Wildlife Research Monographs

Volume 4

Series Editors

The book series on "Wildlife Research Monographs" focuses on all aspects of wildlife biology to fill a vital niche at the intersection of several disciplines. The aim of the series is to summarize knowledge about interaction between wildlife species, their habitats, as well as human activities. Main areas are: wildlife ecology, diseases, toxicology, conservation, management, and the sustainable use of natural resources. The series addresses graduate students, university teachers, scientists and professionals.

More information about this series at http://www.springer.com/series/8826

Catarina C. Ferreira • Cornelya F. C. Klütsch
Editors

Closing the Knowledge-Implementation Gap in Conservation Science

Interdisciplinary Evidence Transfer Across Sectors and Spatiotemporal Scales

Springer

Editors
Catarina C. Ferreira
Department of Conservation Biology
Helmholtz Centre for Environmental
Research - UFZ
Leipzig, Germany

Cornelya F. C. Klütsch
Division of Environment and Natural Resources
Norwegian Institute of Bioeconomy Research
(NIBIO)
Svanvik, Norway

ISSN 2366-8733 ISSN 2366-8741 (electronic)
Wildlife Research Monographs
ISBN 978-3-030-81084-9 ISBN 978-3-030-81085-6 (eBook)
https://doi.org/10.1007/978-3-030-81085-6

This Springer imprint is published by the registered company Springer Nature Switzerland AG.
The registered company address is: Gewerbestrasse 11, 6330 Cham, Switzerland

Acknowledgements

The editors and authors of this book would like to thank the many environmental practitioners and scientists consulted, who all provided valuable insight into knowledge exchange processes and the extent of the knowledge-implementation gap in conservation in various parts of the world. Our thanks go to the following individuals and institutions for contributing to content of this book in various ways, including by sharing data, providing general input and advice, or reviewing earlier draft versions of the chapters.

Alessandro Pititto COGEA srl, Rome, Italy. URL: https://www.linkedin.com/in/apititto

Ana Sofia Soares Cordeiro Fundação para a Ciência e a Tecnologia (FCT), Lisbon, Portugal. URL: https://pt.linkedin.com/in/ana-sofia-soares-cordeiro-50254111

Anna Lukasiewicz Fenner School of Environment & Society, Australian National University, Canberra, Australia. URL: https://iceds.anu.edu.au/dr-anna-lukasiewicz

Anne-Sophie Archambeau Institute of Research for Development, Marseille, France. URL: http://www.patrinat.fr/fr/annuaire/anne-sophie-archambeau-6077

Anne-Sophie Pellier BirdLife International, Cambridge, UK. URL: https://www.linkedin.com/in/anne-sophie-pellier/

Antonio Novellino ETT Solutions Ltd Genova, Italy. URL: https://it.linkedin.com/in/antonio-novellino-7712344

Carly Cook School of Biological Sciences, Monash University, Melbourne, Australia. URL: https://carlycookresearch.wordpress.com/

Casurina Douglas NSW Biodiversity Conservation Trust, Parramatta, New South Wales, Australia. URL: www.bct.nsw.gov.au

Catherine Allan School of Environmental Sciences, Charles Sturt University, New South Wales, Australia. URL: https://www.csu.edu.au/research/ilws/team/profiles/members/catherine-allan

Christopher Cvitanovic College of Science, Australian National University, Canberra, Australia. URL: https://cpas.anu.edu.au/people/academics/dr-christopher-cvitanovic

Claire Wordley Department of Zoology, University of Cambridge, Cambridge, UK. URL: https://clairewordley.com/

Cristina Mendes Da Costa Department of Education and Childhood, University of the West of England, Bristol, UK. URL: https://people.uwe.ac.uk/Person/Cristina2Costa

Edward Lewis United Nations Environmental Programme World Conservation Monitoring Centre, Cambridge, UK. URL: https://www.unep-wcmc.org/employees/edward-lewis

Gemechis Akuma European Commission's Joint Research Center (JRC), Ispra, Italy. URL: https://it.linkedin.com/in/gemechisak

Georg Hanke European Commission's Joint Research Center (JRC), Brussels, Belgium. URL: https://www.researchgate.net/profile/Georg_Hanke

Heather Bingham United Nations Environmental Programme World Conservation Monitoring Centre, Cambridge, UK. URL: https://www.unep-wcmc.org/employees/heather-bingham

Henry Vallius The Geological Surveys of Europa (EuroGeoSurveys), Brussels, Belgium. URL: http://www.eurogeosurveys.org/about-us/organisation/henry-vallius/

Iris Möller Department of Geography, University of Cambridge, Cambridge, UK. URL: https://www.conservation.cam.ac.uk/directory/dr-iris-moeller

Konstantinos Tsiamis European Commission's Joint Research Center (JRC), Brussels, Belgium. URL: https://www.researchgate.net/profile/Konstantinos_Tsiamis

Kylie Durant Holbrook Landcare Network. URL: https://holbrooklandcare.org.au/

Liam Lysaght The National Biodiversity Data Centre, Ireland. URL: https://biodiversityireland.ie/staff/dr-liam-lysaght/

Lorrae van Kerkhoff Fenner School of Environment & Society, Australian National University, Canberra, Australia. URL: https://researchers.anu.edu.au/researchers/van-kerkhoff-le

Natasha Lappin Murray Local Land Services, Albury, New South Wales, Australia. URL: https://www.lls.nsw.gov.au/regions/murray/contact-us/staff-contacts-specialist-services

Nathan Young Faculty of Social Sciences, University of Ottawa, Ontario, Canada. URL: https://uniweb.uottawa.ca/members/973

Nils Ryman Department of Zoology, Stockholm University, Sweden. https://www.su.se/english/profiles/ryman-1.183096

Prue Addison Berks, Bucks and Oxon Wildlife Trust, Berkshire, Buckinghamshire, Oxfordshire, UK. URL: https://www.bbowt.org.uk/about/people

Rachel Morgain NESP Threatened Species Recovery Hub, Australian National University, Canberra, Australia. URL: http://www.nespthreatenedspecies.edu.au/people/rachel-morgain

Rebecca Colvin Resources, Environment and Development, Crawford School of Public Policy, Australian National University, Canberra, Australia. URL: https://researchprofiles.anu.edu.au/en/persons/rebecca-colvin

Rebecca Jarvis Institute for Applied Ecology, New Zealand, School of Science, Auckland University of Technology, New Zealand and Sydney Institute of Marine Science, Sydney, Australia. URL: https://academics.aut.ac.nz/rebecca.jarvis

Robyn Watts Institute for Land, Water, and Society, Charles Sturt University, Albury, New South Wales, Australia. URL: https://www.csu.edu.au/research/ilws/team/profiles/members/robyn-watts

Sergej Olenin Coastal Research and Planning Institute, Klaipeda University, Klaipėda, Lithuania. URL: http://www.mares-eu.org/index.asp?mod=teacher&id=1232

Shadrack Ngene Kenya Wildlife Service, Nairobi, Kenya. URL: https://www.linkedin.com/in/shadrack-ngene-phd-39213511

Tim Wilkinson United Nations Environmental Programme World Conservation Monitoring Centre, Cambridge, UK. URL: http://www.cambridgeconservation.org/person/tim-wilkinson

Tom Spencer Department of Geography, University of Cambridge, Cambridge, UK. URL: http://www.cambridgeconservation.org/person/tom-spencer

Yussuf Wato WWF Kenya, Nairobi, Kenya. URL: https://www.linkedin.com/in/yussuf-wato-196891141

William Sutherland Department of Zoology, University of Cambridge, Cambridge, UK. URL: https://www.zoo.cam.ac.uk/directory/bill-sutherland

Contents

Part I
Introduction

Chapter 1
The Knowledge-Implementation Gap in Conservation Science

Vivian M. Nguyen, Catarina C. Ferreira, and Cornelya F. C. Klütsch

Contents

V. M. Nguyen (✉)
Institute of Environmental and Interdisciplinary Science and Department of Biology, Carleton University, Ottawa, ON, Canada
e-mail: vivian.nguyen@carleton.ca

C. C. Ferreira
Helmholtz-Zentrum für Umweltforschung – UFZ, Leipzig, Germany

C. F. C. Klütsch
Division of Environment and Natural Resources, Norwegian Institute of Bioeconomy Research (NIBIO), Svanvik, Norway
e-mail: cornelya.klutsch@nibio.no

C. C. Ferreira, C. F. C. Klütsch (eds.), *Closing the Knowledge-Implementation Gap in Conservation Science*, Wildlife Research Monographs 4,
https://doi.org/10.1007/978-3-030-81085-6_1

1.1 Introduction

There has never been a greater need for environmental and conservation research to achieve real-world impact with the world facing intractable and complex challenges, such as those associated with climate change and biodiversity loss. Notwithstanding, the impacts of scientific knowledge are overall difficult to trace. They sometimes can have immediate and direct impacts, but most of the time the impacts unfold indirectly over a long time period (Levin 2013; Nguyen et al. 2017). For many decades, scholars particularly in the social sciences, have explored and paved the way for understanding and improving the utilization of scientific knowledge to provide social, environmental, and economic impacts (Fazey et al. 2014).

This challenge in achieving research utilization and impact is widespread in many other applied sciences such as management and organizational science (Pfeffer and Sutton 1999; Starbuck 2006), environmental psychology (McKenzie-Mohr 2000; Sommer 2003), restoration ecology (Higgs 2005), climate science (Meadow et al. 2015), and ecosystem management (McNie 2007). Science for conservation is not alone in facing the challenges of achieving research implementation (Knight et al. 2008).

In this introductory chapter, we provide context on the knowledge-implementation gap in conservation science and discuss our current understanding of the processes, barriers, and enablers for greater implementation of scientific knowledge into conservation practice. Throughout the chapter, we pull on insights and theories from the rich social science literature, as well as within the literature of conservation science and environmental management. The chapter concludes with views of remaining gaps in understanding, enabling and evaluating research implementation, and how this book contributes to filling them.

1.1.1 Overview of Research on the Knowledge-Implementation Gap in Conservation Science

The pursuit of knowledge and its uses are deeply rooted in philosophy dating back to Plato, who described knowledge as "justified true belief." Knowledge has been a central part of human history where rulers believed that *knowledge is power,* evolving to today where knowledge is accessible to almost everyone immediately (Fox 2010). Indeed, knowledge relevant to policy has been exchanged for centuries and is the subject of a vast literature written mainly by historians and scholars in the adjacent disciplines of the social and policy sciences (Fox 2010).

In evidence-based practice and policy-making, rather than merely relying on personal experience or anecdote, practitioners, and policy-makers make decisions and take actions that are informed by systematic and critical analyses of both their own and the world's previous experiences (see Salafsky et al. 2019 for definitions of evidence in conservation practice). This is critical to ensure decisions and actions are timely, relevant, measurable, and have the desired societal impact. In conservation practice and decision-making, collecting, and analyzing appropriate knowledge and evidence is ever more important as it can contribute to solving some of the world's greatest social, economic, and ecological challenges (Sutherland et al. 2004).

In the past couple of decades, there has been increasing interest in the scientific literature in understanding the relationship between knowledge and its use for tackling complex problems across a range of disciplines such as health, education, business, and environmental management (Cvitanovic et al. 2015). Figure 1.1 analyses the results of a search done using Scopus and the keywords "knowledge AND implementation AND gap" OR "knowledge AND action AND gap," by subject area, as well as by country and year. The search retrieved 735 documents between 1978 and 2020, and results by country and year correspond only to the fields of Environmental Sciences, Agricultural and Biological Sciences and Biochemistry, Genetics, and Molecular Biology.

Unsurprisingly, the bulk of the literature between 1978 and 2000 on the knowledge-implementation gap comes from the medical (20.8% of the publications) and social sciences (14.1% of the publications; Fig. 1.1a) with the majority of the papers focused on improving the use of health evidence and technology on primary health care. An increased recognition that scientific knowledge and evidence are pivotal to advance biodiversity conservation was the backdrop for the first studies on this topic developed by environmental scientists in the marine realm (e.g., Melvasalo 2000). These three main fields of expertise (environmental, social, and medical sciences) have contributed to nearly half (48.7%) of all scientific literature produced up to date on the knowledge-implementation gap. This body of knowledge on narrowing the gap between knowledge and action in conservation science grew slowly at first (at a rate of 2.1 papers per year between 2000 and 2009; Fig. 1.1b) and was predominantly linked to the management of water resources, protected areas, and forests. After 2010, this trend shifted with the scientific literature on this topic growing almost exponentially within the field (at a rate of approx. 18 papers per year between 2010 and 2020) culminating in 50 publications in 2020 (Fig. 1.1b). This trend likely reflects an increased interest and necessity in bridging the knowledge sphere with the policy arena and community of practice within the environmental movement. Currently, four countries contribute to over 50% of the scientific literature produced on this topic (USA, United Kingdom, Australia, and Germany), with authors from the Global North also leading this research topic in many other regions of the world (Fig. 1.1c). This creates a clear regional bias in understanding the barriers and enablers of the knowledge-implementation gap that can affect the effectiveness of recommendations to promote the flow between knowledge and conservation action across different sociocultural contexts. This is a serious issue that plagues research done on the gap and ultimately contributes to widening it, because strategies for closing the knowledge-implementation gap developed in the Global North may not always be transferable to local situations in other regions of the world. This can ultimately lead to failure or low impact of conservation initiatives. After all, if conservation science is not leading to actions that effectively

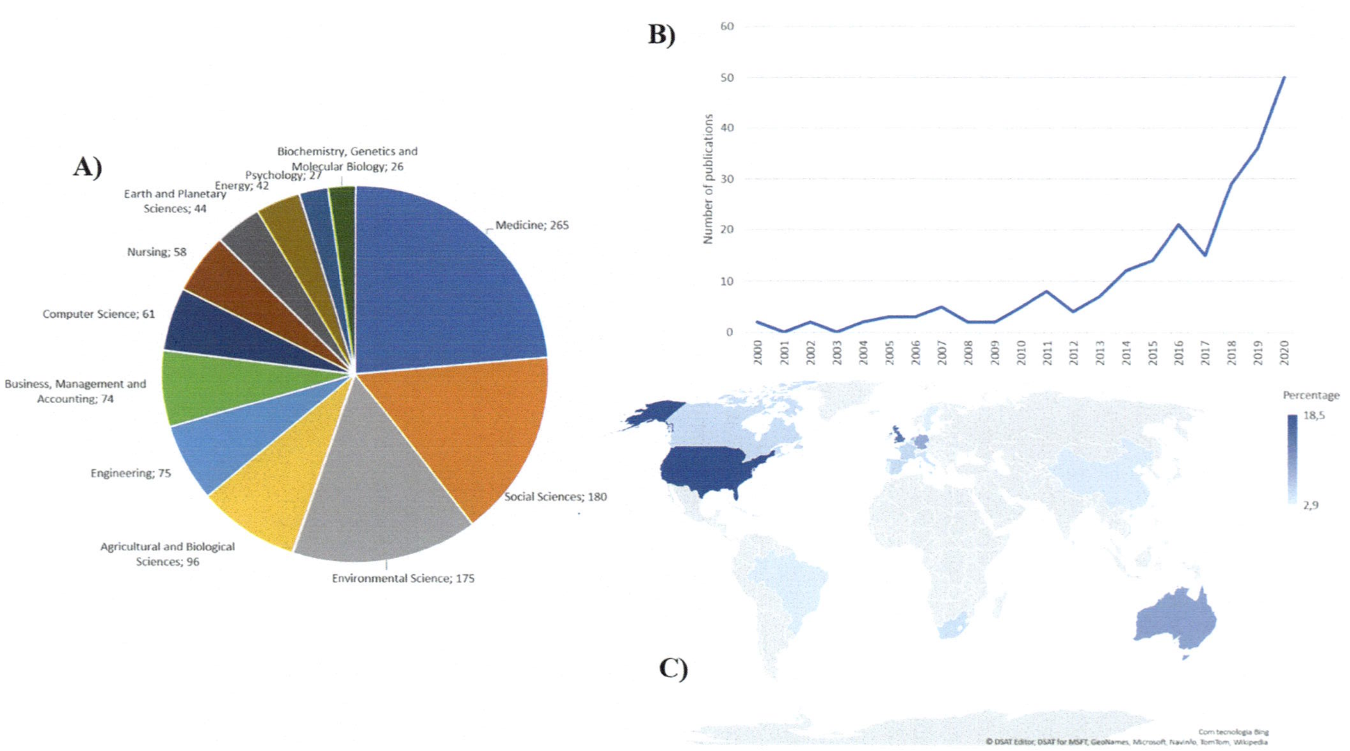

Fig. 1.1 Trends in the scientific literature related to the study of the knowledge-implementation gap as obtained through Scopus (for details about the search parameters please see main text). (**a**) Number of publications by subject area from 1978 to 2020 (total $n = 735$ publications); (**b**) Temporal trends in the number of publications in the fields of Environmental Sciences, Agricultural and Biological Sciences, and Biochemistry, Genetics and Molecular Biology up to 2020 (total $n = 223$ publications); and (**c**). Geographical distribution of the percentage of publications in the fields of Environmental Sciences, Agricultural and Biological Sciences, and Biochemistry, Genetics, and Molecular Biology up to 2020 (total $n = 223$ publications)

conserve nature or stop global extinctions, then what is the point of it? (Whitten et al. 2001). This is an important sentiment shared by many who have begun to question why conservation science has not had its intended impacts. What are the barriers to the implementation of conservation research and how can they be resolved? (e.g., Cooke 2019).

1.1.2 What Is in a Name: Terminologies

The pervasive challenge of producing scientific knowledge that leads to implementation of actions has led to several disciplines coining various terms that all endeavor to discuss the use of knowledge to achieve an outcome. Some terminologies describe the "movement" of knowledge, such as *knowledge transfer*, *knowledge translation*, *diffusion of innovation*, and *knowledge management* (Table 1.1). Each has its own nuances, emphases, and applications and one term is likely not sufficient to capture them all (Fazey et al. 2013).

In the environmental and conservation literature, *knowledge exchange* and *mobilization* are most often used because of its emphasis on two-way exchange and applying knowledge for societal and environmental benefits (Fazey et al. 2013; Cvitanovic et al. 2015). The concept of knowledge exchange encompasses all facets of knowledge production, sharing, storage, mobilization, translation, and use (Best and Holmes 2010). Other terminologies have been coined to label the "gap" between knowledge produced and its intended outcome, such as *knowledge-action* or *knowledge-to-action* gap, *and science-policy*, *science-practice*, or *science-action gap*. In conservation science, the *knowledge-action* or *research or knowledge-implementation* gap is most widely used to specify the particular gap between research and its applicability, and this is the prevalent terminology adopted in this book. Lastly, some terminologies are used to label the "type" of knowledge produced including *actionable* or *usable knowledge* or *science*. Table 1.1 provides descriptions and summaries of the common terminologies found in the literature.

1.2 Barriers to Knowledge Implementation

Efforts have been made to tackle the perceived underuse of scientific knowledge, and ample evidence suggests that creating "useable" scientific knowledge is neither easy nor straightforward (Rosenberg 2007; Young et al. 2014; Nguyen et al. 2018a). Vogel et al. (2007) said it well where the reality is that all too often "*the scientific output is more likely to be mismatched to user requirements, i.e., not what practitioners need; it may not be delivered in time or in appropriate formats; those interacting do not communicate well; scientists feel their credibility is negatively affected by collaborating with practitioners; stakeholders do not feel their legitimate concerns are addressed; and so on*" (Vogel et al. 2007, p. 352).

Table 1.1 Descriptions and sources for the common terminologies found in the literature describing the movement of scientific knowledge to achieving outcomes or the application of scientific knowledge (adapted from Fazey et al. 2013)

Term	Description	References
Knowledge exchange	Implies a two or multiple-path process with reciprocity of the equitable value of the different forms of knowledge being exchanged. The concept of knowledge exchange, therefore, encompasses all facets of knowledge production, sharing, storage, mobilization, translation, and use.	Best and Holmes (2010) Fazey et al. (2013)
Knowledge mobilization	Implies eliciting or spreading knowledge to a wider range of recipients with the intent of increasing application of knowledge for societal and environmental benefit.	Levin (2013) Nguyen et al. (2017)
Knowledge management (in environmental context)	Process of generating, storing, and circulating new knowledge and identifying, bringing together, and applying existing knowledge to achieve a specific objective.	Reed et al. (2014)
Knowledge (or science) translation	Implies communication using a mediated language modified for the audience or recipient.	Curran et al. (2011)
Knowledge (or science) transfer	Implies knowledge is portable, a linear direction, delivery, and reception in a one-way process.	Reviewed in Fazey et al. (2013)
Knowledge sharing	Process whereby individuals mutually exchange their (implicit and explicit) knowledge and jointly create new knowledge. Implies both bringing (or "donating") knowledge and getting (or "collecting" knowledge). Implies a similar process to exchange, but possibly with greater recognition by those involved in the value of the knowledge of those with whom they are sharing.	Van den Hoof and de Ridder (2004) Fazey et al. (2013)
Diffusion of innovation	Study of the how and why innovations are adopted, and the rates and patterns of adoption. Diffusion is the process by which "prior adoption of a trait or practice in a population alters the probability of adoption for remaining non adopted."	Rogers (2003) Mascia and Mills (2018)
Actionable or usable knowledge	Knowledge that prescribes how one should act. Requires propositions that make explicit the causal processes required to produce action. Causality is the key to implementation.	Argyris (2004)
Knowledge or research implementation gap	Disconnect between the knowledge generated by researchers and the information being used to inform policy and practice decisions. This gap impedes effective programming when program planning and implementation proceed with incomplete information and when managers miss opportunities to incorporate relevant knowledge into program decisions.	Cook et al. 2013

The past two decades have seen a call for scientists and researchers to find innovative ways to engage effectively with knowledge users and demonstrate real-world impacts (Lubchenco 1998). So far, most of this documentation has focused on where science has failed to inform policy or practice and on identifying the barriers that prevent the successful use of research (Cvitanovic and Hobday 2018). Common identified barriers range from cultural and institutional differences between functional communities (e.g., science and policy) (e.g., Roux et al. 2006; Cvitanovic et al. 2015; Nguyen et al. 2021), mismatches in scales, timelines and timing (Jacobs et al. 2005; Nguyen et al. 2018b), lack of relevance, and applicability of research to practice (e.g., Weichselgartner and Kasperson 2010), institutional barriers such as poor incentive structures to engage among knowledge producers and users and financial capacity (e.g., Cvitanovic et al. 2015, 2016), competing interests and priorities (Nguyen et al. 2018a), lack of trust and credibility in the knowledge (Cash et al. 2003), lack of relevant skills and competencies (Fazey et al. 2013), and personal perceptions and worldviews (Fazey et al. 2006; Leviston and Walker 2012).

Despite the multiple barriers to the effective uptake of knowledge to inform conservation decisions and action around the world (summarized in Rose et al. 2018), the low priority given to biodiversity conservation is still considered to be most important. Understanding the enablers to mainstreaming science into and communicating it clearly to the decision- and policy-making arenas are therefore instrumental to bridge the knowledge-implementation gap in conservation science (Sutherland et al. 2020).

1.3 Organizing and Structuring What We Are Learning About the Knowledge-Implementation Gap

We have reached the point where we, generally, know and understand what common challenges exist in the implementation of research-informed conservation actions (Rose et al. 2018, 2019). These challenges may be more easily understood by offering a framework to organize and contextualize how knowledge moves. Overall, these challenges may be influenced by the actors involved (e.g., people profile, characteristics, epistemologies, and ontologies; Fazey et al. 2013), the context and environment in which knowledge enters, the relationships between the actors and their environment, the processes, and mechanisms used to mobilize knowledge, as well as the knowledge product itself and its potential outcomes (Nguyen et al. 2017).

In the clinical literature, a BARRIERS scale was developed to improve our understanding of barriers to the utilization of research. The scale organizes barriers into four factors: *characteristics of the adopter/knowledge user*, *characteristics of the organization*, *characteristics of the innovation or knowledge product*, and *characteristics of the communication* (Funk et al. 1991). While this scale is useful to map barriers, it is limited in its use for understanding broader movements of

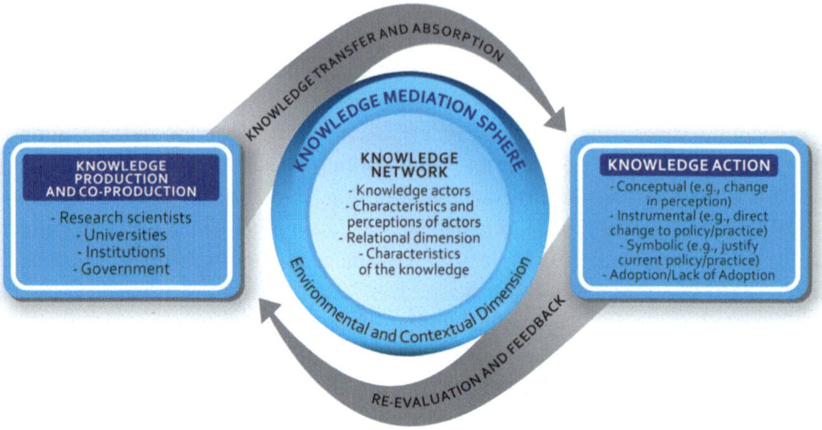

Fig. 1.2 The three main parts of the knowledge-action cycle including knowledge production and co-production, knowledge mediation sphere, and knowledge action/implementation. Reproduced with permission from Nguyen et al. (2017)

knowledge, including potential unintended consequences of knowledge and innovations.

More specific to the conservation and environmental literature, Nguyen et al. (2017) developed a knowledge-action framework, based on a broad literature review, to visualize the components or stages that are important when investigating the movement of knowledge from production to action. The framework is rooted in lessons from the sociology of science and describes the three typical components of knowledge movement: the source of the knowledge (*knowledge production or co-production*), an intermediary (the *mediation sphere*), and an outcome (*knowledge action*) (Fig. 1.2). The novelty of their framework is the emphasis on the social dimension and described nuances of the intermediary, or the "gap," between knowledge and action, which they coined as the "mediation" sphere. The sphere is a representation of the nonlinear ways that knowledge can move through a knowledge (often social) network and interpersonal relationships. In addition to describing the three components of knowledge movement, the knowledge-action framework also illustrates the importance of processes and factors that may influence the movement of knowledge at various scales and levels (Nguyen et al. 2017).

This book follows the knowledge-action framework developed by Nguyen et al. (2017) and its chapters are hence organized within each of these three stages. This framework emphasizes the interdependencies and interconnectedness of knowledge with its environment and with people (Contandriopoulous et al. 2010), which contrasts with the more traditional model in which knowledge is viewed to move linearly and unidirectionally, i.e., a knowledge-deficit or "loading dock" model where scientists, as the producers of knowledge, communicate one-way with the assumption that more information means greater awareness leading to change (Van Kerkhoff and Lebel 2006; Stocklmayer 2013). This, as we know today, is flawed and for the past 50 years, there has been a rapid evolution of science–society interaction thinking, from a linear model characterized by a strong disciplinary-based, basic research focuses to more complex models of science production that embrace interdisciplinary and participatory approaches to address societal problems (Kirchoff et al. 2013). These new models and structures help us understand enabling conditions to mobilize knowledge at various stages including the production, mediation, and action stages.

1.3.1 Narrowing the Knowledge-Implementation Gap at the Knowledge Production Stage

Within the sustainable development literature, Cash et al. (2003) provide a compelling framework for understanding why some science is translated into action whereas other is not. According to the authors, scientific information is likely to have influence if it is perceived by relevant stakeholders to be credible, salient, and legitimate. *Credibility*, in the sense that the information is scientifically adequate (authoritative, believable, and trusted). *Salience* involves the relevance of the information to the needs of the decision makers or knowledge users (relevant to knowledge users and provided when it is needed), and *legitimacy* is whether the information is viewed by the user as respectful of their values and beliefs and produced in an unbiased and fair manner. Without all three elements, research is likely to be ignored by decision makers or conservation practitioners. However, these attributes of information and knowledge are tightly coupled, such that efforts to enhance anyone could incur a cost to the other (Cash et al. 2003). These attributes may also differ in importance depending on the context. For example, Posner et al. (2016) showed that legitimacy of knowledge had a greater impact on decisions, and that activities that enhance the legitimacy of knowledge best explained the impact of ecosystem services science on decision-making.

1.3.2 Narrowing the Knowledge-Implementation Gap at the Knowledge Mediation Stage

Insights from the sociology of science reveal that knowledge is embedded in social relations where people rely on one another to access knowledge as well as judge the credibility and reliability of knowledge. Social connections and networks are thus core to knowledge mobilization and narrowing the knowledge-implementation gap. Knowledge is therefore often mediated through people who interpret it based on their worldviews and shared social constructs (Pohl 2008; Levin 2013; Clark et al. 2016). Building relationships is therefore key to building trust between actors. There are emerging concepts and explanations that help mediate the movement of knowledge into an action/outcome. For example, boundary organizations, knowledge brokers, and knowledge translation are entities or activities that help overcome the gap between research and implementation (e.g., Cook et al. 2013; Nel et al. 2016; Nguyen et al. 2021).

The social sciences also offer a rich body of theory and frameworks for guiding approaches to enhancing conservation science into conservation actions and practice such as transdisciplinary research practice (Reyers et al. 2010), adaptive management (Holness and Biggs 2011), social-ecological systems thinking (Ban et al. 2014), social network analysis (Mills et al. 2014), and social marketing (Wilhelm-Rechmann and Cowling 2013). We describe some of the most common models for enabling knowledge movement at the mediation stage in Table 1.2.

1.3.3 Understanding and Enabling Knowledge Implementation at the Action Stage

When making decisions for conservation actions, it has been suggested that practitioners and decision-makers more often rely on their experiential knowledge than science, which limits the potential success of policy and management decisions and can have downstream consequences for environmental and societal wellbeing and prosperity (Pullin et al. 2004).

Different frameworks and approaches have been developed to evaluate research impacts. For example, eight different types of information use were described by Taylor (1991), which offered a spectrum in ways to think about how information can be used. These eight types of use were further refined into three categories (Oh 1996): (1) *conceptual* information use, where an organization/individual perceives themselves to be better informed about an issue or has changed opinion about the issue; (2) *justification,* where information is used to rationalize or justify a predetermined decision; and (3) *instrumental,* where information is directly used to inform decisions. These three types are most commonly used in conservation science (Beyer 1997; Amara et al. 2004).

One of the most cited studies reviewing the three types of research use is that of Amara et al. (2004) who conducted an empirical survey-based study to evaluate the instrumental, conceptual, and symbolic use of academic research in government agencies. Rudd (2011) further explained that conceptual (for information) impacts may be more pervasive than originally thought and that instrumental impacts should be considered as dependent on conceptual impact. Lastly, Choo (2006) presented three different conceptual models to explain how organizations use information with the lens of why information was sought out in the first place: sense-making in response to a change in their environment, knowledge creating to develop new capabilities or innovations, and decision-making to select alternatives and take a goal-directed action. These elements are important to understand when investigating successes and failures of implementing research into conservation action.

Table 1.2 Common concepts and frameworks for enabling knowledge movement at the mediation stage

Concepts and frameworks	Description	Select references
Deficit model	Stems from the traditional "linear model" in communication that assumes knowledge users pose well-defined questions to which scientists reply by providing credible, legitimate, relevant, and timely knowledge.	Bradshaw and Borchers (2000) Cash (2001) Young et al. (2014)
Co-production or co-evolution of knowledge	A participatory arena that puts researchers, decision-makers, and other users of knowledge on equal footing to work iteratively and interactively toward collaborative learning, shared understanding of key concepts, and co-evolution of common purpose and action. Evidence suggests that when people are closely involved in knowledge production, they are more likely to view the resulting knowledge as credible, salient, and legitimate and to adopt such knowledge for implementation.	Cash et al. (2003) Armitage et al. (2011) Young et al. (2014) Nel et al. (2016) Chapman and Schott (2020) Westwood et al. (2020) Cooke et al. (2020)
Knowledge broker	People or organizations that move knowledge around and create connections between researchers and their various audiences.	Meyer (2010)
Boundary Organization Boundary objects	Organizations that are designed to facilitate collaboration and information flow between the research and knowledge user community. In other words, groups that facilitate the transfer and exchange of knowledge between science and action for informed conservation implementation. Boundary objects are co-produced outputs that are adaptable to different viewpoints yet robust enough to maintain identity across them. Examples are definitions, standards, models, and indicators.	Star and Griesemer (1989) Jacobs et al. (2016)
Boundary spanning	The practice of boundary spanning is to enable exchange between the production and use of knowledge to support evidence-informed decision-making in a specific context. Boundary spanners are "individuals or organizations that specifically and actively facilitate this process."	Nel et al. (2016) Bednarek et al. (2018)
Embedding and sabbaticals	A process of situating a scientist or researcher within the decision-making agency or policy environment, or alternatively, a decision-maker within a research environment. Embedding can range from fixed-term or ongoing appointments.	Gibbons et al. (2008) Jenkins et al. (2012) Cook et al. (2013)
Networks and relationship building	Relationships are the foundation of an organized effort for change. Networks are built based on	Cohen et al. (2012)

(continued)

Table 1.2 (continued)

Concepts and frameworks	Description	Select references
	mutual affiliations or connections between individuals or groups of people or entities. These two concepts are critical to mobilize knowledge and can foster effective collaborations.	Nguyen et al. (2019)
Transdisciplinary research practice	Form of research and practice that synthesizes knowledge from a range of academic disciplines and from the community. Stems from the transfer of knowledge between disciplines, by sharing approaches and assumptions, in dialogue, in order to weave together a new approach to complex issues.	McGregor (2004) Fam et al. (2017)

1.4 Where to Go Next: Current Knowledge Gaps in Knowledge Implementation

Much of the evaluation done of the knowledge-implementation gap in conservation science has been conducted in the western world and democracies (see Fig. 1.1c). This is an issue as some of the most endangered biomes/ecosystems are found in developing worlds and cultures outside of the western world. This geographical bias towards research done in Western and developed nations introduces a risk of erasing vital geographical differences in the personal or collective valuations of material and symbolic goods (Hulme 2010) as well as local knowledge. Since gaps between science and action may arise from cultural and/or social barriers (Amano et al. 2016), in addition to political and institutional factors (Owens 2015; Nguyen et al. 2017), geographical biases can contribute to a misconstruction of challenges and opportunities to bridge these gaps. Hence, assessments of local contexts and the identification of factors that can lead to success or failure of conservation initiatives are important tools in increasing the effectiveness of conservation implementation strategies.

More recently, there has been a call for a shift in the academic study of science–policy–practice interfaces toward the study of bright spots—outliers that perform significantly better than what would be expected (Cvitanovic and Hobday 2018). Cvitanovic and Hobday (2018) argue that the propagation of the failure mantra is counterproductive to improve the relationship between science, policy, and practice and that there is a greater need for documenting successes that help establish a new mantra of optimism that may yield improved strategies of reconciling the knowledge-implementation gap. The authors also suggest that an optimistic environment can also lead to greater creativity linked with innovation and problem solving. Establishing such an optimistic outlook is an important next step for enabling greater research implementation. However, how well these research outliers perform in different local and regional contexts is unknown and needs further research.

In order to address the issues highlighted in this introductory chapter as well as contribute to narrow the knowledge-implementation gap in conservation science and natural resource management in various political, cultural, and social settings, a systematic global review and assessment of various approaches at the different stages of knowledge production, mobilization, and action are warranted. Since all three stages of the knowledge cycle (Fig. 1.2) are pivotal for closing the knowledge-implementation gap, this book was split into three parts representing the three stages (following the knowledge-action framework developed by Nguyen et al. 2017). In each part, chapters focus on the different features of the respective stage and critically review the state of the art, identify problems and gaps, highlight success stories, and offer advice for future directions and improvements in bridging the knowledge-implementation gap in conservation science. By doing so, the book not only synthesizes the status quo of the conservation knowledge-implementation gap but also offers examples and guidelines for successful knowledge implementation.

In the first part, four chapters are dedicated to the main knowledge (co-) production sources. Prominently featured are often neglected and/or emerging knowledge sources like genetics (Chap. 3; Klütsch and Laikre 2021), traditional ecological knowledge (Chap. 5; Ens et al. 2021), and citizen science (Chap. 4; Phillips et al. 2021), all of which are complementary to the more traditionally applied ecological knowledge production (Chap. 2; Ferreira et al. 2021). The goal of this first part is to recognize the differences between these knowledge sources while acknowledging how each source can contribute to narrowing the knowledge-implementation gap, thereby, stressing the significance of a multidisciplinary framework to tackle complex conservation issues. Further, the recognition of traditional ecological knowledge as well as citizen science as sources of knowledge opens up the knowledge (co-)production sphere to often marginalized groups and the general public, which represents an important step in transitioning from a knowledge-deficit model to an inclusive and adaptive knowledge transfer model.

The second part of the book is dedicated to the transfer of knowledge with an emphasis on improving conservation knowledge mobilization. Here, three chapters address the topic from different angles. Chapter 6 (Bixler 2021) describes knowledge networks and identifies the different multilevel actors in the network from group- and institutional-level and how these actors interact (or not) with one another. With the use of case studies, this chapter illustrates how knowledge transfer differs at various scales from local to (inter-)national networks. Chapter 7 (O'Connell and McKinnon 2021) provides guidance on the communication of scientific knowledge to the general public and case studies that offer insight on how impactful science communication by scientists as the major knowledge producers can be achieved. Finally, Chapter 8 (Rose et al. 2021) looks at existing and developing platforms for management and interdisciplinary decision-making to introduce the reader to numerous examples of decision-making tools and how those have the potential to narrow the knowledge-implementation gap.

Finally, in Part 3 of this book, six chapters take a look at the international level and provide a comprehensive overview about the current state of the knowledge-implementation gap and knowledge implementation in different political, societal, and cultural settings. With this approach, diverse initiatives that have worked or failed in different parts of the world are discussed, highlighting the importance of flexible strategies for knowledge transfer. Empirical examples from North and South America (Chaps. 9 and 10; Schwartz et al. 2021 and Josse and Fernandez 2021, respectively), Africa (Chap. 11; Stephenson et al. 2021), Europe (Chap. 12; Araújo et al. 2021), Asia (Chap. 13; Horgan and Kudavidanage 2021), and Oceania (Chap. 14; Knight 2021) offer insight into the reasons why some knowledge implementations fail in specific settings and why some are successful, hence providing recommendations for improving the knowledge flow from knowledge producers to conservation managers and policy/decision-makers as well as the general public.

Knowledge production, transfer, and implementation are inherently linked to a wide range of stakeholders such as scientists, conservation practitioners and managers, policy-makers, local knowledge holders, and the general public. For this reason, it is important to understand the background, interests, and pressures each of these stakeholders has. This book puts a strong emphasis on identifying and describing the roles of diverse stakeholders in the case studies. Consequently, we believe that this book is suitable for most of the above-mentioned stakeholders and other audiences that would like to better understand the dynamics of knowledge transfer in conservation and natural resource management. Students in these fields as well as social science, politics, and law students will find a contemporary global overview of the major factors influencing knowledge mobilization and implementation. Last but not least, this book is intended to be a go-to resource with numerous website links and literature reviews for further reading and study.

References

Amano T, González-Varo JP, Sutherland WJ (2016) Languages are still a major barrier to global science. PLoS Biol 14(12):e2000933

Amara N, Ouimet M, Landry R (2004) New evidence on instrumental, conceptual, and symbolic utilization of university research in government agencies. Sci Commun 26(1):75–106

Araújo R, Cardoso AC, Tornero V, Zampoukas N (2021) The marine conservation landscape in Europe: knowledge support to policy implementation and conservation action. In: Ferreira CC, Klütsch CFC (eds) Closing the knowledge-implementation gap in conservation science – evidence transfer across spatiotemporal scales and different stakeholders. Springer, New York, pp. 223–370

Argyris C (2004) Actionable knowledge. In: Tsoukas H, Knudsen C (eds) The Oxford handbook of organization theory: meta theoretical perspectives. Oxford University Press, Oxford, pp 423–452

Armitage D, Berkes F, Dale A, Kocho-Schellenberg E, Patton E (2011) Co-management and the co-production of knowledge: learning to adapt in Canada's Arctic. Glob Environ Chang 21(3):995–1004

Ban NC, Maxwell SM, Dunn DC, Hobday AJ, Bax NJ, Ardron J, Gjerde KM, Game ET, Devillers R, Kaplan DM, Dunstan PK, Halpin PN, Pressey RL (2014) Better integration of sectoral planning and management approaches for the interlinked ecology of the open oceans. Mar Policy 49:127–136

Bednarek AT, Wyborn C, Cvitanovic C, Meyer R, Colvin RM, Addison PFE, Close SL, Curran K, Farooque M, Goldman E, Hart D, Mannix H, McGreavy B, Parris A, Posner S, Robinson C, Ryan M, Leith P (2018) Boundary spanning at the science–policy interface: the practitioners' perspectives. Sustain Sci 13(4):1175

Best A, Holmes B (2010) Systems thinking, knowledge and action: towards better models and methods. Evid Policy J Res Debate Pract 6(2):145–159

Beyer J (1997) Research utilization: bridging the cultural gap between communities. J Manag Inquiry 6(1):17–22

Bixler RP (2021) The knowledge network: identifying actors and structural dimensions of knowledge transfer. In: Ferreira CC, Klütsch CFC (eds) Closing the knowledge-implementation gap in conservation science – evidence transfer across spatiotemporal scales and different stakeholders. Springer, New York, pp. 143–170

Bradshaw GA, Borchers JG (2000) Uncertainty as information: narrowing the science-policy gap. Ecol Soc 4(1):7

Cash DW (2001) In order to aid in diffusing useful and practical information: agricultural extension and boundary organizations. Sci Technol Human Values 26(4):431–453

Cash DW, Clark WC, Alcock F, Dickson NM, Eckley N, Guston DH, Jäger J, Mitchell RB (2003) Knowledge systems for sustainable development. PNAS 100(14):8086–8091

Chapman JM, Schott S (2020) Knowledge coevolution: generating new understanding through bridging and strengthening distinct knowledge systems and empowering local knowledge holders. Sustain Sci 15(3):931–943

Choo CW (2006) The knowing organization: how organizations use information to construct meaning, create knowledge, and make decisions, 2nd edn. Oxford University Press, Oxford

Clark WC, van Kerkhoff L, Lebel L, Gallopin GC (2016) Crafting usable knowledge for sustainable development. PNAS 113(17):4570–4578

Cohen PH, Evans LS, Mills M (2012) Social networks supporting governance of coastal ecosystems in Solomon Islands. Conserv Lett 5(5):376–386

Contandriopoulos D, Lemire M, Denis JL, Tremblay E (2010) Knowledge exchange processes in organisations and policy arenas: a narrative systematic review of the literature. Milbank Q 88(4):444–483

Cook CN, Mascia MB, Schwartz MW, Possingham HP, Fuller RA (2013) Achieving conservation science that bridges the knowledge–action boundary. Conserv Biol 27(4):669–678

Cooke SJ (2019) From frustration to fruition in applied conservation research and practice: ten revelations. Socio-Ecol Pract Res 1(1):15–23

Cooke SJ, Nguyen VM, Chapman JM, Reid AJ, Landsman SJ, Young N, Hinch SG, Schott S, Mandrak NE, Semeniuk CA (2020) Knowledge co-production: a pathway to effective fisheries management, conservation, and governance. Fisheries 46(2): 89–97

Curran JA, Grimshaw JM, Hayden JA, Campbell B (2011) Knowledge translation research: the science of moving research into policy and practice. J Contin Educ Health Prof 31(3):174–180

Cvitanovic C, Hobday AJ (2018) Building optimism at the environmental science-policy-practice interface through the study of bright spots. Nat Commun 9(1):3466

Cvitanovic C, Hobday AJ, van Kerkhoff L, Wilson SK, Dobbs K, Marshall NA (2015) Improving knowledge exchange among scientists and decision-makers to facilitate the adaptive governance of marine resources: a review of knowledge and research needs. Ocean Coast Manag 112:25–35

Cvitanovic C, McDonald J, Hobday AJ (2016) From science to action: principles for undertaking environmental research that enables knowledge exchange and evidence-based decision-making. J Environ Manag 183(Pt 3):864–874

Ens E, Reyes-García V, Asselin H, Hsu M, Reimerson E, Reihana K, Sithole B, Shen X, Cavanagh V, Adams M (2021) Recognition of Indigenous Ecological Knowledge Systems in Conservation and Their Role to Narrow the Knowledge-Implementation Gap. In: Ferreira CC, Klütsch CFC (eds) Closing the knowledge-implementation gap in conservation science – evidence transfer across spatiotemporal scales and different stakeholders. Springer, New York, pp. 109–139

Fam D, Palmer J, Riedy C, Mitchell C (eds) (2017) Transdisciplinary research and practice for sustainability outcomes. Taylor & Francis, London

Fazey I, Fazey JA, Salisbury JG, Lindenmayer DB, Dovers S (2006) The nature and role of experiential knowledge for environmental conservation. Environ Conserv 33(1):1–10

Fazey I, Evely AC, Reed MR, Stringer LC, Kruijse JHJ, White PCL, Newsham A, Jin L, Cortazzi M, Phillipson J, Blackstock KL, Entwistle N, Sheate WR, Armstrong F, Blackmore C, Fazey JA, Ingram J, Gregson J, Lowe P, Morton S, Trevitt C (2013) Knowledge exchange: a review and research agenda for environmental management. Environ Conserv 40: 19–36

Fazey I, Bunse L, Msika J, Pinke M, Preedy K, Evely AC, Lambert E, Hastings E, Morris S, Reed MS (2014) Evaluating knowledge exchange in interdisciplinary and multi-stakeholder research. Glob Environ Chang 1(25):204–220

Ferreira CC, Stephenson PJ, Gill M, Regan E (2021) Biodiversity Monitoring and the Role of Scientists in the Twenty-first Century. In: Ferreira CC, Klütsch CFC (eds) Closing the knowledge-implementation gap in conservation science – evidence transfer across spatiotemporal scales and different stakeholders. Springer, New York, pp. 23–50

Fox DM (2010) History matters for understanding knowledge exchange. Milbank Q 88(4):484–491

Funk SG, Champagne MT, Wiese RA, Tornquist EM (1991) BARRIERS: the Barriers to research utilization scale. Appl Nurs Res 4(1):39–45

Gibbons P, Zammit C, Youngentob K, Possingham HP, Lindenmayer DB, Bekessy S, Burgman M, Colyvan M, Considine M, Felton A, Hobbs RJ, Hurley K, McAlpine C, McCarthy MA, Moore J, Robinson D, Salt D, Wintle B (2008) Some practical suggestions for improving engagement between researchers and policy-makers in natural resource management. Ecol Manage Restor 9(3):182–187

Higgs E (2005) The two-culture problem: ecological restoration and the integration of knowledge. Restor Ecol 13(1):159–164

Holness SD, Biggs HC (2011) Systematic conservation planning and adaptive management. Koedoe 53(2):1-6

Horgan FG, Kudavidanage EP (2021) Translating research into wildlife conservation actions through multi-sector collaboration in tropical Asia. In: Ferreira CC, Klütsch CFC (eds) Closing the knowledge-implementation gap in conservation science – evidence transfer across spatiotemporal scales and different stakeholders. Springer, New York, pp. 371–411

Hulme M (2010) Global environmental change – problems with making and governing global kinds of knowledge. Glob Environ Chang 20(4):558–564

Jacobs K, Garfin G, Lenart M (2005) More than Just Talk: connecting science and decision making. Environ Sci Policy Sustain Dev 47(9):6–21

Jacobs S, Dendoncker N, Martín-López B, Barton DN, Gomez-Baggethun E, Boeraeve F, McGrath FL, Vierikko K, Geneletti D, Sevecke KJ, Pipart N, Primmer E, Mederly P, Schmidt S, Aragão A, Baral H, Bark RH, Briceno T, Brogna D, Cabral P, De Vreese R, Liquete C, Mueller H, Peh KSH, Phelan A, Rincón AR, Rogers SH, Turkelboom F, Van Reeth W, van Zanten BT, Wam HK, Washbourne CL (2016) A new valuation school: integrating diverse values of nature in resource and land use decisions. Ecosyst Serv 22(Part B): 213–220

Jenkins LD, Maxwell SM, Fisher E (2012) Increasing conservation impact and policy relevance of research through embedded experiences. Conserv Biol 26(4):740–742

Josse C, Fernandez M (2021) Progress and Gaps in Biodiversity Data Mainstreaming and Knowledge Transfer for Conservation in South America. In: Ferreira CC, Klütsch CFC (eds) Closing the knowledge-implementation gap in conservation science – evidence transfer across spatiotemporal scales and different stakeholders. Springer, New York, pp. 255–286

Kirchhoff CJ, Lemos MC, Dessai S (2013) Actionable knowledge for environmental decision making: broadening the usability of climate science. Annu Rev Environ Resour 38:393-414

Klütsch CFC, Laikre L (2021) Closing the conservation genetics gap: integrating genetic knowledge in conservation management to ensure evolutionary potential. In: Ferreira CC, Klütsch

CFC (eds) Closing the knowledge-implementation gap in conservation science – evidence transfer across spatiotemporal scales and different stakeholders. Springer, New York, pp. 51–82
Knight AT (2021) Environmental knowledge exchange in Australia and Oceania: how researchers and practitioners are working together to bring change. In: Ferreira CC, Klütsch CFC (eds) Closing the knowledge-implementation gap in conservation science – evidence transfer across spatiotemporal scales and different stakeholders. Springer, New York, pp. 413–453
Knight AT, Cowling RM, Rouget M, Balmford A, Lombard AT, Campbell BM (2008) Knowing but not doing: selecting priority conservation areas and the research-implementation gap. Conserv Biol 22(3):610–617
Levin B (2013) To know is not enough: research knowledge and its use. Rev Educ 1(1):2–31
Leviston Z, Walker I (2012) Beliefs and denials about climate change: an Australian perspective. Ecopsychology 4(4):277–285
Lubchenco J (1998) Entering the century of the environment: a new social contract for science. Science 279(5350):491–497
Mascia MB, Mills M (2018) When conservation goes viral: the diffusion of innovative biodiversity conservation policies and practices. Conserv Lett 11(3):e12442
McGregor SLT (2004) The nature of transdisciplinary research and practice. Notes from symposium of the Canadian Commission for UNESCO (May 2004)
McKenzie-Mohr D (2000) Promoting sustainable behaviour: an introduction to community-based social marketing. J Soc Issues 56(3):543–554
McNie EC (2007) Reconciling the supply of scientific information with user demands: an analysis of the problem and review of the literature. Environ Sci Policy 10(1):17–38
Meadow AM, Ferguson DB, Guido Z, Horangic A, Owen G, Wall T (2015) . Weather Clim Soc 7(2):179–191
Melvasalo T (2000) Regional marine environmental management and the GPA-LBA: perspectives and the need for scientific support. Ocean Coast Manag 43(8–9):713–724
Meyer M (2010) The rise of the knowledge broker. Sci Commun 32(1):118–127
Mills M, Álvarez-Romero JG, Vance-Borland K, Cohen P, Pressey RL, Guerrero AM, Ernston H (2014) Linking regional planning and local action: towards using social network analysis in systematic conservation planning. Biol Conserv 169:6–13
Nel JL, Roux DJ, Driver A, Hill L, Maherry AC, Snaddon K, Petersen CR, Smith-Adao LB, Van Deventer H, Reyers B (2016) Knowledge co-production and boundary work to promote implementation of conservation plans. Conserv Biol 30(1):176–188
Nguyen VM, Young N, Cooke SJ (2017) A roadmap for knowledge exchange and mobilization research in conservation and natural resource management. Conserv Biol 31(4):789–798
Nguyen VM, Young N, Corriveau M, Hinch SG, Cooke SJ (2018a) What is "usable" knowledge? Perceived barriers for integrating new knowledge into management of an iconic Canadian fishery. Can J Fish Aquat Sci 76(3):463–474
Nguyen VM, Young N, Cooke SJ (2018b) Applying a knowledge-action framework for navigating barriers to incorporating telemetry science into fisheries management and conservation: a qualitative study. Can J Fish Aquat Sci 75(10):1733–1743
Nguyen VM, Young N, Brownscombe JWB, Cooke SJC (2019) Collaboration and engagement produce more actionable science: quantitatively analyzing uptake of fish tracking studies. Ecol Appl 6(29):e01943
Nguyen VM, Delle Palme C, Pentz B, Vandergoot CS, Krueger CC, Young N, Cooke SJ (2021) Overcoming barriers to transfer of scientific knowledge: integrating biotelemetry into fisheries management in the Laurentian Great Lakes. Socio-Ecol Pract Res 3:17–36
O'Connell C, McKinnon M (2021) Communication: The bridge between knowledge and implementation. In: Ferreira CC, Klütsch CFC (eds) Closing the knowledge-implementation gap in conservation science – evidence transfer across spatiotemporal scales and different stakeholders. Springer, New York, pp. 171–197
Oh CH (1996) Issues for the new thinking of knowledge utilization: introductory remarks. Knowl Policy 10(3):3–10

Owens S (2015) Knowledge, policy, and expertise: The UK Royal Commission on Environmental Pollution 1970–2015. Oxford University Press, Oxford

Pfeffer J, Sutton RI (1999) The knowing-doing gap: how smart companies turn knowledge into action, 1st edn. Harvard Business School Press

Phillips TB, Parker A, Bowser A, Hakley M (2021) Publicly generated data: the role of Citizen-Science for knowledge production, action, and public engagement. In: Ferreira CC, Klütsch CFC (eds) Closing the knowledge-implementation gap in conservation science – evidence transfer across spatiotemporal scales and different stakeholders. Springer, New York, pp. 83–107

Pohl C (2008) From science to policy through transdisciplinary research. Environ Sci Pol 11(1):46–53

Posner SM, McKenzie E, Ricketts TH (2016) Policy impacts of ecosystem services knowledge. PNAS 113(7):1760–1765

Pullin AS, Knight TM, Stone DA, Charman K (2004) Do conservation managers use scientific evidence to support their decision-making? Biol Conserv 119(2):245–252

Reed MS, Stringer LC, Fazey I, Evely AC, Kruijsen JHJ (2014) Five principles for the practice of knowledge exchange in environmental management. J Environ Manag 146:337–345

Reyers B, Roux DJ, Cowling RM, Ginsburg AE, Nel JL, Farrell PO (2010) Conservation planning as a transdisciplinary process. Conserv Biol 24(4):957–965

Rogers EM (2003) Diffusion of innovations, 5th edn. Free Press, New York

Rose DC, Sutherland WJ, Amano T, González-Varo JP, Robertson RJ, Simmons BI, Wauchope HS, Kovacs E, Durán AP, Vadrot ABM, Wu W, Dias MP, Di Fonzo MMI, Ivory S, Norris L, Nunes MH, Nyumba TO, Steiner N, Vickery J, Mukherjee N (2018) The major barriers to evidence-informed conservation policy and possible solutions. Conserv Lett 11:e12564

Rose DC, Amano T, González-Varo JP, Mukherjee N, Robertson RJ, Simmons BI, Wauchope HS, Sutherland WJ (2019) Calling for a new agenda for conservation science to create evidence-informed policy. Biol Conserv 238:108222

Rose DC, Despot-Belmonte K, Pollard JA, Shears O, Robertson RJ (2021) Making an impact: how to design relevant and usable decision support systems for conservation. In: Ferreira CC, Klütsch CFC (eds) Closing the knowledge-implementation gap in conservation science – evidence transfer across spatiotemporal scales and different stakeholders. Springer, New York, pp. 199–226

Rosenberg AA (2007) Fishing for certainty. Nature 449(7165):989

Roux DJ, Rogers KH, Biggs HC, Ashton PJ, Sergeant A (2006) Bridging the science–management divide: moving from unidirectional knowledge transfer to knowledge interfacing and sharing. Ecol Soc 11(1):4

Rudd MA (2011) How research-prioritization exercises affect conservation policy. Conserv Biol 25(5):860–866

Salafsky N, Boshoven J, Burivalova Z, Dubois NS, Gomez A, Johnson A, Lee A, Margoluis R, Morrison J, Muir M, Pratt SC, Pullin AS, Salzer D, Stewart A, Sutherland WJ, Wordley CFR (2019) Defining and using evidence in conservation practice. Conserv Sci Pract 1(5):e27

Schwartz MW, Fleishman E, Williamson MA, Williams JN, Morelli TL (2021) The use of boundary-spanning organizations to bridge the knowledge-action gap in North America. In: Ferreira CC, Klütsch CFC (eds) Closing the knowledge-implementation gap in conservation science – evidence transfer across spatiotemporal scales and different stakeholders. Springer, New York, pp. 229–254

Sommer R (2003) Action research and big fuzzy concepts. Hum Ecol Rev 10(2):176–177

Star SL, Griesemer JR (1989) Institutional ecology, 'translations,' and boundary objects: amateurs and professionals in Berkeley's museum of vertebrate zoology, 1907–1939. Social Stud Sci 19(3):387–420

Starbuck WH (2006) The production of knowledge: the challenge of social science research. Oxford University Press, Oxford

Stephenson PJ, Bakarr M, Bowles-Newark N, Kleinschroth F, Mapendembe A, Ntiamoa-Baidu Y, Obura D, Ratsifandrihamanana N, Simaika J, Sitati N, Sumbi P, Teferi T (2021) Conservation science in Africa: mainstreaming biodiversity information into policy and decision-making. In: Ferreira CC, Klütsch CFC (eds) Closing the knowledge-implementation gap in conservation science – evidence transfer across spatiotemporal scales and different stakeholders. Springer, New York, pp. 287–320

Stocklmayer S (2013) Engagement with science: models of science communication. In: Gilbert JK, Stocklmayer S (eds) Communication and engagement with science and technology, issues and dilemmas. Routledge, New York

Sutherland WJ, Pullin AS, Dolman PM, Knight TM (2004) The need for evidence-based conservation. Trends Ecol Evol 19(6):305–308

Sutherland WJ, Brotherton PNM, Davies ZG, Ockenden N, Pettorelli N, Vickery JA (eds) (2020) Conservation research, policy and practice. Cambridge University Press, Cambridge

Taylor RS (1991) Information use environments. Progr Commun Sci 10:217–255

Van den Hoof B, De Ridder JA (2004) Knowledge sharing in context: the influence of organizational commitment, communication climate and CMC use on knowledge sharing. J Knowl Manag 8(6):117–130

Van Kerkhoff L, Lebel L (2006) Linking knowledge and action for sustainable development. Annu Rev Environ Resour 31:445–477

Vogel I, Wendt D, Wolfe R (2007) In the know: conceptualising information interventions in international development, draft report. Institute of Development Studies, Brighton

Weichselgartner J, Kasperson R (2010) Barriers in the science-policy-practice interface: toward a knowledge-action-system in global environmental change research. Glob Environ Chang 20(2):266–277

Westwood AR, Barker NK, Grant S, Amos A, Camfield AF, Cooper K, Dénes FV, Jean-Gagnon F, McBlane L, Schmiegelow FKA, Simpson JI et al (2020) Toward actionable, coproduced research on boreal birds focused on building respectful partnerships. Avian Conserv Ecol 15(1):26

Whitten T, Holmes D, MacKinnon K (2001) Conservation biology: a displacement behavior for academia? Conserv Biol 15(1):1–3

Wilhelm-Rechmann A, Cowling RM (2013) Local land-use planning and the role of conservation: an example analysing opportunities. S Afr J Sci 109(3–4):1–6

Young JC, Waylen KA, Sarkki S, Albon S, Bainbridge I, Balian E, Davidson J, Edwards D, Fairley R, Margerison C, McCracken D, Owen R, Quine CP, Stewart-Roper C, Thompson D, Tinch R, Van den Hove S, Watt A (2014) Improving the science-policy dialogue to meet the challenges of biodiversity conservation: having conversations rather than talking at one-another. Biodivers Conserv 23(2):387–404

Part II
The Knowledge Production or Co-production

Chapter 2
Biodiversity Monitoring and the Role of Scientists in the Twenty-first Century

Catarina C. Ferreira, P. J. Stephenson, Mike Gill, and Eugenie C. Regan

Contents

C. C. Ferreira (✉)
Department of Conservation Biology, Helmholtz Centre for Environmental Research - UFZ, Leipzig, Germany
e-mail: catferreira@gmail.com

P. J. Stephenson
IUCN SSC Species Monitoring Specialist Group, Laboratory for Conservation Biology, Department of Ecology & Evolution, University of Lausanne, Lausanne, Switzerland

M. Gill
NatureServe, Virginia, USA
e-mail: mike.gill@natureserve.org

E. C. Regan
Springer Nature, London, UK

C. C. Ferreira, C. F. C. Klütsch (eds.), *Closing the Knowledge-Implementation Gap in Conservation Science*, Wildlife Research Monographs 4,
https://doi.org/10.1007/978-3-030-81085-6_2

2.1 What Is Biomonitoring and Why Is It Important?

Never, in modern times, has the concept of 'monitoring' been more present and embedded in public discourse than today. Dealing with the implications of an infectious disease spreading globally in 2020 (Covid-19), the media had to juggle terms such as 'data collection', 'tracking numbers', and 'flattening the curve', along with other more complex scientific jargon, in an attempt to make them more palatable to the public. Whilst the world wrestles with the biodiversity crisis, the climate crisis, and a global pandemic with clear links to poor environmental management, these mainstreaming efforts are also important to elevate data-driven and evidence-based decision-making. Although it is relatively straightforward to sensitize the general public to the need for, and value of, developing public health surveillance frameworks, it becomes challenging to get the same level of public attention and buy-in for other gaps between knowledge and action, especially when the focus shifts to monitoring data whose impacts are not always closely felt or understood. Yet, biodiversity is a pivotal component of our health ecosystem as supported by a significant body of knowledge showing the linkages between the emergence of infectious diseases and environmental degradation (e.g., Weiss and McMichael 2004). Further, it is widely acknowledged that all 17 of the United Nations (UN) Sustainable Development Goals (SGDs) which include measures for poverty alleviation, improved health, etc. (UN 2021), are not attainable without achieving the biodiversity-related goals. In other words, effective biodiversity conservation underpins all other human endeavours. Therefore, if we are to establish and implement credible new post-2020 global biodiversity goals, achieve the SDGs, tackle climate change, and reduce mortality from disease, we need to rigorously monitor biodiversity, its drivers, and how they change over space and time, to inform decisions, prioritize efforts, and track progress. Indeed, without quality data, it will not be possible to successfully manage biodiversity and the broader environment.

In this chapter, we look at the knowledge–implementation gap through the lens of biodiversity monitoring (hereafter, biomonitoring) to highlight the role of scientific inquiry as a stream of knowledge production in biodiversity conservation and the ways it can influence the width of the gap. Biomonitoring provides the perfect platform to discuss this. The knowledge it produces fundamentally underpins and informs all aspects of biodiversity conservation and it is one of the few fields in conservation biology most permeable to concurrently employ other sources of ecological knowledge, technological innovation, interdisciplinarity, and collaborative approaches. This means that the scientists leading these efforts likely embody the traits and skills most needed to successfully navigate and close the gap in the future, as they will continue to be integral cogs of the knowledge production machinery in this field. Note, however, that this chapter is not intended to be a thorough review of the topic of biodiversity monitoring nor will we go in-depth in all aspects of this copious field of research (for a good overview on the topic, see, for example, Collen et al. 2013). Instead, we will only highlight the elements of biomonitoring that can influence the width of the knowledge–implementation gap.

For the purposes of this chapter, we adopt the Group on Earth Observations—Biodiversity Observation Network's (GEO BON) definition of biomonitoring as the monitoring, regardless of scale, of biological diversity where the key dimensions of biological diversity include species abundance, distribution and traits, community composition, ecosystem structure and function, and genetic composition, with the end goal of supporting their wise management. Biomonitoring is central to conservation biology as a discipline, allowing the evaluation of species and ecosystem conservation status, biological responses to environmental and policy changes (e.g., Donald et al. 2007; Lepetz et al. 2009; Schmeller et al. 2012; Navarro et al. 2017), and conservation action (including restoration measures). Long-term biomonitoring, in particular, is essential to unveil important ecological patterns and phenomena in the wild that not only enhance our understanding of the natural world but also evoke the deep impact humans have on nature. Given the high temporal variability of many aspects of ecological phenomena, sustained monitoring over time is essential. Indeed, without sustained biomonitoring it would have been impossible to have learned about, *inter alia*:

- The impact of human-induced environmental contamination on wildlife, from the effect of pesticides on the breeding ecology of birds of prey in the United States (U.S.; Hickey and Anderson 1968) to the impact of air pollution on the prevalence of colour morphs of industrial melanic moths in the United Kingdom (U.K.; reviewed in Cook and Saccheri 2013). These studies were significant pieces of the scientific argument used to convince the U.S. government to ban use of the pesticide Dichlorodiphenyltrichloroethane (commonly known as DDT) in the 1970s to protect declining populations of some birds of prey, and to show the effects of environmental legislation like Clean Air Acts in the U.K.;
- The deep transformations ecosystems undergo following apex predator eradication and subsequent reintroduction, as typified by the well-known case of wolves (*Canis lupus*) in Yellowstone National Park in the U.S. (reviewed in Peterson and Peterson 2020) and the trophic cascade of ecological change they enabled, including helping to increase beaver populations and bringing back vegetation as a direct consequence of elk predation;
- The fascinating lives of cyclical species, such as snowshoe hares (*Lepus americanus*) and lemmings in northern Canada, and the impact their population dynamics have on the ecosystems and the human activities they support (e.g., Krebs 2011);
- How soil microbes are accelerating greenhouse-gas production in the permafrost system in the Arctic, the Earth's largest terrestrial carbon sink, as a consequence of rising temperatures caused by climate change (e.g., Brouillette 2021); and
- Fitness costs and the loss of vocal culture in the regent honeyeater (*Anthochaera phrygia*), a critically endangered bird endemic to southeastern Australia, as a result of a severe population decline caused by widespread habitat loss (Crates et al. 2021).

The scientific literature in biodiversity conservation is rich in examples describing natural patterns and/or phenomena detected by well-designed biomonitoring programmes reinforcing their value to better understand nature and how we can most benefit from it without disrupting its balance.

Biological records are 'what, where, when' observations that record, for instance, the presence of a species in a particular place at a particular time, and they can be collected in a variety of ways (Isaac and Pocock 2015). Observations constitute the source datasets that are the basic building blocks of any given biomonitoring system and of which biological variables, like species abundance and distribution, or key aspects and traits of ecosystems, such as structure and function, are derived (Guerra et al. 2019). National and international data centres, such as the Global Biodiversity Information Facility (GBIF; www.gbif.org) and the Ocean Biogeographic Information System (OBIS; https://obis.org/), aggregate and disseminate biological data, with the mission of making biodiversity data available for the scientific community, governments, and non-governmental organizations (NGOs; Gaul et al. 2020). Hence, biomonitoring provides the data needed to make informed decisions about natural resource management and use, whether at local levels (e.g., a local community monitoring the state of a small forest patch being restored), national levels (e.g., governments measuring their contributions towards globally-committed conservation goals), or regional levels (e.g., conservation NGOs monitoring populations of species targeted by recovery programmes). Biomonitoring is also essential for priority setting by helping the scientific and conservation communities identify species and habitats most at risk and in need of conservation action. For businesses, biomonitoring allows sustainability planning and biodiversity performance disclosure. Mascia et al. (2014) analysed some of the most common approaches to conservation monitoring and evaluation to provide a framework to determine which question and approach can best respond to the most common information needs in conservation. They concluded that, despite some commonalities, there is a tension between approaches that (1) by design, are useful in resource-constrained contexts, provide valuable information quickly at a low cost, albeit with less rigour and certainty (because they do not explicitly test—and often assume—the validity of the underlying program logic), and (2) those that require substantial time, expertise (because they attempt to explicitly test or examine the validity of theories of change), and financial investments, which are not always readily available. These different approaches to biomonitoring will have different effects on the magnitude of the knowledge–implementation gap as the level of information accuracy they produce will differ substantially.

Despite their importance to so many stakeholders, evidence suggests biomonitoring programmes were gradually reduced or even cut in recent decades because of financial considerations (e.g., Birkhead 2014) and due to a lack of understanding of the value that biomonitoring brings to decisions affecting human well-being (Lindenmayer and Likens 2018). Moreover, in some instances, biomonitoring became somewhat stigmatized as of lower publication value, frequently ending up in low impact journals or as reports in the 'grey' literature (Robinson et al. 2011). Lately, however, a resurgence has occurred in which the number of papers being published using long-term datasets and analysed with sophisticated statistical methods has grown (e.g., McCarthy et al. 2006). The latter was also encouraged by the creation of specific article types (i.e., data papers) in a mounting number of scientific journals in alignment with recent open data policies and to promote the recycling of ecological time-series collected across taxa, biomes, and scales (e.g., PREDICTS—Hudson et al. 2017; BioTIME—Dornelas et al. 2018). Further, resource managers and researchers are increasingly aware that their knowledge of temporal changes in ecological communities is limited and does not always respond to increasing demands for a greater ability to make predictions about ecosystem changes that result from the rapid transformations caused by global environmental change (Clark et al. 2001; Hanson 2008). For these reasons there has been an upsurge of interest in biomonitoring programmes, particularly those that are long-term, as they offer a better elucidation of eco-evolutionary patterns and increase the probability of finding rare stochastic-pulse, time-lagged, or threshold responses (Franklin 1988). Since long-term biomonitoring programmes grow in value with time, they may also be used to investigate scientific questions that were inconceivable at their inception (Dodds et al. 2012). Designing such biomonitoring schemes and analysing the resulting datasets, however, require a high level of scientific proficiency and sophistication that typically reside in the scholarly community (inside or outside academia). At the same time, they require the foresight and ability to ensure a user-driven design that includes decision- and policy-makers (at relevant scales) where the outputs of the biomonitoring scheme directly and clearly serve key management and policy objectives that will ensure sustained funding (Navarro et al. 2017).

2.1.1 The Importance of Scale in Biomonitoring

Scale is an important feature of biomonitoring schemes that can contribute strongly to widen the knowledge–implementation gap if its impacts are not properly accounted for and translated. The scientific literature on the importance and impacts of scale in biomonitoring is prolific (e.g., Lindenmayer and Likens 2010, 2018; Schmeller et al. 2017; Vellend et al. 2017; Stephenson 2019) and so this chapter will only briefly mention some of the elements of this discussion most relevant to the knowledge–implementation gap. Matching the appropriate scale to the conservation question and context at hand fosters clear expectations about what will be measured, how it will be measured, and the insights for conservation policy and practice that are likely to emerge as a result (Mascia et al. 2014). When it comes to scale, biomonitoring typically considers data collection and measurements across space and time.

Temporal scale is key to biomonitoring as time-series data are needed to differentiate between natural patterns of population variability over time and changes induced by specific human interventions or natural phenomena. The frequency and duration of measurements are a key part of a biomonitoring plan that critically dictate the contribution of the plan to the knowledge–implementation gap. Many conservation projects only monitor biodiversity for a relatively short period of time—often

1–3 years (e.g., Lindenmayer and Likens 2010; Badalotti et al. 2021), even though a change in biodiversity state may take up to ten years or more (Krebs 2011; Stephenson 2019). This is even more critical when the goal of biomonitoring is to measure the impact of a specific intervention. Longer-term biomonitoring schemes are therefore essential to capture these changes and several such initiatives exist around the world. For example, the Long-Term Ecological Research (LTER) Network was started in 1980 by the U.S. National Science Foundation to address the need for research projects that were longer than traditional funding periods. This new model had a commitment to continuous funding for basic ecological research across a network of sites. International recognition of the merits of this funding model has stimulated the establishment of long-term ecological sites in other countries in the last decade (Hanson 2008). Differences in the temporal scale of biomonitoring schemes can, therefore, be a major source of variation that needs to be considered when transferring the knowledge they produce. This is relevant because the detection of (or lack thereof) an impact, pattern, or phenomenon could be the result of an inadequate design or too short of a duration of the scheme instead of a real observation. Time is the only factor that cannot be replicated in ecology, so it is important to understand that the longer and more frequently we look at nature the more we will understand it. Not clearly communicating this to policy- and decision-makers will contribute to widening the gap. Given this, it is essential that biomonitoring schemes go through a structured design process that takes into account the key variables for targeted biomonitoring (Navarro et al. 2017) but also takes advantage of existing data to ensure quick wins that leverage historical time-series when available.

Spatial scale is another key element of biomonitoring programmes that can create ambiguity if not properly translated. The spatial scale of biomonitoring depends very much on the aim of the monitoring scheme, although most biomonitoring occurs at national or sub-national levels (Navarro et al. 2017). As such, biomonitoring has traditionally been largely a decentralized activity due to the great diversity in biomonitoring objectives (Marsh and Trenham 2008), often focused on particular sites (often protected areas) or species (Durant 2013). This can make it difficult to use biodiversity monitoring data on a larger scale (such as national), although this is typically the scale at which countries need to report on and are accountable for. This also means that biodiversity patterns/trends can vary widely depending on the spatial scale they are inspected at. One of the best illustrations of this issue comes from plant community ecology, where meta-analytical studies have shown disparate future trends in plant species diversity and richness at the global versus local, with some nonnative species introductions actually greatly increasing plant species richness in many regions of the world (Vellend et al. 2017). Accounting for scale in analyses is challenging, but detected trends in metrics such as species richness can differ markedly across scales (Chase et al. 2019). This goes to show that experiments conducted at small spatial scales could contradict extinction estimates taken at large scales, contributing to generating confusion if not properly framed and translated to practitioners and policy-makers. Given that many drivers of biodiversity change operate at scales beyond which an individual organization can adequately monitor, it is imperative that biomonitoring efforts are interoperable. Such an approach is being deployed by GEO BON where national, regional, and thematic biodiversity observation networks are being established utilizing the Essential Biodiversity Variables (EBVs) framework (Pereira et al. 2013) thereby allowing effective aggregation and scaling of data across space and time.

Translating knowledge on biodiversity trends and status acquired at different spatial and temporal resolutions is not straightforward and can cause confusion if not adequately communicated to the people who will take action on that knowledge. In this context, a huge point of contention involves effectively communicating scientific uncertainty around these estimates (Rowland et al. 2021), and whilst we will not go into detail on this issue here (but see, for example, Fischhoff and Davis 2014), this is often a major contributor to the knowledge–implementation gap in conservation science.

Many now see us entering a period of 'big data', which is also relevant for biodiversity (Stephenson et al. 2017b). Surveillance approaches to biomonitoring (*sensu* Earth Observation Networks-EON, see Sect. 2.4), which are generally not guided by specific (policy or scientific) questions with a robust experimental design, have become increasingly popular generating large streams of (often real-time) data (Lindenmayer et al. 2018). They also bring additional challenges for storing, processing, and, more importantly, for our capacity to analyse outcomes and communicate them timely and properly. If left unattended, these technological advances could create a widening of the knowledge–implementation gap in biodiversity conservation in the future. Preventing that will require scientists to become more specialized but also interdisciplinary, collaborative, and creative. In this context, Lindenmayer et al. (2018) provide an in-depth analysis of the benefits and limitations of the EON model *versus* the traditional question-driven, experimentally designed environmental monitoring with recommendations for scientists on how to navigate this surge in 'big data' streams in biomonitoring. In a nutshell, the authors argue that there is a strong complementarity between the two approaches but that it is critical to ensure that scientists are (1) intimately involved in guiding data gathering by EONs; (2) have both a broad and deep understanding of their field; (3) are alert to potential discoveries; and (4) are capable of verifying those discoveries by posing focused, high-quality questions, and testing well-formulated hypotheses (Lindenmayer et al. 2018).

Ironically, despite rising efforts to build these large-scale surveillance networks, multiple geographical and taxa biases (Pereira et al. 2012) still persist in biomonitoring, and temporal baselines are generally lacking, often hampering a full understanding of the causal relationships between the incidence of anthropogenic stressors and biodiversity (Mihoub et al. 2017), as well as further contributing to widening the knowledge–implementation gap in conservation science. In the next sections, we will outline the main features of traditional science-based biomonitoring, which will always be a cornerstone of knowledge production in conservation science, with an emphasis on some of the challenges faced by the scientific community that may contribute to widening the knowledge–

implementation gap in the discipline, as well as successful expert-driven initiatives that provide a good template to resolve said challenges.

2.2 Where Do Scientists Fit in Modern Biomonitoring?

In conservation biology, biomonitoring, from scheme design to implementation, has traditionally been carried out by scientists (usually based in academia or governments), with outputs and their communication typically presented following an information-deficit model, a one-way movement of knowledge from experts to the lay public (Haklay 2017; see also Chap. 4; Phillips et al. 2021). However, this is changing fast as the ever-evolving technology allows for a greater engagement of non-scientists (such as citizen scientists) and other forms of knowledge (like local or indigenous) at several development stages of the monitoring schemes (Chap. 5; Ens et al. 2021; Chap. 4; Phillips et al. 2021). Indeed, cross-cultural monitoring systems are becoming more common as they provide an opportunity for indigenous peoples and local communities to apply their traditional ways of knowing, and interpret and act on information they understand, also enabling a degree of knowledge of biodiversity states that both managers and communities can use for decision-making (Lyver et al. 2018; Chap. 5; Ens et al. 2021). Further, indigenous and local monitoring have the advantage of providing, in many cases, intra-annual, cross-seasonal monitoring that is often not feasibly implemented by professional scientists alone.

Nowadays, biomonitoring projects are conducted by academic, government, and regulatory institutions, as well as NGOs and corporations, and have varying degrees of civil society involvement. There is a growing recognition that the equitable participation of different stakeholders, data providers and users, including local communities, is central to adaptive management and leads to better results, buy-in, and sustainability (Jacobson et al. 2009; Danielsen et al. 2014; Stephenson et al. 2017a; Chap. 15; Klütsch and Ferreira 2021). Therefore, a lot of new biomonitoring work involves participatory science practices which represents a departure from traditional models of reporting aggregated study results in ways that are limited to academic settings, such as professional meetings and peer-reviewed publications (Chap. 7; O'Connell and McKinnon 2021). As well as increasing use of citizen scientists in data collection, another major trend is for the growing use of technology in biomonitoring (Stephenson 2020). Rapid improvements in computing, mobile phones, and remote sensing technologies (such as camera traps, acoustic recording devices, next-generation sequencing, environmental DNA, etc.) will augment the amount and quality of biological records gathered (Bohan et al. 2017; Derocles et al. 2018; Kelling 2018). This has huge implications for biomonitoring, conservation science, and understanding ecological systems, as these advances and increased reliance on technology, whilst broadly beneficial, also present multiple caveats for science-based biomonitoring. For example, whilst the remote collection of data using citizen scientists may decrease the need for academics to go to the field, thereby reducing costs of personnel and equipment across a potentially larger monitored area, this poses serious challenges to the integration of the information at multiple levels (across data, schemes, and stakeholders) that have yet to be fully realized and understood (but see Lyver et al. 2018; Alexander et al. 2019; Kühl et al. 2020). Importantly, the challenges and opportunities created by these developments are in themselves a conduit for more scientific research and already support a prolific body of knowledge and the emergence of new subdisciplines, such as conservation technology (Wich and Koh 2018; Stephenson 2020; Wich and Piel 2021).

Moreover, the science for studying and monitoring biodiversity is increasingly a transdisciplinary field that uses tools and theories from many areas of study and different disciplines, from social sciences to information technology to business management. As the scientific community adjusts to this contemporary way of doing biomonitoring and continues to investigate new ways to bring meaning to an ever-expanding volume of biodiversity knowledge, generated in multiple ways at multiple levels, it is important not to weaken the robustness of monitoring schemes or erode the skills and scientific training and professionalism that underpin them. Some scientific disciplines needed for robust biomonitoring, like natural history (Tewksbury et al. 2014) and professional taxonomy, are already considered 'at-risk' in many regions of the world, with documented declines, for example, in Canada, Europe, and New Zealand (The Expert Panel on Biodiversity Science 2010; https://wikis.ec.europa.eu/display/EUPKH/European+Red+List+of+Taxonomists; New Zealand Parliamentary Commissioner for the Environment 2020; Hochkirch et al. 2020). Discovering and characterizing species, and their ecology, as well as observing them, requires scientists empowered, not overshadowed, by leading technologies and collaborations with other stakeholders, to contribute to collections and databases as part of a global biodiversity conservation science (Fischer et al. 2021).

2.3 Challenges Associated with Biomonitoring That Widen the Knowledge–Implementation Gap

A number of major challenges to effective biodiversity conservation, monitoring, and evaluation remain that prevent the full and effective use of biodiversity data in decision- and policy-making processes and widen the knowledge–implementation gap (reviewed in Jones 2013). These include financial constraints (Birkhead 2014, 2018), and lack of technical capacity and tools for collecting, analysing, and interpreting data (Navarro et al. 2017; Addison et al. 2020; Bhatt et al. 2020; Hochkirch et al. 2020; Stephenson 2020; Stephenson et al. 2017a, 2020). Institutional, policy, and legal barriers may also be important challenges to biomonitoring (Collen et al. 2013; Xu et al. 2021).

2.3.1 Biases

As previously mentioned, biomonitoring endeavours suffer from taxonomic and geographic biases (Amano et al. 2016; McRae et al. 2017; Troudet et al. 2017; Fabian et al. 2019; Stephenson and Stengel 2020; Moussy et al. 2021), with large mammals and birds, and Europe and North America, over-represented in the datasets, as well as in the monitoring schemes collecting the data (Pereira et al. 2012; Moussy et al. 2021; Badalotti et al. 2021). What is known about biodiversity change is complicated by these taxonomic, geographical, and temporal biases (Pereira et al. 2012). Overall, trends in processes and species groups that are important for the maintenance of healthy ecosystems and that may be undergoing significant change are missing. For example, there is still little or no information on many taxa important for people, such as invertebrates necessary for pollination and soil health, and trends for most non-commercial species, non-flowering plant species, invertebrates, and smaller organisms, like soil bacteria. The result is that trends for these ecosystem components are normally not reported and hence not considered in policy- or decision-making contributing to the gap in action. Taxonomic expertise and scientific enquiry will be critical to fill in these gaps, and some efforts are already underway. For instance, a global Soil Biodiversity Observation Network is being implemented to address a highly under-represented but extremely important group of organisms (Guerra et al. 2020).

2.3.2 Technical Capacity and Data Needs

Many stakeholders do not have access to the biodiversity data they need and advances in technological tools, like remote sensing and environmental DNA, have further increased this gap by leaving many actors behind (Stephenson 2020). There are also data accessibility issues with many databases (Stephenson and Stengel 2020) and many institutions fail to follow data management best practices (Wilkinson et al. 2016). Where data accessibility is not a barrier, data are frequently scattered, fragmented, of poor quality, and rarely available in the right format at the right time (Nesshöver et al. 2016; Kissling et al. 2018; Stephenson et al. 2017a,b; Stephenson and Stengel 2020). Variability in the spatial and temporal resolution of data, a lack of willingness to share information, and the failure to link risks and dependencies to actions also plague governments and businesses (Walls et al. 2012; Whiteman et al. 2013; Stephenson et al. 2017a). Consequently, many stakeholders, from governments to businesses to conservation NGOs, struggle to identify appropriate indicators for monitoring biodiversity, sources of existing data they can use in their own planning and reports (Walpole et al. 2009; Bubb 2013; Stephenson et al. 2015; Bhatt et al. 2020; Han et al. 2020; Stephenson and Carbone 2021; see also Chap. 11; Stephenson et al. 2021), and the relevant monitoring tools for collecting their own data (Navarro et al. 2017; Addison et al. 2020; Stephenson 2020;

Stephenson et al. 2020; Stephenson and Stengel 2020). Ensuring data providers and users collaborate on producing data and data-derived products in formats that meet decision-makers' needs (in being, for example, brief, understandable, timely, and iterative) is an important first step in increasing willingness to use data (Segan et al. 2011; Sanchirico et al. 2014; see also Chap. 8; Rose et al. 2021). Therefore, solutions proposed so far to unblock the flow of biodiversity data often focus on developing science-policy fora to enhance knowledge transfer between data providers and users (Young et al. 2014; Stephenson et al. 2017a, b). Similar approaches applying data to decision workflows that connect raw data to Essential Biodiversity Variables to indicators and to decisions are being employed at different scales around the world (Turak et al. 2016; Navarro et al. 2017; Schmeller et al. 2017; Muller-Karger et al. 2018).

There is also a strong need to build stakeholders' capacity to collect, use, and share data in easy-to-interpret formats (Tittensor et al. 2014; Stephenson 2019; Stephenson et al. 2015, 2017a, b, 2020; Stephenson and Carbone 2021). However, few concrete solutions have been proposed to date to meet identified user needs (Stephenson et al. 2017b; Fabian et al. 2019; see also Chap. 8; Rose et al. 2021). Various platforms exist for accessing or mapping data (Han et al. 2014; Wilkinson et al. 2016; Stephenson and Stengel 2020), and some efforts have been made to collect tools for certain sectors (e.g., Lammerant et al. 2019) and specifically for biodiversity monitoring (GEO BON 2021) that align needed methods and tools to specific monitoring objectives but more investments in these approaches are needed. The scientific and technological expertise exists; it is just not always available to the right people at the right time to facilitate necessary biomonitoring.

2.3.3 Funding

More than half of the world's species monitoring schemes are government funded (Moussy et al. 2021) which leaves them vulnerable to political sways and changes in priorities. Often, funding structures are designed to serve short-term scientific research needs with few existing funding sources designed to serve the long-term nature of biomonitoring schemes. A recent review of species population monitoring schemes around the globe (Moussy et al. 2021) showed a significant relationship between resource availability and the duration of monitoring: active monitoring projects in high-income countries had been running an average of 21 years, compared to only 10 years in low-income countries. Government funding for biomonitoring was also less in low-income countries. Funding, therefore, plays a major role in the stability and longevity of biomonitoring schemes. Even conservation projects that need to know if they have delivered a result often do not invest a big enough proportion of their budgets in biomonitoring (Stephenson 2019). Further, even when biomonitoring projects are well designed they may not be assigned a budget for communication and dissemination activities to maintain close alignment with the goals of the funding agencies that they serve, resulting in the value of long-term time-series data in driving informed decisions being invisible to policy-makers. Beyond applying more user-driven design approaches to biomonitoring, expanding the diversity of actors with a stake in biomonitoring (e.g., the private sector) will help alleviate, to some degree, these challenges (Chap. 15; Klütsch and Ferreira 2021). Large multinational corporations with large budgets and a desire to contribute to the SDGs and enhance their sustainability can surely fit the bill to conduct more of the necessary biomonitoring.

2.3.4 *Institutional Support*

Institutional arrangements also frequently play a role in how well knowledge flows into implementation. Many biomonitoring programmes become episodic, characterized by short-lived commitments by the agencies involved, and reflecting shifting priorities and competing institutional mandates. Consequently, stakeholders often raise concerns about their transparency, credibility, and influence over decision-making (Cronmiller and Noble 2018). This generates significant uncertainty about the stability of institutional arrangements to support long-term environmental monitoring, and tensions between the need for scientific autonomy for credible science whilst ensuring the pursuit of monitoring questions that are relevant to the day-to-day needs of regulatory decision-makers (Cronmiller and Noble 2018). Regional monitoring programmes require, at a minimum, clear vision and agreed-upon monitoring questions (Lindenmayer and Likens 2010) that are of scientific and management value, meaningful, and balanced stakeholder engagement, with a clear governance process to ensure credibility and influence of biomonitoring results on decision-making (Cronmiller and Noble 2018). Sometimes lack of institutional support for biomonitoring reflects a lack of willingness to identify and share data trends for fear of negative results (Stephenson et al. 2017a). This can only be overcome by introducing a results-based management culture and a safe space to fail, learn, and adapt.

Long-term, standardized, spatially complete, and readily accessible monitoring information, complemented by ecosystem research, provides the most useful findings for policy-relevant assessments of biodiversity status and trends. The lack of this type of information in many areas has hindered progress in incorporating biomonitoring knowledge into conservation action (Lyver et al. 2018).

2.4 Successes in Bridging the Gap

Despite the challenges outlined above, there are now many examples of expert-driven biomonitoring schemes that have successfully circumvented them and are auspiciously filling the knowledge–implementation gap in biodiversity conservation.

There is an increased demand for biodiversity information from a range of decision-makers including from intergovernmental processes, governments, private sector, and NGOs. Examples of this demand include the expanded biodiversity indicators set for the UN SDGs compared to the biodiversity indicators for the Millennium Development Goals. This increased demand has, in some cases, been harnessed in a structured way to design biodiversity monitoring programmes that directly and successfully serve such demand (Navarro et al. 2017).

The best examples of sustained biomonitoring efforts are those that are inclusive, taking a user-driven approach that not only engages scientists, but also key decision-makers (at local, sub-national, and/or national scales), that are flexible in approach utilizing existing data and monitoring capacity and which recognize the need not only for robust design and implementation but on the essential requirements and investment in analytics, data management, and communications. In the following paragraphs, we will highlight a few examples, of many we could list, that illustrate ongoing global initiatives to collect, collate, harmonize, share, and use biodiversity data, and the commonalities across such endeavours that are the key to closing the knowledge–implementation gap.

2.4.1 Indicators, Data, and Databases: Global Initiatives to Monitor Biodiversity

As previously argued, by their very nature, biodiversity monitoring programmes need to be sustained over the long-term in order for their value to be fully realized. In order for them to be supported, they need to be closely aligned with key policy needs and effectively addressing those needs through the sustained production of high-quality biodiversity data (Legg and Nagy 2006). Further, due to the nature of many drivers of biodiversity change operating at scales beyond which a single organization, let alone country, has the capacity to fully detect, understand, and attribute, the most successful biodiversity monitoring approaches are those that take into account interoperability (Pereira et al. 2013; Navarro et al. 2017). This means that observations need to follow certain standards for both data collection and data management in order to facilitate effective data aggregation across scales and integration of these data with other relevant datasets, particularly those concerning stressors on biodiversity (Scholes et al. 2016).

In the last two decades, primarily driven by the need to monitor policy objectives set around the Aichi Biodiversity Targets (https://www.cbd.int/sp/targets) and the SDGs, several global efforts have been made to enhance the collection and use of biodiversity data. We share a few examples here.

The Biodiversity Indicators Partnership (BIP—https://www.bipindicators.net) is a global initiative to promote and coordinate the development and delivery of biodiversity indicators for use by the Convention on Biological Diversity (CBD—https://www.cbd.int/) and other biodiversity-related conventions (e.g., Ramsar

Convention—https://www.ramsar.org, Convention on International Trade in Endangered Species of Wild Fauna and Flora—https://cites.org/eng), the Intergovernmental Science-Policy Platform on Biodiversity and Ecosystem Services (IPBES; https://www.ipbes.net), the SDGs (https://sdgs.un.org/goals), and national and regional agencies. The Partnership successfully spans the knowledge to decision-making gap at the global biodiversity policy scale feeding indicators into numerous intergovernmental processes since its establishment in 2007. Added to this, numerous scientific publications have reviewed the data arising from the biomonitoring of the BIP biodiversity indicators to provide overviews of global biodiversity trends (e.g., Walpole et al. 2009; Butchart et al. 2010; Tittensor et al. 2014; Secretariat of the Convention on Biological Diversity, 2020).

Monitoring of several biodiversity indicators, for the Aichi Targets and SDGs in particular, has been facilitated by global databases that have been established to collate data from national, regional, and global biomonitoring. These include the IUCN Red List of Threatened Species ™ (IUCN 2021—https://www.iucnredlist.org), the Living Planet Index (ZSL 2021—https://www.livingplanetindex.org/home/index), the World Database on Protected Areas (UNEP-WCMC 2021—https://www.protectedplanet.net/en), and the GBIF (GBIF 2021). Several data platforms are also providing access to up-to-date (and sometimes near real-time) data on biodiversity and its threats that are collected by satellite-based remote sensing. For example, the World Resources Institute's Global Forest Watch (http://www.globalforestwatch.org/) provides daily updates on forest loss and active fires around the world. The US National Oceanic and Atmospheric Administration (NOAA) provides data daily on sea temperature, flagging coral reefs where bleaching is most likely, through its Coral Reef Watch (https://coralreefwatch.noaa.gov/satellite/index.php).

Another example of biodiversity monitoring data feeding into international policy processes through the BIP is the Wetland Extent Index. This index was created to give a global picture of trends in wetland extent over time (Dixon et al. 2016) and was used in key global assessments including the IPBES Assessments (e.g., IPBES 2016, 2018a,b), the Ramsar Convention's Global Wetland Outlook (Ramsar Convention on Wetlands 2018), and UN SDG 6 Synthesis Report (UN 2018). The index was developed from over 2000 individual time-series records of change of wetland area from local sites and national trends collated through a systematic literature review (Darrah et al. 2019) and is a notable example of a successful initiative addressing a specific knowledge to decision-making gap.

Other initiatives harmonizing systems and building capacity for data collection and use include the Eye on Earth Alliance (www.eoesummit.org), the IUCN Species Survival Commission's Species Monitoring Specialist Group (www.speciesmonioring.org; Stephenson 2018), and the work of GEO BON and its partners on the EBVs (Pereira et al. 2013; Kissling et al. 2015) and the systematic implementation of national, regional, and thematic biodiversity observation networks (Navarro et al. 2017).

2.4.2 Biodiversity Observation Networks

The GEO BON (https://geobon.org/) is a global network of networks working collaboratively to improve the acquisition, coordination, and delivery of biodiversity observations and related services to users including decision-makers and the scientific community. Since its inception in 2008, GEO BON has developed a global community of practice for biodiversity observations (Scholes et al. 2008, 2012) employing both a top-down and bottom-up approach. The top-down approach includes the development and implementation of the EBVs framework (Pereira et al. 2013) and common data standards and collection protocols (e.g., Darwin Event Core—an extension of Darwin Core Standard (Wieczorek et al. 2012) and Guidelines for Standardized Global Butterfly Monitoring (Van Swaay et al. 2015)) to ensure a consistent structuring of biodiversity observation systems to promote interoperability. The bottom-up approach includes the development and provision of a clearinghouse of state-of-the-art tools for biodiversity data collection, management, analysis, and reporting (BON in a Box; https://boninabox.geobon.org/; Schmeller et al. 2017) and structured guidance on the design and implementation of user-driven, interoperable, and scalable biodiversity observation networks operating at sub-national, national, or regional scales. The BON in a Box online toolkit is designed to lower the threshold for the development or enhancement of a biodiversity observation system through the accelerated transfer of best practice techniques and technology for biodiversity monitoring. The structured yet flexible guidance for designing and implementing a biodiversity observation system involves a nine-step design process that takes into consideration and addresses many of the barriers and common mistakes made in the design and implementation of biodiversity monitoring programmes that often result in their eventual failure (for a full overview of the process, see Navarro et al. 2017). This includes ensuring that the observation network is designed to directly serve user needs, has secured, at the outset, an authorizing environment, and takes an adaptive monitoring approach (Lindenmayer and Likens 2009, 2010) that involves clearly articulating monitoring objectives and questions to be answered and utilizes a conceptualization of the ecosystem to guide the selection of key variables for monitoring. The nine-step design process also incorporates planning techniques to account for data management, analysis, and reporting outputs.

The approach employed by GEO BON has been successfully applied in regions worldwide leading to a number of successfully established and sustained biodiversity monitoring efforts in the Arctic (Circumpolar Biodiversity Monitoring Programme (Gill et al. 2011)), China (China Biodiversity Observation Network (Xu et al. 2017)), and in the Asia-Pacific Region (Asia-Pacific Biodiversity Observation Network (Yahara et al. 2014)). This approach is also being flexibly applied and adapted to local circumstances in Colombia, France, the Tropical Andes, and throughout the Americas.

2.4.3 Businesses and Biomonitoring

Businesses are increasingly aware of biodiversity risks, dependencies, and opportunities for their operations. This growing awareness and action on their part is evident through increased presence at international meetings (e.g., the 2018 CBD Convention Of the Parties 14 Business and Biodiversity Forum in Egypt), increased publishing of biodiversity guidance (*inter alia*, Ekstrom et al. 2015; Hardner et al. 2015; The KBA Partnership 2018; Bennun et al. 2021; Stephenson and Carbone 2021), and increasing corporate biodiversity commitments (De Silva et al. 2019). There is also an increased demand from businesses for biodiversity information to inform their decision-making (*inter alia*, Cooper and Trémolet 2019; IBAT Annual Report 2019).

Bennun et al. (2017) outlined the value of the IUCN Red List of Threatened Species for business decision-making and illustrated how the range of data within the IUCN Red List has many uses across a development project's life cycle. This paper demonstrated the data-information-knowledge-wisdom journey (Baskarada and Koronios 2013) that biodiversity data go through from its raw format as data that are used within the IUCN Red List process to generate a species assessment (i.e., information) to a point in time where a decision-maker interacts with that information to understand the whole biodiversity situation at a location at one point in time in order to make a decision. Within the Integrated Biodiversity Assessment Tool (IBAT; www.ibat-alliance.org), the World Database on Protected Areas and the World Database of Key Biodiversity Areas are, together with the IUCN Red List, disseminated to decision-makers who use these datasets (gathered originally by scientists on the ground) to make various decisions such as screening projects against the World Bank's Environmental and Social Framework (ESF). The Species Threat Abatement Metric will also soon be tested in IBAT to provide businesses in particular with a method for planning their contribution to biodiversity (Mair et al. 2021).

Private sector environmental frameworks such as the World Bank's ESF, the International Finance Corporation's (IFC) Environmental and Social Performance Standards, and the Organisation for Economic Co-operation and Development (OECD) Common Approaches require specific biodiversity information to support decision-making processes within these frameworks. This includes information on protected areas and 'other sensitive locations', such as wetlands and forests, with high biodiversity value for the Common Approaches, and information on 'Critical Habitat' for the World Bank and IFC. Critical Habitat is defined as any area of the planet with high biodiversity value, including (i) habitat of significant importance to Critically Endangered and/or Endangered species; (ii) habitat of significant importance to endemic and/or restricted-range species; (iii) habitat supporting globally significant concentrations of migratory species and/or congregatory species; (iv) highly threatened and/or unique ecosystems; and/or (v) areas associated with key evolutionary processes (IFC 2012).

There is also increasing discussion and guidance on the need for companies themselves to assess and monitor biodiversity (Cousins and Pittman 2021; Stephenson and Carbone 2021). Added to this, there is a global growth in national policies for private sector biomonitoring (zu Ermgassen et al. 2019). Data is also already collected by businesses during Environmental Impact Assessments and legally-obligated monitoring schemes. However, more effort should be made to share such data (see also Chap. 11; Stephenson et al. 2021).

2.4.4 *Commonalities Between Initiatives*

Based on the examples described above, we can identify a number of common traits among efforts to improve biomonitoring that are key ingredients in closing the knowledge–implementation gap in biodiversity conservation. They include:

- Clearly defined users and clearly articulated policy objectives that are used to drive the development of the biodiversity monitoring programme to ensure its outputs are relevant;
- Active participation of policy- and decision-makers alongside scientists in the design and implementation of the biodiversity monitoring programmes to ensure they are designed to directly serve the user needs and policy objectives;
- Early, targeted, and continual outputs that serve to strengthen support for and the profile of the monitoring programme; and
- Adoption of common approaches to facilitate spatial scaling of the monitoring approach.

Such attributes ensure high-quality, continual, and relevant outputs that greatly increase the probability of their sustained operation. When they focus on addressing widely adopted policy frameworks and serving mandated reporting requirements, monitoring efforts become indispensable.

2.5 Concluding Remarks and Recommendations

Whilst it seems logical that scientists will continue to play a pivotal role in closing the knowledge–implementation gap in conservation science, this chapter tried, through the lens of biomonitoring, to highlight aspects of data collection and use that cannot be taken for granted, either because they are under strain or require further development and support. In the field of biomonitoring, there is already a documented erosion of certain skills (for example, taxonomy) and a greater reliance on technology (e.g., remote sensing) and other types of data gatherers (e.g., citizen scientists) that can further distance scientists from nature, increasing society's insulation from ecological realities and natural processes. This trend cripples a vital element of the knowledge production sphere that goes beyond mere data collection and allows for the mechanistic understanding of patterns detected, thereby handicapping the scope of action it can trigger. Hence, it is imperative that a stronger investment is made in supporting the professional skills that are foundational to sustain these activities, including by rising levels of recruitment and retention of young professionals in the field. Moreover, a greater reliance on technology and other sources of knowledge (indigenous and local) will require greater degrees of scientific sophistication and specialization (i.e., a broader and deeper knowledge of the field) but also collaboration to enable a successful integration of these different knowledge streams into impactful conservation action.

Biomonitoring endeavours of the future will therefore require a scientific workforce that is increasingly targeted, inter- and transdisciplinary, collaborative, and integrative. It will require that scientists in this field of research become more comfortable working cross-sectorally, either to engage more effectively with policy- and decision-makers as co-developers of biomonitoring systems (to ensure the definition of users, monitoring objectives, and a clear link to decision-making), or to leverage these partnerships and collaborations to garner adequate long-term funding (which is critical to sustaining commitments to biomonitoring), especially from the private sector. Current challenges in data collection, standardization, sharing, transfer, storage, and mainstreaming will only be successfully addressed through more collaborative, transboundary research.

Knowledge alone will not ensure that the conservation community replicates successes, reforms failures, and avoids repeating the mistakes of the past. Effectively addressing the knowledge–implementation gap in biodiversity conservation will require a transformation of the way scientists view and engage with conservation policy and practice actors, and the development of interpersonal traits and specific expertise that advance knowledge brokerage to inform policy, build capacity for the required biomonitoring (particularly in low-income, high-biodiversity countries), and galvanize a culture of evidence-based decision-making.

Acknowledgments Catarina C. Ferreira was supported by Marie Curie Outgoing International Fellowship for Career Development no. PIOF-GA-2013-621571 within the Seventh Framework Programme of the European Union.

References

Addison PFE, Stephenson PJ, Bull JW, Carbone G, Burgman M, Burgass MJ, Gerber LR, Howard P, McCormick N, McRae L, Reuter KE, Starkey M, Milner-Gulland EJ (2020) Bringing sustainability to life: a framework to guide biodiversity indicator development for business performance management. Bus Strateg Environ 29(8):3303–3313

Alexander SM, Provencher JF, Henri DA, Taylor JJ, Lloren JI, Nanayakkara L, Johnson JT, Cooke SJ (2019) Bridging Indigenous and science-based knowledge in coastal-marine research, monitoring, and management in Canada: a systematic map protocol. Environ Evid 8(1):36

Amano T, Lamming JDL, Sutherland WJ (2016) Spatial gaps in global biodiversity information and the role of citizen science. Bioscience 66(5):393–400

Badalotti A, van Galen L, Vié J-C, Stephenson PJ (2021) Improving the monitoring of conservation programmes: lessons from a grant-making initiative for threatened species. Oryx, 1–7

Baskarada S, Koronios A (2013) Data, Information, Knowledge, Wisdom (DIKW): a semiotic theoretical and empirical exploration of the hierarchy and its quality dimension. Australas J Inf Syst 18(1)

Bennun L, Regan EC, Bird J, van Bochove J-W, Katariya V, Livingstone S, Mitchell R, Savy C, Starkey M, Temple H, Pilgrim JD (2017) The value of the IUCN red list for business decision-making. Conserv Lett 11(1):e12353

Bennun L, van Bochove J, Ng C, Fletcher C, Wilson D, Phair N, Carbone G (2021) Mitigating biodiversity impacts associated with solar and wind energy development. Guidelines for project developers. IUCN, Gland and The Biodiversity Consultancy, Cambridge

Bhatt R, Gill MJ, Hamilton H, Han X, Linden HM, Young BE (2020) Uneven use of biodiversity indicators in fifth national reports to the convention on biological diversity. Environ Conserv 47(1):15–21

Birkhead T (2014) Stormy outlook for long-term ecology studies. Nature 514(7523):405

Birkhead TR (2018) Brighten outlook for long-term seabird monitoring. Nature 563(7729):35–36

Bohan DA, Vacher C, Tamaddoni-Nezhad A, Raybould A, Dumbrell AJ, Woodward G (2017) Next-generation global biomonitoring: large-scale, automated reconstruction of ecological networks. Trends Ecol Evol 32(7):477–487

Brouillette M (2021) How microbes in permafrost could trigger a massive carbon bomb. Nature 591(7850):360–362

Bubb P (2013) Scaling up or down? Linking global and national biodiversity indicators and reporting. In: Collen B, Pettorelli N, Baillie J, Durant S (eds) Biodiversity monitoring and conservation: bridging the gap between global commitment and local action. Wiley-Blackwell and the Zoological Society of London, London, pp 402–420

Butchart SHM, Walpole M, Collen B, van Strien A, Scharlemann JPW, Almond REA, Baillie JEM, Bomhard B, Brown C, Bruno J, Carpenter KE, Carr GM, Chanson J, Chenery AM, Csirke J, Davidson NC, Dentener F, Foster M, Galli A, Galloway JN, Genovesi P, Gregory RD, Hockings M, Kapos V, Lamarque J-F, Leverington F, Loh J, McGeoch MA, McRae L, Minasyan A, Morcillo MH, Oldfield TEE, Pauly D, Quader S, Revenga C, Sauer JR, Skolnik B, Spear D, Stanwell-Smith D, Stuart SN, Symes A, Tierney M, Tyrrell TD, Vié J-C, Watson R (2010) Global biodiversity: indicators of recent declines. Science 328(5982):1164–1168

Chase JM, McGill BJ, Thompson PL, Antão LH, Bates AE, Blowes SA, Dornelas M, Gonzalez A, Magurran AE, Supp SR, Winter M, Bjorkman AD, Bruelheide H, Byrnes JEK, Cabral JS, Elahi R, Gomez C, Guzman HM, Isbell F, Myers-Smith IH, Jones HP, Hines J, Vellend M, Waldock C, O'Connor M (2019) Species richness change across spatial scales. Oikos 128(8):1079–1091

Clark JS, Carpenter SR, Barber M, Collins S, Dobson A, Foley JA, Lodge DM, Pascual M, Pielke R Jr, Pizer W, Pringle C, Reid WV, Rose KA, Sala O, Schlesinger WH, Wall DH, Wear D (2001) Ecological forecasts: an emerging imperative. Science 293(5530):657–660

Collen B, Pettorelli N, Baillie J, Durant S (eds) (2013) Biodiversity monitoring and conservation: bridging the gap between global commitment and local action. Wiley-Blackwell and the Zoological Society of London, London

Cook LM, Saccheri IJ (2013) The peppered moth and industrial melanism: evolution of a natural selection case study. Heredity 110(3):207–212

Cooper G, Trémolet S (2019) Investing in nature: private finance for nature-based resilience. The Nature Conservancy and Environmental Finance, London

Cousins N, Pittman SJ (2021) Guidance for defining ecologically appropriate scales of analysis for marine biodiversity in relation to IFC performance standard 6. Bluedot Associates, Bristol

Crates R, Langmore N, Ranjard L, Stojanovic D, Rayner L, Ingwersen D, Heinsohn R (2021) Loss of vocal culture and fitness costs in a critically endangered songbird. Proc R Soc B 288(1947):20210225

Cronmiller JG, Noble BM (2018) The discontinuity of environmental effects monitoring in the Lower Athabasca region of Alberta, Canada: institutional challenges to long-term monitoring and cumulative effects management. Environ Rev 26(2):169–180

Danielsen F, Jensen PM, Burgess ND, Altamirano R, Alviola PA, Andrianandrasana H, Brashares JS, Burton AC, Coronado I, Corpuz N, Enghoff M, Fjeldså J, Funder M, Holt S, Hübertz H, Jensen AE, Lewis R, Massao J, Mendoza MM, Ngaga Y, Pipper CB, Poulsen MK, Rueda RM, Sam MK, Skielboe T, Sørensen M, Young R (2014) A multicountry assessment of tropical resource monitoring by local communities. Bioscience 64(3):236–251

Darrah SE, Shennan-Farpón Y, Loh J, Davidson NC, Finlayson CM, Gardner RC, Walpole MJ (2019) Improvements to the Wetland Extent Trends (WET) index as a tool for monitoring natural and human-made wetlands. Ecol Indic 99:294–298

De Silva GC, Regan EC, Pollard EHB, Addison PFE (2019) The evolution of corporate no net loss and net positive impact biodiversity commitments: understanding appetite and addressing challenges. Bus Strateg Environ 28(7):1481–1495

Derocles SAP, Bohan DA, Dumbrell AJ, Kitson JJN, Massol F, Pauvert C, Plantegenest M, Vacher C, Evans DM (2018) Biomonitoring for the 21st century: integrating next-generation sequencing into ecological network analysis. Adv Ecol Res 58:1–62

Dixon MJR, Loh J, Davidson NC, Beltrame C, Freeman R, Walpole M (2016) Tracking global change in ecosystem area: the Wetland Extent Trends index. Biol Conserv 193:27–35

Dodds WK, Robinson CT, Gaiser EE, Hansen GJA, Powell H, Smith JM, Morse NB, Johnson SL, Gregory SV, Bell T, Kratz TK, McDowell WH (2012) Surprises and insights from long-term aquatic data sets and experiments. Bioscience 62(8):709–721

Donald PF, Sanderson FJ, Burfield IJ, Bierman SM, Gregory RD, Waliczky Z (2007) International conservation policy delivers benefits for birds in Europe. Science 317(5839):810–813

Dornelas M et al (2018) BioTIME: a database of biodiversity time series for the Anthropocene. Glob Ecol Biogeogr 27(7):760–786

Durant SM (2013) Building sustainable national monitoring networks. In: Collen B, Pettorelli N, Baillie J, Durant S (eds) Biodiversity monitoring and conservation: bridging the gap between global commitment and local action. Wiley-Blackwell and the Zoological Society of London, London, pp 313–334

Ekstrom J, Bennun L, Mitchell R (2015) A cross-sector guide for implementing the mitigation hierarchy. Cross Sector Biodiversity Initiative, London

Ens E, Reyes-García V, Asselin H, Hsu M, Reimerson E, Reihana K, Sithole B, Shen X, Cavanagh V, Adams M (2021) Recognition of Indigenous Ecological Knowledge Systems in Conservation and Their Role to Narrow the Knowledge-Implementation Gap. In: Ferreira CC, Klütsch CFC (eds) Closing the knowledge-implementation gap in conservation science – evidence transfer across spatiotemporal scales and different stakeholders. Springer, New York, pp. 109–139

Fabian Y, Bollmann K, Brang P, Heiri C, Olschewski R, Rigling A, Stofer S, Holderegger R (2019) How to close the science-practice gap in nature conservation? Information sources used by practitioners. Biol Conserv 235:93–101

Fischer EE, Cobb NS, Kawahara AY, Zaspel JM, Cognato AI (2021) Decline of amateur lepidoptera collectors threatens the future of specimen-based research. Bioscience 71(4):396–404

Fischhoff B, Davis AL (2014) Communicating scientific uncertainty. Proc Natl Acad Sci USA 111(4):13664–13671

Franklin JF (1988) Importance and justification of long-term studies in ecology. In: Likens GE (ed) Long-term studies in ecology: approaches and alternatives. Springer, New York, pp 3–19

Gaul W, Sadykova D, Roark E, White HJ, León-Sánchez L, Caplat P, Yearsley JM (2020) What have biological records ever done for us? A systematic scoping review. Front Biogeogr 12(4): e48607

GEO BON (2021) BON in a Box. Website https://geobon.org/bon-in-a-box/ (Accessed 26 April 2021)

Gill MJ, Crane K, Hindrum R, Arneberg P, Bysveen I, Denisenko NV, Gofman V, Grant-Friedman-A, Gudmundsson G, Hopcroft R, Iken K, Labansen A, Liubina OS, Melnikov IA, Moore SE, Reist JD, Sirenko BI, Stow J, Ugarte F, Vongraven D, Watkins J (2011) Arctic marine biodiversity monitoring plan. CAFF monitoring series no. 3, CAFF International Secretariat, Akureyri

Global Biodiversity Information Facility (GBIF) (2021) Global biodiversity information facility. http://www.gbif.org/ (Accessed on 10 May 2021)

Guerra CA, Pendleton L, Drakou EG, Proença V, Appeltans W, Domingos T, Geller G, Giamberini S, Gill M, Hummel H, Imperio S, McGeoch M, Provenzale A, Serral I, Stritih A, Turak E, Vihervaara P, Ziemba A, Pereira HM (2019) Finding the essential: improving conservation monitoring across scales. Global Ecol Conserv 18:e00601

Guerra CA, Heintz-Buschart A, Sikorski J, Chatzinotas A, Guerrero-Ramírez N, Cesarz S, Beaumelle L, Rillig MC, Maestre FT, Delgado-Baquerizo M, Buscot F, Overmann J, Patoine G, Phillips HRP, Winter M, Wubet T, Küsel K, Bardgett RD, Cameron EK, Cowan D, Grebenc T, Marín C, Orgiazzi A, Singh BK, Wall DH, Eisenhauer N (2020) Blind spots in global soil biodiversity and ecosystem function research. Nat Commun 11(1):3870

Haklay M (2017) The three eras of environmental information: the roles of experts and the public. In: Loreto V, Haklay M, Hotho A, Servedio VCP, Stumme G, Theunis J, Tria F (eds) Participatory sensing, opinions and collective awareness. Springer, Cham, pp. 163–179

Han X, Smyth RL, Young BE, Brooks TM, de Lozada AS, Bubb P, Butchart SHM, Larsen FW, Hamilton H, Hansen MC, Turner WR (2014) A biodiversity indicators dashboard: addressing challenges to monitoring progress towards the Aichi biodiversity targets using disaggregated global data. PLoS One 9(11):e112046

Han X, Gill MJ, Hamilton H, Vergara SG, Young BE (2020) Progress on national biodiversity indicator reporting and prospects for filling indicator gaps in Southeast Asia. Environ Sustain Indic 5:1000172

Hanson PC (2008) New ecological insights through the Global Lake Ecological Observatory Network (GLEON). Ecol Sci 27(5):300–302

Hardner J, Gullison RE, Anstee S, Meyer M (2015) Good practices for biodiversity inclusive impact assessment and management planning. Prepared for the multilateral financing institutions biodiversity working group

Hickey JJ, Anderson DW (1968) Chlorinated hydrocarbons and eggshell changes in raptorial and fish-eating birds. Sci New Ser 162(3850):271–273

Hochkirch A, Samways MJ, Gerlach J, Böhm M, Williams P, Cardoso P, Cumberlidge N, Stephenson PJ, Seddon MB, Clausnitzer V, Borges PAV, Mueller GM, Pearce-Kelly P, Raimondo DC, Dijkstra K-DB (2020) A strategy for the next decade to address data deficiency in neglected biodiversity. Conserv Biol 35(2):502–509

Hudson L et al (2017) The database of the PREDICTS (Projecting Responses of Ecological Diversity In Changing Terrestrial Systems) Project. Ecol Evol 7(1):145–188

Integrated Biodiversity Assessment Tool (IBAT) (2019) IBAT annual report 2019. BirdLife International, Conservation International, IUCN, UNEP-WCMC. Available at: https://www.ibat-alliance.org/pdf/ibat-annual-report-2019.pdf

International Finance Corporation (IFC) (2012) Performance standard 6: biodiversity conservation and sustainable management of living natural resources. International Finance Corporation, Washington, DC

International Union for Conservation of Nature (IUCN) (2021) The IUCN Red List of Threatened Species™. Version 2021-1 [website]. Available at: https://www.iucnredlist.org (Accessed 10 May 2021)

IPBES (2016) Global/regional indicator factsheet. Intergovernmental Science-Policy Platform on Biodiversity and Ecosystem Services. IPBES Secretariat, Bonn

IPBES (2018a) Summary for policymakers of the thematic assessment report on land degradation and restoration of the Intergovernmental Science-Policy Platform on Biodiversity and Ecosystem Services. In: Scholes R, Montanarella L, Brainich AN, et al (eds) IPBES Secretariat, Bonn

IPBES (2018b) Summary for policymakers of the regional assessment report on biodiversity and ecosystem services for Europe and Central Asia of the Intergovernmental Science-Policy Platform on Biodiversity and Ecosystem Services. In: Fischer M, Rounsevell M, Torre-Marin Rando A, et al (eds) IPBES Secretariat, Bonn

Isaac NJB, Pocock MJO (2015) Bias and information in biological records. Biol J Linn Soc 115(3):522–531

Jacobson C, Hughey KFD, Allen WJ, Rixecker S, Carter RW (2009) Toward more reflexive use of adaptive management. Soc Nat Resour 22(5):484–495

Jones JPG (2013) Monitoring in the real world. In: Collen B, Pettorelli N, Baillie J, Durant S (eds) Biodiversity monitoring and conservation: bridging the gap between global commitment and local action. Wiley-Blackwell and the Zoological Society of London, London, pp 335–347

Kelling S (2018) Technology developments for biodiversity monitoring and conservation. Biodivers Inform Sci Stand 2:e25833

Kissling WD, Hardisty A, García EA, Santamaria M, De Leo F, Pesole G, Freyhof J, Manset D, Wissel S, Konijn J, Los W (2015) Towards global interoperability for supporting biodiversity research on Essential Biodiversity Variables (EBVs). Biodiversity 16(2–3):99–107

Kissling WD, Ahumada JA, Bowser A et al (2018) Building Essential Biodiversity Variables (EBVs) of species distribution and abundance at a global scale. Biol Rev 93(1):600–625

Klütsch CFC, Ferreira CC (2021) Closing the gap between knowledge and implementation in conservation science: concluding remarks. In: Ferreira CC, Klütsch CFC (eds) Closing the knowledge-implementation gap in conservation science – evidence transfer across spatiotemporal scales and different stakeholders. Springer, New York, pp. 457–473

Krebs CJ (2011) Of lemmings and snowshoe hares: the ecology of northern Canada. Proc R Soc B 278(1705):481–489

Kühl HS, Bowler DE, Bösch L, Bruelheide H, Dauber J, Eichenberg D, Eisenhauer N, Fernández N, Guerra CA, Henle K, Herbinger I, Isaac NJB, Jansen F, König-Ries B, Kühn I, Nilsen EB, Pe'er G, Richter A, Schulte R, Settele J, van Dam NM, Voigt M, Wägele WJ, Wirth C, Bonn A (2020) Effective biodiversity monitoring needs a culture of integration. One Earth 3(4):462–474

Lammerant J, Grigg A, Dimitrijevic J, et al. (2019) Assessment of biodiversity measurement approaches for businesses and financial institutions. Update Report 2. EU Business @ Biodiversity Platform, Brussels

Legg CJ, Nagy L (2006) Why most conservation monitoring is, but need not be, a waste of time. J Environ Manag 78(2):194–199

Lepetz V, Massot M, Schmeller DS, Clobert J (2009) Biodiversity monitoring: some proposals to adequately study species' responses to climate change. Biodivers Conserv 18(12):3185–3203

Lindenmayer DB, Likens GE (2009) Adaptive monitoring: a new paradigm for long-term research and monitoring. Trends Ecol Evol 24(9):482–486

Lindenmayer DB, Likens GE (2010) The science and application of ecological monitoring. Biol Conserv 143(6):1317–1328

Lindenmayer D, Likens GE (2018) Effective ecological monitoring. CSIRO Publishing, Melbourne

Lindenmayer DB, Likens GE, Franklin JF (2018) Earth Observation Networks (EONs): finding the right balance. Trends Ecol Evol 33(1):1–3

Lyver PB, Richardson SJ, Gormley AM, Timoti P, Jones CJ, Tahi BL (2018) Complementarity of indigenous and western scientific approaches for monitoring forest state. Ecol Appl 28(7):1909–1923

Mair L, Bennun LA, Brooks TM, Butchart SHM, Bolam FC, Burgess ND, et al. (2021) A metric for spatially explicit contributions to science-based species targets. Nat Ecol Evol 5:836–844

Marsh DM, Trenham PC (2008) Current trends in plant and animal population monitoring. Conserv Biol 22(3):647–655

Mascia MB, Pailler S, Thieme ML, Rowe A, Bottrill MC, Danielsen F, Geldmann J, Naidoo R, Pullin AS, Burgess ND (2014) Commonalities and complementarities among approaches to conservation monitoring and evaluation. Biol Conserv 169:258–267

McCarthy JM, Hein CL, Olden JD, Vander Zanden MJ (2006) Coupling long-term studies with meta-analysis to investigate impacts of non-native crayfish on zoobenthic communities. Freshw Biol 51(2):224–235

McRae L, Deinet S, Freeman R (2017) The diversity-weighted Living Planet Index: controlling for taxonomic bias in a global biodiversity indicator. PLoS One 12(1):e0169156

Mihoub J-B, Henle K, Titeux N, Brotons L, Brummitt NA, Schmeller DS (2017) Setting temporal baselines for biodiversity: the limits of available monitoring data for capturing the full impact of anthropogenic pressures. Sci Rep 7(1):41591

Moussy C, Burfield IJ, Stephenson PJ, Newton AFE, Butchart SHM, Sutherland WJ, Gregory RD, McRae L, Bubb P, Roesler I, Ursino C, Wu Y, Retief EF, Udin JS, Urazaliyev R, Sánchez-Clavijo LM, Lartey E, Donald PF (2021) A quantitative global review of species population monitoring. Conserv Biol

Muller-Karger FE, Miloslavich P, Bax NJ, Simmons S, Costello MJ, Sousa Pinto I, Canonico G, Turner W, Gill M, Montes E, Best BD, Pearlman J, Halpin P, Dunn D, Benson A, Martin CS, Weatherdon LV, Appeltans W, Provoost P, Klein E, Kelble CR, Miller RJ, Chavez FP, Iken K, Chiba S, Obura D, Navarro LM, Pereira HM, Allain V, Batten S, Benedetti-Checchi L, Duffy JE, Kudela RM, Rebelo L-M, Shin Y, Geller G (2018) Advancing marine biological observations and data requirements of the complementary Essential Ocean Variables (EOVs) and Essential Biodiversity Variables (EBVs) frameworks. Front Mar Sci 5:211

Navarro LM, Fernández N, Guerra C, Guralnick R, Kissling WD, Londoño MC, Muller-Karger F, Turak E, Balvanera P, Costello MJ, Delavaud A, El Serafy G, Ferrier S, Geijzendorffer I, Geller GN, Jetz W, Kim E-S, Kim H, Martin CS, McGeoch MA, Mwampamba TH, Nel JL, Nicholson E, Pettorelli N, Schaepman ME, Skidmore A, Sousa Pinto I, Vergara S, Vihervaara P, Xu H, Yahara T, Gill M, Pereira HM (2017) Monitoring biodiversity change through effective global coordination. Curr Opin Environ Sustain 29:158–169

Nesshöver C, Livoreil B, Schindler S, Vandewalle M (2016) Challenges and solutions for networking knowledge holders and better informing decision-making on biodiversity and ecosystem services. Biodivers Conserv 25(7):1207–1214

New Zealand Parliamentary Commissioner for the Environment (2020) A review of the funding and prioritisation of environmental research in New Zealand (Wellington, New Zealand). Available at: https://www.pce.parliament.nz/publications/environmental-research-funding-review

O'Connell C, McKinnon M (2021) Communication: the bridge between knowledge and implementation. In: Ferreira CC, Klütsch CFC (eds) Closing the knowledge-implementation gap in conservation science – evidence transfer across spatiotemporal scales and different stakeholders. Springer, New York, pp. 171–197

Pereira HM, Navarro LM, Martins IS (2012) Global biodiversity change: the good, the bad and the unknown. Annu Rev Environ Resour 37:25–50

Pereira HM, Ferrier S, Walters M, Geller GN, Jongman RHG, Scholes RJ, Bruford MW, Brummitt N, Butchart SHM, Cardoso AC, Coops NC, Dulloo E, Faith DP, Freyhof J, Gregory RD, Heip C, Höft R, Hurtt G, Jetz W, Karp DS, McGeoch MA, Obura D, Onoda Y, Pettorelli N, Reyers B, Sayre R, Scharlemann JPW, Stuart SN, Turak E, Walpole M, Wegmann M (2013) Essential biodiversity variables. Science 339(6117):277–278

Peterson RO, Peterson TS (2020) Yellowstone wolves are important because they changed science. University of Chicago Press, Chicago

Phillips TB, Parker A, Bowser A, Hakley M (2021) Publicly generated data: the role of Citizen-Science for knowledge production, action, and public engagement. In: Ferreira CC, Klütsch CFC (eds) Closing the knowledge-implementation gap in conservation science – evidence transfer across spatiotemporal scales and different stakeholders. Springer, New York, pp. 83–107

Ramsar Convention on Wetlands (2018) Global wetland outlook: state of the world's wetlands and their services to people. Ramsar Convention Secretariat, Gland

Robinson CT, Doering M, Seelen L (2011) Use of protected areas for freshwater biomonitoring – case studies in Switzerland. Eco mont 3(2):13–22

Rose DC, Despot-Belmonte K, Pollard JA, Shears O, Robertson RJ (2021) Making an impact: how to design relevant and usable decision support systems for conservation. In: Ferreira CC, Klütsch CFC (eds) Closing the knowledge-implementation gap in conservation science – evidence transfer across spatiotemporal scales and different stakeholders. Springer, New York, 199–226

Rowland JA, Bland LM, James S, Nicholson E (2021) A guide to representing variability and uncertainty in biodiversity indicators. Conserv Biol 35(5):1669–1682

Sanchirico JN, Springborn MR, Schwartz MW, Doerr AN (2014) Investment and the policy process in conservation monitoring. Conserv Biol 28(2):361–371

Schmeller DS, Henle K, Loyau A, Besnard A, Henry P-Y (2012) Bird-monitoring in Europe – a first overview of practices, motivations and aims. Nat Conserv 2:41–57

Schmeller DS, Böhm M, Arvanitidis C, Barber-Meyer S, Brummitt N, Chandler M, Chatzinikolaou E, Costello MJ, Ding H, García-Moreno J, Gill M, Haase P, Jones M, Juillard R, Magnusson WE, Martin CS, McGeoch M, Mihoub J-B, Pettorelli N, Proença V, Peng C, Regan E, Schmiedel U, Simaika JP, Weatherdon L, Waterman C, Xu H, Belnap J (2017) Building capacity in biodiversity monitoring at the global scale. Biodivers Conserv 26(12):2765–2790

Scholes RJ, Mace GM, Turner W, Geller GN, Jürgens N, Larigauderie A, Muchoney D, Walther BA, Mooney HA (2008) Toward a global biodiversity observing system. Science 321(5892):1044–1045

Scholes RJ, Walters M, Turak E, Saarenmaa H, Heip CHR, Tuama EÓ, Faith DP, Mooney HA, Ferrier S, Jongman RHG, Harrison IJ, Yahara T, Pereira HM, Larigauderie A, Geller G (2012) Building a global observing system for biodiversity. Curr Opin Environ Sustain 4(1):139–146

Scholes RJ, Gill MJ, Costello M, Sarantankos G, Walters M (2016) Working in networks to make biodiversity data more available. In: Walters M, Scholes RJ (eds) The GEO handbook on biodiversity observations. Springer, New York, pp. 1–17

Secretariat of the Convention on Biological Diversity (2020) Global biodiversity outlook 5. Secretariat of the Convention on Biological Diversity, Montreal

Segan DB, Bottrill MC, Baxter PWJ, Possingham HP (2011) Using conservation evidence to guide management. Conserv Biol 25(1):200–202

Stephenson PJ (2018) A global effort to improve species monitoring for conservation. Oryx 59(3):412–413

Stephenson PJ (2019) The Holy Grail of biodiversity conservation management: monitoring impact in projects and project portfolios. Persp Ecol Conserv 17(4):182–192

Stephenson PJ (2020) Technological advances in biodiversity monitoring: applicability, opportunities and challenges. Curr Opin Environ Sustain 45:36–41

Stephenson PJ, Carbone G (2021) Guidelines for planning and monitoring corporate biodiversity performance. IUCN, Gland

Stephenson PJ, Stengel C (2020) An inventory of biodiversity data sources for conservation monitoring. PLoS One 15(12):e0242923

Stephenson PJ, Burgess ND, Jungmann L, Loh J, O'Connor S, Oldfield T, Reidhead W, Shapiro A (2015) Overcoming the challenges to conservation monitoring: integrating data from in situ reporting and global data sets to measure impact and performance. Biodiversity 16(2–3):68–85

Stephenson PJ, Bowles-Newark N, Regan E, Stanwell-Smith D, Diagana M, Höft R et al (2017a) Unblocking the flow of biodiversity data for decision-making in Africa. Biol Conserv 213(Part B):335–340

Stephenson PJ, Brooks TM, Butchart SHM, Fegraus E, Geller GN, Hoft R, Hutton J, Kingston N, Long B, McRae L (2017b) Priorities for big biodiversity data. Front Ecol Environ 15(3):124–125

Stephenson PJ, Ntiamoa-Baidu Y, Simaika JP (2020) The use of traditional and modern tools for monitoring wetlands biodiversity in Africa: challenges and opportunities. Front Environ Sci 8: 61

Stephenson PJ, Bakarr M, Bowles-Newark N, Kleinschroth F, Mapendembe A, Ntiamoa-Baidu Y, Obura D, Ratsifandrihamanana N, Simaika J, Sitati N, Sumbi P, Teferi T (2021) Conservation science in Africa: mainstreaming biodiversity information into policy and decision-making. In: Ferreira CC, Klütsch CFC (eds) Closing the knowledge-implementation gap in conservation science – evidence transfer across spatiotemporal scales and different stakeholders. Springer, New York, pp. 287–320

Tewksbury JJ, Anderson JGT, Bakker JD, Billo TJ, Dunwiddie PW, Groom MJ, Hampton SE, Herman SG, Levey DJ, Machnicki NJ, del Rio CM, Power ME, Rowell K, Salomon AK, Stacey L, Trombulak SC, Wheeler TA (2014) Natural history's place in science and society. Bioscience 64(4):300–310

The Expert Panel on Biodiversity Science (2010) Canadian taxonomy: exploring biodiversity, creating opportunity. Council of Canadian Academies, Ottawa (Available at: https://cca-reports.ca/wp-content/uploads/2018/10/biodiversity_report_final_e.pdf)

The KBA Partnership (2018) Guidelines on business and KBAs: managing risk to biodiversity. IUCN, Gland

Tittensor DP, Walpole M, Hill SLL, Boyce DG, Britten GL, Burgess ND et al (2014) A mid-term analysis of progress toward international biodiversity targets. Science 346(6206):241–244

Troudet J, Grandcolas P, Blin A, Vignes-Lebbe R, Legendre F (2017) Taxonomic bias in biodiversity data and societal preferences. Sci Rep 7(1):9132

Turak E, Brazill-Boast J, Cooney T, Drielsma M, DelaCruz J, Dunkerley G, Fernandez M, Ferrier S, Gill M, Jones H, Koen T, Leys J, McGeoch M, Mihoub J-B, Scanes P, Schmeller D, Williams K (2016) Using the essential biodiversity variables framework to measure biodiversity change at national scale. Biol Conserv 213(Part B):264–271

United Nations (2018) Sustainable development goal 6 synthesis report 2018 on water and sanitation. New York

United Nations (UN) (2021) Sustainable development goals knowledge platform. Website https://sustainabledevelopment.un.org/ (Accessed 26 April 2021)

United Nations Environment Programme - World Conservation Monitoring Centre (UNEP-WCMC) (2021) Protected planet: the world database on protected areas. Available at: https://www.protectedplanet.net/ (Accessed 15 May 2021)

Van Swaay C, Regan E, Ling M, Bozhinovska E, Fernandez M, Marini-Filho OJ, Huertas B, Phon C-K, K"orösi A, Meerman J, Pe'er G, Uehara-Prado M, Sáfián S, Sam L, Shuey J, Taron D, Terblanche R, Underhill L (2015) Guidelines for standardised global butterfly monitoring group on earth observations biodiversity observation network Leipzig GEO BON Technical Series 1, 32pp

Vellend M, Baeten L, Becker-Scarpitta A, Boucher-Lalonde V, McCune JL, Messier J, Myers-Smith IH, Sax DF (2017) Plant biodiversity change across scales during the anthropocene. Annu Rev Plant Biol 68:563–586

Walls JL, Berrone P, Phan PH (2012) Corporate governance and environmental performance: is there really a link? Strateg Manag J 33(8):885–913

Walpole M, Almond REA, Besançon C, Butchart SHM, Campbell-Lendrum D, Carr GM, Collen B, Collette L, Davidson NC, Dulloo E, Fazel AM, Galloway JN, Gill M, Goverse T, Hockings M, Leaman DJ, Morgan DHW, Revenga C, Rickwood CJ, Schutyser F, Simons S, Stattersfield AJ, Tyrrell TD, Vié J-C, Zimsky M (2009) Tracking progress toward the 2010 biodiversity target and beyond. Science 325(5947):1503–1504

Weiss RA, McMichael AJ (2004) Social and environmental risk factors in the emergence of infectious diseases. Nat Med 10(12):S70–S76

Whiteman G, Walker B, Perego P (2013) Planetary boundaries: ecological foundations for corporate sustainability. J Manag Stud 50(2):307–336

Wich SA, Koh LP (2018) Conservation drones: mapping and monitoring biodiversity. Oxford University Press, Oxford

Wich SA, Piel AK (eds) (2021) Conservation technology. Oxford University Press, Oxford

Wieczorek J, Bloom D, Guralnick R, Blum S, Döring M, Giovanni R, Robertson T, Vieglais D (2012) Darwin core: an evolving community-developed biodiversity data standard. PLoS One 7(1):e29715

Wilkinson MD, Dumontier M, Aalbersberg IJ et al (2016) The FAIR Guiding Principles for scientific data management and stewardship. Scientific Data 3(1):160018

Xu H, Cao M, Wu Y, Cai L, Cao Y, Ding H, Cui P, Wu J, Wang Z, Le Z, Lu X, Liu L, Li J (2017) Optimized monitoring sites for detection of biodiversity trends in China. Biodivers Conserv 26(8):1959–1971

Xu H, Cao Y, Yu D, Cao M, He Y, Gill M, Pereira HM (2021) Ensuring effective implementation of the post-2020 global biodiversity targets. Nat Ecol Evol 5(4):411–418

Yahara T, Ma K, Darnaedi D, Miyashita T, Takenaka A, Tachida H, Nakashizuka T, Kim E-S, Takamura N, Nakano S, Shirayama Y, Yamamoto H, Vergara SG (2014) Developing a regional network of biodiversity observation in the Asia-Pacific Region: achievements and challenges of AP BON. In: Nakano S, Yahara T, Nakashizuka T (eds) Integrative observations and assessments, ecological research monographs. Springer, Tokyo, pp 3–28

Young JC, Waylen KA, Sarkki S, Albon S, Bainbridge I, Balian E, Davidson J, Edwards D, Fairley R, Margerison C, McCracken D, Owen R, Quine CP, Stewart-Roper C, Thompson D, Tinch R, Van den Hove S, Watt A (2014) Improving the science-policy dialogue to meet the challenges of biodiversity conservation: having conversations rather than talking at one-another. Biodivers Conserv 23(2):387–404

Zoological Society of London (ZSL) (2021) Living Planet Index [website]. Available at: www.livingplanetindex.org (Accessed on 10 May 2021)

Zu Ermgassen SOSE, Utamiputri P, Bennun L, Edwards S, Bull JW (2019) The role of "no net loss" policies in conserving biodiversity threatened by the global infrastructure boom. One Earth 1(3):305–315

Chapter 3
Closing the Conservation Genetics Gap: Integrating Genetic Knowledge in Conservation Management to Ensure Evolutionary Potential

Cornelya F. C. Klütsch and Linda Laikre

Contents

C. F. C. Klütsch (✉)
Division of Environment and Natural Resources, Norwegian Institute of Bioeconomy Research (NIBIO), Svanvik, Norway
e-mail: cornelya.klutsch@nibio.no

L. Laikre
Department of Zoology, Division of Population Genetics, Stockholm University, Stockholm, Sweden
e-mail: linda.laikre@popgen.su.se

C. C. Ferreira, C. F. C. Klütsch (eds.), *Closing the Knowledge-Implementation Gap in Conservation Science*, Wildlife Research Monographs 4,
https://doi.org/10.1007/978-3-030-81085-6_3

3.1 Introduction

Genetic diversity is the cornerstone that allows species and populations to survive and adapt to changing environmental conditions. The importance of biological diversity for the functioning of planet Earth is increasingly recognized and included in the United Nations Sustainable Development Goals for 2030 adopted in 2015 (https://sustainabledevelopment.un.org/?menu=1300). Already in 1993, the United Nations adopted the Convention on Biological Diversity (CBD; www.cbd.int)—a sister convention to the widely known climate convention. In the CBD, biological diversity is identified at the levels of ecosystems (i.e., a community of organisms and its environment; Groom et al. 2006), species, and genes (i.e., a particular sequence of DNA nucleotides that is the functional unit of inheritance), and it is agreed that all of these diversity levels should be conserved, monitored, and sustainably used (Fig. 3.1).

The gene level of biodiversity represents variation present in the DNA molecule, which implies that the DNA sequence of a particular gene can vary, and this variation results in gene variants (so-called “alleles”). Genetic diversity is regarded below the species level as genetic differences between individuals within populations and genetic differences between populations. Genetic diversity provides the raw material for evolution (i.e., change of heritable characteristics of populations over generations). It constitutes the basis of all biodiversity, as it is necessary for species’ long-term survival and adaptation to environmental change (Allendorf et al. 2013). Adaptation is the evolutionary process that results in individual characteristics (phenotypes) governed by specific genotypes that are better suited to (local) environmental conditions becoming more common. Large populations can maintain higher levels of genetic diversity and thus have a higher capacity for long-term survival and resilience, whereas small, isolated populations have the lower adaptive capacity, low resilience, and a weak potential for long-term survival (Fig. 3.2).

The consequences of ignoring or downplaying the importance of genetics (i.e., study of heredity, that is, how genes are transmitted from one generation to the next and how those genes affect the progeny) in policy and management can result in large-scale losses of genetic diversity and connectivity of populations. For instance, a recent meta-analysis suggested that approximately 6% of genetic diversity in wild populations has been lost since the industrial revolution (Leigh et al. 2019). In

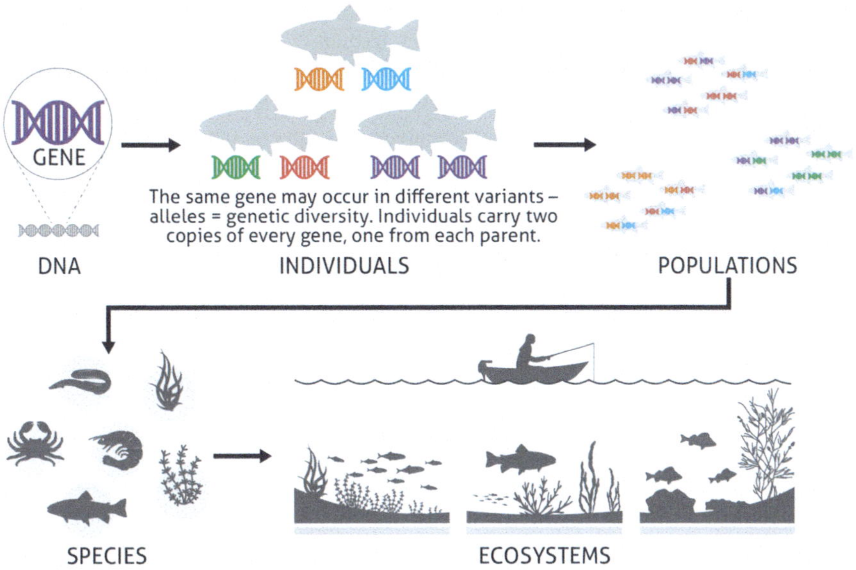

Fig. 3.1 Variation at the genetic level forms the basis of all biodiversity. Illustration: Jerker Lokrantz/Azote (© Jerker Lokrantz 2020. All Rights Reserved)

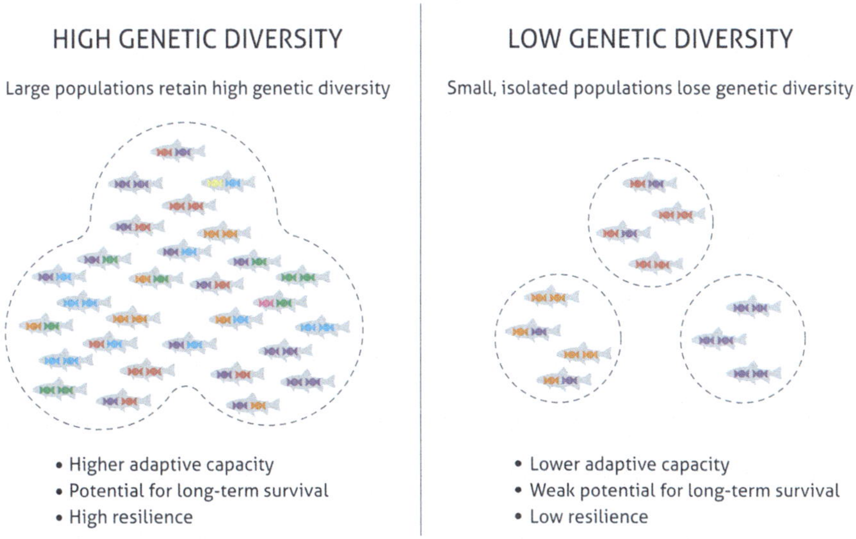

Fig. 3.2 The importance of preserving high genetic diversity in natural populations and the consequences of low genetic diversity. Illustration: Jerker Lokrantz/Azote (© Jerker Lokrantz 2020. All Rights Reserved)

addition, 26% of invertebrates, 29% of vertebrates, and 55% of plant species show insufficient gene flow among populations (Frankham et al. 2019). Thus, it is highly troublesome that increasing evidence shows that protected habitats do not properly safeguard genetic diversity to assure the evolutionary potential of populations (i.e., the ability of a population to evolve and adapt in response to environmental change; Diniz-Filho et al. 2012, 2016; Laikre et al. 2016; Mouillot et al. 2016). Further, evolutionary processes may be triggered by human-induced environmental change (Brodersen and Seehausen 2014; Fitzpatrick and Keller 2015), and genetics can help to develop a firm understanding of these processes.

With the advancement of genetic technologies over the last few decades, genetic methods are widely applied in academic fields such as conservation genetics/genomics, landscape genetics, evolutionary, and ecological genomics, paleogenetics, and phylogenetics (Bowman et al. 2016; Keller et al. 2015; Schwartz et al. 2007). While early studies focused on only small parts of the genetic material of an organism, modern methods study the structure, function, and evolution of large parts of the genome—including many or most of organism's genes (i.e., genomics). Such DNA based methods are frequently utilized to study the consequences of anthropogenic impacts on the genetic composition and evolutionary potential of natural populations (Benestan et al. 2016; Carroll et al. 2014; Harrisson et al. 2014; Mimura et al. 2017; Russello et al. 2015; Shafer et al. 2015; van der Valk et al. 2020). The lag in incorporating this new knowledge in management hinders adaptive and effective biodiversity conservation. In this chapter, we review the global literature documenting the conservation genetics gap. We synthesize reasons proposed for the existence of this gap and recommendations to narrow it down and/or close it.

3.2 The Development of a New Discipline: Conservation Genetics

The importance of individual variation for natural selection driving evolutionary adaptation was first recognized by Charles Darwin (Darwin 1859). Darwin was, however, not aware of the genetic mechanisms behind the inheritance of phenotypic similarities from parents to offspring. Those mechanisms were discovered by Gregor Mendel (Mendel 1866) through his famous experiments with peas (*Pisum sativum*). The importance of Mendel's results was not fully understood until the early twentieth century when Edith Saunders and William Bateson continued studies similar to his (Bateson and Saunders 1902). The field of population genetics was further developed during the 1920s and 1930s by Sir Ronald A. Fisher, John B.S. Haldane, and Sewall Wright integrated the principles of Mendelian genetics with Darwinian Theory on natural selection into mathematical frameworks for evolutionary processes. This work marked the birth of the theoretical basis for modern conservation genetics and included defining and modeling key concepts for understanding mechanisms that much later became paramount for conservation biology and genetics such as inbreeding (i.e., mating and production of offspring from closely related individuals) and genetic drift (i.e., random shifts in allele frequencies due to the stochastic process of allele transfer from one generation to the next, which is particularly pronounced in small populations; Allendorf et al. 2013), and how they relate to the evolutionary forces of migration (genetic exchange among populations, i.e., gene flow), mutation, and selection (Wright 1922, 1931, 1977).

The field of conservation genetics was born as a new research discipline during the 1970s and 1980s when Sir Otto Frankel identified the human responsibility to conserve genetic variation (Frankel 1970, 1974), and when Soulé and Wilcox (1980) and Frankel and Soulé (1981) expanded theoretical population genetic concepts from Sewall Wright and coworkers to provide guidance on how to safeguard genetic variation and evolutionary adaptive potential of populations. Techniques for studying the genetic variation of natural animal and plant populations became available through protein electrophoresis (Allendorf et al. 1976; Utter et al. 1973), and conservation genetic guidelines for natural populations coupled with empirical findings were soon introduced (Ryman 1981; Ryman and Utter 1987; Schonewald-Cox et al. 1983). Katherine Ralls and Jonathan Ballou documented inbreeding depression (i.e., negative effects on fitness following inbreeding; Schonewald-Cox et al. 1983) in a wide range of wild animal species and provided advice for captive breeding programs based on their findings (Ralls and Ballou 1983). The first scientific journal on conservation—Conservation Biology—was published in 1985 with a strong part of published papers focusing on the genetics of conservation (Ehrenfeldt 1995).

The field of conservation genetics continues to develop since these early years. A scientific journal, Conservation Genetics, focusing specifically on the field, has been published since 2000, and a large number of textbooks for university education have been produced (e.g., Allendorf et al. 2013; Ballou et al. 1995; Frankham et al., 2010; Henry 2006). A search in Web of Science for the topic "conservation genetic*" in January 2020 finds nearly 38,000 publications. Clearly, large amounts of scientific information on genetic diversity have been accumulating over the past 50 years. We performed similar searches adding each of the main continents (one at a time) to the topic "conservation genetic*." These searches identified 1264 publications for Africa, 1009 for Asia, 1171 for Australia, 1644 for Europe, 1138 for North America, and 609 for South America, indicating relatively similar levels of knowledge from around the globe, although Europe has a clear lead and South America appears to lag. A marked lag appears to occur for Antarctica, as only 122 studies on conservation genetics focusing on this continent appear to exist (Fig. 3.3).

Thus, a wealth of scientific knowledge on conservation genetics and genetic diversity, including why and how this diversity can be conserved and empirical data from populations of organisms, exists all over the world. This knowledge, together with the international mandate of the CBD, to conserve, monitor, and sustainably use genetic variation, provide a good basis to safeguard and manage genetic diversity actively. So, to what extent has this happened?

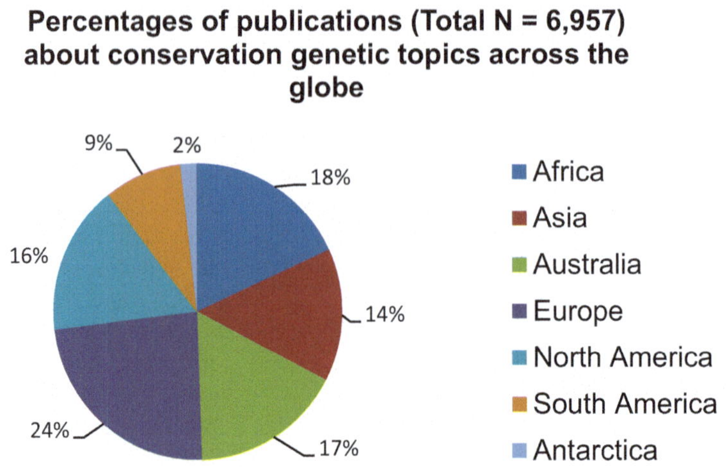

Fig. 3.3 Percentages of publications that deal with conservation genetic topics around the globe, based on a Web of Science search per continent

3.3 The Conservation Genetics Gap

In this section, we describe the lack of connectivity between science, policy, and management that constitutes the conservation genetics gap. This gap is an integral part of the wider knowledge-implementation gap in conservation science that needs to be resolved (Habel et al. 2013; Laurance et al. 2012).

3.3.1 The Conservation Genetics Gap at the International and National Policy Level

The CBD is a great example of a lack of success in implementing international policy and commitments for the preservation of genetic diversity. The majority of the world's nations ratified the CBD during the 1990s, and currently, all countries except the USA and the Holy See (i.e., the Vatican City State) take part in this convention. However, its realization is a struggle (Tittensor et al. 2014). As 2010 approached, scientists and the CBD itself flagged that the world was failing to reach the ambitious goal of halting biodiversity loss by that year (CBD 2001; Mace and Purvis 2008; Mace et al. 2010; Rands et al. 2010; Walpole et al. 2009). In particular, implementing the CBD with respect to genetic biodiversity has been shown to lag at an excruciating rate behind the other levels of biodiversity (Laikre et al. 2010). In 2010, this lag was portrayed by a lack of inclusion of genetic diversity in National Biodiversity Strategies and Action Plans (NBSAPs), absence of targets and indicators for genetic diversity, lack of funding for genetic diversity projects, and lack of inclusion of genetic considerations in species threat status assessments (Laikre

2010). Another issue was the lack of scientists involved in the CBD work that could have aided in integrating scientific evidence regarding the importance of genetic diversity into the convention (Laikre et al. 2008a; van den Hove and Chabason 2009).

Further, signatories to the CBD were also found to be failing to meet commitments related to maintenance and use of genetic diversity (Laikre et al. 2010). This failure was present at the national CBD execution level and several of the major international platforms for CBD implementation (e.g., UN Environment Program's World Conservation Monitoring Center (UNEP-WCMC), IUCN Red List of Threatened Species, the Global Biodiversity Information Facility (GBIF), the Global Environmental Facility (GEF), the Global Biodiversity Indicators and Partnership; see Laikre 2010 and Santamaría and Méndez 2012 for more examples). Similarly, genetic considerations were missing when assessing conservation status within the European Union (Laikre et al. 2009).

When this chapter was being finalized, a new Global Biodiversity Framework for 2021–2050 was being drafted (CBD/WG2020/2/3; https://www.cbd.int/article/2020-01-10-19-02-38). A major omission of this initial draft (the zero draft) is that it does not recognize that genetic diversity should be protected in all species and not only in cultivated plants and domesticated animals, their wild relatives, and other socioeconomically and culturally valuable species (Laikre et al. 2020). A recent review of national policy for CBD implementation again finds a lag concerning the protection of genetic diversity (Hoban et al. 2021a). Further, the draft plan does not address specifically enough which measures should be taken to prevent further genetic erosion and loss of intra-specific adaptive potential. Scientists are proposing specific modifications that are needed in this respect (Hoban et al. 2020, 2021b).

The lack of recognition of the importance of genetic diversity in even recently drafted international policy documents emphasizes that more work is needed to integrate genetic concepts better into policy. A lack of clear guidelines in policy documents on how to preserve genetic diversity has been identified as a major reason for the deficient implementation of international and national policy agreements at regional and local levels (Hoban et al. 2013a, b; Laikre et al. 2016; Ottewell et al. 2016; Sandström et al. 2016, 2019; Vernesi et al. 2008). As a first step to achieve this goal, quantification of how often conservation policy documents use genetic and evolutionary knowledge is required.

In 2017, Cook and Sgrò did a Web of Science search for policy documents using keywords associated with genetic and evolutionary concepts (e.g., inbreeding, adaptation, etc.) in three large countries (i.e., Australia, Canada, and South Africa) across governance levels. The same keywords were used to scope the scientific literature to see whether there was a difference in the use of these concepts between policy documents and scientific literature. They found that the term "genetic diversity" was significantly more often stated than other terms in international policy documents (like the CBD). At the national level, <50% of Australian national conservation documents contained the term genetic diversity, followed by <40% in Canadian and ~20% in South African documents. At the regional level, genetic diversity was still most commonly used, but additional terms like "adaptation" and "evolution" were used more often than at the national and international levels, indicating that provincial documents have a more precise language to describe different evolutionary phenomena. However, the most commonly used term "genetic diversity" was largely used in general contexts without linking it to concrete management actions or recommendations. Similarly, the discussion of other evolutionary and genetic concepts, if mentioned at all, was limited to emphasizing their importance. In contrast, in the scientific literature, keywords were more commonly mentioned and also more critically discussed. This study by Cook and Sgrò (2017), however, did not assess whether the scientific literature included clear recommendations for conservation actions (but see Sect. 3.3.2).

In a similar study, Pierson et al. (2016) assessed how often genetic erosion associated with small population size (e.g., genetic diversity loss, inbreeding) are integrated into at-risk species recovery plans in Europe, North America, and Australia. This study also investigated whether genetic data were included or whether instructions to collect genetic data were given to practitioners. Genetics as a risk factor was built-in in 63% of the USA, 55% of Australian, and 33% of European recovery plans. Within Europe, there were differences between countries with France including genetic aspects more often than other European countries. Globally, the UK and Luxemburg included them less often than North America and Australia. The majority of US recovery plans (82%) included recommendations to collect and include genetic data, while only 52% of Australian and 17% of European recovery plans did do so. However, only 46% of North American, 12% of Australian, and 11% of European recovery plans did actually include any type of genetic data; thus, it appears that there is a large discrepancy between the recommendation to generate and include genetic data and the actual use of these data in recovery plans. Furthermore, recovery plans for animals included genetic factors (57%) significantly more often than those for plants (43%), and the latter included fewer recommendations to collect genetic data (Pierson et al. 2016). In agreement with Cook and Sgrò (2017), this study showed "genetic variation," to be the most commonly considered factor in recovery plans and that more specific concepts were clearly mentioned less frequently. Despite this seeming improvement, genetic-driven conservation action is still incipient.

In Indonesia, genetic diversity is mentioned in several national policy documents and the term "genetic resource" is about to be introduced in an upcoming Act (Ragamustari and Sukara 2019). The definitions are in line with the CBD; however, they remain imprecise on the use of genomic data and the meaning of the term "genetic resource" (Ragamustari and Sukara 2019). Several other studies have discussed the conservation genetics gap at more regional scales and are unanimous in finding a general lack of integration of conservation genetics into regional practice (e.g., Laikre et al. 2016; Rodriguez-Clark et al. 2015; Taylor et al. 2017; see also sections below).

Although these studies are not directly comparable because they investigated partially different document types and different regions, they provide *important insights and trends about the consideration and use of genetic concepts and data in conservation and policy documents*:

- International and national policy documents either fail to recognize the importance of genetic diversity and evolutionary processes for the protection of biodiversity, or they are vague in language. Better integration of genetic concepts and clear recommendations for the conservation of genetic diversity are required.
- The surveys suggest that <50% of national conservation policy documents and recovery plans worldwide include genetic concepts and data, with some regions and taxa clearly lagging even further behind (Cook and Sgrò 2017). Despite this, recovery plans do seem to mention genetic concepts more often (Pierson et al. 2016).
- There is a bias towards animal taxa (mainly vertebrates), while plants are far less often considered in terms of genetic diversity assessments (Pierson et al. 2016). This means that the vast majority of taxonomical biodiversity is largely excluded in the context of evolutionary potential and genetic diversity considerations for conservation planning, which leads to knowledge deficiencies in species assessments (e.g., Wilson et al. 2019).
- The conservation genetics gap is a global problem.

3.3.2 What About Recommendations for Conservation Action in Scientific Articles?

Although the field of conservation genetics was born from a need to provide scientific evidence to conserve genetic diversity and evolutionary processes, several studies suggest that this evidence is not systematically translated into practical management (Vernesi et al. 2008). A meta-analysis of 300 conservation genetic publications from 2006 to 2017 hints that this could partly be due to a lack of clear recommendations for action and policy in these studies (only <40% of this literature explicitly included such recommendations; Britt et al. 2018). In another study, focusing on the conservation genetics gap in Latin America that reviewed >500 conservation genetic articles, only approx. half (49%) of the articles contained concrete recommendations for conservation and management actions (Torres-Florez et al. 2018). Similarly, approx. 30% of publications on genetic diversity of Baltic Sea species published during 2010–2015 provided concrete management advice for the species studied (Wennerström et al. 2013). In line with Pierson et al. (2016), the majority of conservation genetic research papers dealt with animals (~65%; Torres-Florez et al. 2018).

These studies provide clear evidence that scientific articles on conservation genetics often do not provide recommendations concrete enough for action to inform conservation managers properly. This is underpinned by a disconnect between the scientific community agenda and the real needs for conservation on the ground.

3.3.3 The Conservation Genetics Gap Between Science and Management

The lack of inclusion of genetic diversity in conservation management has become increasingly acknowledged and recognized (Fig. 3.4; Cook et al. 2016, Cook and Sgrò 2018, 2019a, b; Sandström et al. 2019; Taylor et al. 2017; Holderegger et al. 2019). The conservation genetics gap in the context of science versus management is inferred by the fact that scientific knowledge on genetic biodiversity, including how it is essential for biodiversity conservation and how it can be included in practical management, has long been available (Sect. 3.2), but this knowledge largely remains in the academic sphere and seldom transposes into management (Cook et al. 2013). Although the main part of the conservation genetic gap is between science and management (Fig. 3.4)—and this is the primary focus of the following sections—it is important to keep in mind that there is also deficient implementation of policy in management plans as outlined in Sect. 3.3.1.

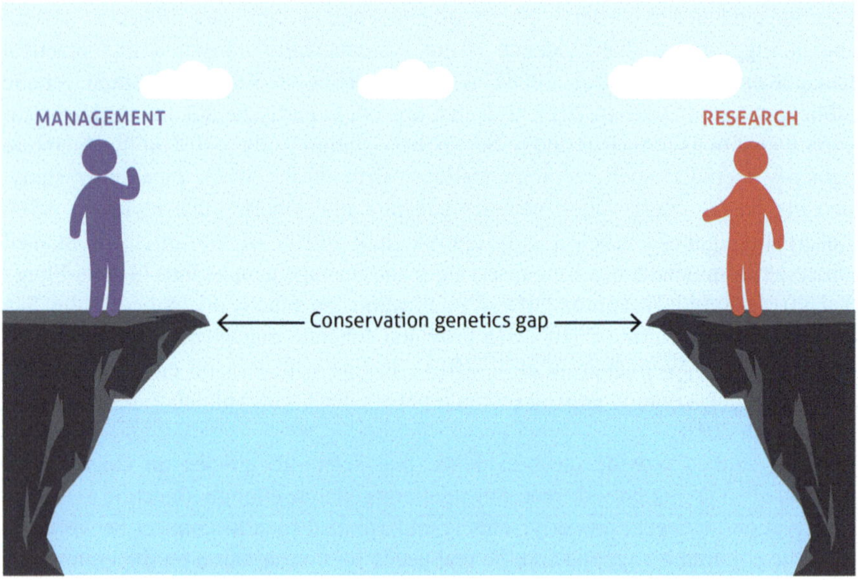

Fig. 3.4 The conservation genetics gap leads to a disconnect between research and applied management of genetic diversity. Illustration: Jerker Lokrantz/Azote (© Jerker Lokrantz 2020. All Rights Reserved)

3.4 Barriers to Integration of Genetics in Conservation Management and Policy

Numerous reasons for the lack of inclusion and application of genetic and evolutionary information into policy and management have been pointed out, including educational, institutional, communication, and societal barriers (Cook and Sgrò 2017; Cook et al. 2016; Haig et al. 2016; Lundmark et al. 2017, 2019; Sandström et al. 2016, 2019; Shaffer et al. 2015; Taylor et al. 2017). Here, we summarize some of the main issues identified thus far.

3.4.1 Lack of Genetic Knowledge Among Managers

A major barrier to the integration of genetic concepts and theory into natural resource policy and management is a lack of understanding of genetic concepts by practitioners and policymakers (Cook and Sgrò 2018, 2019a, b; Frankham 2010; Kinnison et al. 2007; Sandström et al. 2016; Taylor et al. 2017). Practitioners often do not have formal training to interpret and assess the strengths (or weaknesses) of genetic results (Haig et al. 2016). Empirical population genetics is a relatively young scientific discipline and a smaller field as compared to, for example, ecology. This may result in even many university-trained managers lacking population genetics competency.

Cook and Sgrò (2019a) found that scientists have mostly a better understanding of evolutionary concepts than practitioners. General concepts like genetic diversity and evolution were better understood than more specialized concepts (e.g., "gene flow") among practitioners. This coincided with a general lack of formal training in genetics and evolutionary biology for practitioners. But although practitioners demonstrated that they had problems defining concepts such as gene flow (Cook and Sgrò 2019a), they often proposed some type of genetic exchange to aid small and isolated populations (Cook and Sgrò 2019b).

3.4.2 Lack of Platforms for Knowledge Transfer and Communication

The general need for institutional meeting points for managers and researchers where genetic knowledge can be transferred and practical issues can be discussed been identified in many studies (e.g., Cook et al. 2013; Hoban et al. 2013a, b; Sandström et al. 2016, 2019; Shafer et al. 2015; Taylor et al. 2017, Holderegger et al. 2019). Insufficient and imprecise communications, as well as the use of jargon by geneticists, can pose a problem when communicating genetic and evolutionary concepts and results to policymakers and conservation practitioners (Garner et al. 2016;

Hoban et al. 2013a, b; Laurence et al. 2012; Pierson et al. 2016; Shafer et al. 2015). A certain cultural distance is perceived by practitioners in relation to geneticists that translates into uncertainty on how to bridge these two communities. However, when this bridge is crossed, interactions are usually reported as very positive (Taylor et al. 2017). In addition, practitioners typically voice interest in genetic information if it is readily available to them (Hoban et al. 2013a, b; Lundmark et al. 2017, 2019; Sandström et al. 2016; Taylor et al. 2017), so the demand exists for such knowledge and interaction.

3.4.3 Lack of Clarity of Policy

As discussed in Sect. 3.3.1, policy documents are often vague and insufficient when it comes to the assessment and integration of genetic diversity (e.g., Cook and Sgrò 2017). For instance, the CBD Aichi target 13 that was applicable until 2020 focused on the genetic diversity of cultivated plants and domesticated animals, their wild relatives and other socioeconomically and culturally valuable species. However, genetic diversity of natural animal and plant species, in general, are not mentioned, despite the pressing need to conserve genetic diversity in the wild to maximize adaptive potential in the light of rapid climate change and other habitat modifications (Allendorf et al. 2013; Allendorf 2017; Harrison et al. 2014; Laikre et al. 2020). Stronger and clearer formulated goals and recommendations at the international, national, and regional/local levels will be critical to facilitate exchange between research and practice.

3.4.4 Perceived Lack of Applicability of Conservation Genetic Research

A perception that genetics is not suitable to tackle particular conservation problems appears to exist, and hence the applicability of genetic and evolutionary research in management and policy is alleged to be low (Haig et al. 2016; Taylor et al. 2017). Genetic dynamics are often seen as long-term issues and, therefore, receive lower priority in conservation plans (Cook et al. 2016; Smith et al. 2014). Further, within the research field of conservation biology, some leading ecologists debated the role of genetics in conservation in the 1990s, and although their skepticism has now been proven erroneous, the effects of these views have lingered on (Sarre and Georges 2009). It has also been proposed that practitioners are reluctant to change established management practices if they are uncertain about the benefits of adding new knowledge/tools as there is a risk involved in shifting methods (Byrne et al. 2011; Cook and Sgrò 2018; Frankham et al. 2010; Shafer et al. 2015). From the scientists' standpoint, this means that there is a lack of clear and repeated communication of how genetic knowledge can be beneficial for conservation management and why it should receive high priority in the policy.

3.4.5 *Misalignment Between Priorities for Management* Versus *Research*

Several studies have suggested a mismatch between the conservation needs by practitioners and research priorities of conservation geneticists (Britt et al. 2018; Habel et al. 2013; Hoban et al. 2013a, b; Shafer et al. 2015).

Scientists frequently have the pressure to do cutting-edge science implementing the newest technologies and programs that lack immediate applicability to practitioners (Hoban et al. 2013a, b). One example of this is probably the current transition from genetic to genomic tools in many of the relevant, recently re-named, sub-disciplines like conservation genomics, landscape genomics, phylogenomics, etc. (Allendorf 2017; Holderegger et al. 2019; Russello et al. 2015; Shafer et al. 2015; Taylor et al. 2017). While science is moving on to the latest and greatest gadgets and producing an ever-growing amount of data (i.e., big data science), the conservation genetics gap seems to be widening because of added complexity resulting from the introduction of new genomic approaches and concepts without adequate knowledge transfer to stakeholders outside the academic bubble which are often advertised as being the target audience for these tools (Taylor et al. 2017). This creates a cultural gap that potentially leads to an even slower uptake of new information. Moreover, scientists are regularly pressured to produce scientific articles presenting novel results to secure funding and/or job promotion and therefore applied conservation genetic questions can have a low priority for them.

3.4.6 *Lack of Access to Scientific Publications*

Managers often lack access to scientific journals, making it difficult for them to follow results from conservation genetics research (Fuller et al. 2014; Hogg et al. 2017; Sandström et al. 2016). Managers also often lack work time devoted to upholding and broadening competency and keeping in touch with research (Sandström et al. 2016). Open access publications and other efforts, such as translations of English texts into local languages, to make conservation science research broadly accessible are highly warranted (Fuller et al. 2014; Sandström et al. 2016, 2019).

3.4.7 Lack of Practical Decision-Support Tools for Conservation Managers

Conservation managers and practitioners often mention that there are no practical decision-support tools for them to integrate genetic knowledge into conservation management plans (Carroll et al. 2014; Cook and Sgrò 2017; Frankham 2010; Ottewell et al. 2016). Thus, managers and practitioners ask for clearer guidelines from scientists on how to integrate genetic results into management plans. One example of how this issue might be solved is given in case study 2 below (Sect. 3.7).

3.4.8 Lack of Strategic Funding for Cross-Sectorial Efforts

Targeted funding for multi- and interdisciplinary research involving both researchers and practitioners to bridge the existing gap appears scarce but is pivotal (Bromham et al. 2016; Shafer et al. 2015). The pressures experienced by scientists to "publish or perish" have led to a frenetic pursuit of novel research, a trend that is additionally fueled by funding priorities of national research councils and foundations, the main funding sources available to researchers. In order to be competitive, scientists need to habitually demonstrate excellence and novelty in research applications and explain why species are good model systems to answer certain research questions. However, many conservation genetics questions need to be answered individually for different species as transferability of results is generally low because of differences in life histories, habitat, and local adaptations (Wennerström et al. 2013; Shafer et al. 2015). From a management perspective, (conservation genetic) research is often perceived as costly (Vernesi and Bruford 2009), and there is little agreement among practitioners on how to fund it (Taylor et al. 2017). Hence, institutional barriers like inadequate and poorly designed governmental funding schemes that do not take these realities into account are hampering progress to involve scientists in more applied conservation genetic research questions. Similarly, practitioners depend on government funding that can cease if changes at the political level occur.

3.5 Opportunities to Close the Conservation Genetics Gap

Ways to bridge the conservation genetics gap, which is aggravated by the barriers previously laid out in Sect. 3.4, have been proposed by several researchers (e.g., Britt et al. 2018; Hoban et al. 2013a, b; Laikre 2010; Ottewell et al. 2016; Sandström et al. 2019; Shafer et al. 2015; Taylor et al. 2017). Institutional, economical, and organizational changes and clearer policy and guiding principles are needed to close the gap. Here we synthesize the main measures needed.

3.5.1 Platforms for Continuous Knowledge Exchange and Science-Management Exchanges

The most frequently identified measure to bridge the conservation genetics gap is to promote close collaboration between scientists and managers (Britt et al. 2018; Hoban et al. 2013a, b; Lundmark et al. 2017, 2019; Sandström et al. 2019; Shafer et al. 2015; Ottewell et al. 2016; Taylor et al. 2017). Platforms for learning and knowledge transfer need to be funded, and it is vital that such efforts are secured over time and not only linked to short-term research projects (Lundmark et al. 2019; Sandström et al. 2019). National conservation genetic centers/hubs will need to become cornerstone infrastructures if modern conservation biology is to tackle this gap (Haig et al. 2016; Sjögren-Gulve et al. 2009; Taylor et al. 2017).

Lundmark et al. (2017) tested the efficiency of two different communication approaches for knowledge transfer on genetic biodiversity: lectures and deliberative group discussions. They showed that both efforts could positively affect public manager's perceptions of genetic diversity, but the effects are not long-lasting (Lundmark et al. 2019), thus advocating for long-term knowledge communication platforms (Sandström et al. 2019).

3.5.2 Scientists Become More Involved in Practical Conservation and Policy

Calls for conservation scientists to take time to engage in policy and practical management have occurred (Laikre et al. 2008a), and with the establishment of the Intergovernmental Science-Policy Platform on Biodiversity and Ecosystem Services (IPBES; www.ipbes.net), the international science-policy interface has certainly been strengthened (Schmeller and Bridgewater 2016). The importance of genetics is highlighted in the first global IPBES assessment (IPBES 2019). Clearly, it is vital that conservation geneticists are involved in the IPBES work, but involvement in policy work is needed at all levels—from international to national, regional, and local (Sandström et al. 2019). Funding agencies, universities, and developers of metrics to assess scientific impact, innovation and excellence need to become better aware of this gap and take action to make such involvement possible.

3.5.3 Knowledge on Genetics Increases Among Managers

Strengthening managers´ knowledge of population and conservation genetics is an important part of closing the gap. Here, authorities at local, regional, national, and international levels need to assure that such knowledge is available among or provided to their staff (Cook et al. 2013). Competency must be maintained through, for example, access and worktime devoted to scientific literature or specific training (Lundmark et al. 2017; Sandström et al. 2016).

3.5.4 Collaboration Between Scientists and Managers in Conservation Research

The close collaboration of scientists and managers through research projects, from early study design to implementation of actions, has been proposed as an effective strategy to close the knowledge-implementation gap in conservation science (Hulme 2014). This strategy may alleviate several of the aforementioned parts of the conservation genetics gap, including increasing manager's knowledge, strengthen researchers' understanding of practical management, decision-support tools, etc., and can lead to timely identification and alignment of research and management questions that can be supported by genetic knowledge. Further, more realistic expectations on both sides concerning pressures around time frames and goals may develop with this approach, which will aid in keeping project progress on track and avoiding stakeholder frustrations. In addition, practitioners can provide direct feedback on the feasibility of implementation of recommendations by scientists as they possess the "on-the-ground" expertise (Ottewell et al. 2016).

Scientists can work in government organizations to bring their expertise closer to decision-makers (Ridley and Alexander 2016). In some countries, like Canada, offices of government agencies like Environment Canada and Climate Change or the Ministry of Natural Resources and Forestry have been successfully integrated on university campuses to foster collaborations between scientists and practitioners. Embedding scientists in government agencies might help in compiling scientific information and making it more accessible to decision-makers. Haig et al. (2016) suggested the establishment of a National Center for Small Population Biology in the USA to study and bridge the gap between conservation geneticists and managers, and Sjögren-Gulve et al. (2009) proposed a similar center for genetic diversity in Sweden. None of these suggestions have yet been realized, though, but represent a promising development to narrow the gap.

3.5.5 Institutional and Financial Support Systems

Funding agencies, as well as universities and research institutes, need to facilitate interdisciplinary and cross-sectorial collaboration to enable research that will lead to biodiversity policy implementation and management. Important steps include:

- Funding agencies should give priority to multi- and interdisciplinary research projects aiming to assist the integration of different lines of evidence for genetic diversity conservation (Shafer et al. 2015).

- Funding agencies and universities need to increase incentives for researchers to engage in activities in policy-management interactions. Until such work is not promoted and rewarded, researchers will give it a low priority.
- Funding agencies could actively support networks, multi- and interdisciplinary workshops, and retreats to foster local and regional initiatives. The European **Co**operation in **S**cience and **T**echnology (i.e., COST—https://www.cost.eu/cost-actions/what-are-cost-actions/) actions is one example of such a funding scheme.

3.5.6 Accessibility of Conservation Genetic Knowledge and Data

Making conservation genetic research widely accessible, for instance, through open access publication, is highly important for closing the conservation genetic gap (Fuller et al. 2014; Sandström et al. 2016). Further, review articles summarizing the state of the art of subdisciplines and concepts in conservation genetics are important as such articles are often seen as of high quality by decision-makers and their condensed format appeals better to policymakers and managers to get a quick overview and improve their overall understanding about a particular topic (Ridley and Alexander 2016). Some initiatives to inform and integrate genetic knowledge in management and practice are summarized in Table 3.1.

Making genetic data accessible and publicly available to allow use in subsequent studies and management is needed. There are many publicly available databases for storing genetic data (e.g., Dryad (http://www.datadryad.org/), GenBank (https://www.ncbi.nlm.nih.gov/genbank/), the Barcode of Life Database (http://www.barcodinglife.com/), and scientific journals usually require deposition of genetic data in one of these platforms upon publication (Pope et al. 2015). However, a considerable proportion of articles (40%) does not report any geographic information associated with genetic data or temporal information, making the re-use of these data difficult (Pope et al. 2015).

Similarly, biological collections that can be used for DNA extraction to map and monitor genetic biodiversity are very important (Jackson et al. 2012; Laikre et al. 2008b). Such collections are currently often widespread among museums and universities where individual scientists may keep collections (Posledovich et al. 2021a, b). Typically, there is a lack of coordination among such initiatives, and it is often difficult to find supporting documentation about which samples are available and where. Further, in developing countries like Indonesia, there is currently a lack of infrastructure for comprehensive national gene banks dedicated to natural genetic diversity and digital collection facilities for genetic data storage, despite Indonesia and other developing countries having high biodiversity levels (Ragamustari and Sakaru 2019).

Table 3.1 Initiatives to inform and educate the public and integrate genetics and/or evolution of natural populations into conservation practice and management

Name of Organization (geographical scope)	Link	Focus
Educational		
EvoKE (Evolutionary Knowledge for Everyone; international, based in Europe)	https://evokeproject.org/	To increase evolution literacy among the general public and ensure informed decision-making by society
Euroscitizen (Europe)	http://www.euroscitizen.eu/	EU-funded COST action that aims to increase scientific literacy in evolution
Evolution in Action (Finland)	https://www.evolutioninaction.fi/home.html	A Finnish initiative that uses art to teach evolution
Melanogaster Catch The fly (Europe)	http://melanogaster.eu/?page_id=581	European Citizen science project to teach high school students and teachers about adaptive genomics by participation in scientific experiments.
Dissemination forum		
IUCN SSC Conservation Genetics Specialist Group (CGSG; international)	https://www.iucn.org/commissions/ssc-groups/disciplinary-groups/conservation-genetics	To increase the use of genetics in conservation management and decision-making, to assist the Species Survival Commission in applying genetics to threatened species and to lead the development and analysis of genetic data in conservation.
The Society for Conservation Biology's Conservation Genetics Working Group (SCB CGWG; international)	https://conbio.org/groups/working-groups/conservation-genetics-working-group/	To promote the use of genetics in conservation by having a community of conservation scientists that are aware of how genetics can contribute to conservation decisions
GEO BON—Genetics Composition working group (international)	https://geobon.org/ebvs/working-groups/genetic-composition/	To develop, test, and improve approaches for assessing and interpreting genetic diversity
The Wildlife Society—Molecular Ecology Working Group (USA)	https://wildlife.org/mewg/	To promote scientific advancement applying molecular techniques to wildlife ecology, management, and conservation. The group seeks to enhance awareness of molecular ecology and genetic applications to wildlife biology

(continued)

Table 3.1 (continued)

Name of Organization (geographical scope)	Link	Focus
Knowledge exchange in practice		
COST Action Genomics Biodiversity Knowledge for Resilient Ecosystems (G-BIKE, Europe)	https://www.cost.eu/actions/CA18134 https://www.facebook.com/gbikecost/ https://g-bikegenetics.eu/en	To assist scientists and practitioners in the use of standard and routine tools for assessing, monitoring, and managing the genetic resilience and related adaptive potential of wild and captive populations
Conserving Genetic Resources for Effective Species Survival (ConGRESS, Europe)	www.ConGRESSgenetics.eu	To provide information and resources for biodiversity managers and policymakers in the nature conservation sector to encourage the use of genetic data and studies on species and populations in biodiversity projects
BaltGene—Baltic Sea Genetics for Managers (Baltic Europe)	https://bambi.gu.se/baltgene https://www.gu.se/en/cemeb-marineevolutionary-biology/management-conservation/baltgene	Knowledge communication from several research projects focusing on conservation genetics in the Baltic Sea
NCEAS/NESCent Working Group on Genetic Monitoring (GeM: USA)	https://www.nceas.ucsb.edu/workinggroups/genetic-monitoring-development-tools-conservation-and-management	Provide guidelines on how genetic monitoring can be used in management. Guidelines from GeM being adopted by the US Fish and Wildlife Service https://www.fws.gov/r7/gem/principles_I.htm

3.5.7 Guidelines and Decision-Support Tools

Managers need specific information on how genetic diversity can be included in conservation practice. Efforts to provide guidelines and platforms for solid information include the Conservation Genetic Resources for Effective Species Survival (ConGRESS; Table 3.1; Hoban et al. 2013b). Key components of this platform include:

1. A "*knowledge pack*" in different languages that explains genetic concepts, genetic data types, and best practices for conservation actions based on genetic data in mostly jargon-free plain language. This pack addresses several issues, including the removal of language barriers, making genetic information more easily accessible for non-specialists, and policymakers and conservation managers in all the 23 EU languages.
2. A "*publication database*" to facilitate easy access and search functions for researchers, managers, and policymakers.

3. A "*decision-making tool*" providing guidance for practitioners and policymakers for the identification of appropriate conservation genetic tools for common management issues.
4. A "*sample planning tool*" to assist in study design (e.g., testing the effectiveness of sampling schemes) and estimation of funding needed for managers and researchers to assess study feasibility.
5. A "*forum*" provides direct contact to discuss questions online with scientists and practitioners (searchable by keywords and expertise).

Future add-ons include "gray literature" like reports and next-generation sequencing information as the transition from classic genetic markers to genomic markers progresses (Hoban et al. 2013b). Extensions for the planning tools to include phylogenetic, forensics, and environmental DNA are desirable. However, Hoban et al. (2013b) emphasize that this platform approach has its challenges because frequent updates are required and therefore, involvement of the community and continuous funding.

Similar platforms have been established by the US Fish and Wildlife Services in the USA (Stetz et al. 2011; https://www.fws.gov/r7/gem/mainPage_1.htm; http://www.fws.gov/ConservationGeneticsCOP/), and for the Baltic Sea region in Europe (https://bambi.gu.se/baltgene). Challenges include keeping platforms such as these alive after funding for their construction has ceased (Hoban et al. 2013b; Sandström et al. 2019).

In the following section, we present two case studies that illustrate solutions to some of the issues raised in this chapter. More specifically, the first case study looks at a multidisciplinary framework that aimed to understand why genetic diversity is underrepresented in policy and management documents. The second case study deals with the provision of clear guidelines on how to interpret genetic research results for policy and conservation managers.

3.6 Case Study 1: Framing Management for Baltic Sea Species by Understanding Their Evolutionary Potential

The Baltic Sea is a semi-enclosed brackish water body in northern Europe experiencing rapid environmental change, and with nine countries bordering this Sea, governance structure is complex (Reusch et al. 2018). Genetic diversity has been identified to be particularly important to conserve and monitor in the Baltic Sea because of the rapid environmental changes, the generally low genetic diversity of many marine species that have adapted to the brackish water, and the genetic uniqueness of many populations inhabiting this area (Johannesson et al. 2011).

A large research project focused on various aspects of genetic diversity and adaptive potential of key ecological species in the Baltic Sea in the context of climate change (www.bambi.gu.se). Within this project, a multidisciplinary team represented by the political, educational, and natural sciences addressed governance structures, policy instruments, and management measures concerning genetic diversity. Specifically, the role of genetic diversity in marine protected areas (MPAs) was investigated, and the work followed a three-stage approach (Sandström et al. 2019).

Stage 1: How is genetic biodiversity considered in international and national policies, and how are these policies reflected in Baltic Sea MPA management plans? To answer this question a quantitative and qualitative textual analysis was done of 240 policy documents representing these levels using four focal countries (Estonia, Finland, Sweden, and Germany). The findings show that international and national policies do identify genetic biodiversity as important to protect—particularly in regards to protected areas—but policies directed towards marine environments do so to a lesser extent than other conservation policies. MPA plans poorly considered genetic diversity—only about 20% included some mentions of genetic biodiversity that were, according to the qualitative analysis, very shallow and primarily concerned terrestrial species. Thus, even though international and national policies express goals for conserving genetic biodiversity, these goals are not reflected in MPA management plans (Laikre et al. 2016).

Stage 2: What factors explain the minor role given to genetic biodiversity in Baltic Sea MPA management? Explanations for the neglect of the genetic component in MPA management include a multi-level governance system with numerous goals, actors, and interests. This leads to ambiguity and considerable room for interpretation at the regional level. Therefore, the research team turned directly to the regional conservation managers focusing on Sweden and Finland. In-depth interviews confirmed the minor role of genetic biodiversity in MPA management. Genetic biodiversity is not a key aspect when identifying new protected areas or governing existing ones. The reasons are found in the policy framework itself, in the resources available for on-the-ground implementation, and in managers' own knowledge and views on genetic biodiversity. Even though there is no consensus about what policies actually govern their work, the common view is that these documents do not include guidelines on genetic biodiversity. Moreover, the conservation managers stated lacking resources (time, money, and knowledge) to translate the few guidelines into actual conservation action properly. Finally, even though managers expressed that genetic biodiversity is important, they hesitated to explain how important it is and how it should be handled in MPA management (Fig. 3.5; Sandström et al. 2016).

Stage 3: Does knowledge transfer influence conservation managers' views on genetic biodiversity, and are some forms of communications more efficient than others?

The perceived need for more knowledge motivated the third stage of this research, in which the effectiveness of different channels of communication was assessed on managers' views on genetic biodiversity. Altogether 72 managers from nine different regional authorities participated in the study; half of them joined a traditional lecture on genetic biodiversity in the Baltic Sea, and the other half joined a deliberative discussion organized on site. The managers′ self-assessed knowledge and views on genetic diversity prior to the communication, directly after, and 3-4 months later were assessed through a survey. The findings showed that both

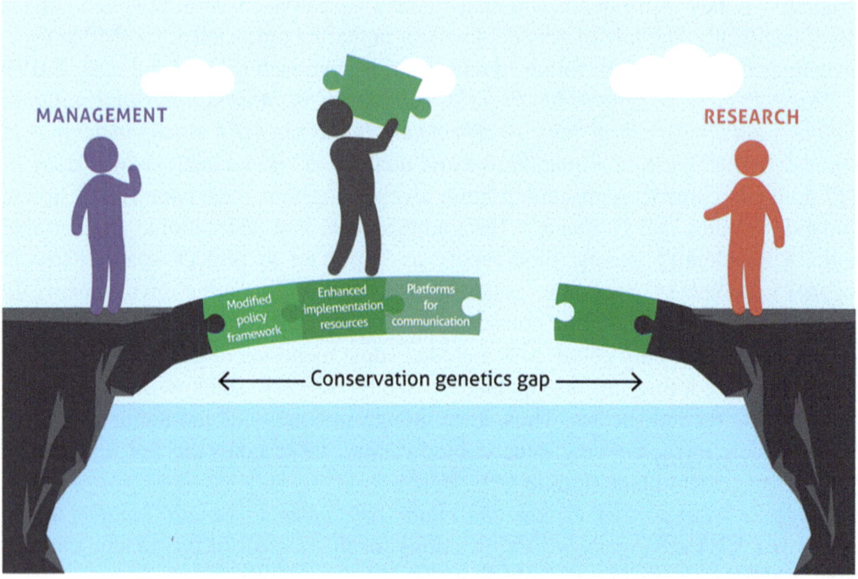

Fig. 3.5 Key elements identified that could bridge the conservation genetics gap identified in the case study focusing on Baltic Sea genetic diversity. Illustration: Jerker Lokrantz/Azote (© Jerker Lokrantz 2020. All Rights Reserved)

the lecture and the discussion had positive and remaining effects on managers' knowledge on genetic biodiversity in the Baltic Sea. The observation made directly after the session implied that the lecture was more effective than group deliberations in transforming the managers' views; however, this difference disappeared over time. Thus, the results suggest that continuity, rather than format, is important when setting up platforms for genetic knowledge exchange between scientists and managers (Figs. 3.5, 3.6; Lundmark et al. 2017, 2019).

The final stage of the project also included reviewing existing knowledge on the genetic diversity of Baltic Sea species to provide recommendations for conservation action. These were collated and made accessible to managers in a review paper, a website and through a stakeholder conference and networking activity (Fig. 3.6; Wennerström et al. 2013; https://bambi.gu.se/baltgene; https://bambi.gu.se/activities/baltgene-2017). The outcomes of this case study are summarized in Fig. 3.6 (see also Sandström et al. (2019)). Spin-off effects of the policy-management-research interactions during this project include plans for a national program to monitor genetic diversity of some selected Baltic Sea species, as well as of some freshwater species, by the Swedish Agency for Marine and Water Management.

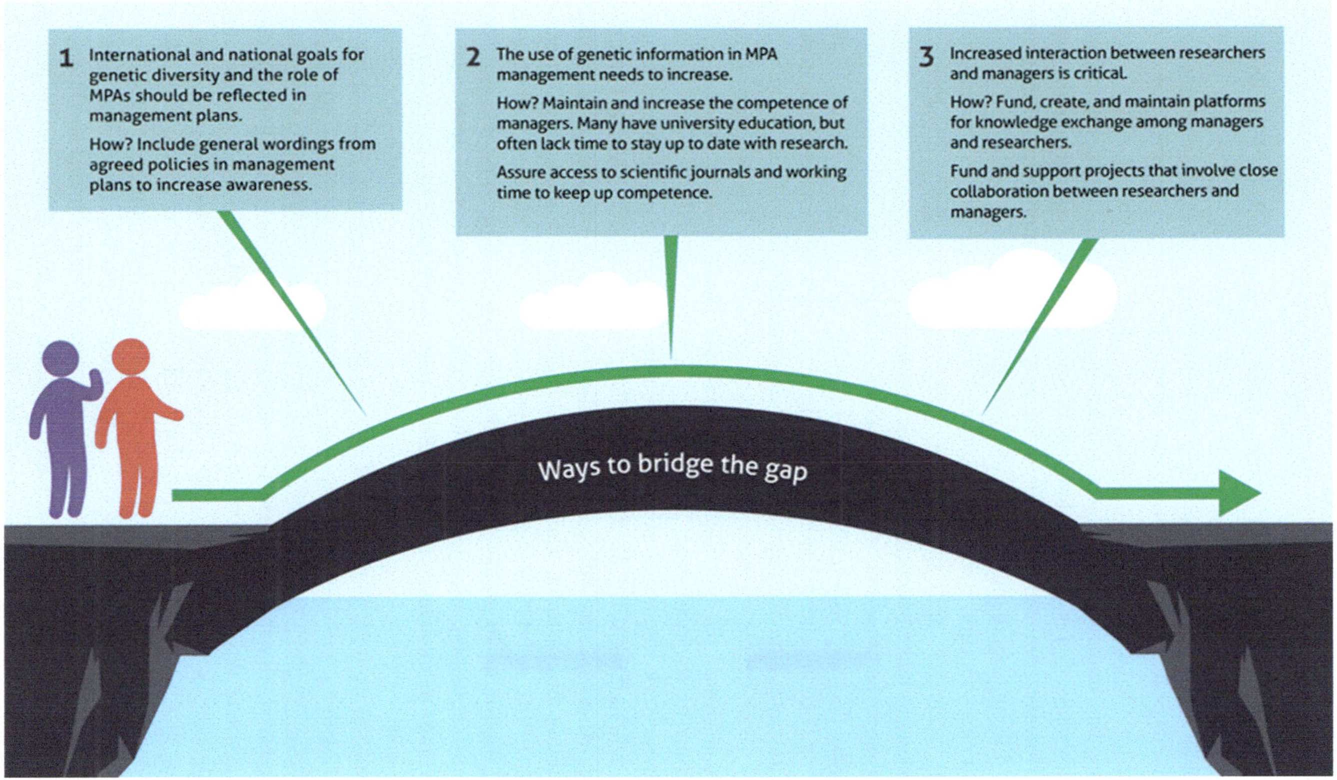

Fig. 3.6 Key recommendations identified for the management of Baltic Sea genetic diversity in marine protected areas (MPAs). Illustration: Jerker Lokrantz/ Azote (© Jerker Lokrantz 2020. All Rights Reserved)

3.7 Case Study 2: A Genetic Assessment Framework for Plant Conservation

Ottewell et al. (2016) developed a plain language genetic decision framework for population-level conservation prioritization of plant species using key genetic indicators, namely genetic differentiation, genetic diversity, and inbreeding (Fig. 3.7). Key elements of this decision framework that was designed to help practitioners and conservation managers are clear guidelines on how to interpret single genetic indicators. The framework addresses a fundamental difficulty in the interpretation of genetic results, that is, those different indicators can give seemingly contradictory signals. However, Ottewell et al. (2016) suggest that the combination of indicators captures different population genetic and evolutionary histories sufficiently well to provide a clear decision tree, with associated guidelines for action, under a wide range of scenarios and species that can help conservation managers and policymakers to make better use of genetic results.

For instance, let us assume we have a species with high genetic differentiation and high genetic diversity within populations. In addition, high inbreeding coefficients are observed. Using the decision tree in Fig. 3.7, we would end up with strategy 6. This strategy is aimed at populations that are genetically diverse but are divergent and show some degree of inbreeding (Ottewell et al. 2016). The recommendation would be to ensure the breeding of unrelated individuals to maintain or increase genetic diversity (Ottewell et al. 2016).

The framework was tested using two examples that led to clarification of management actions (Ottewell et al. 2016). First, the genetic diversity of Mount Compass Oak-bush (*Allocasuarina robusta*) in Australia was studied. This species is listed as endangered in the Australian Environmental Protection and Biodiversity Conservation Act (EPBC Act 1999). Genetic analyses showed that genetic diversity was similarly high in all locations studied, inbreeding was low in all populations, and genetic differentiation was low. Therefore, the conservation recommendations included the maintenance of gene flow (e.g., corridors) to prevent future diversity loss.

The second example demonstrated how additional knowledge could refine the recommendations based on the decision framework. The genetic diversity and differentiation of geographically separated remnant and restocked populations of the endangered Monarto Mintbush (*Prostanthera eurybioides*) in Australia were assessed. Genetic differentiation between the disconnected populations was low, and genetic diversity was low in one of the populations but high in the other. Similarly, inbreeding was high in the first population but low in the second. Both populations rely on restocking efforts, but the population with higher inbreeding coefficients was likely of smaller population size. One could come to the conclusion that strategy 1 may also apply in this instance. However, there are different climatic conditions for the two populations that are 160 km apart, and the species is dependent on autumn rainfall for germination. Further, ecological threats like grazing are thought to be responsible for low recruitment rather than genetic issues. Hence, strategy 5 was

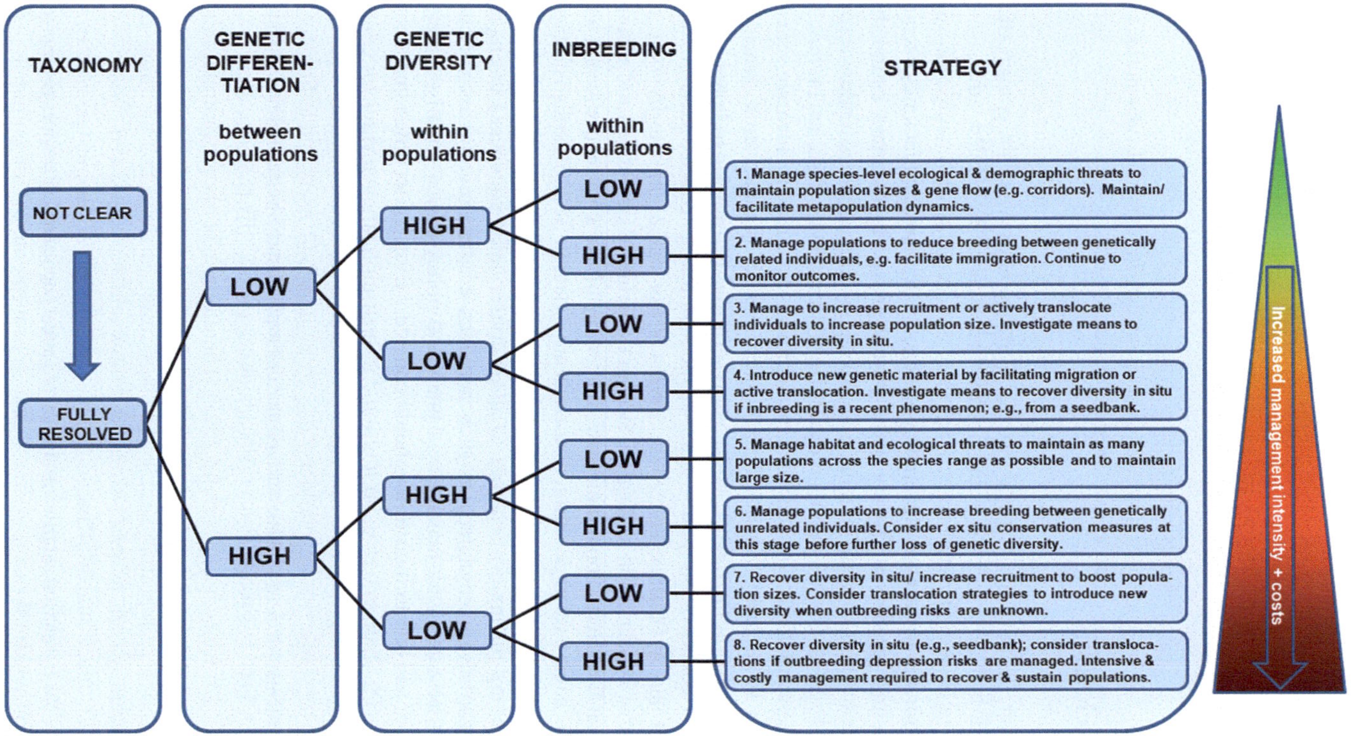

Fig. 3.7 Decision tree for finding genetic conservation strategies at the population-level for plants. The three indicators assessed are population differentiation, genetic diversity, and inbreeding. These are rated either high or low. Depending on the path chosen, one of eight strategies is selected. The taxonomic status of the species should be clear before using this decision framework. Adapted from Ottewell et al. (2016) with permission

more appropriate because it included the management of ecological threats, and more attention was paid to the fact that one location suffered from genetic erosion.

Particularly the second example shows that the flexible decision framework requires the active participation of both scientists and practitioners to guide the decision process. It has yet to be shown if this approach will be broadly up taken by practitioners and how widely applicable it is.

3.8 Conclusions

In this chapter, we have summarized the current state and extent of the conservation genetics gap, including its causes and potential remedies. This review emphasizes that conservation geneticists need to be more involved in communication and knowledge transfer to decision- and policymakers. Such interaction includes becoming more active in advisory panels and working groups to facilitate knowledge mobilization and uptake of evolutionary knowledge by different stakeholders across geographical scales. Clearer communication by conservation geneticists on how to use genetic research results in policy and management is required to integrate genetic diversity into biodiversity conservation initiatives better. Conservation managers and policymakers, on the other hand, have to increase their efforts to better incorporate genetic concepts into policy and management that go beyond general statements about the importance of genetic diversity but make genetic diversity an integral part of conservation strategies. Stronger articulated objectives and recommendations at the international, national, and regional/local governance levels will be crucial to enable exchange between research and practice. To accomplish efficient research policy management collaboration and exchange, structural changes are needed including in funding schemes to allow conservation geneticists to engage (and be rewarded for engaging) in such activities continuously to foster better alignment of research with management goals and policy cycles. Finally, government agencies and research institutes should consider hiring conservation geneticists with a least a Ph.D. degree to build multi-and transdisciplinary teams for long-term continuous and effective conservation genetic research and management.

Acknowledgements Dr. Linda Laikre received financial support from Swedish Research Council grant 2019-05503 and Swedish Research Council Formas grant FR-2016/0005.

References

Allendorf FW (2017) Genetics and the conservation of natural populations: allozymes to genomes. Mol Ecol 26(2):420–430

Allendorf F, Ryman N, Stennek A, Ståhl G (1976) Genetic variation in Scandinavian brown trout (*Salmo trutta* L.): evidence of distinct sympatric populations. Hereditas 83(1):73–82

Allendorf FW, Luikart GH, Aitken SN (2013) Conservation and the genetics of populations, 2nd edn. Wiley-Blackwell, West Sussex
Ballou JD, Gilpin M, Foose TJ (eds) (1995) Population management for survival and recovery. Analytical methods and strategies in small population conservation. Columbia University Press
Bateson W, Saunders ER (1902) The facts of heredity in the light of Mendel's discovery. Reports to the Evolution Committee of the Royal Society, I. pp 125–160
Benestan LM, Ferchaud A-L, Hohenlohe PA, Garner BA, Naylor GJP, Baums IB, Schwartz MK, Kelley JL, Luikart G (2016) Conservation genomics of natural and managed populations: building a conceptual and practical framework. Mol Ecol 25(13):2967–2977
Bowman J, Greenhorn JE, Marrotte RR, McKay MM, Morris KY, Prentice MB, Wehtje M (2016) On applications of landscape genetics. Conserv Genet 17:753–760
Britt M, Haworth SE, Johnson JB, Martchenko D, Shafer ABA (2018) The importance of non-academic coauthors in bridging the conservation genetics gap. Biol Conserv 218:118–123
Brodersen J, Seehausen O (2014) Why evolutionary biologists should get seriously involved in ecological monitoring and applied biodiversity assessment programs. Evol Appl 7(9):968–983
Bromham L, Dinnage R, Hua X (2016) Interdisciplinary research has consistently lower funding success. Nature 534:684–687
Byrne M, Stone L, Millar MA (2011) Assessing genetic risk in revegetation. J Appl Ecol 48(6):1365–1373
Carroll SP, Jørgensen PS, Kinnison MT, Bergstrom CT, Denison RF, Gluckman P, Smith TB, Strauss SY, Tabashnik BE (2014) Applying evolutionary biology to address global challenges. Science 346(6207):1245993
CBD (2001) Global Biodiversity Outlook 1. Secretariat of the Convention on Biological Diversity Montreal 2001. https://www.cbd.int/gbo1/
Cook CN, Sgrò CM (2017) Aligning science and policy to achieve evolutionarily enlightened conservation. Conserv Biol 31(3):501–512
Cook CN, Sgrò CM (2018) Understanding managers' and scientists' perspectives on opportunities to achieve more evolutionarily enlightened management in conservation. Evol Appl 11(8):1371–1388
Cook CN, Sgrò CM (2019a) Conservation practitioners' understanding of how to manage evolutionary processes. Conserv Biol 33(5):993–1001
Cook CN, Sgrò CM (2019b) Poor understanding of evolutionary theory is a barrier to effective conservation management. Conserv Lett 12(2):e12619
Cook CN, Mascia MB, Schwartz MW, Possingham HP, Fuller RA (2013) Achieving conservation science that bridges the knowledge–action boundary. Conserv Biol 27(4):669–678
Cook CN, de Bie K, Keith DA, Addison PFE (2016) Decision triggers are a critical part of evidence-based conservation. Biol Conserv 195:46–51
Darwin C (1859) On the origin of species by means of natural selection. John Murray, London
Diniz-Filho JAF, Melo DB, de Oliveira G, Collevatti RG, Soares TN, Nabout JC, Lima JS, Dobrovolski R, Chaves LJ, Naves RV, Loyola RD, de Campos Telles MP (2012) Planning for optimal conservation of geographical genetic variability within species. Conserv Genet 13: 1085–1093
Diniz-Filho JAF, Diniz JVBPL, de Campos Telles MP (2016) Exhaustive search for conservation networks of populations representing genetic diversity. Genet Mol Res GMR 15(1): gmr.15017525
Ehrenfeld D (ed) (1995) Readings from conservation biology. Genes, populations, and species. The Society for Conservation Biology and Blackwell Science, The Sheridan Press, Hanover, Pennsylvania
Fitzpatrick MC, Keller SR (2015) Ecological genomics meets community-level modelling of biodiversity: mapping the genomic landscape of current and future environmental adaptation. Ecol Lett 18(1):1–16
Frankel OH (1970) Variation – the essence of life. Proc Linn Soc New South Wales 95(2):158–169
Frankel OH (1974) Genetic conservation: our evolutionary responsibility. Genetics 78(1):53–65

Frankel OH, Soulé ME (1981) Conservation and evolution. Cambridge Univ. Press, Cambridge
Frankham R (2010) Challenges and opportunities of genetic approaches to biological conservation. Biol Conserv 143:1919–1927
Frankham R, Ballou JD, Briscoe DA (2010) Introduction to conservation genetics, 2nd edn. Cambridge University Press, Cambridge
Frankham R, Ballou JD, Ralls K, Eldridge M, Dudash MR, Fenster CB, Lacy RC, Sunnucks P (2019) A practical guide for genetic management of fragmented animal and plant populations. Oxford University Press, Oxford
Fuller RA, Lee JR, Watson JEM (2014) Achieving open access to conservation science. Conserv Biol 28(6):1550–1557
Garner BA, Hand BK, Amish SJ, Bernatchez L, Foster JT, Miller KM, Morin PA, Narum SR, O'Brien SJ, Roffler G, Templin WD, Sunnucks P, Strait J, Warheit KI, Seamons TR, Wenburg J, Olsen J, Luikart G (2016) Genomics in conservation: case studies and bridging the gap between data and application. Trends Ecol Evol 31(2):81–83
Groom MJ, Meffe GK, Carroll CR (2006) Principles of conservation biology, 3rd edn. Sinauer Associates, Sunderland, MA
Habel JC, Gossner MM, Meyer ST, Eggermont H, Lens L, Dengler J, Weisser WW (2013) Mind the gaps when using science to address conservation concerns. Biodivers Conserv 22:2413–2427
Haig SM, Miller MP, Bellinger R, Draheim HM, Mercer DM, Mullins TD (2016) The conservation genetics juggling act: integrating genetics and ecology, science and policy. Evol Appl 9(1):181–195
Harrisson KA, Pavlova A, Telonis-Scott M, Sunnucks P (2014) Using genomics to characterize evolutionary potential for conservation of wild populations. Evol Appl 7(9):1008–1025
Henry RJ (ed) (2006) Plant conservation genetics. CRC Press, ISBN 9781560229971
Hoban SM, Hauffe HC, Pérez-Espona S, Arntzen JW, Bertorelle G, Bryja J, Frith K, Gaggiotti OE, Galbusera P, Godoy JA, Hoelzel AR, Nichols RA, Primmer CR, Russo I-R, Segelbacher G, Siegismund HR, Sihvonen M, Verensi C, Vilà C, Bruford MW (2013a) Bringing genetic diversity to the forefront of conservation policy and management. Conserv Genet Resour 5: 593–598
Hoban S, Arntzen JW, Bertorelle G, Bryja J, Fernandes M, Frith K, Gaggiotti O, Galbusera P, Godoy JA, Hauffe HC, Hoelzel AR, Nichols RA, Pérez-Espona S, Primmer C, Russo I-RM, Segelbacher G, Siegismund HR, Sihvonen M, Sjögren-Gulve P, Vernesi C, Vilà C, Bruford MW (2013b) Conservation Genetic Resources for Effective Species Survival (ConGRESS): bridging the divide between conservation research and practice. J Nat Conserv 21(6):433–437
Hoban S, Bruford M, D'Urban Jackson J, Fernandes-Lopes M, Heuertz M, Hohenlohe PA, Sjögren-Gulve P, Segelbacher G, Vernesi C, Aitken S, Bertola LD, Bloomer P, Breed M, Rodríguez-Correa H, Funk WC, Grueber CE, Hunter ME, Jaffe R, Liggins L, Mergeay J, Moharrek F, O'Brien D, Ogden R, Palma-Silva C, Paz-Vinas I, Oierson J, Ramakrishnan U, Simo-Droissart-M, Tani N, Waits L, Laikre L (2020) Genetic diversity targets and indicators in the CBD post-2020 Global Biodiversity Framework must be improved. Biol Conserv 248:108654
Hoban S, Campbell CD, da Silva JM, Ekblom R, Funk WC, Garner BA, Godov JA, Kershaw F, MacDonald AJ, Mergeay J, Minter M, O'Brien D, Paz Vinas I, Pearson SK, Pérez-Espona S, Potter KM, Russo I-RM, Segelbacher G, Vernesi C, Hunter ME (2021a) Genetic diversity is considered important but interpreted narrowly in country reports to the convention on biological diversity: current actions and indicators are insufficient. Biol Conserv 261:109233
Hoban S, Bruford MW, Funk WC, Galbusera P, Griffith MP, Grueber CE, Heuertz M, Hunter ME, Hvilsom C, Kalamujic Stroil B, Kershaw F, Khoury CK, Laikre L, Lopes-Fernandes M, MacDonald AJ, Mergeay J, Meek M, Mittan C, Mukassabi TA, O'Brien D, Ogden R, PALMA-SILVA C, Ramakrishnan U, Segelbacher G, Shaw RE, Sjögren-Gulve P, Veličković N, Vernesi C (2021b) Global commitments to conserving and monitoring genetic diversity are now necessary and feasible. Bioscience 71(9):964–976

Hogg CJ, Grueber CE, Pemberton D, Fox S, Lee AV, Ivy JA, Belov K (2017) "Devil Tools & Tech": a synergy of conservation research and management practice. Conserv Lett 10(1):133–138

Holderegger R, Balkenhol N, Bolliger J, Engler JO, Gugerli F, Hochkirch A, Nowak C, Segelbacher G, Widmer A, Zachos FE (2019) Conservation genetics: linking science with practice. Mol Ecol 28(17):3848–3856

Hulme PE (2014) Bridging the knowing-doing gap: know-who, know-what, know-why, know-how and know-when. J Appl Ecol 51(5):1131–1136

IPBES (2019) Summary for policymakers of the global assessment report on biodiversity and ecosystem services of the Intergovernmental Science-Policy Platform on Biodiversity and Ecosystem Services. Díaz S, Settele J, Brondízio ES, Ngo HT, Guèze M, Agard J, Arneth A, Balvanera P, Brauman KA, Butchart SHM, Chan KMA, Garibaldi LA, Ichii K, Liu J, Subramanian SM, Midgley GF, Miloslavich P, Molnár Z, Obura D, Pfaff A, Polasky S, Purvis A, Razzaque J, Reyers B, Chowdhury R, Shin YJ, Visseren-Hamakers IJ, Willis KJ, Zayas CN (eds.). IPBES secretariat, Bonn, Germany. 56 pages

Jackson JA, Laikre L, Baker CS, Kendall KC, The Genetics Monitoring Group (2012) Guidelines for collecting and maintaining archives for genetic monitoring. Conserv Genet Recourses 4: 527–536

Johannesson K, Smolarz K, Grahn M, André C (2011) The future of Baltic Sea populations: local extinction or evolutionary rescue? Ambio 40:179–190

Keller D, Holderegger R, van Strien MJ, Bolliger J (2015) How to make landscape genetics beneficial for conservation management? Conserv Genet 16:503–512

Kinnison MT, Hendry AP, Stockwell CA (2007) Contemporary evolution meets conservation biology II: impediments to integration and application. Ecol Res 22:947–954

Laikre L (2010) Genetic diversity is overlooked in international conservation policy implementation. Conserv Genet 11:349–354

Laikre L, Larsson LC, Palmé A, Charlier J, Josefsson M, Ryman N (2008a) Potentials for monitoring gene level biodiversity: using Sweden as an example. Biodivers Conserv 17:893–910

Laikre L, Jonsson B-G, Ihse M, Marissink M, Dock Gustavsson A-M, Ebenhard T, Hagberg L, Ståhl P-O, von Walter S, Wramner P (2008b) Wanted: scientists in the CBD process. Conserv Biol 22(4):814–815

Laikre L, Nilsson T, Primmer CR, Ryman N, Allendorf FW (2009) Importance of genetics in the interpretation of favourable conservation status. Conserv Biol 23(6):1378–1381

Laikre L, Allendorf FW, Aroner LC, Baker CS, Gregovich DP, Hansen MM, Jackson JA, Kendall KC, McKelvey K, Neel MC, Olivieri I, Ryman N, Schwartz MK, Bull RS, Stetz JB, Tallmon DA, Taylor BL, Vojta CD, Waller DM, Waples RS (2010) Neglect of genetic diversity in implementation of the convention on biological diversity. Conserv Biol 24(1):86–88

Laikre L, Lundmark C, Jansson E, Wennerström L, Edman M, Sandström A (2016) Lack of recognition of genetic biodiversity: international policy and its implementation in Baltic Sea marine protected areas. Ambio 45:661–680

Laikre L, Hoban S, Bruford MW, Segelbacher G, Allendorf FW, Gajardo G, Rodríguez AG, Hedrick PW, Heuertz M, Hohenlohe PA, Jaffé R, Johannesson K, Liggins L, MacDonald AJ, Wengel PO, Reusch TBH, Rodríguez-Correa H, Russo IRM, Ryman N, Vernesi C (2020) Post-2020 goals overlook genetic diversity. Science 367(6482):1083–1085

Laurance WF, Koster H, Grooten M, Anderson AB, Zuidema PA, Zwick S, Zagt RJ, Lynam AJ, Linkie M, Anten NPR (2012) Making conservation research more relevant for conservation practitioners. Biol Conserv 153:164–168

Leigh DM, Hendry AP, Vázquez-Domínguez E, Friesen VL (2019) Estimated six per cent loss of genetic variation in wild populations since the industrial revolution. Evol Appl 12(8):1505–1512

Lundmark C, Andersson K, Sandström A, Laikre L (2017) Effectiveness of short-term knowledge communication on Baltic Sea marine genetic biodiversity to public managers. Reg Environ Chang 17:841–849

Lundmark C, Sandström A, Andersson K, Laikre L (2019) Monitoring the effects of knowledge communication on conservation managers' perception of genetic biodiversity – a case study from the Baltic Sea. Mar Policy 99:223–229

Mace GM, Purvis A (2008) Evolutionary biology and practical conservation: bridging a widening gap. Mol Ecol 17(1):9–19

Mace GM, Cramer W, Díaz S, Faith DP, Larigauderie A, Le Prestre P, Palmer M, Perrings C, Scholes RJ, Walpole M, Walther BA, Watson JEM, Mooney HA (2010) Biodiversity targets after 2010. Curr Opin Environ Sustain 2(1-2):3–8

Mendel G (1866) Versuche über Pflanzenhybriden, Verhandlungen des naturforschenden Vereines in Brünn, Bd. IV für das Jahr, 1865, Abhandlungen: 3–47

Mimura M, Yahara T, Faith DP, Vázquez-Domínguez E, Colautti RI, Araki H, Javadi F, Núñez-Farfán J, Mori AS, Zhou S, Hollingsworth PM, Neaves LE, Fukano Y, Smith GF, Sato Y-I, Tachida H, Hendry AP (2017) Understanding and monitoring the consequences of human impacts on intraspecific variation. Evol Appl 10(2):121–139

Mouillot D, Parravicini V, Bellwood DR, Leprieur F, Huang D, Cowman PF, Albouy C, Hughes TP, Thuiller W, Guilhaumon F (2016) Global marine protected areas do not secure the evolutionary history of tropical corals and fishes. Nat Commun 7:10359

Ottewell KM, Bickerton DC, Byrne M, Lowe AJ (2016) Bridging the gap: a genetic assessment framework for population-level threatened plant conservation prioritization and decision-making. Divers Distrib 22(2):174–188

Pierson JC, Coates DJ, Oostermeijer JGB, Beissinger SR, Bragg JG, Sunnucks P, Schumaker NH, Young AG (2016) Genetic factors in threatened species recovery plans on three continents. Front Ecol Environ 14(8):433–440

Pope LC, Liggins L, Keyse J, Carvalho SB, Riginos C (2015) Not the time or the place: the missing spatio-temporal link in publicly available genetic data. Mol Ecol 24(15):3802–3809

Posledovich D, Ekblom R, Laikre L (2021a) Mapping and monitoring genetic diversity in Sweden – suggestions for pollinating species. Report to the Swedish Environmental Protection Agency, 6958. ISBN 978-91-620-6958-2. Available from: https://www.naturvardsverket.se/Om-Naturvardsverket/Publikationer/ISBN/6900/978-91-620-6958-2/

Posledovich D, Ekblom R, Laikre L (2021b) Mapping and monitoring genetic diversity in Sweden. A proposal for species, methods and costs. The Swedish Environmental Protection Agency, Report 6959, 77 pp. ISBN 978-91-620-6959-9

Ragamustari SK, Sukara E (2019) Strengthening the genetic diversity conservation narrative in Indonesia: challenges and prospects. Biodivers Conserv 28:1647–1665

Ralls K, Ballou J (1983) Extinction: lessons from zoos. In: Schonewald-Cox CM, Chambers SM, MacBryde B, Thomas L (eds) Genetics and conservation: a reference for managing wild animal and plant populations. Benjamin/Cummings, Menlo Park, CA, pp 164–184

Rands MRW, Adams WM, Bennun L, Butchart SHM, Clements A, Coomes D, Entwistle A, Hodge I, Kapos V, Scharlemann JPW, Sutherland WJ, Vira B (2010) Biodiversity conservation: challenges beyond 2010. Science 329(5997):1298–1303

Reusch TBH, Dierking J, Andersson HC, Bonsdorff E, Carstensen J, Casini M, Czajkowski M, Hasler B, Hinsby K, Hyytiäinen K, Johannesson K, Jomaa S, Jormalainen V, Kuosa H, Kurland S, Laikre L, MacKenzie BR, Margonski P, Melzner F, Oesterwind D, Ojaveer H, Refsgaard JC, Sandström A, Schwarz G, Tonderski K, Winder M, Zandersen M (2018) The Baltic Sea as a time machine for the future coastal ocean. Sci Adv 4(5):eaar8195

Ridley CE, Alexander LC (2016) Applying gene flow science to environmental policy needs: a boundary work perspective. Evol Appl 9(7):924–936

Rodríguez-Clark KM, Oliveira-Miranda MA, Meneses MA, Martino Á, Méndez MA, Miyaki C, Montiel-Villalobos MG, de Oliveira-Miranda RM, Poulin E, Ruzzante D, Solé-Cava A (2015)

Finding the "conservation" in conservation genetics – progress in Latin America. J Hered 106 (S1):423–427
Russello MA, Waterhouse MD, Etter PD, Johnson EA (2015) From promise to practice: pairing non-invasive sampling with genomics in conservation. PeerJ 3:e1106
Ryman N (1981) Fish gene pools: preservation of genetic resources in relation to wild fish stocks. Ecol Bullet Stockholm 34:111 pp
Ryman N, Utter F (eds) (1987) Population genetics and fishery management. Washington Sea Grant Publications, University of Washington Press, Seattle, Washington. Reprinted 2009 by The Blackburn Press, Caldwell, New Jersey
Sandström A, Lundmark C, Jansson E, Edman M, Laikre L (2016) Assessment of management practices regarding genetic biodiversity in Baltic Sea marine protected areas. Biodivers Conserv 25:1187–1205
Sandström A, Lundmark C, Andersson K, Johannesson K, Laikre L (2019) Understanding and bridging the conservation-genetics gap in marine conservation. Conserv Biol 33:725–728
Santamaría L, Méndez PF (2012) Evolution in biodiversity policy–current gaps and future needs. Evol Appl 5(2):202–218
Sarre SD, Georges A (2009) Genetics in conservation and wildlife management: a revolution since Caughley. Wildl Res 36(1):70–80
Schmeller DS, Bridgewater P (2016) The Intergovernmental Platform on Biodiversity and Ecosystem Services (IPBES): progress and next steps. Biodivers Conserv 25:801–805
Schonewald-Cox CM, Chambers SM, MacBryde B, Thomas WL (eds) (1983) Genetics and conservation: a reference for managing wild animals and plant populations. Benjamin-Cummings Publishing, Menlo Park, CA
Schwartz MK, Luikart G, Waples RS (2007) Genetic monitoring as a promising tool for conservation and management. Trends Ecol Evol 22(1):25–33
Shafer ABA, Wolf JBW, Alves PC, Bergström L, Bruford MW, Brännström I, Colling G, Dalén L, De Meester L, Ekblom R, Fawcett KD, Fior S, Hajibabaei M, Hill JA, Hoelzel AR, Höglund J, Jensen EL, Krause J, et al. (2015) Genomics and the challenging translation into conservation practice. Trends Ecol Evol 30(2):78–87
Sjögren-Gulve P et al. (2009) Suggestion for a national action program to conserve genetic diversity of wild plants, animals and fungi. Report of a mission to the Swedish Government. Swedish Environmental Protection Agency Dnr 05-404-06 Nv. (Förslag till Nationellt handlingsprogram för bevarande av genetisk variation hos vilda växter, djur och svampar Redovisning av ett regeringsuppdrag. Naturvårdsverket Dnr 05-404-06 Nv; In Swedish)
Smith TB, Kinnison MT, Strauss SY, Fuller TL, Carroll SP (2014) Prescriptive evolution to conserve and manage biodiversity. Annu Rev Ecol Evol Syst 45:1–22
Soulé ME, Wilcox B (1980) Conservation biology: an evolutionary-ecological perspective. Sinauer Associates, Sunderland, MA
Stetz JB, Kendall KC, Vojta CD, Genetic Monitoring (GeM) Working Group (2011) Genetic monitoring for managers: a new online resource. J Fish Wildl Manag 2(2):216–219
Taylor HR, Dussex N, van Heezik Y (2017) Bridging the conservation genetics gap by identifying barriers to implementation for conservation practitioners. Global Ecol Conserv 10:231–242
Tittensor DP, Walpole M, Hill SLL et al (2014) A mid-term analysis of progress toward international biodiversity targets. Science 346(6206):241–244
Torres-Florez JP, Johnson WE, Nery MF, Eizirik E, Oliveira-Miranda MA, Galetti PM Jr (2018) The coming of age of conservation genetics in Latin America: what has been achieved and what needs to be done. Conserv Genet 19:1–15
Utter FM, Hodgins HO, Allendorf FW, Johnson AG, Mighell JL (1973) Biochemical variants in Pacific salmon and rainbow trout: their inheritance and application in population studies. In: Schröder JH (ed) Genetics and mutagenesis of fish. Springer, Berlin, pp 329–339
Van den Hove S, Chabason L (2009) The debate on an Intergovernmental Science Policy Platform on Biodiversity and Ecosystem Services (IPBES): exploring gaps and needs. IDDRI – Idées pour le débat:1–24

van der Valk T, Gonda CM, Silegowa H, Almanza S, Sifuentes-Romero I, Hart TB, Hart JA, Detwiler KM, Guschanski K (2020) The genome of the endangered dryas monkey provides new insights into the evolutionary history of the vervets. Mol Biol Evol 37(1):183–194
Vernesi C, Bruford MW (2009) Recent developments in molecular tools for conservation. In: Bertorelle G, Bruford MW, Hauffe HC, Rizzoli A, Vernesi C (eds) Population genetics for animal conservation. Cambridge University Press, Cambridge, pp 321–344
Vernesi C, Bruford MW, Bertorelle G, Pecchioli E, Rizzoli A, Hauffe HC (2008) Where's the conservation in conservation genetics? Conserv Biol 22(3):802–804
Walpole M, Almond REA, Besançon C et al (2009) Tracking progress toward the 2010 biodiversity target and beyond. Science 325(5947):1503–1504
Wennerström L, Laikre L, Ryman N, Utter FM, Ab Ghani NI, André C, DeFaveri J, Johansson D, Kautsky L, Merilä J, Mikhailova N et al (2013) Genetic biodiversity in the Baltic Sea: species-specific patterns challenge management. Biodivers Conserv 22:3045–3065
Wilson B, Beech E, Window J, Allen DJ, Rivers M (2019) European Red List of selected endemic shrubs. Brussels and Cambridge: IUCN. Available at: https://portals.iucn.org/library/node/48438
Wright S (1922) Coefficients of inbreeding and relationship. Am Nat 56(645):330–338
Wright S (1931) Evolution in Mendelian populations. Genetics 16(2):97–159
Wright S (1977) Evolution and the genetics of populations, vol. 3. University of Chicago Press, Chicago

Chapter 4
Publicly Generated Data: The Role of Citizen Science for Knowledge Production, Action, and Public Engagement

Tina B. Phillips, Alison Parker, Anne Bowser, and Muki Haklay

Contents

T. B. Phillips (✉)
Center for Engagement in Science and Nature, Cornell Lab of Ornithology, Cornell University, Ithaca, NY, USA
e-mail: tina.phillips@cornell.edu

A. Parker
The Science and Technology Innovation Program, The Wilson Center, Washington, DC, USA
e-mail: alison.parker@wilsoncenter.org

A. Bowser
NatureServe, Arlington, VA, USA
e-mail: Anne_Bowser@natureserve.org

M. Haklay
Extreme Citizen Science (ExCiteS) Research Group, Department of Geography, University College London (UCL), London, UK
e-mail: m.haklay@ucl.ac.uk

C. C. Ferreira, C. F. C. Klütsch (eds.), *Closing the Knowledge-Implementation Gap in Conservation Science*, Wildlife Research Monographs 4,
https://doi.org/10.1007/978-3-030-81085-6_4

4.1 Introduction: What Is Citizen Science?

As the list of environmental challenges grows, the field of conservation biology needs to expand its portfolio to include long-term, viable, and evidence-based solutions that maximize impact (Chandler et al. 2017; Sullivan et al. 2017). Research has identified a temporal gap between knowledge generation and the time it can be acted upon, which inhibits efficient use of knowledge (Knight et al. 2008; Cook et al. 2013). According to Cook et al. (2013), minimizing the knowledge gap requires that knowledge be relevant, trustworthy, timely, and actionable. In this chapter, we describe how citizen science can be utilized to generate new knowledge, as well as to narrow the knowledge–implementation gap to solve some of the world's most challenging environmental issues.

The term "citizen science," first coined by Alan Irwin, was used to describe public involvement in decision making around scientific issues through dialogue, civil discourse, and scientific research (Irwin 1995). A year later, the term gained traction in the USA to mean public participation in scientific research through engagement with the scientific process via activities such as data collection, asking questions, and analyzing and interpreting data (Bonney 1996). Initially, this latter form of citizen science had less emphasis on dialogue and decision making, and more emphasis on helping scientists answer important biological questions (Bonney et al. 2014; Cooper and Lewenstein 2016). Since then, the term has been expanded to include multiple forms of public engagement in varying contexts (Newman et al. 2011; Eitzel et al. 2017). For example, projects that occur completely online and for the main purpose of classifying data are often referred to as crowd-science, networked science, mass collaboration, cyberscience, or collaboratories. Other projects that are locally generated and often environmentally focused are referred to as participatory or community science or volunteer monitoring.

In 2014, the Oxford English Dictionary (OED) offered the first formal definition for citizen science: "scientific work undertaken by members of the general public, often in collaboration with or under the direction of professional scientists and scientific institutions." This definition acknowledges the prominence of citizen science in society and is sufficiently broad to encompass a wide range of activities with varying goals and in diverse settings.

Today, citizen science-generated data are being used to study everything from analyzing the blocked neural pathways associated with Alzheimer's Disease (www.eyesonalz.com) to mapping marine protected areas (Cigliano et al. 2015). Thanks to the scope, scale, and resolution of information shared by volunteers, citizen science data offer an unprecedented level of accessibility to new forms of information and knowledge creation by multiple stakeholders (Chandler et al. 2017). This paradigm shift toward co-ownership of knowledge creation and use is at the core of what makes citizen science uniquely different from traditional approaches. Hence, there is growing interest in the use of citizen science data in the knowledge to action pipeline (Aoki et al. 2008; Sullivan et al. 2017). But how, and under what conditions, can citizen science influence action? Is citizen science a paradigm shift that can help narrow the knowledge–action gap faster than research conducted through other means? This chapter will explore these and other relevant questions to understand the cycle of knowledge production via citizen science, and its uptake by conservation practitioners, researchers, and decision makers.

We begin with a brief summary of the diverse nature and context of citizen science and why it is a useful tool for knowledge production. Then we describe how citizen science produces valid knowledge that leads to action and highlight existing case studies. We end the chapter with considerations about where the field of citizen science is headed next and its potential to re-imagine the nature of science knowledge and science–society relationships toward solutions to some of our greatest environmental challenges.

4.2 Why Citizen Science?

Although citizen science is not a new phenomenon—some have argued that it has existed for millennia (Miller-Rushing et al. 2012)—in the last decade it has become a valid and reliable source of knowledge creation (Bonney et al. 2014; Haklay 2017). Through increased availability of the Internet and mobile technologies (Bonney et al. 2014; Baker 2016; Davies et al. 2016; Kullenberg and Kasperowski 2016), citizen science leverages the collective crowd to gather data at geographic and temporal scales unattainable through traditional methods, addressing diverse topics like astronomy, public health, and zoology (Dickinson and Bonney 2012; Bonney et al. 2014; Theobald et al. 2015; Baker 2016; McKinley et al. 2017). Beginning in the early 2000s, the exponential growth of peer-reviewed publications using citizen science data resulted in a more than ten-fold jump over the previous decade (Follett and Strezov 2015), highlighting the tremendous potential of volunteer collected data to answer important scientific questions. Globally, environmental issues such as climate change, biodiversity loss, natural resource management, and air, soil, and water pollution are being addressed through citizen science that directly impact conservation and natural resource management (McKinley et al. 2017).

Many citizen science projects also try to influence participant learning outcomes such as interest, efficacy, and motivation for science and the environment, skills of science inquiry, science literacy, and environmental stewardship (Phillips et al. 2012, 2018; Stepenuck and Green 2015; Bonney et al. 2016). Theobald et al. (2015) estimate that around the world approximately two million people participate in biodiversity projects annually. Such participation may have cumulative impact, especially in localized or residential areas where natural resource issues require community involvement (Cooper et al. 2007). Citizen science participants vary in their age and skills, ranging from children to adults, in projects in or out of school with very simple to very complex protocol requirements. Therefore, researchers and educators committed to citizen science often provide training and educational materials for supporting multiple volunteer communities. However, when designed for science and educational outcomes, the potential to engage individuals in the process of science and enhance science literacy is great (Bonney et al. 2016; Phillips 2017).

Citizen science-generated data can be collected and used at local, regional, national, and global scales (DeVictor et al. 2010). Temporally, projects may last a day, as is the case for BioBlitzes (Roger and Klistorner 2016), or go on for decades like the 100+ year Christmas Bird Counts (Dunn et al. 2005). The temporal robustness of citizen science datasets is particularly useful for detecting changes in abundance and distribution of species, geographic and environmental conditions, and climate change patterns. Increasingly, the use of citizen science also is expanding to less obvious issues such as natural disaster risk reduction, management of energy security, public health, and even affordable health care (Haklay 2017). Efforts are underway to link citizen science to the UN's Sustainable Development Goals (SDGs)—seventeen targets for ensuring global health and well-being by addressing issues like climate change, food security, and nutrition for every person on the planet.

There also is a growing body of professional associations around the world (Europe, North America, and Australia each has their own citizen science associations) that are building capacity for citizen science project design, management, implementation, and evaluation (Storksdieck et al. 2016). Researchers and practitioners in Asia, Africa, and South America have expressed interest in or are establishing associations, and an emerging organization called the Citizen Science Global Partnership (citizenscienceglobal.org) seeks to help support and network these initiatives. Such efforts have moved citizen science from a budding new research tool to a worldwide network of professionals and volunteers representing citizens, academic institutions, government agencies, non-government organizations, and businesses in the pursuit of advancing and applying scientific knowledge.

Thus, in all its varied forms, citizen science is a growing global movement that facilitates research at broad temporal and geographic scales. With a wide array of potential issues to study, flexible design choices, cost-effectiveness, and enhanced value through public engagement, citizen science is an appealing and practical approach for the twenty-first century knowledge generation (Cooper et al. 2007; Bonney et al. 2009a; Wiggins and Crowston 2011; Dickinson and Bonney 2012).

4.3 How Is Knowledge Generated in Citizen Science?

There is a growing desire to understand the history of knowledge production within the realm of citizen science. Haklay (2017) outlines three "Eras of Environmental Information" that describe how environmental information comes into being and by whom, who uses it and to what ends, and what the direct contribution to decision-making processes may be. The first era began with the modern environmental movement at the end of the 1960s, when environmental information was produced and used largely by experts and scientists. This era is analogous to the *public education or deficit model*, which assumes that scientific knowledge is universal, objective, and deficient on behalf of the public (Pouliot 2009; Davies et al. 2016). This *deficit model* considers knowledge production and communication a one-way movement from experts to the lay public (Davies et al. 2016).

The second era (early 1990s–2010) is marked by the opening up of environmental information to the public while maintaining the paradigm of information production by experts and scientists (Haklay 2017). This era, sometimes referred to as the *public debate model* takes a slightly more subjective approach to knowledge, suggesting that the public is a legitimate source of knowledge that should be considered in larger, expert-driven knowledge production systems (e.g., through consultation).

The third era, which we are now experiencing within citizen science, is characterized by opening up the information production process. Both the production and consumption of environmental information are undertaken by the public, experts, and scientists (Davies et al. 2016; Haklay 2017). The *co-production of knowledge model* suggests that professional scientists and the "lay" public are equally valued as knowledge sources, and both empowered to contribute to knowledge production.

As quoted in Nguyen et al. (2017, p. 791), "the effectiveness of knowledge on conservation practices and natural resource management depends on how knowledge moves, how it is exchanged, how it is used, and how it interacts with the social world (p. 791)." When knowledge is co-produced, particularly around issues of relevance, affected individuals are often more motivated to take action on their own behalf (Kincheloe 2005; Nold and Francis 2017). Whereas traditional scientific approaches moved knowledge via deficit or public education models, citizen science operates in multi-directional co-production model for acquiring, synthesizing, and using knowledge that acknowledges contributions from a range of professional and public stakeholders (Haywood and Besley 2014). For example, while citizen scientists may gather and submit information to scientists for their use, these data are often made publicly available, thereby providing broad access to individuals and communities to use those data to ask and answer their own questions. Another example of expanded stakeholder engagement that seeks to break down traditional, dominant power structures is the integration of Traditional Ecological Knowledge (TEK) with newer, western ways of knowing (see Chap. 5; Ens et al. 2021). Thus, linking knowledge production to action requires understanding of the social context of how it is created and used. And whereas in traditional science there is a disconnect between the natural and social sciences (Norgaard 2008), citizen science has evolved with a more parallel acknowledgement of socio-ecological systems and their respective epistemologies (Jordan et al. 2015).

In describing the knowledge to action process within citizen science, it is also important to understand the various typologies used to describe the nature and character of different forms of citizen science, particularly where governance is concerned. Governance typologies are important to knowledge production because they address the roles that the various stakeholders are likely to play (Table 4.1).

Despite the apparent differences between the governance structures highlighted in Table 4.1, one is not necessarily better than the other, but each offers constraints and affordances. For example, top-down contributory models tend to be better resourced and able to accept, maintain, curate, and disseminate large volumes of data. When

Table 4.1 Governance structures within citizen science

Governance structure	Categorized by	Example	Source
Increasing involvement and ownership in research by volunteers	How participants are involved with the scientific process	Contributory, Collaborative, and Co-created	Wilderman (2005), Bonney et al. (2009b), Shirk et al. (2012)
	Participation levels within Volunteered Geographic Information (VGI)	Crowdsourcing, Distributed Intelligence, Participatory Science, and Extreme Citizen Science	Haklay (2013)
Governance aligned to project goals and activities	Project goals and organizational structure	Action, Conservation, Investigation, Virtual and Education	Wiggins and Crowston (2011)
	Types of activities	Data collection, data classification, curriculum, and community science	Bonney et al. (2016)

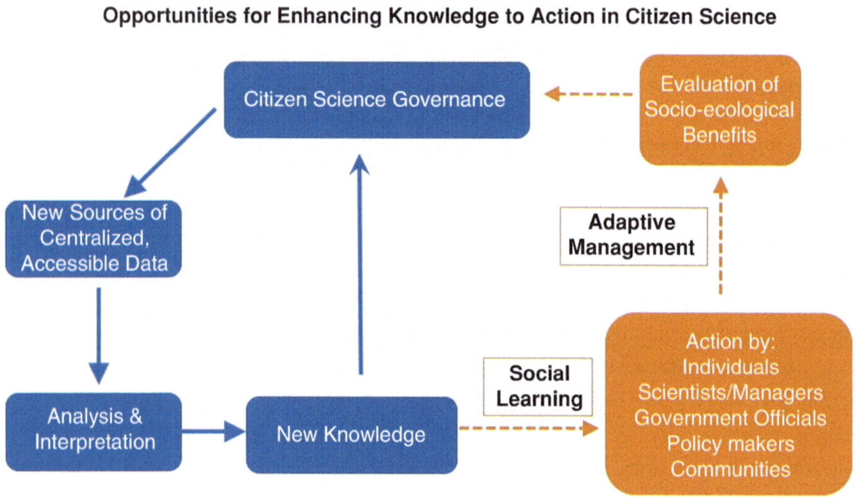

Fig. 4.1 Depiction of traditional citizen science models (blue) and enhanced opportunities for knowledge to action via social learning and adaptive management (orange)

these data are publicly available, they can have multiple uses by diverse stakeholders (Sullivan et al. 2017). In bottom-up or co-created structures, opportunities for communities or individuals developing or co-developing the question of interest are greater, thereby enhancing relevance to communities, which may influence impact at a local scale (Danielsen et al. 2005). To illustrate how knowledge is generated and used in citizen science, we refer to Fig. 4.1, which presents the "typical" citizen science approach in blue, and "enhanced opportunities" for knowledge to action in orange.

Regardless of governance models or project types, a unifying characteristic of citizen science is the technological infrastructure to house, curate, and disseminate data (Newman et al. 2012). Increasingly, databases are web-based, enabling participants to enter their observations directly into a relational database. Such databases require significant investment to develop and maintain, and many smaller projects turn to organizations, such as citsci.org, to facilitate project start-up, protocol development, management of volunteers, and building and maintaining databases. Data management and continuity are basic pillars of credible citizen science and best practices for the handling, curation, and delivery of data are available (Wiggins et al. 2013; Haklay 2015).

Another hallmark of citizen science knowledge production is the open access data repositories where data are made publicly available. However, web-based data-out tools vary greatly in their sophistication; some projects with large volumes of data, such as eBird, provide numerous data-out tools to analyze and interpret data in real time via tables, graphs, and maps. eBird also allows users to request full access to the raw data tables. Other datasets are quite limited, providing "canned" versions or updates of data on a daily, weekly, or monthly basis.

There also is a general expectation among all stakeholders that data are being analyzed and interpreted and project findings communicated. This is a critical function of the knowledge to action pipeline: that data are used—at a minimum by the scientists housing the data—and ideally, by the larger community of volunteers (Sullivan et al. 2017). However, data users should be acutely aware of the unique characteristics and limitations of citizen science-generated data, and use it accordingly (Haklay 2015). For example, the production of information by a large group of participants who are contributing as part of their leisure can mean that data will be more frequent during weekends, or that it is more likely to be concentrated in urban areas and popular spots. There are also advantages and limitations in the places that participants can access—for example, their backyard, which can be challenging to access by professionals (Cooper et al. 2012). At the same time, the limitation of participant recruitment will mean that the pattern of backyards to which access is granted will be irregular and not necessarily systematic. Each of these challenges requires appropriate statistical methodologies that address inherent biases in the data (e.g., Bird et al. 2014).

4.4 How Can Citizen Science Enhance Knowledge to Action?

The combination of free, accessible, and timely information results in multiple opportunities for the production of new knowledge by multiple actors in the system. But knowledge in and of itself is not enough, it must be utilized by diverse stakeholders in order to benefit science and society. We posit that the intake and utilization of information likely occurs through social learning mechanisms or "the collaborative or mutual development and sharing of knowledge by multiple stakeholders (both people and organizations) through learning-by-doing" (Armitage et al. 2009). Citizen science has great potential to foster social learning and build social capital by increasing the capacity of community members to monitor, communicate, and share information about environmental conditions with one another. Such connections build and maintain social capital through "trust that facilitate coordination and cooperation for mutual benefit" (Putnam 1995). Communities with programs that support the kind of collaborative social learning that strengthens social capital are better able to stimulate collective action to help address environmental and social problems (Bowles and Gintis 2002; Teorell 2003; Overdevest et al. 2004). Social learning can stimulate actions such as decision making, communication, resource mobilization, and conflict resolution by many actors including individuals, scientists, government officials, communities, and policy makers.

For instance, individual actions may include: transfer of knowledge to non-participants, increased involvement in citizenship and civic engagement (voting, attending town halls, commenting on environmental impact assessments), conservation-oriented behavior, using data to back a scientific claim, becoming leaders or activists, and training or recruiting new participants (Phillips 2017). There is also an increasing desire for citizens to have open access to academic publications as a means of recognition of their efforts and to support their continued individual and collective learning (Haklay 2015).

Scientists' actions typically involve utilization of statistically viable and actionable data often resulting in peer-reviewed publications that advance scientific knowledge (Wiggins et al. 2018). As in traditional science, new knowledge often leads to new research questions, methodologies, and technologies. This can spur increased institutional support for continued engagement with the public. Occasionally, scientists may become activists in support of a particular issue, species, or community. However, long-standing institutional structures and cultural norms around objectivity and rigor make it difficult for scientists to tread from public engagement to advocacy, despite a growing trend (Stilgoe 2009).

Actions by government officials tend to revolve around using citizen science data to implement, evaluate, or enforce legislation or regulations that directly influence program planning, natural resource management, habitat restoration, and environmental protections (McKinley et al. 2017). Stepenuck (2013) estimates that volunteer water quality data have been used to develop or enact policy at every level of government, including in North American, municipal, state, federal, and tribal jurisdictions. Policymakers and other decision makers act on citizen science-generated knowledge to recommend, amend, or pass legislation or policy, set priorities, and allocate resources. Key examples citing the use of citizen science data include the USA's Endangered Species Act, the Migratory Bird Treaty, and the National Environmental Policy Act (McKinley et al. 2017) and "illustrate the emergence of citizen science into the decision-making sphere" (Chari et al. 2017).

Finally, by leveraging the social capital generated from citizen science data and knowledge, communities act to coordinate, empower, and build capacity for representation in the decision-making process. This includes voicing their concerns and presenting data as evidence to scientists, government agencies, industry, and legislators. Without such knowledge, communities are often left out of the decision-making process and continually marginalized; with such knowledge communities can work in concert with decision makers to find solutions and resolve conflicts that build trust and credibility between stakeholders (Chari et al. 2017). Here, citizen science has the greatest potential to mitigate the science–society relationship and display a democratization of science indicative of a paradigm shift.

In closing the knowledge to action gap, affected stakeholders should also undertake evaluation of the socio-ecological benefits of the entire system but this step is often overlooked or minimized (Knight et al. 2008; Margoluis et al. 2013). A lack of reflection on measuring successes and failures serves to reinforce the cycle of decisions based on assumptions and past experiences, rather than evidence (Margoluis et al. 2013). Conservation biology has been arguing for evaluation of approaches through adaptive co-management that considers the domains related to institutions and processes, socioeconomics and livelihoods, and the ecological system (Armitage et al. 2009; Plummer and Armitage 2007). However, there are few examples of this occurring in citizen science (see Fernandez-Gimenez et al. 2008 and Jordan et al. 2016 for exceptions). Interestingly, in a meta analysis describing citizen science in natural resource management settings, Aceves-Bueno et al. (2015) conclude (with caveats) that citizen science can help correct two major shortcomings of adaptive management; namely insufficient monitoring and low stakeholder involvement. Community level citizen science programs are especially well aligned to employ adaptive management approaches that consider evidence of effectiveness to guide collaborative decision making in a continual process of feedback and iteration (Armitage et al. 2009; Jordan et al. 2016; Conrad and Hilchey 2011). Thus, we argue that an evaluative approach to understanding the risks and benefits of integrating adaptive management into citizen science would greatly enhance our understanding of how to maximize the knowledge to action pipeline for environmental conservation.

4.5 Citizen Science in Action

There is a growing literature documenting action by scientists/researchers, by communities, and by individuals through citizen science projects and data. McKinley et al. (2017) provide eight case studies illustrating the influence of citizen science on conservation science, natural resource management, and environmental protection, including two distinct pathways to inform policy adoption and implementation: one through building scientific knowledge, and the other through public engagement and public input. Newman et al. (2017) provide various examples of how a focus on place can improve conservation decision making, increase participation, and improve community resilience. Haklay (2015) additionally focuses on place by exploring how citizen science contributes to policy and shapes decision making at local, regional, and national scales. McElfish et al. (2016) document numerous examples of citizen science informing government decision making, from iNaturalist supplanting information generation needs, to the Louisiana Bucket Brigade informing enforcement action. Through an analysis of 104 examples of environmental monitoring, Danielsen et al. (2010) found that scientist-led monitoring can inform decisions at the regional and national level over the long term, but that community monitoring involving local people is more effective at achieving local impact quickly.

To further illustrate the process of knowledge production to action described above and in Fig. 4.1, we present several examples below and in Table 4.2. Each example highlights how citizen science can motivate action by broadening the decision-making conversation to include multiple perspectives from scientists/researchers, individual volunteers, community members, and policy makers.

4.5.1 eBird

eBird is a global project run by the Cornell Lab of Ornithology that gathers data on sightings of bird species worldwide. Begun in 2001, eBird has amassed more than a billion bird observations and includes several data quality checks. For instance, eBird validates volunteer observations as they are being entered with algorithms that match sightings with known species ranges and abundances (Wood et al. 2011; Fig. 4.2). Data that fall outside the expected algorithm indices are sent to local expert reviewers for confirmation or are flagged as questionable. The resulting dataset is openly available for use in research, education, and conservation. Observations from eBird are an important and widely-used data source on bird occurrence and provide 20% of all data on bird occurrence in the Global Biodiversity Information Facility (GBIF). In a survey of eBird data users, Sullivan et al. (2017) documented 159 tangible conservation actions from data use across six categories: research and monitoring, conservation planning, habitat and species management, site/habitat protection, and law, policy, and regulation. Data users were mostly from North America and included private individuals, students, government employees, NGO employees, academic researchers, and teachers.

4.5.1.1 Stakeholder: Scientists

Scientists from the Cornell Lab of Ornithology and from academic, NGO, and governmental institutions have used eBird data extensively in more than 180 scientific publications and to support conservation, biological, social, and interdisciplinary science. (https://ebird.org/science/publications). For example, Hurlbert and Liang (2012) used eBird data to identify the characteristics of bird species that shifted their range most drastically with climate change.

Table 4.2 Examples of citizen science projects and data that informed or motivated government decisions at the local, state, or national level

Project	Geographic scope	Action	Action taken by whom	Citation or more information
Wildbook for Whale Sharks	Worldwide	Published studies based on citizen science data from whaleshark.org were cited in the IUCN revision of the whale shark from vulnerable to endangered.	International Union for Conservation of Nature (IUCN)	Whaleshark.org Pierce and Norman (2016)
Bumble Bee Watch www.BumbleBeeWatch.org	North America	Citizen science data used in Endangered Species Act listing of the Rusty patched bumble bee (*Bombus affinis*)	US Fish and Wildlife	https://www.fws.gov/midwest/endangered/insects/rpbb/
Queensland Wader Study Group	Queensland, Australia	Citizen science data contributed to a long-term dataset used to list several shorebirds as critically endangered	Academic researcher, students, and volunteers	Hansen (2018)
Riverfly Partnership/Anglers' Monitoring Initiative	United Kingdom	Data from this initiative provided an early warning system for river pollution incidents. The data are fed directly into reporting by the UK Environment Agency	The UK Environment Agency and the UK Government	Ballard et al. (2017)
Bayesian Belief Network	North Queensland, Australia	Managers use citizen science datasets to perform dynamic sensitivity and scenario analyses to make decisions about fire management in conservation reserves in north Queensland, Australia	Managers	Smith et al. (2008)
CyberTracker	Worldwide	Data on the presence and subsequent absence of the lowland gorilla in Gabon and the Republic of Congo collected by non-literate indigenous people alerted authorities of Ebola outbreaks	Health authorities	Liebenberg (2016)

4.5.1.2 Stakeholder: Government/Decision Makers

Government employees such as the US Fish and Wildlife Service used eBird as one of several primary data sources to understand seasonal patterns of occurrence for the

Fig. 4.2 Ebird participants observing birds in the field and gathering data on their abundance and distribution. (Photo credit: Christopher Wood)

rufa Red Knot (*Calidris canutus rufa*) because traditional data sources (consistent long-term surveys) were only available for a few locations and most breeding and nonbreeding areas were not adequately covered. The Fish and Wildlife Service acknowledged the value of this dataset and the rigorous nature of the quality control protocols in the final listing rule (Sullivan et al. 2017, Federal Register Vol. 79, No. 238, p. 73731).

4.5.1.3 Stakeholder: Individual Data Users

Sullivan et al. (2017) found that unaffiliated individuals accounted for 30% of the eBird data use audience, downloading data more than any other type of data users. These users utilized the data at the local or regional level, often to support the protection of local parks or natural areas or argue against development projects (Sullivan et al. 2017).

4.5.2 Flint Water Crisis

In 2015, a year after Flint's water supply was switched from Lake Huron to the Flint River, LeAnne Walters, a mother in Flint, Michigan, noticed changes in the quality of the tap water and suspected a link between the water and health issues in her family. After her concerns were not acknowledged by government officials, Walters reached out to Marc Edwards, a researcher at Virginia Polytechnic Institute who worked on water quality issues. They began a collaboration based on community data collection, developing and implementing protocols for the collection and testing of water samples. This project led to the broad exposure and acknowledgement of the Flint water crisis—a public health crisis caused by widespread contamination of the city's water supply after a series of problems with improper treatment and oversight (Pieper et al. 2018).

4.5.2.1 Stakeholder: Community Members

Since the partnership began, there is evidence that the partnership and collaborative research effort has led to additional opportunities for community-led action (Soleri et al. 2016). Scientists, students, residents, NGO's, and others continue to collaborate on the Flint water study (http://flintwaterstudy.org/) and have even developed a course (free to community residents) to address the complex problem through open dialogue (Selig and Sahli 2016).

4.5.2.2 Stakeholder: Policy Makers

Data originating with citizen science ultimately led to the recognition of the Flint water crisis, which in turn has led to some action at local, state, and national levels (Ottinger 2009; Chari et al. 2017; Edwards and Walters 2017). Most notably, in 2017, Congress allocated nearly \$100 million to remove and replace lead pipes in thousands of affected homes (https://www.epa.gov/flint).

4.5.3 Locally-Based Monitoring in the Philippines

As part of a new monitoring scheme established by the Philippine government, 97 rangers and 350 community volunteers monitored one million hectares of protected areas over 2.5 years. This monitoring scheme included four methods: focus group discussions with community members and monitoring groups consisting of forest product gatherers, hunters, and fishers; protocols for observations of wildlife and resource use (the field diary method); photography of fixed points; and line transect surveys. Monitoring focused on a list of specific taxa and resource use identified by staff and community members. Prior to this monitoring scheme, collaboration between community members and park staff was minimal, and local people were very rarely involved in park management. Danielsen et al. (2005) identified decisions and actions resulting from the monitoring scheme between 1998 and 2001, focusing primarily on conservation management interventions. Interventions included raising awareness about resource management, bylaws at the tribal, village, or municipal level, cooperation between authorities, enforcement of existing regulations, extraction method bans, livelihood assistance, Protected Area Management Board resolutions, permit system establishments, sanctuary establishments, seasonal closures, and size-limits for harvesting.

4.5.3.1 Stakeholder: Managers and Staff

Prior to this new scheme, monitoring by park staff and managers was limited to extracted timber. Following its implementation, protected area staff conducted 83% of conservation management interventions (of a total of 156 interventions) ranging from raising awareness to enforcement of laws (Danielsen et al. 2005). The vast majority of these were accomplished without external support, indicating that park managers and staff were self-sufficient in implementing interventions.

4.5.3.2 Stakeholder: Community Members

Focal group discussions were particularly effective in initiating action and decisions by community members. As a result of involvement in the monitoring scheme, community members implemented 70% of the 89 management interventions. Focus group discussions also initiated joint government/community member interventions including enforcement activities, which were often combined with activities involving local people, potentially leading to more socially-acceptable enforcement.

4.5.3.3 Stakeholder: Policy Makers

More than half of the interventions were conducted by policy makers in local government and community institutions; these included local bylaws governing resource use established by indigenous people, villages, and municipalities. Additionally, observation protocols and focus group discussions resulted in protected area councils issuing resolutions or making other policies. Although some policies were issued to protect local species or habitats, many focused on the supply of resources for local people and addressed fishing, hunting, and gathering of forest products (Danielsen et al. 2005).

4.5.4 Biodiversity in Southern Africa

Citizen science species atlases and monitoring projects in southern Africa such as the Protea Atlas Project, the Southern African Bird Atlas Project, the Custodians of Rare and Endangered Wildflowers (CREW) project, and MyBirdPatch have developed protocols for the generation of biodiversity data. Over the last 15 years, these data have been fed into national platforms that have provided additional quality control and allowed for analysis, modeling, and synthesis of datasets. These datasets now fill some critical gaps in biodiversity and climate change monitoring over large scales. Trends data, graphs, and summary statistics are generated, allowing for the development of policy support tools such as Red List Data conservation status assessment. The South African National Biodiversity Institute (SANBI) supports this work and facilitates the integration of professional and citizen science datasets. Together, these integrated datasets provide early warning systems for biodiversity in southern Africa by translating data into narratives and clear messages for decision making (Barnard et al. 2017).

4.5.4.1 Stakeholder: Scientists

Statistical ecology is essential for early warning of conservation issues, allowing for the "distinction of signal from noise" in a timely and defensible way. However, scientists are relatively rare in this new field in Africa. The SANBI system focuses on concrete interactions with centers of statistical ecology in order to provide conservation guidance to relevant stakeholders, especially scientists (Barnard et al. 2017).

4.5.4.2 Stakeholder: Policy Makers

Early warning systems communicate signals of dangerous fragmentation or species population declines to policy makers. For example, South Africa's Custodians of Rare and Endangered Wildflowers (CREW) provide data on populations and threat information for threatened plants that directly support the National Red List database, providing data on 1147 species in over 5000 sites since 2003. Other outputs such as detailed species checklists for environmental impact assessments are reported to national and international policy makers. In South Africa, data from a set of citizen science projects like CREW make up a significant component of iterative biodiversity assessments that are conducted every seven years to support national planning and management (Barnard et al. 2017).

4.6 Next Steps

We see a number of next steps for advancing citizen science knowledge generation and closing the knowledge–action gap, including within conservation science and related fields. Bonney et al. (2014) and Ellwood et al. (2017) recommend three actions for citizen science to be effective to conservation science through: (1) coordinating and communicating insights, (2) creating interdisciplinary teams utilizing citizen science as one of many tools to tackle wicked problems, and (3) improving coordination among investments in citizen science. Here, we offer two additional recommendations: (4) continued study of public engagement in scientific research, particularly through meta studies and community-based monitoring, and (5) expansion of citizen science beyond project-based approaches. Each of these recommendations is discussed below.

Through networked professional associations coordinating citizen science on multiple scales, communication of findings, best practices, and insights are happening in myriad ways. For example, the open access, peer-reviewed journal *Citizen Science: Theory and Practice*, publishes articles from scientists, practitioners, educators, evaluators, and technologists on topics that help to advance citizen science and bolster a global community of practice. Another example of communicating insights is the burgeoning Citizen Science Global Partnership (CSGP), a consortium of existing networks launched during a 2017 United Nations Science-Policy-Business Forum on the Environment that seeks to promote citizen science globally to support a sustainable world. In addition to helping engage 1 billion coordinated citizen science contributions around Earth Day 2020, the Citizen Science Global Partnership hopes to influence sustainability initiatives by establishing systems that support interoperable data for use in local-to-global research and decision making, and help understand the contributions of citizen science to the sustainable development goals. These resources will help coordinate and raise awareness of ongoing initiatives and best practices and create information products to narrow the knowledge to action gap.

With regard to tackling wicked problems, in Europe in particular, there is growing coordination between teams of researchers and data sources to address invasive species and their impact on local flora and fauna (Tollington et al. 2017). Understanding and managing invasive species is a challenging problem that requires pulling together disparate data sources as well as inspiring policy-based actions and individual control strategies. Citizen science can help advance invasive species monitoring by supporting data collection on new spatial and temporal scales, including in areas such as private property not otherwise accessible. This process can also create an early warning network for the spread of new invasive species, all while raising public awareness of the problem and (in cases such as the Sparrow Swap Project in the USA) inviting volunteers to contribute directly to invasive species control (Larson et al. 2015).

Climate change is another wicked problem that citizen science is uniquely poised to address (Dickinson and Bonney 2012; Bonney et al. 2014). New technologies and statistical approaches allow for the assimilation of data from multiple sources such as traditional monitoring schemes and remote sensor data that can better predict climate change impacts. Such advances are also being made in long-term ecological research sites and protected areas, where traditional ecological knowledge and community-based monitoring are likely to occur, providing opportunities for increased connections between communities, scientists, and technological resources (Chandler et al. 2017).

Beyond targeted efforts around specific issues, new infrastructure is being developed to help solve a range of known and unknown problems. In Australia, The Atlas of Living Australia (ALA) was created as an open access infrastructure portal for housing and disseminating information about Australia's biodiversity from multiple data sources. The ALA is designed to be used by scientists, decision makers, and industry to address issues related to biodiversity conservation, sustainable ecosystem development, natural resource management, and environmental impact assessments on various scales. In addition to the infrastructure portal, ALA supports the BioCollect tool (https://www.ala.org.au/biocollect/), a customizable mobile application for data collection designed to fit a range of citizen science use cases. These types of infrastructures and flexible platforms minimize the knowledge to action gap by allowing citizen scientists to easily mobilize around relevant issues in real time.

While funding and investments in citizen science are limited, a few examples provide hope for future endeavors. Since 2007, the UK lottery has supported citizen science programs with over £10m, which created the Open Air Laboratories (OPAL—https://www.opalexplorenature.org/), focused mostly on terrestrial environmental observations, and Capturing Our Coast (Co-Coast https://www.capturingourcoast.co.uk/), focused on Marine species. While these investments are time limited, they have contributed to awareness of citizen science in the UK, with research funders starting to take notice.

There is evidence that citizen science is having an impact in the U.S. federal government as well. In September 2015, President Barack Obama's Science Advisor John Holdren noted in a memo to federal agency heads that "citizen science and crowdsourcing projects can enhance scientific research and address societal needs... in recognition of these potential benefits, this memorandum encourages the use, where appropriate, of citizen science and crowdsourcing by Federal agencies" (Holdren 2015). In 2016, the United States Congress authorized the Crowdsourcing and Citizen Science Act of 2016 (15 USC 3724) with bipartisan support, which allows heads of multiple Federal science agencies to use appropriated funds to support citizen science efforts. These formal acknowledgements support the work of more than 300 federal employees in integrating citizen science and crowdsourcing across 60 federal agencies.

With regard to our fourth recommendation, we encourage continued social science research on why and the ways in which people engage in citizen science worldwide, in different cultures, and in different contexts, as well as the socio-ecological outcomes resulting from citizen science. To date, much of what we know about citizen science engagement comes from studies of single projects, which are often difficult to translate to other contexts. A recent surge in "meta studies" has revealed some key insights about motivation and engagement in citizen science more broadly. For example, Aceves-Bueno et al. (2015) looked at 83 peer-reviewed articles on citizen science in natural resource management and found that sense of place, technology, and action-related motivations correlated with data use for management purposes. Phillips (2017) found that participants in co-created projects (where volunteers can engage in all aspects of the science process) were more likely to have extrinsic motivations such as fear or concern and less likely to engage in environmental stewardship practices than participants in contributory projects (where volunteers engage mainly in data collection), and who were more intrinsically motivated. Continued efforts such as these are critical to advance our understanding of best practices for effective recruitment and retention of volunteers, program design and implementation, and the factors that influence socio-ecological impacts.

Our last recommendation suggests looking beyond the traditional citizen science approaches and governance structures, toward emerging, complementary paradigms that challenge the project-based approach. These novel approaches provide additional venues for expanding citizen science as a form of co-created knowledge production. For example, researchers have begun to link citizen science with "smart cities" approaches to urban development (Craglia and Granell 2014). These approaches use networked technology to facilitate public participation in urban development and environmental activism and raise the quality of life. An illustration of this can be found in Europe's Living Labs, which suggests that innovation research should be moved from artificial *in vitro* to real world, or *in vivo,* settings (Dutilleul et al. 2010). Living labs were initially conceived as an approach to product development where citizens collaborated with technologists to exchange ideas and develop proofs-of-concept in real-world settings. Living labs now convene a range of public and "professional" stakeholders to collaborate on smart city innovation, including identification of new technologies to improve urban living, conducting background research, developing prototypes and proof-of-concepts, gathering data where appropriate, and evaluating the technologies produced. Living labs are increasingly considered innovation incubators that empower communities to work together to identify and solve a range of real-world challenges (Veeckman et al. 2013). As such, living labs have (at least conceptually) transcended the public debate model of knowledge production to support the full co-creation of knowledge (Callon 1999). The living labs model also narrows the knowledge to action gap by bringing the public together with decision makers including government representatives and private sector technology developers to collaboratively take action to develop new technologies.

Other novel approaches to citizen science seek to further emphasize the Irwin (1995) model of involving citizens in the initial phases of deliberation of science and technology issues. In this vein, ECAST, The Expert and Citizen Assessment of Science and Technology, is a "distributed network of institutions for peer-to-peer deliberation," which seeks to engage the public in dialogue about science and technology, early and often, not just in response to extant crises. Typically conducted in collaboration with public science museums in North America, ECAST has hosted public deliberations on issues related to biodiversity, technology assessment, genetically modified algae, and community resilience planning for extreme weather conditions. According to the Consortium for Science, Policy and Outcomes at the Arizona State University, "the method of public engagement in governing emerging technologies developed by ECAST addresses the tension between democratic and expert-led decision making. It offers the opportunity for mutual learning, integrating formal expertise, local knowledge, and public engagement. It provides transparency and offers a form of informed consent. Finally, engagement can build trust among experts, stakeholders, and public audiences." Initiatives such as ECAST allow citizens to be represented in the *prioritization* of research agendas at the federal/national scale, something that has largely been the realm of scientists and government agencies alone. To strengthen the knowledge to action pipeline, more initiatives like ECAST that allow the public full access to the decision-making process for setting the research agenda should be encouraged.

4.7 Conclusion

In this chapter we have argued for the important role that citizen science has and should continue to play in scientific knowledge production. However, there is still much to do and much to be learned before we concede that citizen science also represents a paradigm shift for how knowledge is produced. For one thing, it is likely that a majority of scientists have not heard of citizen science, have not envisioned their work utilizing citizen science, or just do not find it to be a credible source of knowledge production. Additionally, in co-created or participatory projects where the potential paradigm shift is the greatest, communities are often challenged with finding a scientist who finds relevance in their issues. To that end, a new international working group has been formed within the Society for Conservation Biology, the Participatory and Citizen Science (PaCS) working group, whose main goals include: (1) Raising awareness within SCB, other conservation scientists, practitioners, and the public of the validity of using citizen science to advance biodiversity conservation; (2) Promote the ways that citizen science contributes to conservation biology, natural resource management, climate change adaptation, and policy; and (3) Increase knowledge exchange for citizen science within the field of conservation biology.

To continue its global advancement, citizen science must also address other challenges. Debates about the quality of volunteer collected data continue despite numerous studies that have found volunteer collected data to be comparable to that collected by professionals (Danielsen et al. 2014; Lewandowski and Specht 2015). Data quality is enhanced by designing easy to use protocols that are appropriate for the research question, including data validation filters, employing statistical approaches that account for temporal, spatial, and species bias, and having documented plans for data management, continuity, and dissemination (Chandler et al. 2017). Additionally, international working groups are formulating recognized standards for acquiring and managing data and metadata (Bowser et al. 2017).

As citizen science expands to other cultures and contexts, it also raises legal and ethical dilemmas such as privacy, liability, physical and intellectual property, and institutional cultural change. The engagement of indigenous groups in citizen science introduces issues of data sovereignty, ownership, consent, and reciprocity. Continued debate on such matters will explore the legal, policy, and organizational challenges that need to be overcome so that citizen science can better operate in bureaucratic systems and inform public policy more seamlessly. Along with these newer challenges, the traditional challenges still exist, namely limited funding and resources, motivating unengaged publics, retaining existing volunteers, keeping pace with technological advances, and balancing sometimes competing agendas of scientists, educators, activists, and project developers. Addressing these and other issues will take time, effort, and coordination, but are critical for citizen science to maintain itself as a credible, trustworthy, and relevant source of knowledge, to scientists, individuals, communities, and decision makers.

Despite these challenges, we are confident that citizen science has a valid place in the knowledge to action pipeline. A decade ago, we did not envision the expansion of citizen science into every imaginable discipline, but driven by innovation and the prospect of discovery, we argue that like traditional science, the role of citizen science in knowledge production is boundless. Unlike traditional science however, the structure of citizen science seeks to minimize barriers toward the democratization of science, and perhaps in time, we will witness a full paradigm shift in knowledge production. In the meantime, as citizen science continues to evolve, we will be faced with other uncertainties, new wicked problems, and increasing shortages of time and resources. We contend, however, that the growing body of expertise both by scientists and the public will add much needed capacity for coping with these issues, and in turn, spur the knowledge to action pipeline at scales that are relevant and impactful for science and society.

References

Aceves-Bueno E, Adeleye AS, Bradley D et al (2015) Citizen science as an approach for overcoming insufficient monitoring and inadequate stakeholder buy-in in adaptive management: criteria and evidence. Ecosystems 18(3):493–506

Aoki PM, Honicky RJ, Mainwaring A, Myers C, Paulos E, Subramanian S, Woodruff A (2008) Common sense: mobile environmental sensing platforms to support community action and citizen science. Adjunct Proceedings Ubicomp

Armitage DR, Plummer R, Berkes F, Arthur RI, Charles AT, Davidson-Hunt IJ et al (2009) Adaptive co-management for social–ecological complexity. Front Ecol Environ 7(2):95–102

Baker B (2016) Frontiers of citizen science: explosive growth in low-cost technologies engage the public in research. Bioscience 66(11):921–927

Ballard HL, Robinson LD, Young AN, Pauly GB, Higgins LM, Johnson RF, Tweddle JC (2017) Contributions to conservation outcomes by natural history museum-led citizen science: examining evidence and next steps. Biol Conserv 208:87–97

Barnard P, Altwegg R, Ebrahim I, Underhill LG (2017) Early warning systems for biodiversity in southern Africa – how much can citizen science mitigate imperfect data? Biol Conserv 208:183–188

Bird TJ, Bates AE, Lefcheck JS, Hill NA, Thomson RJ, Edgar GJ et al (2014) Statistical solutions for error and bias in global citizen science datasets. Biol Conserv 173:144–154

Bonney R (1996) Citizen science: a lab tradition. Living Bird 15(4):7–15

Bonney R, Cooper CB, Dickinson J, Kelling S, Phillips T, Rosenberg KV, Shirk J (2009a) Citizen science: a developing tool for expanding science knowledge and scientific literacy. Bioscience 59(11):977–984

Bonney R, Ballard H, Jordan R, McCallie E, Phillips T, Shirk J, Wilderman CC (2009b) Public participation in scientific research: defining the field and assessing its potential for informal science education. A CAISE Inquiry Group Report. Washington, DC: Center for Advancement of Informal Science Education (CAISE). Retrieved from http://eric.ed.gov/?id=ED519688
Bonney R, Shirk JL, Phillips TB, Wiggins A, Ballard HL, Miller-Rushing AJ, Parrish JK (2014) Next steps for citizen science. Science 343(6178):1436–1437
Bonney R, Phillips TB, Ballard HL, Enck JW (2016) Can citizen science enhance public understanding of science? Public Underst Sci 25(1):2–16
Bowles S, Gintis H (2002) Social capital and community governance. Econ J 112(483):F419–F436
Bowser A, Brenton P, Stevenson R, Newman G, Schade S, Bastin L, Parker A, Oliver J (2017) Citizen science association data & metadata working group: report from CSA and future outlook. Workshop report, Woodrow Wilson International Center for Scholars, Washington, DC
Callon M (1999) The role of lay people in the production and dissemination of scientific knowledge. Sci Technol Soc 4(1):81–94
Chandler M, See L, Copas K, Bonde AMZ, López BC, Danielsen F et al (2017) Contribution of citizen science towards international biodiversity monitoring. Biol Conserv 213:280–294
Chari R, Matthews LJ, Blumethal MS, Edelman AF, Jones T (2017) The Promise of Community Citizen Science. PE-256-RC. RAND Corporation, Santa, Monica, CA, p 2. www.rand.org/pubs/perspectives/PE256.html
Cigliano JA, Meyer R, Ballard HL, Freitag A, Phillips TB, Wasser A (2015) Making marine and coastal citizen science matter. Ocean Coastal Manage 115:77–87
Conrad CC, Hilchey KG (2011) A review of citizen science and community-based environmental monitoring: Issues and opportunities. Environ Monit Assess 176(1–4):273–291
Cook CN, Mascia MB, Schwartz MW, Possingham HP, Fuller RA (2013) Achieving conservation science that bridges the knowledge–action boundary. Conserv Biol 27(4):669–678
Cooper CB, Lewenstein BV (2016) Two meanings of citizen science. In: Cavalier D (ed) The rightful place of science: citizen science. Arizona State University Press, Tempe, AZ, pp 51–62
Cooper CB, Dickinson J, Phillips T, Bonney R (2007) Citizen science as a tool for conservation in residential ecosystems. Ecol Soc 12(2):11
Cooper CB, Hochachka WM, Dhondt AA (2012) Opportunities and challenges of citizen science as a tool for ecological research. In: Dickinson JL, Bonney R (eds) Citizen science: public participation in environmental research. Comstock Publishing Associates, Ithaca, NY, pp 99–113
Craglia M, Granell C (2014) JRC technical reports: citizen science and smart cities. Ispra, Italy: European Commission Joint Research Centre (JRC)
Danielsen F, Burgess ND, Balmford A (2005) Monitoring matters: examining the potential of locally-based approaches. Biodivers Conserv 14:2507–2542
Danielsen F, Burgess ND, Jensen PM, Pirhofer-Walzl K (2010) Environmental monitoring: the scale and speed of implementation varies according to the degree of peoples involvement. J Appl Ecol 47(6):1166–1168
Danielsen F, Jensen PM, Burgess ND, Altamirano R, Alviola PA, Andrianandrasana H, Brashares JS, Burton AC, Coronado I, Corpuz N, Enghoff M, Fjeldsa J, Funder M, Holt S, Hubertz H, Jensen AE, Lewis R, Massao J, Mendoza MM, Ngaga Y, Pipper CB, Poulsen MK, Rueda RM, Sam MK, Skielboe T, Sorensen M, Young R (2014) A multicountry assessment of tropical resource monitoring by local communities. Bioscience 64:236–251
Davies L, Fradera R, Riesch H, Lakeman-Fraser P (2016) Surveying the citizen science landscape: an exploration of the design, delivery and impact of citizen science through the lens of the Open Air Laboratories (OPAL) programme. BMC Ecol 16(1):17
Devictor V, Whittaker RJ, Beltrame C (2010) Beyond scarcity: citizen science programmes as useful tools for conservation biogeography. Divers Distrib 16(3):354–362
Dickinson JL, Bonney R (2012) Citizen science: public participation in environmental research. Cornell University Press, Ithaca, NY

Dunn EH, Francis CM, Blancher PJ, Drennan SR, Howe MA, Lepage D et al (2005) Enhancing the scientific value of the Christmas Bird Count. Auk 122(1):338–346
Dutilleul B, Birrer FA, Mensink W (2010) Unpacking European living labs: analysing innovation's social dimensions. Central Eur J Publ Policy 4(1):60–85
Edwards M, Walters L (May 18, 2017) Keynote Presentation, Citizen Science Association Conference, St. Paul, Minn. As of April 6, 2018: https://csa2017.sched.com/event/9qpw/keynote-presentation-dr-marc-edwards-department-of-civil-environmental-engineering-virginia-tech-leeanne-walters-coalition-for-clean-water
Eitzel MV, Cappadonna JL, Santos-Lang C, Duerr RE, Virapongse A, West SE, Kyba CCM, Bowser A, Cooper CB, Sforzi A, Metcalfe AN, Harris ES, Thiel M, Haklay M, Ponciano L, Roche J, Ceccaroni L, Shilling FM, Dörler D, Heigl F, Kiessling T, Davis BY, Jiang Q (2017) Citizen science terminology matters: exploring key terms. Citizen Sci Theory Pract, 2(1):1–20
Ellwood ER, Crimmins TM, Miller-Rushing AJ (2017) Citizen science and conservation: recommendations for a rapidly moving field. Biol Conserv 208:1–4
Ens E, Reyes-García V, Asselin H, Hsu M, Reimerson E, Reihana K, Sithole B, Shen X, Cavanagh V, Adams M (2021) Recognition of indigenous ecological knowledge systems in conservation and their role to narrow the knowledge-implementation gap. In: Ferreira CC, Klütsch CFC (eds) Closing the knowledge-implementation gap in conservation science – evidence transfer across spatiotemporal scales and different stakeholders. Springer, New York, pp 109–139
Fernandez-Gimenez ME, Ballard HL, Sturtevant VE (2008) Adaptive management and social learning in collaborative and community-based monitoring: a study of five community-based forestry organizations in the Western USA. Ecol Soc 13(2):4
Follett R, Strezov V (2015) An analysis of citizen science based research: usage and publication patterns. PLoS One 10(11):e0143687
Haklay M (2013) Citizen science and volunteered geographic information—overview and typology of participation. In: Sui D, Elwood S, Goodchild M (eds) Crowdsourcing geographical knowledge: Volunteered Geographic Information (VGI) in theory and practice. Springer Publishing, New York, NY
Haklay M (2015) Citizen science and policy: a European perspective. Washington, DC: Woodrow Wilson International Center for Scholars. https://www.wilsoncenter.org/sites/default/files/Citizen_Science_Policy_European_Perspective_Haklay.pdf
Haklay M (2017) The three eras of environmental information: the roles of experts and the public. In: Loreto V, Haklay M, Hotho A, Servedio VCP, Stumme G, Theunis J, Tria F (eds) Participatory sensing, opinions and collective awareness. Springer, Cham, pp. 163–179
Hansen BD, Clemens RS, Gallo-Cajiao E, Jackson MV, Maguire GS, Maurer G, Milton D, Rogers DI, Weller DR, Weston MA, Woehler EJ, Fuller RA (2018) Shorebird monitoring in Australia: a successful long-term collaboration between citizen scientists, governments and researchers. In: Legge S, Lindenmayer D, Robinson N, Scheele B, Southwell D, Wintle B (eds) Monitoring threatened species and ecological communities. CSIRO, Canberra, pp 100–120
Haywood BK, Besley JC (2014) Education, outreach, and inclusive engagement: towards integrated indicators of successful program outcomes in participatory science. Public Underst Sci 23(1):92–106
Holdren J (2015) Addressing societal and scientific challenges through citizen science and crowdsourcing. September 30 memorandum to the heads of executive departments and agencies. Washington, DC: Office of Science and Technology Policy, Executive Office of the President
Hurlbert AH, Liang Z (2012) Spatiotemporal variation in avian migration phenology: citizen science reveals effects of climate change. PLoS One 7(2):e31662
Irwin A (1995) Citizen science: a study of people, expertise and sustainable development. Routledge, London
Jordan R, Crall A, Gray S, Phillips T, Mellor D (2015) Citizen science as a distinct field of inquiry. Bioscience 65(2):208–211

Jordan R, Gray S, Sorensen A, Newman G, Mellor D, Newman G, Hmelo-Silver C, LaDeau S, Biehler D, Crall A (2016) Studying citizen science through adaptive management and learning feedbacks as mechanisms for improving conservation. Conserv Biol 30(3):487–495

Kincheloe JL (2005) Critical constructivism primer. P. Lang, New York, NY

Knight AT, Cowling RM, Rouget M, Balmford A, Lombard AT, Campbell BM (2008) Knowing but not doing: selecting priority conservation areas and the research implementation gap. Conserv Biol 22:610–617

Kullenberg C, Kasperowski D (2016) What is citizen science? – a scientometric meta-analysis. PLoS One 11(1):e0147152

Larson LR, Cooper CB, Hauber ME (2015) Emotions as drivers of wildlife stewardship behavior: examining citizen science nest monitors' response to invasive house sparrows. Human Dimens Wildl 21(1):18–33

Lewandowski E, Specht H (2015) Influence of volunteer and project characteristics on data quality of biological surveys: Data Quality of Volunteer Surveys. Conserv Biol 29(3):713–723

Liebenberg L (2016) The CyberTracker Story

Margoluis R, Stem C, Swaminathan V, Brown M, Johnson A, Placci G, Salafsky N, Tilders I (2013) Results chains: a tool for conservation action design, management, and evaluation. Ecol Soc 18(3):22

McElfish J, Pendergrass J, Fox T (2016) Clearing the path: citizen science and public decision making in the United States. Washington, DC: Woodrow Wilson International Center for Scholars. www.wilsoncenter.org/sites/default/files/clearing_the_path_eli_report.pdf

McKinley DC, Miller-Rushing AJ, Ballard HL, Bonney R, Brown H, Cook-Patton SC et al (2017) Citizen science can improve conservation science, natural resource management, and environmental protection. Biol Conserv 208:15–28

Miller-Rushing A, Primack R, Bonney R (2012) The history of public participation in ecological research. Front Ecol Environ 10(6):285–290

Newman G, Graham J, Crall A, Laituri M (2011) The art and science of multiscale citizen science support. Eco Inform 6(3–4): 217–227

Newman G, Wiggins A, Crall A, Graham E, Newman S, Crowston K (2012) The future of citizen science: emerging technologies and shifting paradigms. Front Ecol Environ 10(6):298–304

Newman G, Chandler M, Clyde M, McGreavy B, Haklay M, Ballard H et al (2017) Leveraging the power of place in citizen science for effective conservation decision making. Biol Conserv 208: 55–64

Nguyen VM, Young N, Cooke SJ (2017) A roadmap for knowledge exchange and mobilization research in conservation and natural resource management. Conserv Biol 31(4):789–798

Nold C, Francis L (2017) Participatory sensing: recruiting bipedal platforms or building issue-centred projects? In: Loreto V, Haklay M, Hotho A, Servedio VCP, Stumme G, Theunis J, Tria F (eds) Participatory sensing, opinions and collective awareness. Springer, Cham, pp. 213–235

Norgaard RB (2008) Finding hope in the millennium ecosystem assessment. Conserv Biol 22(4):862–869

Ottinger G (2009) Buckets of resistance: standards and the effectiveness of citizen science. Sci Technol Human Values 35(2):244–270

Overdevest C, Orr CH, Stepenuck K (2004) Volunteer stream monitoring and local participation in natural resource issues. Hum Ecol Rev 11(2):177–185

Phillips TB (2017) Engagement and learning in environmentally-based citizen science: a mixed methods comparative case study. Doctoral Dissertation, Cornell University. Ithaca NY

Phillips T, Bonney R, Shirk J (2012) What is our impact? Toward a unified framework for evaluating outcomes of citizen science participation. In: Dickinson JL, Bonney R (eds) Citizen science: public participation in environmental research. Cornell University Press, Ithaca, NY, pp 82–95

Phillips TB, Bonney R, Porticella N, Constas M (2018) Articulating and measuring individual learning outcomes from citizen science: a conceptual model. Citizen Sci Theory Pract 3(2):1–19.

Pieper K, Martin RL, Tang M, Walters L, Parks J, Roy S, Devine CL, Edwards MA (2018) Evaluating water lead levels during the Flint Water Crisis. Environ Sci Technol 52(15): 8124–8132

Pierce SJ, Norman B (2016) *Rhincodon typus*. The IUCN Red List of Threatened Species 2016: e. T19488A2365291

Plummer R, Armitage D (2007) A resilience-based framework for evaluating adaptive co-management: linking ecology, economics and society in a complex world. Ecol Econ 61(1):62–74

Pouliot C (2009) Using the deficit model, public debate model and co-production of knowledge models to interpret points of view of students concerning citizens' participation in socioscientific issues. Int J Environ Sci Educ 4(1):49–73

Putnam RD (1995) Tuning in, tuning out: the strange disappearance of social capital in America. Polit Sci Polit 28(4):664–684

Roger E, Klistorner S (2016) BioBlitzes help science communicators engage local communities in environmental research. J Sci Commun 15(3):A06

Selig S, Sahli M (2016) University-community collaboration to address Flint water crisis. Eur J Publ Health, 26 (suppl_1):ckw171.002

Shirk JL, Ballard HL, Wilderman CC, Phillips T, Wiggins A, Jordan R et al (2012) Public participation in scientific research: a framework for deliberate design. Ecol Soc 17(2):29

Smith C, Felderhof L, Bosch OJH (2008) Adaptive management: making it happen through participatory systems analysis. Syst Res Behav Sci 24(6):567–587

Soleri D, Long J, Ramirez-Andreotta M, Eitemiller R, Pandya R (2016) Finding pathways to more equitable and meaningful public-scientist partnerships. Citizen Sci Theory Pract 1(1):9, 1–11

Stepenuck KF (2013) Improving understanding of outcomes and credibility of volunteer environmental monitoring programs. Doctoral Dissertation. University of Wisconsin, Madison. Available at: http://aquadoc.typepad.com/files/stepenuck_dissertation-finalpdf

Stepenuck K, Green L (2015) Individual-and community-level impacts of volunteer environmental monitoring: a synthesis of peer-reviewed literature. Ecol Soc 20(3):19

Stilgoe J (2009) Citizen Scientists: reconnecting science with civil society. Demos, London. http://creias.ipleiria.pt/files/2010/08/Citizen_Scientists_-_web.pdf

Storksdieck M, Shirk JL, Cappadonna J, Domroese M, Goebel C, Haklay M, Miller-Rushing AJ, Roetman P, Sbrocchi C, Vohland K (2016) Associations for citizen science: regional knowledge, global collaboration. Citizen Sci Theory Pract 1(2):10

Sullivan BL, Phillips TB, Dayer AA, Wood CL, Farnsworth A, Iliff MJ, Davies IJ, Wiggins A, Fink D, Hochachka W, Rodewald AD, Rosenberg KV, Bonney R, Kelling S (2017) Using open access observational data for conservation action: a case study for birds. Biol Conserv 208:5–14

Teorell J (2003) Linking social capital to political participation: voluntary associations and networks of recruitment in Sweden. Scand Polit Stud 26(1):49–66

Theobald EJ, Ettinger AK, Burgess HK, DeBey LB, Schmidt NR, Froehlich HE et al (2015) Global change and local solutions: tapping the unrealized potential of citizen science for biodiversity research. Biol Conserv 181:236–244

Tollington S, Turbe A, Rabitsch W, Groombridge JJ, Scalera R, Essl F, Shwartz A (2017) Making the EU legislation on invasive species a conservation success. Conserv Lett 10(1):112–120

Veeckman C, Schuurman D, Leminen S, Westerlund M (2013) Linking living lab characteristics and their outcomes: towards a conceptual framework. Technol Innov Manag Rev 3:6–15

Wiggins A, Crowston K (2011) From conservation to crowdsourcing: a typology of citizen science. In 2011 44th Hawaii International Conference on System Sciences (HICSS) (pp. 1–10)

Wiggins A, Bonney R, Graham E, Henderson S, Kelling S, Littauer R, LeBuhn G, Lotts K, Michener W, Newman G, Russell E, Stevenson R, Weltzin J (2013) Data management guide for public participation in scientific research. Albuquerque, NM, DataONE

Wiggins A, Bonney RE, LeBuhn G (2018) A science products inventory for citizen science planning and evaluation. Bioscience 68(6):436–444

Wilderman C (2005) Portrait of a watershed: Shermans Creek. Pennsylvania Department of Environmental Protection, Harrisburg, PA

Wood C, Sullivan B, Iliff M, Fink D, Kelling S (2011) eBird: engaging birders in science and conservation. PLoS Biol 9(12):e1001220

Chapter 5
Recognition of Indigenous Ecological Knowledge Systems in Conservation and Their Role to Narrow the Knowledge-Implementation Gap

Emilie Ens, Victoria Reyes-García, Hugo Asselin, Minna Hsu, Elsa Reimerson, Kiri Reihana, Bevyline Sithole, Xiaoli Shen, Vanessa Cavanagh, and Michael Adams

Contents

E. Ens (✉)
Department of Environmental Sciences, Macquarie University, North Ryde, NSW, Australia
e-mail: emilie.ens@mq.edu.au

V. Reyes-García
Institució Catalana de Recerca i Estudis Avançats (ICREA), Barcelona, Spain

Institut de Ciència i Tecnologia Ambientals (ICTA), Universitat Autònoma de Barcelona, Bellatera, Barcelona, Spain
e-mail: victoria.reyes@uab.cat

H. Asselin
Université du Québec en Abitibi-Témiscamingue 445, Boulevard de l'Université, Rouyn-Noranda, Québec, Canada
e-mail: hugo.asselin@uqat.ca

M. Hsu
IUCN Cultural Practices and Ecosystem Management thematic group (Commission on Ecosystem Management), Gland, Switzerland
e-mail: minnahsu@gmail.com

E. Reimerson
Department of Political Science, Arcum – Arctic Research Centre and Várdduo – Centre for Sámi Research, Umeå University, Umeå, Sweden
e-mail: elsa.reimerson@umu.se

K. Reihana
Ngāpuhi, Te Whakatōhea, Ngai Tūhoe; Manaaki Whenua, Hamilton Office, Hamilton, New Zealand
e-mail: reihanak@landcareresearch.co.nz

C. C. Ferreira, C. F. C. Klütsch (eds.), *Closing the Knowledge-Implementation Gap in Conservation Science*, Wildlife Research Monographs 4,
https://doi.org/10.1007/978-3-030-81085-6_5

5.1 Introduction

Since the 1980s international directives for nations to include Indigenous knowledge (IK) in conservation have occurred in parallel with initiatives aiming to address Indigenous rights. In 1982, the *Declaration of the Rights of Indigenous Peoples* (IP) was initiated by the United Nations (UN) Economic and Social Council, although it took 25 years to ratify (in 2007) due to complexities in regional acceptance and aspirations of Indigenous Peoples. In 2000, the UN established the *Permanent Forum on Indigenous Issues* to coordinate global Indigenous leaders and facilitate the reassertion of Indigenous rights and roles in the global discourse. Similarly, international conservation agreements, such as The Aichi Targets of the *Convention on Biological Diversity* (2011–2020) and *Intergovernmental Panel on Biodiversity and Ecosystem Services* (IPBES) also clearly mandate Indigenous involvement in global conservation efforts. These international directives are increasingly being observed at national and, to varying degrees, local levels. However, as the regional case studies below show, the recognition of IK systems in on-ground conservation programs often requires a willingness of decision-makers and conservation leaders to make clear statements and practical plans that define who, where, and how IK systems are included in conservation actions.

Progress and challenges toward supporting IK recognition in conservation are summarized below. Sections are oriented to capture enabling and disabling factors that allow the recognition of IK in conservation, although due to the high diversity of

B. Sithole
Research Institute for Environment and Livelihoods at Charles Darwin University, Casuarina, NT, Australia
e-mail: bev.sithole@gmail.com

X. Shen
State Key Laboratory of Vegetation and Environmental Change, Institute of Botany, The Chinese Academy of Sciences, Beijing, China
e-mail: xlshen@ibcas.ac.cn

V. Cavanagh · M. Adams
School of Geography and Sustainable Communities, University of Wollongong, Wollongong, NSW, Australia
e-mail: vanessa@uow.edu.au; madams@uow.edu.au

Indigenous Peoples, across the world and the differing socio-political climates in which they live, the task is challenging. Despite some great achievements in recognizing and respecting IK systems in some regions, many Indigenous groups continue to be marginalized, contributing to variable knowledge-implementation gaps.

5.1.1 Africa

In some contemporary African societies, politically and economically motivated contestation around Indigenous Peoples hides voluminous local knowledge and experience that could enhance sustainable land use practice. While there is ongoing skepticism regarding the value and extent of IK in conservation in some parts of Africa (Briggs 2005), examples exist where IK is being successfully and increasingly applied. Gichuki et al. (1999) stated that IK is dynamic and continues to evolve as long as the ethnic group lives in, and has access to land to practice the knowledge. For example, Ocholla et al. (2016) found that the Samburu, among other pastoral communities in northern Kenya, have a vast knowledge of habitat and behavior of wildlife that is useful to conservation. They found that the cultural rules of the Samburu ensure they do not hunt wild animals for income or food except during serious drought and famine where specific animals provide their sustenance. Similarly, Eyong (2007) outlined that the Ba'Aka who maintain a traditional lifestyle in the Dzanga Sangha National Park, central Africa, retain the rights to hunt and gather medicinal plants, fruits, and other wild foods but do not hunt certain species protected under national legislation, such as elephant, gorilla, and chimpanzee. Moreover, the Ba'Aka has developed a strong awareness of the seasons and moods of the forest and how they influence plants and animals, and this knowledge is being put to use in managing the reserve and its wildlife (Eyong 2007).

Up until the mid-2000s, the exclusion of Indigenous Peoples from Protected Areas occurred in many African countries, such as parts of Ethiopia and Botswana (Adams and Hutton 2007). In the mid-2000s, Botswana declared illegal the exclusion of the San Bushmen from the Central Kalahari Game Reserve. About the same time, the *Land Reforms* of Zimbabwe (where farms of white Zimbabweans of European ancestry were "redistributed" back to black subsistence farmers) saw some of the most aggressive devolutions of power, land ownership, and control seen in any colonized nation in recent times (Jones and Murphree 2004). Corresponding with increases in Indigenous Peoples land ownership and recognition of rights, the value of IKs and experience in conservation has been growing (Mwangi 1988; Lalonde 1993).

In the same line, in sub-Saharan Africa attempts to reduce resource use in conservation areas saw "community conservation" programs evolve in the 1970s and 1980s. Community conservation initiatives developed out of safari hunting revenue and were designed to include Indigenous People in conservation and deter them from hunting and overexploiting conservation areas, as these practices threatened to undermine the safari hunting industry (Barrow and Murphree 2001; Turner

2004; DeGeorges and Reilly 2009; Jones 2015). An early example of community conservation was the Communal Areas Management Program for Indigenous Resources (CAMPFIRE) in Zimbabwe, a program that devolved some control and management of natural resources to local communities and was eventually funded by USAID (DeGeorges and Reilly 2009).

Inclusive conservation was recognized as the preferred conservation practice at the Durban World Parks Congress in 2003 and the first People and Parks Forum in 2004. Now, across Africa, community involvement in conservation takes many forms in different locations and in different socio-political and biophysical contexts (Barrow and Murphree 2001). Community participation in conservation has opened up spaces for Indigenous Peoples to participate more directly and apply their IKs. In different African countries, examples of community-based approaches occur in wildlife, rangeland, water, forests, inland fisheries, and coastal and marine resources management. Further examples, many of which are co-funded by international conservation and development organizations, are discussed by DeGeorges and Reilly (2009) including *Living in a Finite Environment* in Namibia (run through World Wildlife Fund and funded by USAID), the *Selous Conservation Program* in Tanzania (funded by the German funding agency GTZ) and the *Luangwa Integrated Resource Development Project* in Zambia (funded by the Norwegian Agency for International Development).

Another example of IK being increasingly incorporated in conservation is the case of fire management. According to Shaffer (2010), Malian farmers use seasonally targeted annual burning regimes to manage savanna landscapes for hunting and agricultural production. Increased biodiversity accompanying the habitat patchiness created by the fires is desirable to communities and to conservation as it facilitates a wider range of potential foods, medicines, construction material, and herd forage to communities. In some government-run conservation areas, such as Kruger National Park, South Africa, Greater St. Lucia Wetlands National Park, South Africa, and Reserva Especial de Maputo (REM), Mozambique, scientists, and managers are re-establishing Indigenous fire regimes (Shaffer 2010). Eriksen (2007) suggested that despite the historical importance of fire as a savanna management tool, there is still much controversy about the use and sustainability of Indigenous land management practices in contemporary African savannas. Eriksen (2007) described that the farming systems in Zambia have largely moved away from a shifting cultivation system that incorporated fires (locally known as chitemene), to the prevalence of more permanent and intensive agricultural systems (e.g., ifibunde and impunta).

Importantly, some researchers note that, as a result of colonization, population growth and a shift in livelihood economies, not all IK systems or management regimes are still intact, or arguably relevant. Often the conditions under which these knowledge systems developed have changed or the communities holding the knowledge have been relocated to landscapes and locations which possibly have rendered the information irrelevant (Little and Brokensha 1989; Cunningham 2014). Mapira and Mazambara (2013) suggested that in Zimbabwe, it might be hard for some communities to go back to the old knowledge systems and practices, while some communities may not be willing to share their knowledge. Therefore, attempts to enhance IKs in conservation in Africa should recognize the power structures or hierarchies within which knowledge systems are embedded and the aspirations and abilities for Indigenous Peoples to provide the knowledge (Agrawal 2002; Gaventa and Cornwall 2008).

5.1.2 Asia

Two-thirds of the world's estimated 370-400 million Indigenous Peoples reside in Asia. Amidst the inter-ethnic and cross-border tensions that characterize many areas across Asia, there is ongoing conflict between supporters of Indigenous Peoples (and their knowledge) and state policies. Numerous names are used to refer to Indigenous Peoples throughout Asia, such as "ethnic minorities" (China and Laos), "aboriginal tribes" (Taiwan), "cultural minorities" (Philippines), "hill tribes" (Thailand), "natives" (Malaysian Borneo), "scheduled tribes" (India), or "ethnic nationalities" (Burma).

Indigenous Peoples and their knowledge systems are often isolated from major land use and conservation decisions in Asia and therefore, their contribution to narrow the conservation-implementation gap is particularly incipient in this region. Indeed, many governments are reluctant to recognize the collective rights of Indigenous Peoples, particularly with respect to land rights and use of natural resources. Dramatic socio-environmental transformations due to development have also devalued and, in some cases eliminated, IKs and practices in Asia. For example, medicinal plants are an integral part of traditional medical practices and are highly valued in both folk medicine and codified medical systems such as Chinese traditional medicine and Ayurveda (Sivarajan and Balachandran 1994; Lama et al. 2001). However, medicinal plant species growing in the wild are being rapidly exploited via the international expansion of medicinal plant industries. In the Nepalese Himalaya, Indigenous ethnoecological knowledge, monitoring, and management practices regarding medicinal plants greatly benefit conservation initiatives, especially around knowledge of phenology, species habitat requirements, and adaptive management approaches (Ghimire et al. 2004; Uprety et al. 2010). Proper management of high-value and high-priority medicinal plants could serve as a sustainable source of income for communities, in turn helping to generate incentives for biodiversity conservation and thus ensuring the long-term availability of medicinal plants for household and commercial uses (Uprety et al. 2010).

In Asia, like in Africa, there is increasing adoption of Community-Based Natural Resource Management (CBNRM) programs as well as an increasing recognition of Indigenous rights to own and manage land. Nevertheless, in areas such as Cambodia, China, India, Indonesia, Laos, Nepal, the Philippines, Thailand, and Vietnam tensions between achieving more equitable distribution of resources and greater livelihood and natural resource outcomes occur (Mahanty et al. 2006; Shivakoti et al. 2016). Based on a case study in Indigenous Atayal territory, Taiwan, Lu et al. (2012) demonstrated how, despite the legal recognition in terms of land tenure of co-managed protected areas (CMPA), commensurate legislation must also encourage resource management agencies to devolve their power and share management responsibilities with local communities. Yet, although the most successful examples of CBNRM are often those built around customary institutions that give scope for Indigenous peoples' knowledge, skills, and rights (Li 2002; Colchester 2004; Tang and Tang 2010), officials often remain to relinquish authority over resources to local users (Sudtongkong and Webb 2008). In Asia, community-conserved areas are often not recognized by national laws, although the establishment of small protected areas by individuals, private sectors, and local communities is permitted, even encouraged, by provincial governments, with co-management existing at the local level (Li 2014).

In Thailand, the Thung Yai Naresuan Wildlife Sanctuary and World Heritage Site present examples of how conflicts can emerge in the process of designating protected areas. In this particular case, the "voluntary" resettlement of Karen ethnic peoples within the Thung Yai area involved soldiers and forest rangers demolishing villages and prohibiting agricultural activities (Buergin 2015). In a study of communities on the Andaman coast of Thailand, Bennett and Dearden (2014) found that locals perceived marine protected areas as having a negative impact on livelihoods by "undermining access to or lacking support for the development of cultural, social, political, financial, natural, human and physical assets." Contrastingly, the Annapurna Conservation Area in Nepal has been cited as a positive case study for protected area management, because it "explicitly recognize[d] local settlement land use and resource management within [the protected area and made] co-management and local management the basis of regional conservation and development" (Stevens 1997: 258). These examples suggest that recognition of the value of IKs and pressure on decision-makers to devolve power and allow space for Indigenous Peoples and their knowledge is required to advance the inclusion of IK in conservation in many parts of Asia.

In China, international NGOs began advocating for the inclusion of traditional practices and local communities in contemporary conservation in the late 1980s. They did so through participatory programs, such as the *Community Forestry Project* of the Ford Foundation and *Integrated Conservation Development Project* of WWF. This heralded increased interest from the scientific community, formation of local NGOs for cultural revival and environmental protection, and efforts to promote the legal recognition of community-based conservation (Li 2014). Many recent studies demonstrate the contributions of traditional practices to conservation in China, especially in the Tibetan regions (e.g., Xu et al. 2005; Brandt et al. 2013; Shen et al. 2015). Since the 2000s, the *Sacred Land Project* and *Conservation Concession Program* implemented by Conservation International and Shanshui Conservation Center promoted the legal recognition of Tibetan traditional practices and local institutions in government conservation plans (Shen and Tan 2012). Although community-conserved areas are yet to be recognized by national laws, small protected areas have been encouraged by provincial governments since the 1990s, as a model to include local communities in conservation in densely populated regions. With the establishment of the National Park System in 2015, the function of protected areas in China has shifted from sole conservation to the inclusion of multiple goals, including nature-based education and recreation needs, as well as supporting the livelihood of local people. These objectives are clearly stated in the "Master Plan on Establishing China's National Park System" and reflected by the functional zones of national parks (e.g., core protection, ecological conservation, recreation, and traditional utilization zones). A multidimensional governance framework designed to involve different stakeholders including local communities, private sectors, and the general public in conservation and nature resource management is advocated and actively under exploration both by the government and the civil society, for example, the Chinese Civil Protected Area Network launched by 23 international and domestic NGOs and foundations in 2017, such as the SEE Foundation, The Nature Conservancy, and Shanshui Conservation Centre, aiming to protect 1% land area of China.

Widely diverse histories, cultures, climates, and political contexts have contributed to the varied Indigenous knowledge systems in Asia. There has been a gradual paradigm shift in policies and practices where Indigenous communities are increasingly being recognized as an integral part of local conservation initiatives. However, further recognition at the national level is required to better incorporate the richness of IK systems into conservation in Asia to narrow the knowledge-implementation gap (Persoon et al. 2003; Roth 2004).

5.1.3 Australia

The territories of over 250 Indigenous language groups span the Australian continent. Since colonization, the maintenance of IK systems has varied, generally depending on proximity to the initial colony—areas in southeastern Australia being earliest and are most affected, while parts of central and northern Australia were colonized later and have generally speaking, been less affected by colonization (Ens et al. 2015). The use of IK in conservation tends to reflect this pattern. For example, in remote central and northern Australia, Indigenous Peoples still often use fire in customary ways to care for their ancestral estates. However, in the more densely populated and colonized parts of the southeast, cultural use of fire must be negotiated within state agencies' regulatory controls (Neale et al. 2019). While a recent revival and acknowledgment of IK, language, and practice are occurring in southeastern Australia, it remains marginalized.

Some Indigenous people now use employment as an avenue to gain access to lands and maintain customary knowledge and practice for conservation of natural and cultural resources. Historically, Aboriginal people in Australia were able to access and, to varying degrees, manage their lands through participation in the pastoral industry (Goodall 2001; Harrison 2004), as well as working in government departments such as the NSW National Parks and Wildlife Service, which first employed an Aboriginal man, Ray Kelly, in 1973 (Kijas et al. 2005). During the 1980s, the community-driven Aboriginal Ranger movement began to gain traction. At this time, Aboriginal Rangers were also being employed to work in government-

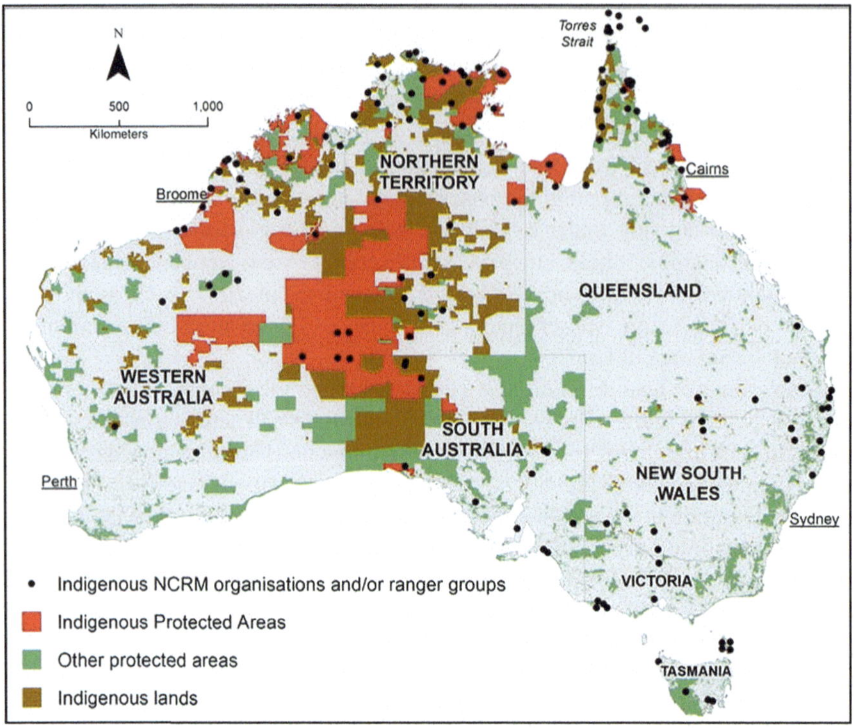

Fig. 5.1 Location of Australian Indigenous Protected Areas, other protected areas, Indigenous lands, and Indigenous Ranger groups across Australia in 2017 (Source: Ens and Grech 2018)

run National Parks across Australia (Smyth et al. 1985). Ranger's work is a propitious niche for Indigenous Peoples (Moorcroft 2015) that complements IK systems and cultural practice. The converging interests and objectives of Aboriginal communities and government conservation departments have resulted in the rapid growth and community support of Aboriginal "Caring for Country" programs and the Ranger movement across Australia (Hill et al. 2013).

In 1997, the Australian Government formally supported Indigenous conservation with the establishment of the *Indigenous Protected Area* (IPA) program. In 2021, over 74 million hectares of Indigenous-owned and managed land has been voluntarily declared by Indigenous Traditional Owners as part of Australia's National Reserve System to be managed under IUCN Categories, iii, iv, v, or vi (Australian Government 2021), the latter of which allows for not just environmental conservation but also traditional land uses. In 2021, IPAs made up 46% of Australia's National Reserve System (Fig. 5.1) and have become a significant part of Australia's conservation agenda and compliance with the international *Convention on Biological Diversity* (1999). Additionally, in 2007, the Australian Government established the *Working on Country* (Aboriginal Ranger) program. In 2017, over 777 Aboriginal Ranger jobs (full-time to casual) were supported by the Australian

Government (Fig. 5.1). However, much of this investment has been in central and northern Australia, while the majority of Indigenous Australians live in southeast Australia, where conventional Western qualifications are prioritized over cultural knowledge to gain employment within government conservation agencies and departments.

Despite these inequities, the revitalization of Indigenous conservation has evolved with increased awareness of the benefits of IK. Indigenous people across Australia are increasingly using fire following ancestral methods, accessing bush food and medicine species, maintaining cultural sites as well as dance, language, art, and story. Modern Australian Indigenous fire management is encouraging other Indigenous Peoples around the globe to revitalize fire management and indeed other culturally motivated conservation practices, such as in North America (see below). Aboriginal Ranger groups were the first to capitalize on their traditional use of fire in contemporary carbon offset markets. The West Arnhem Land Fire Abatement scheme emerged as the world's first Indigenous-led carbon offset program where high carbon-emitting companies fund Aboriginal Ranger groups to implement early dry season low-intensity patch burns to prevent late dry season fires and, hence, reduce carbon emissions (Russell-Smith et al. 2009).

Indigenous land and sea management is a rapidly growing part of Australia's conservation agenda (Hill et al. 2013). Aboriginal Ranger groups tend to blend "traditional" and "new" ways of understanding and managing Country.[1] Many Ranger groups use Geographic Information System (GIS), electronic mapping tools, and databases to record traditional and new knowledge (e.g., of threatened species) and engage in "fee for service work" like quarantine inspections, fishery patrols, and revegetation works. Although these work programs are driven by the market, they concomitantly serve to create employment for Aboriginal people on-Country and generate resources to access Country and maintain place-based IK and practice (Muller 2008; Greiner 2010; Concu 2013). However, despite the significant progress in recognizing IKs in mainstream conservation in Australia, several fundamental challenges remain. Equitable funding, Indigenous leadership, Indigenous control, and prioritization of IK and preferred methods can conflict with Western approaches. In some places, Aboriginal people have the necessary skills to self-manage conservation projects in the modern context, with strong Western literacy, organizational, and computer skills. In more remote areas, where English literacy is not as strong, employment of non-local people is often needed to fulfil roles to meet particular funding and associated administrative requirements. Simultaneously, we note the rise in *Country Based Plans* and *Healthy Country Planning* initiatives that aim to combine Indigenous and Western approaches to conservation planning to narrow the knowledge-implementation gap (Moorcroft et al. 2012).

[1]Country is a modern term used by Australian Aboriginal and Torres Strait Islander Peoples that refers to more than just a geographical area by encompassing the associated values, places, resources, stories, spirits, and cultural obligations (see Smyth 1994).

A fundamental challenge to respecting and recognizing IK in conservation in Australia remains in the enormous disparity of funding between state-operated protected areas, which have high-level resources guaranteed, and Indigenous conservation programs, which often run on shoestring budgets while delivering conservation benefits on nearly half of Australia's protected area estate (Hill et al. 2013). Additionally, despite progress since the 1960s in Indigenous voting rights, land claims, and self-determination, there are many examples where subjugation and suppression of rights and aspirations still occur. For example, since 1985, following a successful land claim, Uluru-Kata Tjuta National Park, and World Heritage Area (an iconic site in Australia) has been jointly managed by the Australian Government and *Anangu* Traditional Owners. While this central desert area is often presented as a showcase of Aboriginal culture, Anangu continues to struggle to establish appropriate recognition of their right to care for the Country (Adams 2014). While the Australian Government is to be commended for a commitment to closing "the climb" (tourists climbing Uluru/Ayers Rock), it has taken 34 years of agitation by Anangu to achieve this.

Although Australia has made great progress in incorporating Indigenous Peoples, knowledge, and Country into its broader conservation agenda, many gaps and opportunities remain. Aboriginal people want to have full control of their lands and seas; however, funding, monitoring, and reporting requirements often inhibit this aspiration. Not only does the capacity of Aboriginal people to operate within Western governance systems need to grow, but the capacity of non-Aboriginal decision-makers to respond appropriately to the needs, aspirations, preferred methods, and priorities of Aboriginal cultures needs to occur (Ens et al. 2012, 2014). Until these two-way capacity and understanding gaps close, there will be ongoing tensions in recognizing IK in contemporary Australian conservation hampering its use to bridge the knowledge-action gap in the region.

5.1.4 South America

Indigenous peoples and local communities in South America hold legal titles over large shares of forest. For example, the territories of the 375 Indigenous groups cover 25.3% of the so-called "Global Amazon" (Van Dam 2011). When adding Protected Areas, many of which overlap with Indigenous territories, the percentage of land in the continent under Indigenous and local communities management is 41.2% (Toledo 2001; Sunderlin et al. 2005; Porter-Bolland et al. 2012).

Indeed, the complex landscapes and rich biodiversity in South America result, not only from natural conditions in the area, but also from centuries of Indigenous management (Denevan 1966, 1992; Chazdon 2003; Heckenberger 2003; Heckenberger et al. 2008; Macía 2008), which has often safeguarded and sometimes enhanced natural resources availability (Posey 1985; Dufour 1990; Wiersum 1997; Paneque-Gálvez et al. 2018). For example, slash-and-burn agriculture, as practiced by Amazonian small-scale societies, increases landscape biodiversity through the creation of habitat mosaics (Peters 2000; Wiersum 2004). Similarly, Amazonian swidden cultivation-fallow management systems can be considered sustainable agroforestry systems with significant ecological and economic benefits (Coomes et al. 2000). Such local agricultural systems could promote food production while contributing to conservation goals.

Contemporary South American Indigenous Peoples continue to extensively manage many useful species (Macía 2008) and to use biodiversity enhancing techniques, including fire to manage savanna and forest landscapes, which have been shown to enhance both domesticated and wild biodiversity (Mistry et al. 2016). Indeed, several authors have proposed that Indigenous management practices (e.g., burning or tree cultivation) generate an intermediate level of disturbance that does not affect landscape heterogeneity (Wiersum 2004; Pérez-Llorente et al. 2013; Paneque-Gálvez et al. 2018). For example, the Quechua-speaking communities of the Tunari (Bolivia) privilege integrated and diversified use of their territory through mixed agriculture, pastoralism, small-scale forestry, and off-farm labor that leads to high habitat heterogeneity.

However, like Protected Areas, areas inhabited by Indigenous Peoples in South America are subject to new threats, irrespective of whether or not Indigenous Peoples rights to these lands are recognized. Such threats include pressures from nearby urban centers (like the Bolivian Altiplano or northeastern Brazil); the rapid growth of Indigenous populations (McSweeny and Arps 2005); illegal logging (Reyes-Garcia et al. 2013; Pacheco et al. 2016); livestock farmers and agricultural companies attracted by grain or biofuel projects (Butler and Laurance 2009); mining and oil interests (Finer and Orta-Martínez 2010; Orta and Finer 2010); illegal activities such as coca cultivation (Fernández-Llamazares et al. 2018); or simply, the replacement of traditional biodiversity management practices by Western practices, as illustrated by the replacement of *chinampas* (raised beds) by subsidized plastic greenhouses in Mexico (Merlin-Uribe et al. 2013). Recognizing Indigenous peoples' rights to land, benefit sharing and control of their own institutions could help alleviate these threats and thus meet local and global conservation goals (Garnett et al. 2018).

Moreover, throughout South America, Indigenous Peoples face governance challenges that can affect their management of lands and natural resources. Importantly, unresolved contradictions between national legislation and Indigenous rights, as in the case of non-timber forest products in some Andean countries, for example, make it difficult to regulate for sustainable resource use and management (de la Torre et al. 2011). In the absence of other institutional support, adverse perceptions of traditional practices combined with policies that do not protect sustainable local practices work against Indigenous Peoples. The downgrading of protected status for multiuse protected areas and the shrinking borders of Indigenous territories present an uncertain future for Indigenous Peoples and their resource management. In 2012, Brazil reduced the size of the Amazon National Park by 47,080 hectares, an area already threatened by colonial agricultural policies and cattle ranching for hydropower generation (Laue and Arima 2016). In 2017, Bolivia reduced protections in the Isiboro-Sécure National Park and Indigenous Territory, the ancestral lands of three Indigenous groups and national biodiversity hotspot, in order to build a road through the center of the park (Fernández-Llamazares et al. 2018).

Despite these challenges, in the last 20 years, Indigenous communities have implemented different projects and initiatives oriented to the sustainable management of their forest in order to reduce the knowledge-implementation gap. Many of these initiatives have been implemented in coordination with conservation or development organizations (Van Dam 2000) and they represent a variety of potential ways in which IKs can be used to achieve conservation goals. In South America, initiatives oriented to the sustainable management of resources and incorporating Indigenous knowledge, to varying degrees, range from community-based conservation projects (Reyes-Garcia et al. 2013), to projects inspired by the industrial forestry model, such as the utilization of non-timber forest products, agroforestry, forest-grazing, or small farming proposals. The most common has been to combine community forms of organization with production for the market, promoting small-scale forest management and use, which is generally called community forest management, a very popular type of project in Mexico (Valdez et al. 2012). While these initiatives face challenges related to centralized forest policies and the taxation system, the adaptability of the decision-making structures based on traditional knowledge and institutions make these enterprises economically sustainable and simultaneously contribute to conservation goals.

Often, Indigenous Peoples have serious difficulties in designing long-term management plans that suit conservation scientists and this contributes to the persistence of the knowledge-implementation gap in the region (Van Dam 2011). Notable exceptions include the few *Indigenous Territorial Management* (ITM) experiences promoted by the Confederación de Pueblos Indígenas de Bolivia, and some *Life Plans* (or Planes de Vida) developed by Indigenous organizations from Ecuador and Colombia to envision their community development. There are also difficulties in bringing IK and management into conservation legislation. However, in Brazil, where 13.3% of national lands are quilombo (i.e., Indigenous territories), Indigenous Peoples values have been included in the *National Biodiversity Strategy for 2020*. Recognizing and supporting the contributions of Indigenous Peoples to biodiversity conservation could be done through many mechanisms already present in conservation legislation. Enhanced integration of IK in conservation could be achieved through increased recognition of the *Indigenous and Community Conserved-Area* program or through the promotion of "other effective area-based conservation measures" (OECMs), to which Indigenous people's lands are a substantial subset in South America.

5.1.5 North America

In North America as in other parts of the World, Indigenous cultures and—to various extents—livelihoods are intimately linked to the land (Asselin 2015). Traditional beliefs, knowledge, and practices are continually enriched and transmitted through generations (Berkes 2004), so that cultural landscapes are perpetuated (Watson et al. 2011). The myth of the pristine pre-contact North American wilderness has been debunked (Denevan 1992) and it is now recognized that Indigenous People have shaped and sustainably used the environment for millennia before the first Europeans set foot on the east coast. For example, Indigenous People increased the percentage of acorn- and nut-bearing trees in the landscape (*Quercus*, *Carya*, and *Juglans*) in order to benefit from easily accessible high-protein food (Abrams and Nowacki 2008). Fire was used for a variety of purposes such as to create habitat for large game (e.g., bison), control pests, and increase production of berries, mushrooms, and medicinal plants (Kimmerer and Lake 2001). While several Indigenous groups in North America were nomadic or seminomadic, some were practicing an agriculture centered on the cultivation of native crops, including the so-called "three sisters," i.e., corn, squash, and beans (Asch and Hart 2004).

Following European colonization, North American Indigenous Peoples were forced to cede traditional lands and therefore lost access to vital resources. Traditional practices were either prohibited (e.g., prescribed burning) or dramatically reduced (Trosper et al. 2012). This resulted in "invisible losses" (*sensu* Turner et al. 2008), in terms of culture, identity, health, knowledge, and self-determination. To this day, Indigenous Peoples in North America have little control over land and play a minor, if any, role in natural resources governance. Nevertheless, traditional knowledge continues to be transmitted and its role in environmental planning, conservation, management, and restoration is increasingly being recognized (Berkes 2004; Watson et al. 2011; Uprety et al. 2012; Asselin 2015). For example, the Canadian Committee on the Status of Endangered Wildlife in Canada (COSEWIC) recognizes that incorporating traditional knowledge into assessments of species at risk improves the process by "bringing information and perspectives on wildlife species that are not available in published scientific literature." Therefore, COSEWIC has an Aboriginal Traditional Knowledge Subcommittee whose role is to decide how IK will be incorporated into the process of assessing species at risk that ultimately leads to conservation action.

Considering themselves as stewards of the land (Parlee et al. 2005), North American Indigenous Peoples follow ethical principles such as reciprocity, sharing, respecting seasonal cycles, and avoiding waste (LaRiviere and Crawford 2013). Inspired by the Aboriginal Ranger Program from Australia, the Indigenous Guardians Program was developed by the Indigenous Leadership Initiative to empower communities in Canada to promote intergenerational knowledge sharing, create land-use and marine-use plans, monitor ecological health, maintain cultural sites and protect sensitive areas and species. Indeed, Indigenous People possess deep knowledge of wildlife and plant habitats, especially those of cultural keystone species (Garibaldi and Turner 2004; Colombi 2012; Uprety et al. 2013; Emery et al. 2014; Tendeng et al. 2016). As in Australia, GIS technology is increasingly used to facilitate monitoring in community-based observation networks (Herrmann et al. 2014; Alessa et al. 2016).

Similar to other parts of the world, protected areas are an alien concept to Indigenous cultures in North America (Gladu et al. 2003). Their creation has often led to population displacements and disruption of socioeconomic systems (Notzke 1994; Healy 2007). Indigenous People were forcibly removed from several well-known protected areas (e.g., Glacier, Mesa Verde, Olympic, Yellowstone, and Yosemite National Parks in the United States, and Quetico and Algonquin Provincial

Parks in Canada), and traditional activities were prohibited in others (e.g., Wood Buffalo and Prince Albert National Parks in Canada) (Gladu et al. 2003; Healy 2007). However, in line with global initiatives to increase Indigenous participation in conservation over recent decades, there are growing examples where Indigenous People are playing a role in formal conservation efforts in North America. According to Gladu et al. (2003), increased acceptability of conservation by Indigenous People will ensue if: they are involved in park establishment and management; their rights are respected; traditional activities are allowed within the park; it provides employment and economic development opportunities; and if cultural awareness is paramount.

A good example is the Gwaii Haanas National Park Reserve in British Columbia (Canada) which is co-managed by the Haida Nation and Parks Canada, where traditional activities are allowed, and which financially supports the Indigenous Guardians Program. Mixed-use management is also adopted in the Great Bear Rainforest (Price et al. 2009), which covers 6.4 million hectares of coastal temperate rain forest in British Columbia (Canada). About one-third of the area is protected (two million hectares), whereas the rest is under extensive management for timber and other ecosystem services, using ecological thresholds and natural variability to establish management targets. This area is under shared responsibility by the provincial government and Indigenous People. While co-management is gradually gaining importance in conservation practices in North America (Martin 2016), it is also paving the way for community-based conservation (Berkes 2004), emphasizing "the return of 'inalienable possessions' by the federal government, rather than the extension of limited harvest rights under the current 'subsistence' regime" (Thornton 2010).

5.1.6 Aotearoa, New Zealand

Mātauranga Māori (Indigenous Māori knowledge) has its origins in Polynesia going back at least ~5000 years. It flourished and developed in New Zealand (Aotearoa in Māori) within a holistic Te Ao Māori worldview and is a living and dynamic knowledge system.

Reference is given from most tribes (iwi/hapū) to great explorers such as Kupe (Ngapuhi-northern tribes of Aotearoa) who brought much knowledge to Aotearoa (Smith 1913; Te Rito 2007). In contrast to other parts of the world, traditional knowledge has been maintained as a basis for modern IK, for example, in culturally sustainable harvesting and natural resource management practices (Lyver et al. 2008; Lyver and Moller 2012). Coupling the traditional with the contemporary Western science paradigms has created a new model of state of the environment health practice in New Zealand (Huntington 2000; Moller et al. 2004; Harmsworth et al. 2016).

Mātauranga or Māori knowledge can be summarized as the body of knowledge that seeks to explain the biophysical, metaphysical, and social harmony that exists as lore in the Māori world. Whakapapa (Ancestral linkages) is the transferral mechanism for traditional knowledge, via oral traditions through waiata (songs), pūrakau (stories), mōteatea (traditional chants), karakia (sacred incantations), and waiata tangi (laments) (Marsden and Henare 1992; Moeke-Pickering 1996).

In 1840, the Treaty of Waitangi was signed by over 500 Māori iwi/hapū chiefs from around the country to bring Māori under British sovereignty and rule. However, once signed, subsequent legislations excluded and marginalized Māori from their resources and land. The *Native Land Act* (cc 1861) forced surveyed boundaries, land confiscation, and individualized title and ownership on all land, dispositions which transgressed Māori values and belief systems around collective ownership and decision-making (MfCH 2017a). Under British sovereignty, legislation was introduced to support the cultural assimilation of Māori into the Western culture. Successive legislation encouraged cultural displacement, which allowed confiscation of Māori land by the Government (MfCH 2017b). The disenfranchising of Māori from their land fractured their cultural knowledge base, although the basic cultural tenets remained strong (Glover et al. 1996; Harmsworth et al. 2016).

The *Resource Management Act* 1991 (RMA) and its recent (2017) amendments are the primary overarching environmental conservation legislation in New Zealand. These include local government requirements to recognize and take account of Māori culture and values. Since the passing of the RMA, national policy standards such as one for freshwater (NPSFM 2014) have mandated that local Māori should be involved in collaborative decision-making processes, planning, policy, and environmental monitoring to better integrate Indigenous knowledge that will help to narrow the knowledge-implementation gap. Further, the *Environmental Reporting Act* (2015) reinforces the need to include Māori values and perspectives in state of the environmental assessment and reporting.

In line with iwi/hapū needs and requirements and in application to various national and regional legislation and policy, a large number of mātauranga Māori-based assessment approaches have been developed in New Zealand. One of the first frameworks to incorporate Māori values and perspectives was the Cultural Health Index (CHI) for stream and waterways (Tipa and Teirney 2003, 2006). In terms of legislation in respect to Treaty of Waitangi claims, one of the best examples of local and regional mātauranga Māori-based values and concepts influencing modern legislative conservation thinking is the Whanganui River that in 2017 was granted legal status for its personhood. A new legal entity has been created *Te Awa Tupua* referring to "an indivisible and living entity which spans from the mountains to the sea, incorporating the Whanganui River and all of its physical and metaphysical elements" (Te Awa Tupua Whanganui River Settlement Act, section 13(b)). This is the epitome of Mātauranga Māori in conservation action.

Indigenous knowledge is applied to address many modern conservation issues and challenges in New Zealand. Mātauranga can provide knowledge and guidance through innovation and technology such as mobile applications and web pages. For example, a holistic mātauranga-based cultural framework and method was developed in the Waikato region, North Island, with several iwi/hapū groups in response to the NPSFM National Objectives Framework (NOF), where Māori can assess, monitor, and report on mahinga kai-customary resources as part of a wider regional and national freshwater environmental health scheme. The framework and design of

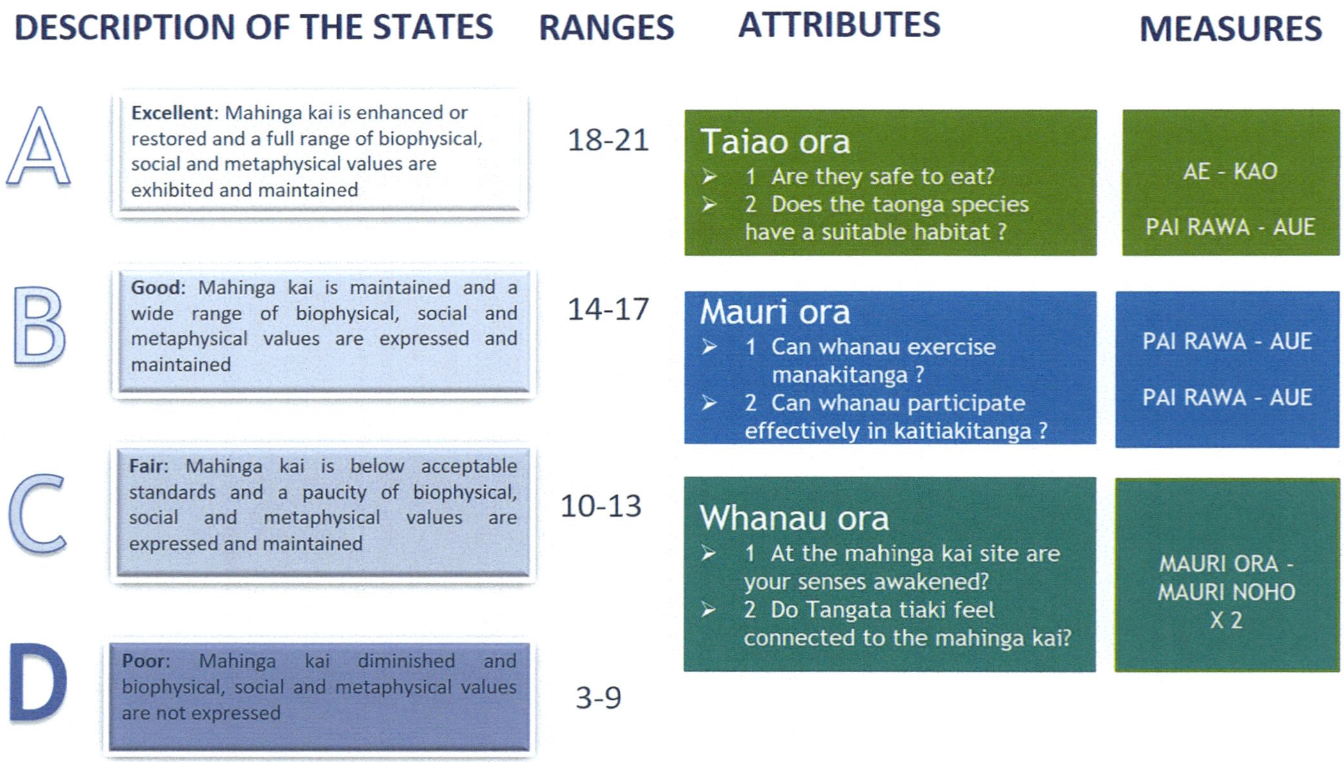

Fig. 5.2 Descriptions of the attribute assessments and their ranking measures of the "Wai Ora Wai Maori" Freshwater Assessment Tool. Once these assessments scores are aggregated, a final ranking is then selected from the States table on the left. These prioritization states to aid in restoration and management decision-making for iwi, the data is assessed alongside freshwater science data to create a more holistic picture of the health of the mahinga kai

(food harvesting site). Figure Glossary: Mahinga Kai—is a traditional food harvesting site on a river/stream; Taiao ora—Flourishing nature; Taonga—precious, treasured, sacred or revered; Mauri ora—The essence of vitality; Whanau—Family, extended family; Manakitanga—Hospitality, kindness, generosity, support, the process of showing respect, generosity, and care for others; Kaitiakitanga—Guardianship, stewardship, trusteeship; Whanau ora—Thriving families; Tangata tiaki—people of the land who are guardians; Ae—yes, Kao—No; Pai rawa—Excellent, a resource in very good condition; Auē—expression of distress—low; Mauri ora—Life essence to support human wellbeing—being invigorated; Mauri noho—life essence at a place—dormant; Kaupapa—Māori ideology, Māori based

Fig. 5.3 Iwi (Ngāti Tahu Ngāti Whaoa) kaitiaki (iwi/hapū environmental guardians) trailing the Framework and mobile application in the field. Photo credit: Kiri Reihana

the kaupapa Māori assessment tool "Wai Ora Wai Maori" is presented in Fig. 5.2 showing the NOF bands from excellent to poor for freshwater assessment within a Te Ao Māori knowledge band framework. Figure 5.2 shows mātauranga-based descriptions, attributes, states, and measures with Māori language terms (Te Reo Māori) and numeric rankings and ranges (Awatere et al. 2017). The New Zealand national freshwater standards (Hellenic Ministry of the Environment, Energy and Climate Change 2014) have determined "bottom lines or limits" for attributes and measures of water quality health, these are known as bands (A, B, C, D), which are descriptors of the state of health of a water body. Our framework aligns to the national standards but is ascribed to a Te Ao Maori worldview. After the conceptual framework was validated with kaitiaki (iwi natural resource managers) in the field, it was then transferred into a mobile phone application. The mobile application was also trialled by kaitiaki before it was launched as an additional survey tool (Fig. 5.3).

These various mediums allow for mātauranga assessment data to be used alongside scientific data to provide a "richer picture of understanding and assessment" in contemporary conservation. A mātauranga design process model (Fig. 5.4) was developed to capture the process steps for replication and transferral of the framework to other iwi (tribes).

The integration of mātauranga into conservation science and natural resource management shown by the Wai ora case study demonstrates that the knowledge-implementation gap can be closed in the field. By using technology, mātauranga becomes a living dynamic entity to strengthen the fabric of Māori society, improve the articulation of Māori values, and at the same time keeps the integrity of past knowledge and traditions in modern society.

5.1.7 Europe

While European countries are home to numerous minority peoples, officially recognized Indigenous Peoples are few (Åhrén 2016). The focus is most often placed on

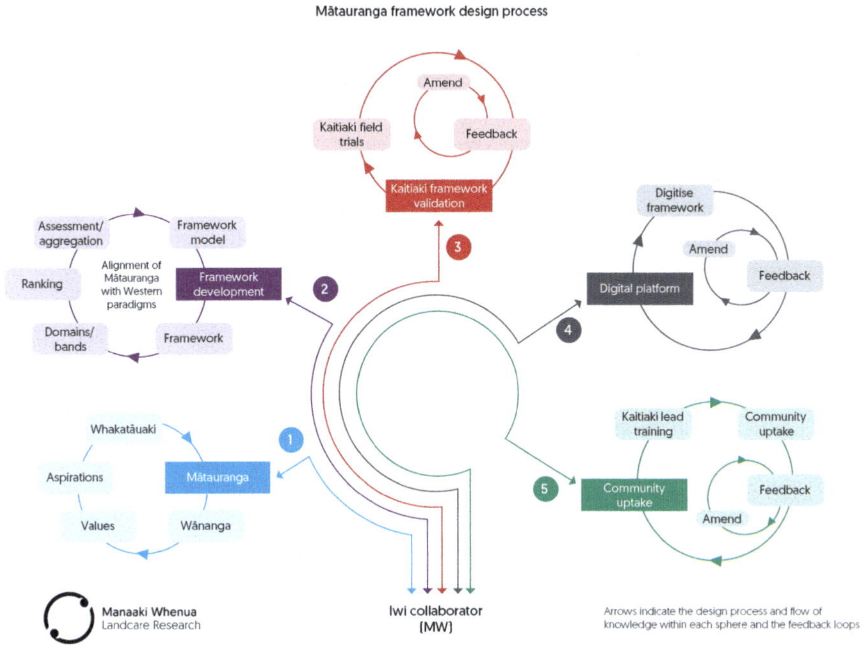

Fig. 5.4 Mātauranga design process. Figure Glossary: Iwi—tribe; Kaitiaki—Māori resource manager; MW—Manaaki Whenua; Wānanga—to meet discuss, deliberate, consider; Mātāpono—principles, values; Wawata—aspirations; Whakataukī—proverb by a known author

the Arctic Indigenous Peoples, including the Inuit (Kalaallit) of Greenland; the Sámi of Norway, Sweden, Finland, and Russia; and the Indigenous Peoples of the Russian North (see e.g., Broch Hansen et al. 2017). In discussions on Indigenous Traditional Ecological Knowledge in a European context, the Sámi people and the situation in Norway, Sweden, and Finland is often the focus (Hernández-Morcillo et al. 2014).

The Sámi have lived in and used the lands of northern Fennoscandia since time immemorial. Their traditional lands, known as Sápmi,[2] include large parts of Norway, Sweden, northern Finland, and the Russian Kola Peninsula. The emerging Scandinavian nation-states' attempts to colonize Sápmi began in the Middle Ages. Up until the eighteenth century, the states generally recognized the Sámi as a distinct people and acknowledged their rights to land and natural resources (Korpijaakko-Labba 2005). From the mid-1700s, the states intensified their efforts to colonize and increase their control over Sápmi. By the end of the nineteenth century, the states claimed ownership of Sámi lands and their policies became openly racist and colonial (Minde 2003; Lantto and Mörkenstam 2007). The relationships between

[2]*Sápmi* is the name for the Sámi territory and people in Northern Sámi, the most widely spoken of the Sámi languages.

the Sámi and the nation-states have evolved in partly different ways, and recognition of Sámi rights today differs between countries (Allard 2011).

Over the past century, Sámi activism and legal battles have led to a strengthening of the status of the Sámi and their rights to land (Minde 2001; Henriksen 2008; Lantto 2010; Allard 2015). However, the effects of colonialism in Sápmi are ongoing (Mörkenstam 2005; DO 2008; Hansen et al. 2008; UNHRC 2011). Sámi self-determination, recognition and legal protection of rights, and adequate protection of Sámi culture, language, and livelihoods are still not fully satisfied (Henriksen 2008; Josefsen 2010; Allard and Skogvang 2015).

Sámi use of land and natural resources is under continuous and increasing pressure from competing land use interests. These include industrial activities, such as mining, logging, wind and hydropower plants, and conservation efforts. Nature conservation in Sápmi has historically followed the "Western" European norm of strict protection and top-down governance, with limited or no Sámi influence. Recently, following general international trends, there is an increasing preference for co-management models, focus on local stakeholders, and emphasis on sustainable resource use as well as a growing interest in Sámi knowledge as a potential contributor to environmental goals that aim at reducing the knowledge-implementation gap (see e.g., Bay-Larsen 2010; Hovik et al. 2010; Heinämäki et al. 2014; Holmgren et al. 2017).

Sámi scholar Dr. Jelena Porsanger defines *árbediehtu,*[3] Sámi traditional knowledge, as "the collective wisdom, practical skills and theoretical competence evolved and acquired by Sámi people through centuries in order to subsist economically, socially and spiritually" (Porsanger 2011, p. 242). In parallel to work undertaken to build up the documentation and study of *árbediehtu* (see, e.g., Porsanger and Guttorm 2011b; Tunón et al. 2015) and to use *árbediehtu* to achieve Sámi environmental goals (see, e.g., Swedish Sámi Parliament 2009), some recent policy developments provide interesting examples of opportunities and challenges of incorporating Sámi traditional knowledge in conservation governance and management. These include a reform of Norwegian Protected Area management and the establishment of a novel management arrangement for a World Heritage site in Swedish Sápmi.

In 2010, Norway introduced a new management model for Protected Areas that allows the government to delegate management authority to local National Park Boards (NPBs; Fig. 5.5). Part of the justification for this reform was Norway's obligations under the Convention on Biological Diversity to protect and promote Sámi traditional knowledge (Risvoll et al. 2014; Reimerson 2017). The main form for Sámi inclusion and participation in protected area governance in Norway is through consultations with the Sámi Parliament, as regulated through agreements between the Norwegian Government and the Sámi Parliament (Procedures for Consultations between State Authorities and the Sámi Parliament 2005; Ministry

[3]*Árbediehtu* is a Northern Sámi word, meaning "inherited knowledge", that has become an established term to denote Sámi traditional knowledge (Porsanger and Guttorm 2011a).

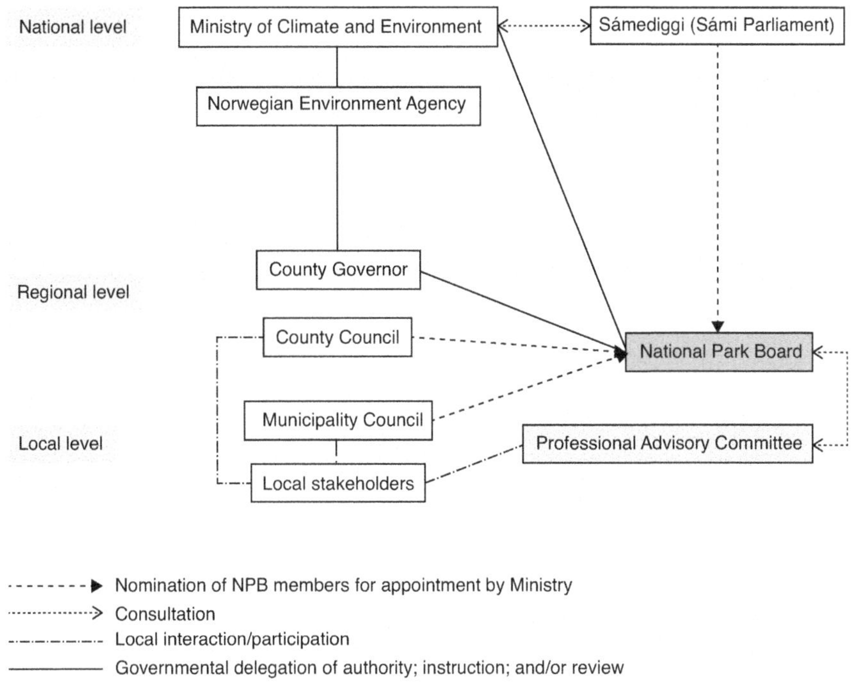

Fig. 5.5 Protected area governance and management in Norway (adapted from Risvoll et al. 2014)

of the Environment and Sámediggi – Sametinget 2007). The reform and its introduction of Sámi representation in NPBs facilitate influence also on specific management practices. In areas of importance for Sámi culture and industries, Sámi representation on NPBs is a requirement (Fauchald and Gulbrandsen 2012; Overvåg et al. 2016). In addition, local stakeholders, including reindeer herders or representatives of other Sámi interests, may participate in Professional Advisory Committees appointed by and invited to advise and collaborate with the NPBs. However, County Governors have the right to appeal decisions made by the NPBs, and the Ministry may revoke the NPBs' delegated authority and mandate, if it finds an NPB's decisions or activities to be inconsistent with relevant legislation or regulations (Prop. 1 S (2009–2010)).

In 2012, the Swedish Government authorized a newly established nonprofit organization to take over parts of the management of the Laponia World Heritage site in northern Sweden. This organization, Laponiatjuottjudus,[4] was formed by municipal, state, and Sámi stakeholders and its statutes stipulate Sámi majority in the organization's decision-making bodies. The living cultural heritage of the reindeer herding Sámi in the area was an important part of the justification for the

[4]*Tjuottjudus* is a term for management or administration in Lule Sámi, one of the Sámi languages spoken in the area where Laponia is situated.

inscription of Laponia on the World Heritage List, and the nine reindeer herding communities in the area had used this in their strategies to negotiate influence over the management of the site (Green 2009).

The Sámi reindeer herding communities' position as holders of traditional knowledge has worked to strengthen their demands for inclusion in, and influence over, the management of Laponia. Laponiatjuottjudus also explicitly engages with *árbediehtu* in the management of Laponia. It uses Sámi concepts and terminology to describe and structure its organization and work methods, and it engages in Sámi language revitalization and documentation of Sámi traditional knowledge in relation to the protection and conservation of Laponia (Reimerson 2016).

However, the level of Sámi influence and control achieved in Laponia is still an exception on the Swedish side of Sápmi. To a much greater extent, Norway has institutionalized Sámi rights in the general political system, and the 2010 reform applies to all larger protected areas. This provides avenues for Sámi influence that could serve as an example for conservation management on Indigenous lands elsewhere—but it also calls attention to other potential challenges. Norway's Protected Area management under the reform is organized largely in accordance with conventional Norwegian principles, similar to other state or municipal arrangements, and without explicit engagement with traditional Sámi organizational practices. The protection and promotion of Sámi traditional knowledge is a key element of the reform, but it encases Sámi participation in existing structures that are not necessarily conducive to alternative approaches—for example, based on Sámi traditional knowledge (Reimerson 2017).

5.2 Summary and Conclusions

Indigenous knowledge systems are best applied and practiced by IK custodians on land controlled or owned by them. The summaries above allude to several key enabling factors for enhanced integration of IK in conservation thereby contributing to close the knowledge-implementation gap:

- Recognition of Indigenous Peoples and their rights.
- Formal recognition of Indigenous-owned lands.
- The state valuing IKs for their conservation benefits.
- Devolution of decision-making power by the state to allow for Indigenous voices, knowledge systems, and priorities.

There has been some progress in the legal re-recognition and support of Indigenous land ownership and autonomous management in many parts of the world, such as Australia, parts of Africa, South America, and Norway. Although, as the situation in South America attests, pressure on Indigenous lands is likely to continue. Often, a fundamental tension persists in the very different underlying beliefs, methods, and priorities between Western conservation approaches founded on command-and-control attitudes, and Indigenous caring for land approaches founded on the agency of place and an ethic of respecting and encouraging life in all its forms. Nevertheless, in many situations, these knowledge types can also be considered complementary rather than in opposition (Asselin 2015), which can result in mutual benefits to science and Indigenous Peoples. Examples of such mutual benefits include fire management in places like Australia, Asia, and North America and freshwater management in New Zealand. The examples discussed in this chapter suggest that IKs can be applied in a spectrum from full deployment of traditional knowledge, such as in parts of Africa and Brazil, to blends of traditional and new knowledge, especially where Indigenous people have been forced, encouraged, or willingly adopted the ways of colonists. Partnerships and the creation of locally controlled programs such as co-management agreements, community-based natural resource management, and Indigenous Protected Areas are contributing to the rise in acceptance of the value and importance of IK in conservation (i.e., closing the knowledge-implementation gap). More generally, the detailed local biodiversity knowledge of IK holders opens opportunities for the implementation of decentralized and tailor-made conservation policies and actions on the ground. This is an important aspect of impactful conservation action because specific local, social, and cultural conditions vary considerably even at small geographical scales and there is not a "one size fits all" solution to conservation. Further, traditional and alternative Indigenous Peoples' approaches to resource management practices may lead to diversification of management regimes and methods. Finally, the holistic approach to nature conservation by Indigenous Peoples acknowledges the importance of species–species interactions, relational values (Russell et al. 2020) and the importance of the ecosystem as a whole. Conservation management plans could benefit more from this type of knowledge to transition from a species-centric model to an ecosystem model of conservation protection and stewardship.

External factors have also played a role in closing the knowledge-implementation gap. The signing of the *Convention on Biological Diversity* (1999) clearly had an impact on some countries in their dedication to inclusion of Indigenous people, knowledge, and land in conservation agendas. For example, Australia established their Indigenous Protected Area system (1997) and Sweden accepted Sámi co-management of Laponia. In continents such as Africa, Asia, and South America, international environmental NGOs are playing an important role in asserting the rights of Indigenous Peoples in broader conservation efforts. However, there have been notable conflicts between environmental and Indigenous approaches that can serve to further disempower and displace Indigenous Peoples, as occurred in some Protected Areas in Africa, Asia, and North America. New technologies such as digital devices, GPS (Global Positing Systems), GIS (Geographic Information Systems), and the Internet can also enhance and raise awareness of IK and its value in contemporary conservation (Herrmann et al. 2014; Ramirez-Gomez et al. 2016).

Despite international recognition of the value of Indigenous Peoples, knowledge, and lands significant gaps in the implementation of IK systems in "mainstream" conservation continue to exist. The pressures of economic development (e.g., from mining, farming, and forestry), population growth, the infiltration and dominance of Western practice, the lingering colonial perceptions of IK as myth, and the Western construct of Protected Areas often continue to widen the IK-implementation gap. Constant pressure from international to local organizations is required to maintain the invaluable and proven methods of Indigenous Peoples' sustainable environmental knowledge worldwide.

References

Abrams MD, Nowacki GJ (2008) Native Americans as active and passive promoters of mast and fruit trees in the eastern USA. The Holocene 18(7):1123–1137

Adams MJ (2014) Pukulpa pitjama Ananguku ngurakutu-Welcome to Anangu land: World Heritage at Uluru-Kata Tjuta National Park. In: Disko S, Tugendhat H (eds) World Heritage Sites and Indigenous Peoples' Rights. IWGIA, Forest Peoples Programme and Gundjeihmi Aboriginal Corporation, Copenhagen, Denmark, pp 289–311

Adams WM, Hutton J (2007) People, parks and poverty: political ecology and biodiversity conservation. Conserv Soc 5(2):147–183

Agrawal A (2002) Indigenous knowledge and the politics of classification. Int Soc Sci J 54(173):287–297

Åhrén M (2016) Indigenous peoples' status in the international legal system. Oxford University Press, New York

Alessa L, Kliskey A, Gamble J, Fidel M, Beaujean G, Gosz J (2016) The role of Indigenous science and local knowledge in integrated observing systems: moving toward adaptive capacity indices and early warning systems. Sustain Sci 11(1):91–102

Allard C (2011) The Nordic countries' law on Sámi territorial rights. Arc Rev Law Politics 2(2):159–183

Allard C (2015) Some characteristic features of Scandinavian laws and their influence on Sami matters. In: Allard C, Skogvang SF (eds) Indigenous rights in Scandinavia: autonomous Sami Law. Routledge, London, pp 49–64

Allard C, Skogvang SF (eds) (2015) Indigenous rights in Scandinavia: autonomous Sami Law. Routledge, London

Asch DL, Hart JP (2004) Crop domestication in prehistoric eastern North America. In: Goodman RM (ed) Encyclopedia of plant and crop science. Routledge, New York, pp 314–319

Asselin H (2015) Indigenous forest knowledge. In: Peh K, Corlett R, Bergeron Y (eds) Routledge handbook of forest ecology. Routledge, New York, pp 586–596

Australian Government (2021) Indigenous protected areas. https://www.environment.gov.au/land/indigenous-protected-areas. Accessed 6/8/2021

Awatere S, Robb M, Taura Y, Reihana K, Harmsworth G, Te Maru J, Watene-Rawiri E (2017) Wai Ora Wai Māori–a kaupapa Māori assessment tool. Landcare Research, Maaki Whenua, Policy Brief no 19

Barrow E, Murphree M (2001) Community conservation: from concept to practice. In: Hulme D, Murphree M (eds) African wildlife and livelihoods: The promise and performance of community conservation. James Currey, Oxford, pp 24–37

Bay-Larsen I (2010) The conservationists' concerns: on national administration response to integrated use and protection planning. Local Environ 15(4):357–371

Bennett NJ, Dearden P (2014) Why local people do not support conservation: community perceptions of marine protected area livelihood impacts, governance and management in Thailand. Mar Policy 44:107–116

Berkes F (2004) Rethinking community-based conservation. Conserv Biol 18(3):621–630

Brandt JS, Wood EM, Pidgeon AM, Han L-X, Fang Z, Radeloff VC (2013) Sacred forests are keystone structures for forest bird conservation in southwest China's Himalayan Mountains. Biol Conserv 166:34–42

Briggs J (2005) The use of indigenous knowledge in development: problems and challenges. Prog Dev Stud 5(2):99–114
Broch Hansen K, Jepsen K, Leiva Jacquelin P (eds) (2017) The Indigenous World 2017. The International Work Group for Indigenous Affairs (IWGIA), Copenhagen
Buergin R (2015) Contested rights of local communities and indigenous peoples in conflicts over biocultural diversity: the case of Karen communities in Thung Yai, a world heritage site in Thailand. Mod Asian Stud 49(6):2022–2062
Butler RA, Laurance WF (2009) Is Oil Palm the Next Emerging Threat to the Amazon? Trop Conserv Sci 2(1):1–10
Chazdon RL (2003) Tropical forest recovery: legacies of human impact and natural disturbances. Perspect Plant Ecol Evol Syst 6(1–2):51–71
Colchester M (2004) Indigenous peoples and communal tenures in Asia. Land Reform 1:28–45
Colombi BJ (2012) Salmon and the adaptive capacity of Nimiipuu (Nez Perce) culture to cope with change. Am Indian Q 36(1):75–97
Concu N (2013) Indigenous development through payments for environmental services in Arnhem Land, Australia: a critical analysis. In: Muradian R, Rival L (eds) Governing the provision of ecosystem services. Springer, New York, pp. 171–189
Coomes OT, Grimard F, Burt GJ (2000) Tropical forests and shifting cultivation: secondary forest fallow dynamics among traditional farmers of the Peruvian Amazon. Ecol Econ 32(1):109–124
Cunningham AB (2014) Applied ethnobotany: people, wild plant use and conservation. Routledge, New York
de la Torre L, Valencia R, Altamirano C, Ravnborg HM (2011) Legal and administrative regulation of palms and other NTFPs in Colombia, Ecuador, Peru and Bolivia. Bot Rev 77(4):327–369
DeGeorges PA, Reilly BK (2009) The realities of community based natural resource management and biodiversity conservation in Sub-Saharan Africa. Sustainability 1(3):734–788
Denevan W (1966) The aboriginal cultural geography of the Llanos de Mojos of Bolivia. Ibero-Americana 48
Denevan WM (1992) The pristine myth: the landscape of the Americas in 1492. Ann Assoc Am Geogr 82(3):369–385
DO (2008) Discrimination of the Sami – the rights of the Sami from a discrimination perspective (DO report no. 2008:1 eng). Stockholm
Dufour DL (1990) Use of tropical rainforests by native Amazonians. BioScience 40(9):652–659
Emery MR, Wrobel A, Hansen MH, Dockry M, Moser WK, Stark KJ, Gilbert JH (2014) Using traditional ecological knowledge as a basis for targeted forest inventories: Paper birch (*Betula papyrifera*) in the US Great Lakes region. J For 112(2):207–214
Ens E, Grech A (2018) Indigenous ranger programs are working in Queensland – they should be expanded. The Conversation. https://theconversation.com/indigenous-ranger-programs-are-working-in-queensland-they-should-be-expanded-89766
Ens EJ, Finlayson M, Preuss K, Jackson S, Holcombe S (2012) Australian approaches for managing 'country'using Indigenous and non-Indigenous knowledge. Ecological Management Restoration 13(1):100–107
Ens E, Burns E, Russell-Smith J, Sparrow B, Wardle G (2014) The cultural imperative: broadening the vision of long-term ecological monitoring to enhance environmental policy and management outcomes. Biodiversity and environmental change: Monitoring, challenges and direction 83
Ens EJ, Pert P, Clarke PA, Budden M, Clubb L, Doran B, Douras C, Gaikwad J, Gott B, Leonard S, Locke J, Packer J, Turpin G, Wason S (2015) Indigenous biocultural knowledge in ecosystem science and management: review and insight from Australia. Biol Conserv 181:133–149
Eriksen C (2007) Why do they burn the 'bush'? Fire, rural livelihoods, and conservation in Zambia. Geogr J 173(3):242–256
Eyong CT (2007) Indigenous knowledge and sustainable development in Africa: Case study on Central Africa. In: Boon EK, Hens L (eds) Indigenous knowledge systems and development: Relevance for Africa. Kamla-Raj Enterprises, New Delhi, pp. 121–139

Fauchald OK, Gulbrandsen LH (2012) The Norwegian reform of protected area management: a grand experiment with delegation of authority? Local Environ 17(2):203–222

Fernández-Llamazares A, Helle J, Eklund J, Balmford A, Moraes RM, Reyes-García V, Cabeza M (2018) New law puts Bolivian biodiversity hotspot on road to deforestation. Curr Biol 28(1): R15–R16

Finer M, Orta-Martínez M (2010) A second hydrocarbon boom threatens the Peruvian Amazon: trends, projections, and policy implications. Environ Res Lett 5:014012

Garibaldi A, Turner N (2004) Cultural keystone species: implications for ecological conservation and restoration. Ecol Soc 9(3):1

Garnett ST, Burgess ND, Fa JE, Fernández-Llamazares Á, Molnár Z, Robinson CJ, Watson JEM, Zander KK, Austin B, Brondizio ES et al (2018) A spatial overview of the global importance of Indigenous lands for conservation. Nat Sustain 1(7):369–374

Gaventa J, Cornwall A (2008) Power and knowledge. In: Reason P, Bradbury H (eds) The Sage handbook of action research: participative inquiry and practice. SAGE, Thousand Oaks, pp 172–189

Ghimire SK, McKey D, Aumeeruddy-Thomas Y (2004) Heterogeneity in ethnoecological knowledge and management of medicinal plants in the Himalayas of Nepal: implications for conservation. Ecol Soc 9(3):6

Gichuki NN, Adams NJ, Slotow RH (1999) The contribution of indigenous knowledge to bird conservation. Johannesburg: BirdLife South Africa. Durban: Proc 22:1345–1350

Gladu JP, Brubacher D, Cundiff B, Baggio A, Bell A, Gray T (2003) Honouring the promise: aboriginal values in protected areas in Canada. National Aboriginal Forestry Association

Glover RB, Hunt TM, Severne CM (1996) Ohaaki Ngawha; Ohaaki Pool. In Proceedings 18th NZ Geothermal Workshop, pp 77–84

Goodall H (2001) Renewing country: aboriginal people and their lands in rangeland environments. Rangeland J 23(1):99–115

Green C (2009) Managing Laponia: a world heritage site as arena for Sami ethno-politics in Sweden. Thesis. Uppsala University, Uppsala, Sweden

Greiner R (2010) Payments for environmental services (PES): contribution to Indigenous livelihoods. In: Aravossis K, Brebbia CA (eds) Environmental economics and investment assessment Iii. Wit Press, Southampton, pp 163–174

Hansen KL, Melhus M, Høgmo A, Lund E (2008) Ethnic discrimination and bullying in the Sami and non-Sami populations in Norway: the SAMINOR study. Int J Circumpolar Health 67(1):97–113

Harmsworth G, Awatere S, Robb M (2016) Indigenous Māori values and perspectives to inform freshwater management in Aotearoa-New Zealand. Ecol Soc 21(4):9

Harrison R (2004) Shared landscapes: archaeologies of attachment and the pastoral industry in New South Wales. UNSW Press

Healy RG (2007) Parks and people in North America—one hundred and thirty five years of change. Policy Matters 15:261–272

Heckenberger MJ, Kuikuro A, Kuikuro UT, Russell JC. Schmidt M, Fausto C, Franchetto B (2003) Amazonia 1492: Pristine forest or cultural parkland? Science 301(5640):1710–1714

Heckenberger MJ, Russell JC, Fausto C, Toney JR, Schmidt MJ, Pereira E, Franchetto B, Kuikuro A (2008) Pre-Columbian Urbanism, Anthropogenic Landscapes, and the Future of the Amazon. Science 321(5893):1214–1217

Heinämäki L, Herrmann TM, Neumann A (2014) The protection of the culturally and spiritually important landscapes of arctic indigenous peoples under the convention on biological diversity and first experiences from the application of the Akwé: Kon guidelines in Finland. Yearbook Polar Law Online 6(1):189–225

Hellenic Ministry of the Environment, Energy and Climate Change (2014) National biodiversity strategy and action plan

Henriksen JB (2008) The continuous process of recognition and implementation of the Sami people's right to self-determination. Camb Rev Int Aff 21(1):27–40

Hernández-Morcillo M, Hoberg J, Oteros-Rozas E, Plieninger T, Gómez-Baggethun E, Reyes-García V (2014) Traditional ecological knowledge in Europe: status Quo and insights for the environmental policy agenda. Environ Sci Policy Sustain Dev 56(1):3–17
Herrmann TM, Sandström P, Granqvist K, D'Astous N, Vannar J, Asselin H, Saganash N, Mameamskum J, Guanish G, Loon J-B, Cuciurean R (2014) Effects of mining on reindeer/caribou populations and indigenous livelihoods: community-based monitoring by Sami reindeer herders in Sweden and First Nations in Canada. Polar J 4(1):28–51
Hill R, Pert P, Davies J, Robinson C, Walsh F, Falco-Mammone F (2013) Indigenous Land Management in Australia: extent, scope, diversity, barriers and success factors. CSIRO Ecosystem Sciences, Cairns
Holmgren L, Sandström C, Zachrisson A (2017) Protected area governance in Sweden: new modes of governance or business as usual? Local Environ 22(1):22–37
Hovik S, Sandström C, Zachrisson A (2010) Management of protected areas in Norway and Sweden: challenges in combining central governance and local participation. J Environ Policy Plan 12(2):159–177
Huntington HP (2000) Using traditional ecological knowledge in science: methods and applications. Ecol Appl 10(5):1270–1274
Jones BT (2015) Community conservation in Africa-successes and failures – lessons for the current poaching crisis. In: Conference on Conservation and Communities in Africa, organized by the Centre d'Estudis Africans (CEA) in Barcelona, Barcelona
Jones BTB, Murphree MW (2004) Community-based natural resource management as a conservation mechanism: Lessons and directions. In: Child B (ed) Parks in Transition - Biodiversity, Rural Development and the Bottom Line. Routledge, London. pp. 63–103.
Josefsen E (2010) The Saami and the national parliaments - channels for political Influence. Inter-Parliamentary Union (IPU) and United Nations Development Programme (UNDP). http://archive.ipu.org/splz-e/chiapas10/saami.pdf
Kijas J, NSW Department of Environment and Climate Change, and NSW Department of Environment and Conservation (2005) Revival, renewal & return: Ray Kelly & the NSW Sites of Significance Survey. NSW Department of Environment and Conservation, Hurstville
Kimmerer RW, Lake FK (2001) The role of Indigenous burning in land management. J For 99(11):36–41
Korpijaakko-Labba K (2005) Samernas rätt till mark och vatten: utvecklingen från 1500-talets mitt fram till våra dagar. In: Bladh G, Kuvaja C (eds) Svenskt i Finland – finskt i Sverige. 1, Dialog och särart: människor, samhällen och idéer från Gustav Vasa till nutid 2005. Svenska litteratursällskapet i Finland, Helsingfors, pp. 320–374
Lalonde A (1993) African indigenous knowledge and its relevance to sustainable development. In: Inglis J (ed) Traditional ecological knowledge: concepts and cases. International development Research Centre, Ottawa, pp 55–62
Lama Y, Ghimire S, Thomas Y (2001) Medicinal plants of Dolpo: Amchis' knowledge and conservation. WWF Nepal Program, Kathmandu
Lantto P (2010) Borders, citizenship and change: the case of the Sami people, 1751–2008. Citizsh Stud 14(5):543–556
Lantto P, Mörkenstam U (2007) Sami rights and Sami challenges. Scand J Hist 33(1):26–51
LaRiviere CM, Crawford SS (2013) Indigenous principles of wild harvest and management: an Ojibway community as a case study. Hum Ecol 41(6):947–960
Laue JE, Arima EY (2016) Spatially explicit models of land abandonment in the Amazon. J Land Use Science 11(1):48–75
Li TM (2002) Engaging simplifications: community-based resource management, market processes and state agendas in upland Southeast Asia. World Dev 30(2):265–283
Li S (2014) Management and outside interventions of community-conserved areas. Peking University Press, Beijing

Little PD, Brokensha DW (1989) Local institutions, tenure and resource management in East Africa. In: Anderson D, Grove RH (eds) Conservation in Africa: people, policies and practice. Cambridge University Press, Cambridge, pp 193–209

Lu D-J, Chueh H-C, Kao C-W (2012) Why they cannot work together: a study of the co-management of natural resources with indigenous people in Taiwan. Soc Nat Resour 25(1):105–112

Lyver POB, Moller H (2012) An alternative reality: Māori spiritual guardianship of New Zealand's Native Birds. In: Tidemann S, Gosler A (eds) Ethno-ornithology: birds, indigenous peoples. Culture and Society. Routledge, New York, pp 241–264

Lyver POB, Taputu TM, Kutia ST, Tahi B (2008) Tūhoe Tuawhenua mātauranga of kererū (Hemiphaga novaseelandiae novaseelandiae) in Te Urewera. N Z J Ecol 32(1):7–17

Macía MJ (2008) Woody plants diversity, floristic composition and land use history in the Amazonian rain forests of Madidi National Park, Bolivia. Biodivers Conserv 17(11):2671–2690

Mahanty S, Fox J, Nurse M, Stephen P, McLees L (eds) (2006) Hanging in the balance: equity in community-based natural resource management in Asia. Regional Community Forestry Training Center for Asia and the Pacific, Bangkok

Mapira J, Mazambara P (2013) Indigenous knowledge systems and their implications for sustainable development in Zimbabwe. J Sustain Dev Africa 15(5):90–106

Marsden M, Henare TA (1992) Kaitiakitanga: a definitive introduction to the holistic world view of the Māori. Ministry for the Environment

Martin T (2016) Beyond the protection of the Land, National Parks in the Canadian Arctic: a way to actualized and institutionalized aboriginal cultures in the global. In: Herrmann TM, Martin T (eds) Indigenous peoples' governance of land and protected territories in the Arctic. Springer, New York, pp 167–187

McSweeny K, Arps S (2005) A "demographic turnaround": the rapid growth of indigenous populations in Lowland Latin America. Lat Am Res Rev 40(1):3–29

Merlin-Uribe Y, Gonzalez-Esquivel CE, Contreras-Hernandez A, Zambrano L, Moreno-Casasola P, Astier M (2013) Environmental and socio-economic sustainability of chinampas (raised beds) in Xochimilco, Mexico City. Int J Agric Sustain 11(3):216–233

MfCH (2017a) Treaty events 1800–1849. In: Ministry for Culture and Heritage

MfCH (2017b) Treaty events 1850–99. In: Ministry for Culture and Heritage, editor

Minde H (2001) Sami land rights in Norway: a test case for Indigenous peoples. Int J Minor Group Rights 8:107–125

Minde H (2003) Assimilation of the Sami – implementation and consequences. Acta Borealia 20(2):121–146

Ministry of the Environment and Sámediggi - Sametinget (2007) Avtale mellom Sametinget og Miljøverndepartementet om retningslinjer for verneplanarbeid etter naturvernloven i samiske områder. Ministry of the Environment, Oslo

Mistry J, Bilbao BA, Berardi A (2016) Community owned solutions for fire management in tropical ecosystems: case studies from Indigenous communities of South America. Phil Trans R Soc B Biol Sci 371(1696):20150174

Moeke-Pickering TM (1996) Maori identity within whanau: A review of literature. University of Waikato, Hamilton, New Zealand

Moller H, Berkes F, O'Brian Lyver P, Kislalioglu M (2004) Combining science and traditional ecological knowledge: monitoring populations for co-management. Ecol Soc 9(3):2

Moorcroft H (2015) Paradigms, paradoxes and a propitious niche: conservation and Indigenous social justice policy in Australia. Local Environ 21(5):591–614

Moorcroft H, Ignjic E, Cowell S, Goonack J, Mangolomara S, Oobagooma J, Karadada R, Williams D, Waina N (2012) Conservation planning in a cross-cultural context: the Wunambal Gaambera Healthy Country Project in the Kimberley, Western Australia. Ecol Manage Restor 13(1):16–25

Mörkenstam U (2005) Indigenous peoples and the right to self-determination: the case of the Swedish Sami People. Canadian J Native Stud XXV:433–461

Muller S (2008) Indigenous payment for environmental service (PES) opportunities in the Northern Territory: negotiating with customs. Aust Geogr 39(2):149–170
Mwangi E (1988) Indigenous knowledge and environmental conservation in east Africa: annotated bibliography. School of Public and Environmental Affairs, Indiana University
Neale T, Carter R, Nelson T, Bourke M (2019) Walking together: a decolonising experiment in bushfire management on Dja Dja Wurrung country. Cultural Geogr 26(3):341–359
Notzke C (1994) Aboriginal peoples and natural resources in Canada. Captus Press, Concord (ON)
Ocholla GO, Mireri C, Muoria PK (2016) Application of indigenous knowledge systems in wildlife management: a case study of the Samburu Pastoral Community in Kenya. Int J Appl Sci Technol 6(1):72–80
Orta M, Finer M (2010) Oil frontiers and indigenous resistance in the Peruvian Amazon. Ecol Econ 70(2):207–218
Overvåg K, Skjeggedal T, Sandström C (2016) Management of mountain areas in Norway and the persistence of local–national conflicts. J Environ Plan Manag 59(7):1186–1204
Pacheco P, Mejía E, Cano W, De Jong W (2016) Smallholder forestry in the Western Amazon: outcomes from forest reforms and emerging policy perspectives. Forests 7(9):193
Paneque-Gálvez J, Pérez-Llorente I, Luz AC, Guèze M, Mas J-F, Macía MJ, Orta-Martínez M, Reyes-García V (2018) High overlap between traditional ecological knowledge and forest conservation found in the Bolivian Amazon. Ambio 47(8):908–923
Parlee B, Berkes F, Teetl'it Gwich'in Renewable Resources Council (2005) Health of the land, health of the people: a case study on Gwich'in berry harvesting in northern Canada. EcoHealth 2(2):127–137
Pérez-Llorente I, Paneque-Gálvez J, Luz AC, Guéze M, Pino J, Macía MJ, Domínguez-Gómez JA, Reyes-García V (2013) Changing indigenous cultures, economies and landscapes. The case of the Tsimane', Bolivian Amazon. Landsc Urban Plan 120:147–157
Persoon G, Van Est DME, Sajise PE (2003) Co-management of natural resources in Asia: a comparative perspective. Nias Press
Peters C (2000) Precolumbian sylviculture and indigenous management of Neotropical forests. In: Lentz D (ed) Imperfect balance: landscape transformations in the Precolumbian Americas. Columbia University Press, New York, pp 203–223
Porsanger J (2011) The problematisation of the dichotomy of modernity and tradition in Indigenous and Sami contexts. In: Porsanger J, Guttorm G (eds) Working with traditional knowledge: communities, institutions, information systems, law and ethics. Sámi allaskuvla, Guovdageaidnu, pp 225–252
Porsanger J, Guttorm G (2011a) Building up the field of study and research on Sami traditional knowledge (árbediehtu). In: Porsanger J, Guttorm G (eds) Working with traditional knowledge: communities, institutions, information systems, law and ethics. Sámi allaskuvla, Guovdageaidnu, pp 13–57
Porsanger J, Guttorm G (eds) (2011b) Working with traditional knowledge: communities, institutions, information systems, law and ethics. Writings from the Árbediehtu pilot project on documentation and protection of Sami traditional knowledge. Sámi allaskuvla, Guovdageaidnu
Porter-Bolland L, Ellis EA, Guariguata MR, Ruiz-Mallén I, Negrete-Yankelevich S, Reyes-García V (2012) Community managed forest and forest protected areas: An assessment of their conservation effectiveness across the tropics. Forest Ecol Manage 268:6–17
Posey DA (1985) Indigenous management of tropical forest ecosystems: the case of the Kayapó indians of the Brazilian Amazon. Agrofor Syst 3(2):139–158
Price K, Roburn A, MacKinnon A (2009) Ecosystem-based management in the Great Bear Rainforest. For Ecol Manag 258(4):495–503
Procedures for Consultations between State Authorities and the Sami Parliament (2005)
Prop. 1 S (2009–2010) Proposisjon til Stortinget (forslag til stortingsvedtak) for budsjettåret 2010, Miljøverndepartementet

Ramirez-Gomez SOI, Brown G, Verweij PA, Boot R (2016) Participatory mapping to identify indigenous community use zones: Implications for conservation planning in southern Suriname. J Nat Conserv 29:69–78
Reimerson E (2016) Sami space for agency in the management of the Laponia World Heritage site. Local Environ 21(7):808–826
Reimerson E (2017) International arenas, local space for agency and national discourses as mediator: protected areas in Swedish and Norwegian Sápmi. In: Elenius L, Allard C, Sandström C (eds) Indigenous rights in modern landscapes: nordic conservation regimes in global context. Routledge, London, pp 167–184
Reyes-Garcia V, Ruiz-Mallen I, Porter-Bolland L, Garcia-Frapolli E, Ellis EA, Mendez M-E, Pritchard DJ, Sanchez-Gonzalez M-C (2013) Local understandings of conservation in South-eastern Mexico and their implications for community-based conservation as an Alternative Paradigm. Conserv Biol 27(4):856–865
Risvoll C, Fedreheim GE, Sandberg A, BurnSilver S (2014) Does pastoralists' participation in the management of national parks in Northern Norway contribute to adaptive governance? Ecol Soc 19(2):71
Roth R (2004) Spatial organization of environmental knowledge: conservation conflicts in the inhabited forest of northern Thailand. Ecol Soc 9(3):5
Russell S, Ens E, Rangers NY (2020) Connection as country: relational values of billabongs in indigenous northern Australia. Ecosyst Serv 45:101169
Russell-Smith J, Whitehead P, Cooke P (2009) Culture, ecology and economy of fire management in north Australian savannas: rekindling the Wurrk tradition. CSIRO, Collingwood
Shaffer LJ (2010) Indigenous fire use to manage savanna landscapes in southern Mozambique. Fire Ecol 6(2):43–59
Shen X, Tan J (2012) Ecological conservation, cultural preservation, and a bridge between: the journey of Shanshui conservation center in the Sanjiangyuan region, Qinghai-Tibetan Plateau, China. Ecol Soc 17(4):38
Shen X, Li S, Wang D, Lu Z (2015) Viable contribution of Tibetan sacred mountains in southwestern China to forest conservation. Conserv Biol 29:1518–1526
Shivakoti G, Pradhan U, Helmi H (2016) Redefining diversity and dynamics of natural resources management in Asia. In: Sustainable natural resources management in dynamic Asia, vol 1. Elsevier, London
Sivarajan V, Balachandran I (1994) Ayurvedic drugs and their plant sources. Oxford and IBH Publishing
Smith SP (1913) The Lore of the Whare-wananga. New Plymouth 2
Smyth D (1994) Understanding country: the importance of land and sea in Aboriginal and Torres Strait Islander societies. Australian Government Publishing Service, Canberra
Smyth D, Taylor P, Willis A (eds) (1985) Aboriginal ranger training and employment in Australia. Australian National Parks and Wildlife Service, Canberra
Stevens S (1997) Annapurna conservation area: empowerment, conservation, and development in Nepal. In: Stevens S (ed) Conservation through cultural survival: Indigenous peoples and protected areas. Island Press, Washington DC, pp 237–261
Sudtongkong C, Webb EL (2008) Outcomes of state-vs community-based mangrove management in southern Thailand. Ecol Soc 13(2):27
Sunderlin WD, Angelsen A, Belcher B, Burgers P, Nasi R, Santoso L, Wunder S (2005) Livelihoods, forests, and conservation in developing countries: an overview. World Dev 33(9):1383–1402
Swedish Sami Parliament. 2009. The Sami Parliament's Living Environment Program: Eallinbiras. Kiruna
Tang C-P, Tang S-Y (2010) Institutional adaptation and community-based conservation of natural resources: the cases of the Tao and Atayal in Taiwan. Hum Ecol 38(1):101–111
Te Rito JS (2007) Whakapapa: a framework for understanding identity. MAI Review LW 1:10

Tendeng B, Asselin H, Imbeau L (2016) Moose (*Alces americanus*) habitat suitability in temperate deciduous forests based on Algonquin traditional knowledge and on a habitat suitability index. Ecoscience 23(3–4):77–87

Thornton TF (2010) A Tale of Three Parks: Tlingit Conservation, Representation, and Repatriation in Southeast Alaska's National Parks. Hum Organ 69(2):107–118

Tipa G, Teirney LD (2003) A cultural health index for streams and waterways: indicators for recognising and expressing Māori values. Ministry for the Environment

Tipa G, Teirney LD (2006) A cultural health index for streams and waterways: a tool for nationwide use. Ministry for the Environment Wellington

Toledo VM (2001) Indigenous peoples and biodiversity. In: Levin SA (ed) Encyclopedia of biodiversity. Academic Press, New York, pp 1181–1197

Trosper RL, Clark F, Gerez-Fernandez P, Lake F, McGregor D, Peters CM, Purata S, Ryan T, Thomson A, Watson AE, Wyatt S (2012) North America. In: Parrotta JA, Trosper RL (eds) Traditional forest-related knowledge: sustaining communities, ecosystems and biocultural diversity. Springer Science & Business Media, New York, pp 157–201

Tunón H, Kvarnström M, Malmer P (2015) Report from the project: indigenous and Local Knowledge in a Scoping Study for a Nordic IPBES Assessment. Uppsala

Turner S (2004) A crisis in CBNRM? Affirming the commons in southern Africa. In: 10th IASCP Conference Oaxaca, pp 9–13

Turner NJ, Gregory R, Brooks C, Failing L, Satterfield T (2008) From invisibility to transparency: identifying the implications. Ecol Soc 13(2):7

UNHRC (2011) Report of the Special Rapporteur on the rights of indigenous peoples, Addendum: The situation of the Sami people in the Sápmi region of Norway, Sweden and Finland. A/HRC/18/35/Add.2. United Nations Human Rights Council

Uprety Y, Asselin H, Boon EK, Yadav S, Shrestha KK (2010) Indigenous use and bio-efficacy of medicinal plants in the Rasuwa District, Central Nepal. J Ethnobiol Ethnomed 6(1):3

Uprety Y, Asselin H, Bergeron Y, Doyon F, Boucher J-F (2012) Contribution of traditional knowledge to ecological restoration: practices and applications. Ecoscience 19(3):225–237

Uprety Y, Asselin H, Bergeron Y (2013) Cultural importance of white pine (*Pinus strobus* L.) to the Kitcisakik Algonquin community of western Quebec, Canada. Can J For Res 43(6):544–551

Valdez GVV, Hansen EN, Bliss J (2012) Factors impacting marketplace success of community forest enterprises: the case of TIP Muebles, Oaxaca, México. Small-Scale For 11(3):339–363

Van Dam C (2000) Two decades of participatory forest development...but how participatory? Forest Trees People 42:11–17

Van Dam C (2011) Indigenous territories and REDD in Latin America: opportunity or threat? Forests 2(1):394–414

Watson A, Matt R, Knotek K, Williams DR, Yung L (2011) Traditional wisdom: protecting relationships with wilderness as a cultural landscape. Ecol Soc 16(1):36

Wiersum KF (1997) Indigenous exploitation and management of tropical forest resources: an evolutionary continuum in forest-people interactions. Agric Ecosyst Environ 63(1):1–16

Wiersum KF (2004) Forest gardens as an 'intermediate' land-use system in the nature-culture continuum: characteristics and future potential. Agrofor Syst 61:123–134

Xu J, Ma ET, Tashi D, Fu Y, Lu Z, Melick D (2005) Integrating sacred knowledge for conservation: cultures and landscapes in southwest China. Ecol Soc 10(2):7

Part III
The Knowledge Mediation Sphere

Chapter 6
The Knowledge Network: Identifying Actors and Structural Dimensions of Knowledge Transfer

R. Patrick Bixler

Contents

6.1 Introduction

This chapter addresses barriers, opportunities, and strategies for transferring conservation knowledge across spatiotemporal scales and stakeholders by highlighting four key concepts: knowledge, scale, networks, and bridging organizations. The

R. P. Bixler (✉)
The LBJ School of Public Affairs, University of Texas at Austin, Austin, TX, USA
e-mail: rpbixler@utexas.edu

C. C. Ferreira, C. F. C. Klütsch (eds.), *Closing the Knowledge-Implementation Gap in Conservation Science*, Wildlife Research Monographs 4,
https://doi.org/10.1007/978-3-030-81085-6_6

conceptual framework presented is developed through empirical research in North America, yet the identification of these key variables, as well as the lessons learned, are generalizable to the challenges of knowledge transfer in conservation science.

Knowledge is at once the most crucial and elusive aspect of a sustainable future. The structures and processes that generate, disseminate, and implement scientific knowledge of the environment into policy and practice are not well understood. Rather than reflecting an objective reality, scientific knowledge itself needs to be understood as embedded in a "social" context, i.e., social practices, identities, norms, conventions, discourses, instruments, and institutions (Jasanoff 2004). Scientific knowledge also needs to be understood as embedded in social structures by which it is created, disseminated, and absorbed, i.e., knowledge networks. By doing so, we can better understand the opportunities and barriers to the knowledge-implementation gap in social–ecological systems and conservation science.

Social–ecological connections are complex involving multiple and competing resource values (e.g., recreation, resource extraction, esthetic), experiential (e.g., actions such as moving and living on the land), cognitive (e.g., beliefs, norms, attitudes), emotional and ontological. Knowledge systems are neither purely abstract nor purely embedded in place but go beyond the local level to include multilevel actors, institutions, and networks (Bixler et al. 2016), thus requiring governance arrangements to integrate a plurality of interests, opinions, and values with regard to human–nature connections at different scales (Duraiappah et al. 2014).

Collaborative, polycentric, and network approaches are increasingly theorized in the environmental governance literature, generally summarized as approaches to adaptive governance (Bixler 2014; Bixler et al. 2018). This conceptual framework commonly emphasizes an integrative approach to governance that transcends land tenure and jurisdictional boundaries as well as the constraints of spatial and temporal scales (Folke et al. 2005). Adaptive governance encourages collaboration that builds social capital and innovation and multilevel learning and networks that promote effective governance and adaptation (Pahl-Wostl 2009; Pahl-Wostl et al. 2012). These approaches promise improved decision-making under uncertainty, collective learning and action, long-term experimentation, and inclusive decision-making and governance (Folke et al. 2005).

Although scholars have contributed to the theoretical and conceptual framings of the links between governance and knowledge strategies, our empirical understanding remains limited (Fischer et al. 2012; Holley et al. 2013). Little is known about how, and under what conditions, governance arrangements or attributes lead to effective knowledge transfer across spatiotemporal scales and/or different stakeholders. Many have questioned the link between knowledge, social learning, and collaborative governance with social or environmental outcomes (Cundill and Rodela 2012; Ojha et al. 2013). This chapter advances the conversation on knowledge networks and scale and will proceed as follows. First, the theoretical and conceptual framing of knowledge and scale will be presented. This will be followed by two aspects of governance—boundary organizations and knowledge networks—that guide navigation of the scale and knowledge challenge. This is followed by three empirical case studies in the U.S. intermountain west region known as the Crown of the Continent. The first case study focuses on invasive species management; the second is focused on grizzly bear (*Ursus arctos horribilis*) conservation and the third looks at climate adaptation. The case studies provide some insight into the barriers, opportunities, and strategies for transferring conservation knowledge across spatio-temporal scales and stakeholders.

6.2 Theoretical and Conceptual Underpinnings of Knowledge and Scale in Conservation Science

In order to make sense of the empirical examples presented below, some theoretical and conceptual background is required. The following background will include literature from four thematic areas: knowledge, scale, knowledge networks, and bridging organizations. Considered cross-cutting themes of importance to closing the knowledge-implementation gap across spatiotemporal scales, they form a two-by-two matrix, diagramed in Fig. 6.1. Based on extensive research, this chapter will discuss a number of empirical examples that fit into the different quadrants (found in Fig. 6.1).

6.2.1 The Case for Socially Embedded Knowledge in Conservation Science

One common characteristic in the collaborative and adaptive forms of governance is the role that knowledge, knowing, and learning plays in being "collaborative" and "adaptive". The production and legitimatization of objective knowledge vis-à-vis science is an important topic of scholarly interest. The traditional notion of science removes objects from their specific contexts in order to make parts meaningful independent of wholes (Jasanoff 2010). Scientific knowledge focuses on relationships and phenomena that do not vary across space and time, is reductionist and looks at relationships between specific variables. To effectively do so, the production of science tends to erase specificity and transcend the subjective and contingent

Fig. 6.1 Conceptual matrix for closing the knowledge-implementation gap across spatiotemporal scales

	Knowledge	Scale
Bridging Organization		
Knowledge network		

circumstances of the local. In this way, "facts" may float freely and carry legitimacy through objectivity (Latour 1987). Perhaps more than any other field of scientific inquiry, conservation science is socially embedded in everyday activities, which highlights that science is "co-produced" by various sectors of society, and separation of "facts" and "values" cannot be achieved (Pielke 2004). Social practices, for example, private landowner ranching activities in a private–public land landscape (see grizzly bear case below), shape ecosystem dynamics under scientific investigation and conservation policy prescriptions often aim to restrict social practices deemed problematic. Yet, frequently local knowledge constitutes valuable stores of knowledge that bridge the knowledge-implementation gap since local knowledge can frequently be applied to local conservation issues.

How can we think about conservation knowledge in this context? When we transition from considering knowledge of "what is" to "what should be done" the boundary between expertise and knowledge on the one hand and sentiments and values on the other become blurred (Carolan 2006). Tensions arise when impersonal and apolitical science comes into conflict with the subjective, situated, and normative individuals interacting with nature (Jasanoff 2010). How does a local knowledge system embedded in "public sentiments and values" inform conservation science and policy? The linear model leading from "more science" to "less uncertainty" to "political action" is inherently flawed (Sarewitz 2004). The box enclosing this push and pull of legitimate knowledge and the appropriate way to do science has been "opened up," argue some, in the discussions of post-normal science (Funtowicz and Ravetz 1993), trans-science (Weinberg 1972), and the democratization of science (Carolan 2006). Yet, scientific knowledge of environmental change remains paramount to solving conservation problems of today. Moving from knowledge to implementation requires an understanding of how knowledge—scientific and local—moves through networks of various stakeholders and across scales in different contexts.

6.2.2 Scale and Cross-Scale Linkages in Conservation Science

The connection between (spatiotemporal) scale and knowledge transfer is important for conservation science. While much debate surrounds the degree to which "scale" is a social construction versus an objective observation, there is broad normative agreement that "cross-scale linkages" are critical for bridging the knowledge-implementation gap. Cross-scale linkages are defined as "social, institutional, or ecological connections among individuals or organizations. Such connections may be horizontal (e.g., across geographical space) or vertical (e.g., across different levels of organization)" (Armitage et al. 2009: p. 96). The literature contends that cross-scale linkages that link multiple levels (e.g., community institutions with other levels of organization) are critical for adaptive governance (Carlsson and Sandström 2008);

however, less empirically understood is how knowledge networks mitigate the problem of "institutional fit" in conservation science to fit the scale of the problem and solution (Young 2006). Challenges arise because these social and ecological systems operate on fundamentally different temporal and spatial scales complicating the design and implementation of institutions for effective conservation, particularly conservation issues that require consideration at multiple levels (Cash and Moser 2000; Sayre 2005).

Cash and Moser (2000) note three challenges of scale that are relevant for conservation science:

1. Matching scales of biogeophysical systems with scales of management systems (an institutional fit problem)
2. Avoiding scale discordance (matching the scale of ecological assessment with the scale of management)
3. Accounting for cross-scale dynamics (understanding the linkages between scales, and how they affect decision-making information flows, and the integration of information into the decision-making process)

These challenges have been well documented in the literature (Cumming et al. 2006; Ekstrom and Young 2009; Ostrom 1990). An inherent tension exists between scale and levels of governance in conservation science because of differences in social organization, in how information is generated, shared, and applied, and the ways that power is exercised to coordinate conservation action. Who should undertake what activities at which level and how organizations network together collectively across complex jurisdictional, cultural, and ecological boundaries are enduring questions in environmental governance (Lemos and Agrawal 2006). There is a "paradox of scale" at play, where on the one hand, there is the normative need to address large-scale global or regional conservation and on the other hand, a primacy on community-scale collaboration (Wyborn and Bixler 2013). Few studies have produced a grounded analysis in the way that cross-scale linkages bridge the knowledge-implementation gap in conservation science. Two mechanisms to negotiate the interplay of knowledge and scale in conservation science are bridging organizations and knowledge networks.

6.2.3 *Knowledge Networks in Conservation Science*

Networks have been evoked as both an explanatory tool and as an outcome of a broad spectrum of social processes and have been described as an essential feature of effective conservation. The network perspective assumes that:

1. Relationships among actors are important.
2. Actors are interdependent rather than autonomous.
3. A relationship between two actors represents a flow of material or nonmaterial resources.

4. Network structures enhance or inhibit actors' ability to act (Wasserman and Faust 1995).

More specifically, knowledge networks are conceptualized as a set of nodes—individuals or higher-level collectives that serve as heterogeneously distributed repositories of knowledge and agents that search for, transmit, and create knowledge—that are interconnected by social relationships that enable and constrain efforts to acquire, transfer, and create knowledge (Phelps et al. 2012). Knowledge networks are inherently multilevel and influence the processes of production, diffusion, and absorption of knowledge as well as the efficacy and efficiency by which knowledge is accessed and applied (Phelps et al. 2012). Innovation is the result of the interaction among several actors belonging to diverse sectors and engaged in reciprocal, preferential, and supportive actions (Powell 1990). Interaction in networks is an important means of gaining and transferring new knowledge, gathering relevant information about new organizations, and finding external support and services.

Some aspects of knowledge networks that will be discussed in this chapter include weak ties (Granovetter 1973, 1983) and structural holes (Burt 1992, 2004). Granovetter's Strength of Weak Ties (SWT) argument asserts that our acquaintances (weak ties) are less likely to be socially involved with one another than are our close friends (strong ties) (Granovetter 1973). In network terms, the SWT theory translates to say that people tend to have stronger ties with people who are similar to themselves (McPherson et al. 2001). Strong ties characterize a dense cluster of actors who are all mutually connected to each other. Since this subcluster of strongly connected actors is likely to interact frequently, much of the information circulating in this social system is redundant. Conversely, weak ties are characterized by the relative infrequency of interaction between focal organization and contacts and enable the discovery of opportunities because they serve as bridges to potential sources of novel ideas and innovation (Borgatti and Halgin 2011; Granovetter 1973). The connecting medium of bridging weak ties lubricates the exchange of ideas and innovation by connecting cohesive subgroups that are not too cohesive to be entirely closed off (Crona and Hubacek 2010). Weak ties are less redundant and more flexible than strong ties and can bridge longer distances within a network, and thus, provide new information and knowledge for the network.

Burt extends the SWT argument by asserting that it is not so much the strength or weakness of a tie that determines its information potential, but rather whether a structural hole exists between a focal organization's strong tie contacts (Burt 1992, 2004). Burt claims that individuals near the "holes" in a social structure have a higher likelihood of good ideas because people that connect across groups are exposed to alternative ways of thinking which give them more options to select from and synthesize (Burt 2004). Being positioned between structural holes presents the opportunity to receive more nonredundant information at any given time, which is the key to good ideas and social capital (ibid). Individuals can exploit structural holes to act as brokers and connect otherwise disconnected groups and thus promote innovation and learning, providing opportunities for emergent leadership and collaborative innovation (Crona and Parker 2012).

6.2.4 *Bridging Organizations in Conservation Science*

The bridging organization, as a formal, third party entity distinct from the individuals or organizations it connects, provides an arena for knowledge co-production, trust building, sense making, learning, vertical and horizontal collaboration, and conflict resolution (Berkes 2009). They assist collaborative governance by gathering and interpreting technical information or providing legal, financial, or simply moral support. In these capacities, such organizations are thought to lower the transaction costs associated with multiparty collaboration and provide social incentives to modify behavior or enhance participation in resource governance (Hahn et al. 2006). In a structural sense, they might provide a stabilizing role in social networks, buffering disturbances such as changes in leadership (Olsson et al. 2004), or they could serve as catalysts for multiple ad hoc collaborations that arise in response to specific issues, with varying degrees of betweenness-centrality for each (Hahn et al. 2006). Bridger's need to be able to understand the science, communicate with policy- and decision-makers, and incorporate community values into practice. Figure 6.2 illustrates the position of a bridge in a network and across scales.

Organizations that play a bridging role are able to connect individuals/organizations across boundaries (Sternlieb et al. 2013), across science and policy (White et al. 2008), and across different nodes of expertise (Berkes 2009; Biggs et al. 2010; Crona and Parker 2012).

6.3 Comparing Three Cases of Knowledge Transfer in the U.S. Intermountain West: Managing Noxious Weed Species, Grizzly Bear Conservation, and Climate Adaptation

6.3.1 *Data Collection*

The author began the collection of the empirical data for the three case studies in September 2012 with participant observation, and then 4 months of on-site fieldwork from January 2013 to April 2013. For this project, data collection and analytical techniques combined qualitative field research methods and social network analysis (SNA, Wasserman and Faust 1995). In total, 62 interviews were conducted with consenting participants. While the sampling was purposive, it also followed a rough quota framework that sought proportional representation from community-based practitioners and landscape-level practitioners working in the Crown of the

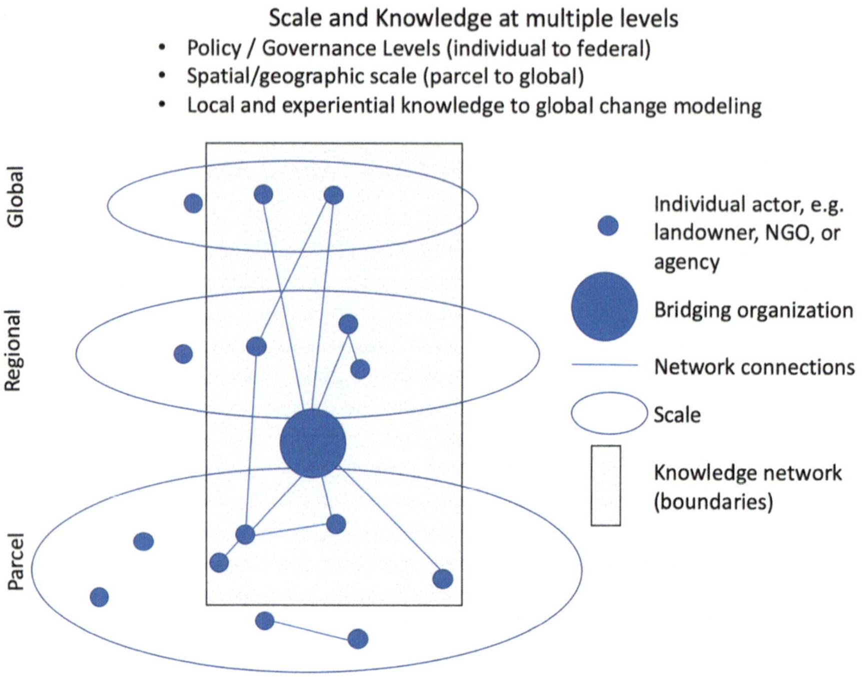

Fig. 6.2 Knowledge networks and boundary organization operating at different scales. Bridging organization connects actors across scales and knowledge systems. Knowledge networks link individual actors and enable/constrain efforts to acquire and transfer knowledge

Continent landscape in northwest Montana, southeast British Columbia, and southwestern Alberta (Fig. 6.3).

In addition to the interviews, a social network survey was sent to the 62 interview respondents. Data regarding their social relationships for both knowledge sharing and conservation implementation was collected through this survey. The survey focused on two of the three case studies: invasive species and climate adaptation. The grizzly bear conservation network was constructed entirely from an analysis of the qualitative data. This process entails coding for network data while at the same time coding for larger themes.

All of the quotes presented below were collected using permission and informed consent protocols as part of the qualitative research and are anonymized for confidentiality. Below, a number of organizations will be referenced and used to illustrate specific points related to knowledge, scale, networks, and bridging organizations. Table 6.1 lists those organizations, provides an external web address and includes a list of organizational attributes.

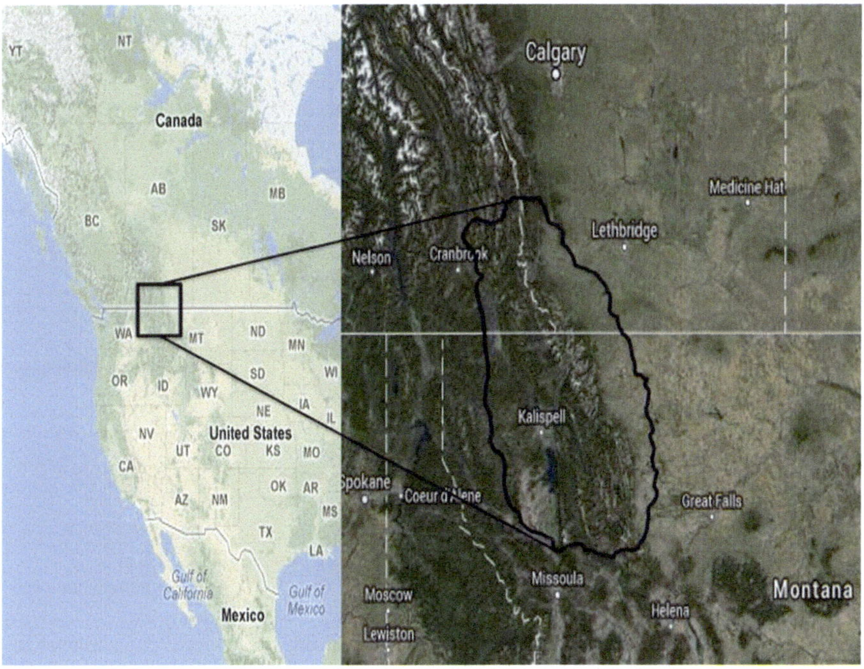

Fig. 6.3 Map of North America (left) with inset of the Crown of the Continent region (right)

6.3.2 Invasive Species in the Crown of the Continent

Invasive species management, primarily noxious weed control, is not a flashy attention grabber like charismatic megafauna or climate change. Yet, knowledge regarding invasive species management is particularly salient and relevant across all kinds of stakeholder and organization types, across multiple levels of conservation activity (all the way up to a national priority for this specific landscape), and provides a great case of knowledge networks and the role of bridging organizations across scales and stakeholders.

Invasive species are recent (voluntary or accidental) introductions of nonnative, exotic, or nonindigenous species that are (or have the potential to become) successfully established or naturalized, spreading into new localized natural habitats or ecoregions with the potential to cause economic or environmental harm (Lodge et al. 2006). The "invasion" of noxious weeds threatens the viability of native plants and grasses, which are valuable resources that support the types of ranching livelihoods found in many parts of the Crown of the Continent landscape. Invasive weedy species pose a threat to both public and private landowners working to maintain native range and grasslands and broader ecosystem health in the region.

Table 6.1 Organizations of the Crown of the Continent, in order of mention in the text

Organization name	Stakeholder-type	Acronym	Type of organization and Focal scale of work	Web address
Clearwater Resource Council	Community-based organization	CRC	Local, watershed	http://crcmt.org/
Roundtable on the Crown of the Continent	Regional NGO	The Roundtable	Regional, Landscape-scale; Transboundary (U.S.-Canada)	http://www.crownroundtable.net/
Rocky Mountain Front Weed Roundtable	Regional NGO	RMFWR	Landscape-scale	http://www.rmfweedroundtable.org/
Working Lands Council	Community-based, National NGO	WLC	Links local landowners to Federal agency leadership in Washington, DC	–
Blackfoot Challenge	Community-based organization	The Challenge	Local, watershed	http://www.blackfootchallenge.org/
Great Northern Landscape Conservation Cooperative	Federal agency collaborative	GNLCC	Science-based organization, landscape-scale	https://greatnorthernlcc.org/
Interagency Grizzly Bear Committee	Federal agency collaborative	IGBC	Science-based organization, landscape-scale	http://igbconline.org/
Swan Ecosystem Center	Community-based organization	SEC	Local, watershed	https://www.swanecosystemcenter.org/
Confederated Salish and Kootenai	Sovereign tribal nation, community-based organizations	CSKT	Local	http://tribalnations.mt.gov/cskt
Greater Yellowstone Coordination Committee	Federal agency collaborative	GYCC	Landscape-scale	https://www.fedgycc.org/
Teton Valley Land Trust	Regional NGO	TVLT	Landscape-scale	http://tetonlandtrust.org/
The Nature Conservancy—High Divide	National NGO	TNC	Landscape-scale	https://www.nature.org/en-us/get-involved/how-to-help/places-we-protect/high-divide-headwaters/
Heart of the Rockies	Regional NGO	HOR	Landscape-scale	https://heart-of-rockies.org/

(continued)

Table 6.1 (continued)

Organization name	Stakeholder-type	Acronym	Type of organization and Focal scale of work	Web address
Vital Ground	Regional NGO	Vital Ground	Landscape-scale	https://www.vitalground.org/
Crown Managers Partnership	Federal Agency collaborative	CMP	Landscape-scale	https://www.crownmanagers.org/

6.3.2.1 Knowledge

Knowledge of invasive species, particularly noxious weeds, is grounded at the local scale and frequently based on experiential knowledge. Frequently, knowledge regarding the presence and intensity of noxious weed distribution on any particular parcel of land is generated by observation, which is dependent on access to the land. This results in an isolated and fragmented network of knowledge regarding the type, intensity, and distribution of noxious weeds across a landscape.

Research participants discussed "being grounded," of being on the ground and interacting with the landscape on a daily basis. A local weed management business owner reflects, "When you say grounded, I think it revolves around your sense of place. I think I'm pretty grounded. I'm on it [the ground] all the time; I see that stuff every day."

There are interesting implications of the local nature of knowledge with invasive species management. There exist highly contextual meanings associated with noxious weeds that often emerge from local places that include local knowledge and everyday ways of doing things. To share and organize this knowledge, community-based (NGO) and county-based (public sector) committees have been formed across the landscape. These groups focus on sharing knowledge and coordinated action between government agencies and nongovernmental organizations, including government-funded initiatives or contracts (Ansell and Gash 2008). One community-based organization, the Clearwater Resource Council (CRC) saw a coordination gap and developed a weed management plan that included establishing (1) high priority weed-free areas, (2) weed control areas, (3) and widespread invader infested areas. In the high priority areas, for example, resources are spent to aggressively and immediately eradicate "invaders" from delineated weed-free areas, and infested areas are designated non-priority given resource constraints (although individual control efforts of landowners and agencies are still encouraged in those areas). Through this project, the CRC was able to build a strong network of local-level actors sharing knowledge and working in a coordinated effort against weeds.

The weed management business owner reflects, "It is interesting for me to go and hear what everyone is doing, what everyone is planning." Noxious weed management opens up the door for many informal ways of interacting and sharing knowledge.

6.3.2.2 Scale

Invasive species management makes an important cross-scale case for a number of reasons. The way local groups are organized illustrates the importance of this topic for those "on-the-ground." Yet, officials at the highest level of the U.S. Department of Interior have also emphasized invasive species coordination and alignment of technical information in the Crown of the Continent through America's Great Outdoors demonstration landscape project. Momentum to focus on invasive species management is occurring at many different levels with a strong emphasis on cross-scale linkages and knowledge exchange.

A quote from the research articulates the cross-scale nature of this issue in surprisingly clear network terms while referring to how different place-based networks link together to have a landscape-level effect:

> You're starting to find the dots [referring to the Roundtable on the Crown of the Continent]. Now find more dots and make them connected. Nobody is doing that in the Crown. The invasive species stuff, you start knitting that together. Just yesterday the North Fork called and I said you're one of the folks I can reach to. In the next round of grants we should include you. So now we are doing weed work in the Blackfoot, Swan, Front, and then the North Fork. We are starting to get a larger landscape focus to some similarities. Maybe we can stop the invaders [noxious weeds]. We are beginning to have a landscape level effect based on local modules of local people with their partners getting stuff done...

This multi-scale organizational context highlights the challenges and opportunities of multilevel governance as these groups, on the one hand, develop local networks to engage in local action, while on the other hand recognize that local networks are embedded within larger social and ecological systems in ways that affect the success at all scales of organization. In this sense, community-based groups recognize a strategic interdependence that can be achieved in stitching together local efforts at invasive species management.

6.3.2.3 Networks

Through the networks of noxious weed management, community-based groups are linking together to coordinate action to have a larger effect on noxious weed management. Through sharing of knowledge, public and private landowners have formed a shared vision of how their actions (or lack of action) influence conservation efforts at the landscape level, which thus creates an incentive for participation in the larger knowledge network. With noxious weed management, there is an "increasing return of conservation" through the strategic interdependence of the various stakeholders because the local costs of weed mitigation go down as the scale of coordinated cross-boundary noxious weed management increases. If it is possible to "push out the boundaries" or keep the "invaders" from entering in the first place then there is value-added for local-level efforts by "scaling" the efforts outward. Some research has discussed this process in terms of innovation, where benefits flow through the network as a consequence of increasing returns; with exponential increases in output

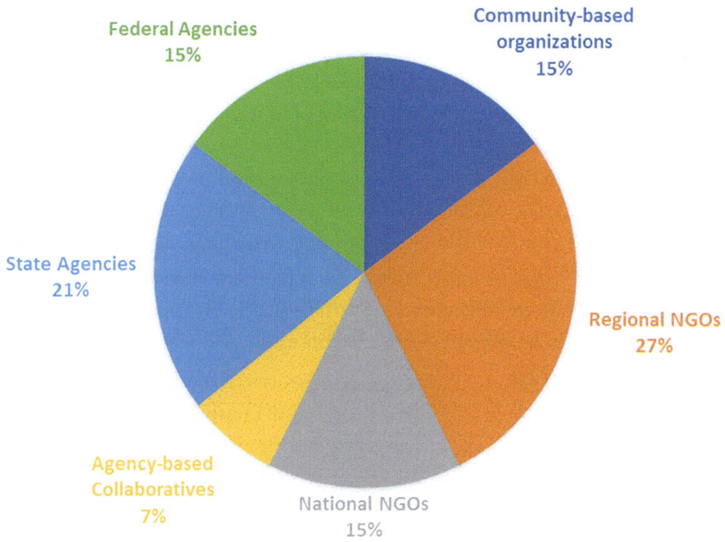

Fig. 6.4 Types of actors involved in Invasive Species Network in the Crown of the Continent

(and rewards or wealth) that spread throughout a network of relationships (Urry 2011). Increasing returns is an example of a positive feedback mechanism, where improved knowledge sharing and coordination between organizations spread nonlinear gains and benefits.

Empirically, two key variables of the knowledge network are: network heterogeneity (or diversity) and the presence of regionally-focused NGOs that play the role of bridging organizations. For example, based on analyzing the diversity of actors in the "network" of invasive species, approximately 27% of the core actors are regional NGOs, 21% are state agencies, 15% are community-based organizations, 15% are national NGOs, 15% are federal agencies, and 7% are agency-driven collaboratives (Fig. 6.4).

The knowledge network is socially and institutionally diverse yet balanced in terms of the proportion of different stakeholder types: state and federal agencies, and NGOs that operate at different scales. Diversity has been related to the function of resource mobilization, performance, and innovation (Carlsson and Sandström 2008), the ability to facilitate problem-solving and adaptiveness (Prell et al. 2009), and the ability to engage with a broader set of issues and challenges by contributing novel approaches to solving problems, and the flexibility in the governance process (Baird et al. 2019).

6.3.2.4 Bridging Organizations

Governmental and nongovernmental organizations have "stitched" together to make overlapping and complementary contributions toward invasive species management in the Crown of the Continent landscape. Indeed, bridging organizations actually navigate the network to bridge the knowledge-implementation gap. Two examples illustrate bridging organizations in this case: the Rocky Mountain Front Weed Roundtable (RMFWR) and the Working Lands Council (WLC).

The RMFWR formed in 2010 and implemented a cost-effective, broad-scale, integrated weed management model. The project's goal was to change weed management from less effective treatment of established weed patches to cost-effective, integrated weed management at the landscape scale using all appropriate techniques. In addition to "enhancing the ecological health" of the landscape, maintaining "economic opportunities" of the Rocky Mountain Front was a clear objective of the initiative. According to their website, in 2012, the RMFWR achieved:

- The collection of 300,000 leafy spurge flea beetles from field insectaries
- Distributed 25,000 biological control insects for knapweed across four counties
- Hosted nine cooperative events for pulling and learning about weeds, and
- Hosted a "Sun Canyon Weed Whacker Rodeo" where over 500 pounds of knapweed were pulled

By navigating and coordinating private and public landowners in the landscape, the RMFWR connected different stakeholders and served as a hub of knowledge. Yet, leaders of the RMFWR knew that a multi-scale approach was necessary to have a greater impact.

Working with community-based leaders of four watersheds in the Southern Crown of the Continent, they formed an umbrella organization to bridge across geographical space called the Working Lands Council. The WLC is an unincorporated collective of private landowners representing the Southern Crown, including the Rocky Mountain Front, the Blackfoot, Clearwater, and Swan Valleys. According to the declaration of partnership, the WLC is focused on community-driven working lands conservation delivered on a landscape scale working in partnership with the US Department of Interior (DOI) and the US Department of Agriculture (USDA) with the intent to establish transparent and regular communications between working landowner interests and USDA and DOI leadership. The plan for this partnership was to have two meetings every year—one in the Crown of the Continent and one in Washington DC—to discuss important landowner issues in the region.

In many ways, this is a clear example of a "cross-scale" linkage. The coordinator for the WLC notes: "Right now, people in the Rocky Mountain Front, the Blackfoot, and the Swan have been working together to talk about more of the southern part of the Crown and whether and what they can work on that makes sense across this landscape. . .Just getting them talking and having those conversations is valuable. That is an attempt to scale up. You have to find that balance, what is a crown-wide issue that can really be worked on and solved *versus* just getting people together to have more political power."

A compelling narrative around invasive species management has led these Crown groups to intersect. This illustrates the power (political and practical) that occurs when a bridging organization connects knowledge, scale, and networks across a specific conservation issue.

6.3.3 Grizzly Bear Conservation

Grizzlies once inhabited the entire north-south trajectory of the North American continent, from Alaska to Mexico, and from the West Coast to nearly the center of the continent. The North American range of grizzly bears has contracted in the past century and a half because of human-caused mortality, habitat loss, and population fragmentation (Proctor et al. 2012). In the conterminous United States, 98% of their range has been lost, and two major populations of grizzly bears remain, one in the Yellowstone region and another in the Crown of the Continent into the Canadian Rockies. Grizzly bears are a key umbrella species—if you protect habitat in a way that grizzly populations persist, you have protected almost every other species (Angelstam and Roberge 2004).

In some ways similar to invasive species, grizzly bears move across the landscape that includes multiple ownerships, and across those varying ownerships are varying attitudes and behaviors toward wildlife. Grizzly bears move out of parks and public lands into adjacent ranches and communities, damaging property and crops or threatening human lives, and as a result, face an increasing risk of anthropogenic mortality. Often, private lands in valley bottoms and foothills adjacent to public lands are problematic zones, especially when available bear attractants coincide with occupied grizzly bear habitat. Specifically, in the Crown of the Continent, conflicts or incidents include bears killing livestock, destroying beehives, foraging for garbage close to homes, or, in rare cases, threatening human safety (Wilson 2011).

6.3.3.1 Knowledge

Grizzly bears, who navigate corridors between habitat patches as they move across the landscape, are the quintessence subject of the science of connectivity conservation and offer an interesting case of bridging conservation science to implementation through networks. For example, in 2012, a multi-year study was published in Wildlife Monographs identifying remaining "biological bridges" between "islands" of grizzly bear populations and threats, such as crowding and habitat depletion, to sustainable populations of grizzly bears in the Rocky Mountain region of North America (Proctor et al. 2012). Another study tracked 388 radio-collared grizzly bears and found that people killed 77–85% of the 99 grizzly bears known or suspected to have died while radio-collared (McLellan et al. 1999). Half of those 99 grizzly bears were killed for being too close to human habitation (while the other half was permitted hunting and legal harvesting). Because of their reputation as dangerous carnivores, grizzly bears often experience higher rates of anthropogenic mortality in human-dominated landscapes than can be sustained by their low reproductive rates (Garshelis et al. 2005).

As grizzly bears re-expand their range onto private lands in the Crown of the Continent (Wilson et al. 2014), the chances for conflicts or incidents with humans increase significantly. Recognizing this, the Blackfoot Challenge—a community-based conservation organization based in the Blackfoot Valley in Montana—brought together the rural landowners, wildlife agencies, and conservation groups to determine exactly what the problem was and how best to address it. Not surprisingly, there were multiple perspectives or definitions of what was the "problem." As Wilson et al. (2014) reflect, some people felt that there were simply too many bears, some celebrated new grizzly bear activity, some defined the problem as primarily one of risk to human safety, and some linked the increased grizzly activity to an erosion of personal rights and freedoms exacerbated by the regulatory burdens of the Endangered Species Act.

Through a process of engaging key stakeholders, that officially began in 2002, the Blackfoot Challenge implemented a participatory GIS mapping program that mapped land use practices, bear attractants, and other relevant features and took that information back to the community to collectively re-frame the problem. Recognizing that the traditional practice of dumping dead livestock carcasses in "bone yards" was attracting bears onto ranches and driving much of the human-grizzly conflict, the Blackfoot Challenge started a carcass removal program. In the past 3 years, an average of 633 carcasses per year was removed, and the program now covers 70–80 ranches over 607,000 ha (Wilson 2011). In the Blackfoot watershed from 2003 to 2009, grizzly bear–human conflicts decreased 93% (Wilson et al. 2014).

In addition to the locally focused and generated information on grizzly bears, science organizations focused on grizzly bear research and conservation are critical pieces to generating knowledge on the ecology of grizzlies. In particular, two organizations—the Great Northern Landscape Conservation Cooperative (an initiative of the U.S. Department of Interior) and the Interagency Grizzly Bear Committee—are involved in all aspects of research on grizzly bears in the region. A conservation practitioner from a community-based conservation organization reflects that they get better and better data from agencies, such as the Interagency Grizzly Bear Committee.

Because of the data generated by organizations such as the Interagency Grizzly Bear Committee on key habitats and corridors for grizzly bears, the knowledge to implementation strategy itself has been transformed from a "buckshot approach" to one that is more strategic. A key part of this strategy is connecting the information to local practitioners and stakeholders.

6.3.3.2 Scale

Many organizations working across a variety of levels of governance are engaged in grizzly bear conservation in the Crown of the Continent. Many "place-based

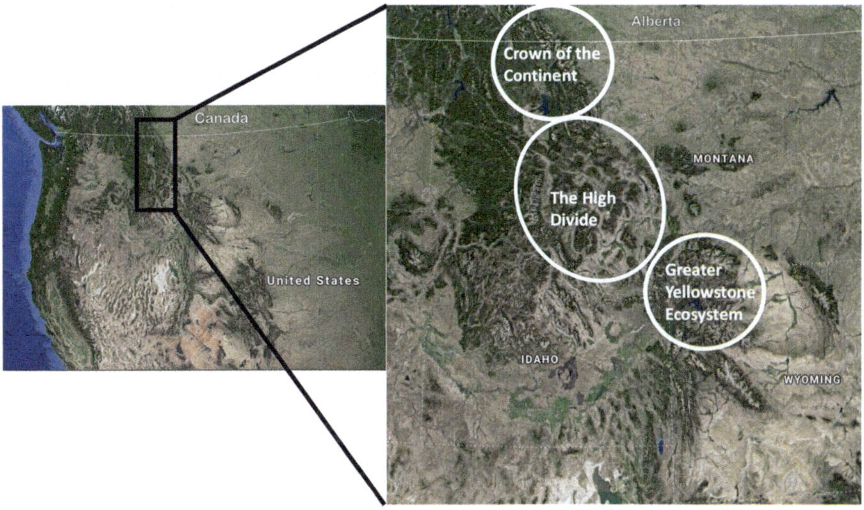

Fig. 6.5 The stretch of the Rocky Mountains that include the Greater Yellowstone Ecosystem, the High Divide, and the Crown of the Continent is prime Grizzly Bear habitat

networks" or community-based conservation organizations, such as the Blackfoot Challenge previously mentioned, work with private landowners and local communities to generate and share local knowledge of the grizzly bear activity, behavior, and movements at locally relevant scales. Across the Crown of the Continent, place-based networks or organizations focus on particular valleys or communities. Other organizations include the Swan Ecosystem Center, CRC (also involved in the invasive species case study), and the tribal nation Confederated Salish and Kootenay. These local groups are an important interface between science and the community. Data on grizzly bears is often collected, analyzed, and reported in the scientific community, but conservation implementation is applied in local communities.

Much attention has been given to the scale at which to manage grizzly bears and transferring effective conservation practices horizontally to other locations. Grizzlies in the Greater Yellowstone Ecosystem were threatened earlier than populations in the Crown of the Continent and thus important cross-scale connections were developed that connect practitioners and scientists in the Greater Yellowstone to the Crown of the Continent. Relatedly, both the Greater Yellowstone Ecosystem and the Crown of the Continent are nested within a larger geographical region of the Northern Rockies connected by a region known as the "high divide" (Fig. 6.5). Different public entities (Great Northern Landscape Conservation Cooperative) and NGOs (Yellowstone to Yukon) have drawn boundaries around this larger system and work to coordinate science and conservation that is relevant to this scale of the grizzlies' entire range. In order to generate information and translate that information to conservation action, a multilevel network is required to connect the various actors across scales.

6.3.3.3 Networks

Like the majority of large carnivores, Grizzly bears move across boundaries (political, administrative, managerial, and institutional) and in the process connect a variety of actors at different scales to a broader sense of the landscape. That is, local-level action and habitat conservation is linked to landscape-scale science assessment and conservation planning. Grizzly bear conservation in the Crown of the Continent, as a network of environmental governance, highlights several positive features of effective networks, namely that they bridge, integrate, translate, and generate knowledge. Particular organizations, such as Vital Ground, play an important role in bridging place-based land trusts and other community-based groups to the science of grizzly bear conservation. In doing so, they often translate "data" to meaningful information to those that can use it for conservation purposes and integrate community values into those conservation efforts.

The ties that bind the different actors are heterogeneous, some connections are strong while others are weak. Some are driven by geography (place-based actors connected by watershed) and others are driven by values (a desire for grizzly bear protection). Some ties, importantly, bridge "structural holes" that are critical to connecting knowledge to implementation (Fig. 6.6).

There were three organizations that bridged that structural hole and established relationships to both the science organizations and the community organizations,

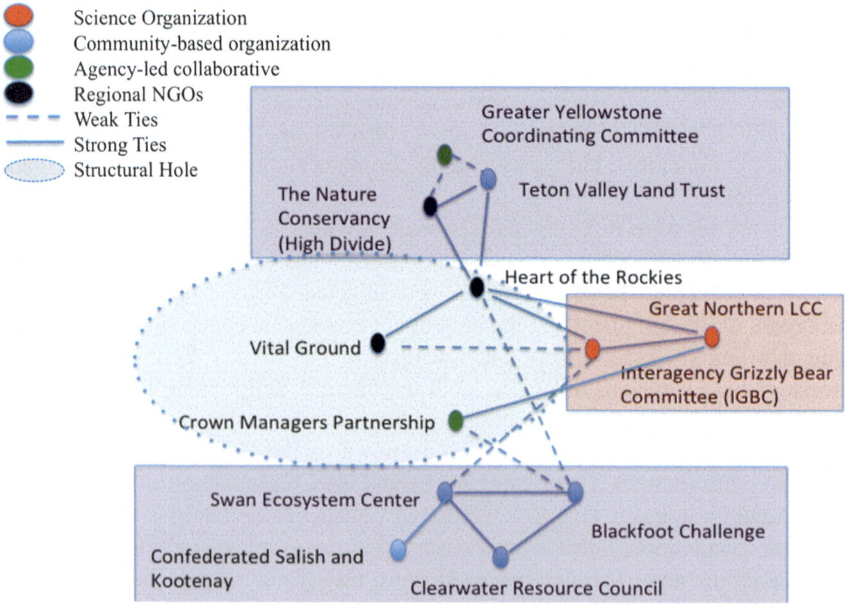

Fig. 6.6 Grizzly bear network and organizations that bridge structural information holes

namely the Heart of the Rockies (HOR), Vital Ground, and the Crown Managers Partnership (CMP).

6.3.3.4 Bridging Organizations

As Fig. 6.6 illustrates, there are certain organizations that bridge those structural holes and in doing so acquire informational advantages they use to advance conservation. In network terms, these ties are often weak as indicated by the dotted lines in Fig. 6.6. Through informal networks and weak ties, information on grizzly bear habitat conservation flows from organization to organization.

Having this type of partnership is about information benefits (Burt 2004), but also about receiving a valuable piece of information and knowing how to use it (Burt 2005). Frequently, research identifies critical grizzly bear habitat on private lands, and one of the primary tools to conserve private lands is a conservation easement (i.e., voluntary legal agreement between a landowner and a land trust or government agency that permanently limits uses of the land in order to protect its conservation values). This raises critical questions about the role of local networks in issues like grizzly bear conservation, where, unlike invasive species management, most of the scientific assessment, planning, and information is derived from supra-local places. Importantly, this highlights the importance of "bridging organizations" that can build rapport in local community networks. Bridging structural holes is not just connecting scientific information across scales but being able to cross cultural boundaries is ways that connect scientific information to local implementation contexts. Crossing these boundaries and building these relationships are necessary to bridge structural "informational" holes.

6.3.4 Climate Adaptation

Climate change adaptation is of central concern for multilevel governance networks because much of the science occurs at global scales yet many of the impacts, and adaptation, will occur locally. Multi-scale environmental problems, such as climate change, often extend across spatiotemporal scales, giving rise to complex patterns of social–ecological interdependencies, which make knowledge networks critical for bridging this knowledge-implementation gap. Climate adaptation is an intersection of the science of downscaled climate impacts on regional natural resources, stakeholder-driven vulnerability assessments (the analysis of the extent to which a species, habitat, ecosystem, or human system is susceptible to harm from climate change impacts) and adaptive governance. Significant advances have been made in synthesizing the projected effects of climate change on ecosystems and describing the strategies for adapting current resource management practices to sustain these evolving ecosystems and the social, economic, and environmental services they provide (Sample et al. 2016). However, we still have much to figure out when translating this scientific knowledge into climate adaptation on the ground.

As has emerged in the Crown of the Continent, the people and organizations charged with the conservation and sustainable management of the regions' natural resources are at the forefront of efforts to understand and address these challenges (Bixler et al. 2016; Bixler et al. 2018). To promote adaptation, networks must help stimulate innovation and creativity across multiple sectors: state and federal agencies, NGOs, and communities.

6.3.4.1 Knowledge

Science suggests that the challenge of climate change in forested landscapes is unprecedented and staggering in its scope, pace, and complexity (Sample et al. 2016). Yet, as scientific information on climate change becomes more ubiquitous it also becomes more available to a wide range of practitioners searching for ways to meet more specific needs of on-the-ground conservation. The amount of data, research, and findings regarding the drivers and consequences continues to increase and be downscaled for particular regions and communities. A key aspect of connecting knowledge to implementation for climate adaption is matching the scale of knowledge generated to that for which strategies are implemented. The mismatch between the scale of knowledge and implementation is one challenge, yet this mismatch is driven by how knowledge and scale are framed. For example, an employee from The Nature Conservancy in the Crown of the Continent remarks:

> I think we are going to have to deal with climate change. It is a huge learning issue. Our scientists tell us people come in and ask what we're doing for climate change and what we're saying is we're not affecting things at a global scale but we're doing the best we can which is to work at a large-landscape scale so that if there is movement plants and animals we've got elevational gradient and we've got room for things to change and so the scale is a lot better than if we were looking at a smaller scale.

In this comment, he shifts the focus from the "global scale" to focus on the "landscape scale." This is an important shift because it makes impacts and adaptations more tangible and highlights the role of scale.

6.3.4.2 Scale

Scale is a critical variable with bridging the knowledge-implementation gap in climate adaptation. Effective climate adaptation work will employ cross-scale linkages and the ability to translate the science of climate change at a global scale to implementation at a local scale will require a scale-dependent framing of knowledge and activities (Wyborn and Bixler 2013). One research participant that works with communities in the Crown of the Continent illustrates this point: "I kind of feel like there is enough common ground here that I think all we need to do is work on stuff that communities do want to work on you can really accomplish a lot of good conservation work. It is just the way you frame it. Take the climate change issue. My feeling is if you go into a community and say we are going to help you put together a climate adaptation program that will better help you adapt to changes that are happening. Climate change, that is a tough sell to people. But on the other hand, if you go into a community and say what are your development goals and what community goals do you have in the pipeline then they'll start talking about connecting open space that is exactly what the climate change professionals will tell you should do. It is no longer a climate change adaptation project but it is now a community project."

Despite concerted efforts, there remains a tension between the science of climate change and the relevance of that information for work "on-the-ground." In this case, certain network characteristics may inhibit knowledge to implementation.

6.3.4.3 Networks

A number of knowledge network strategies are used in attempts to bridge the knowledge divides between research, policy formulation, management decisions, and implementation on this topic. Strengthening the role of community-based conservation may be a fundamental key to broader systemic adaptation efforts for both local livelihoods as well as natural resources.

In contrast to the core of the invasive species knowledge network, the core of the climate adaptation network is dominated by federal agencies and science organizations. Figure 6.7 compares the stakeholder composition of the three networks.

The composition and diversity of the networks are remarkably different. The organizations in the climate adaptation network increase in their scale of focus (less community-based organizations and more regional NGOs) and include a much higher number of science-based organizations than do any of the other two networks. While perhaps not surprising given the fundamentally different nature of these conservation issues, the comparison provides insight into bridging the science-implementation gap and suggests that climate adaptation knowledge networks need to build and support local scale organizations to truly bridge the structural informational hole.

6.3.4.4 Bridging Organizations

To resolve the gap between producers and potential users of climate information, some have suggested sustained interaction between scientists and decision-makers (Cross et al. 2013; Littell et al. 2011; Raymond et al. 2013). Science-management partnerships have created (and continue to create) formal opportunities for sharing information such as workshops and conferences; but also, flexible opportunities and space for regular information exchange between parties, including sharing of experiences, discussion of new ideas, and joint problem-solving (Cross et al. 2013; see also Chap. 9; Schwartz et al. 2021).

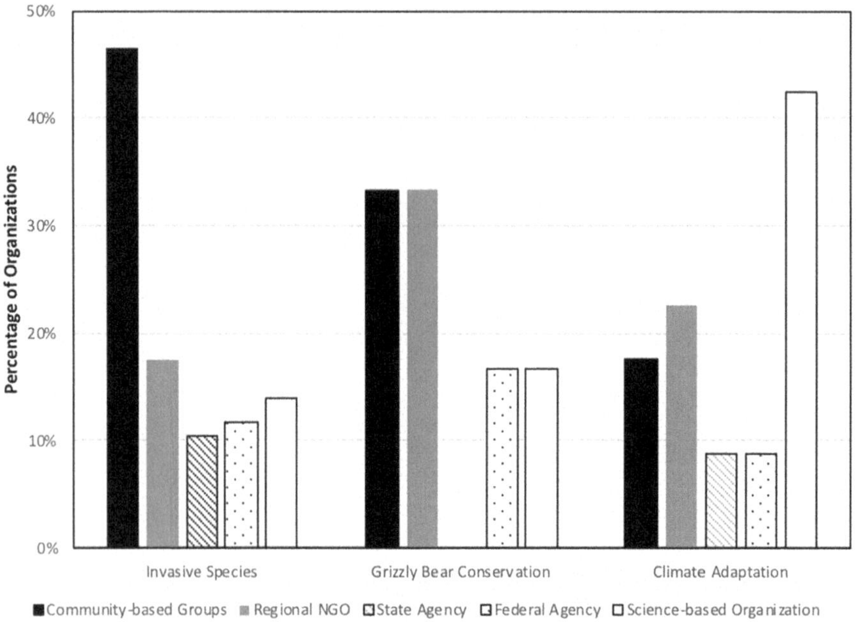

Fig. 6.7 Comparison of organizational diversity in the invasive species management, Grizzly Bear conservation, and Climate Adaptation Knowledge Networks in the Crown of the Continent

Particularly when dealing with climate change, these strategies are increasing. For example, Littell et al. (2011) document science-management partnerships on the Olympic National Forest in Washington State and the Tahoe National Forest in California and Raymond et al. (2013) in North Cascadia. Through this process, climate change scientists provide the scientific knowledge base on which adaptations could be based, and resource managers develop adaptation options based on their understandings of ecosystem management. These science-management partnerships typically involve iterative sharing of climate and climate effects information by scientists, and of local climate, ecological, and management information by managers and have become a forum for conducting vulnerability assessments and developing adaptation plans at both the strategic and tactical levels (Halofsky et al. 2018; Littell et al. 2011). Therefore, developing this model to include a broader range of stakeholder types is one mechanism to start bridging the knowledge-implementation gap in climate adaptation.

6.3.5 *Summary of the Case Studies*

As this chapter has outlined, knowledge networks and bridging organizations that can navigate the interplay between networks, knowledge, and scale can help bridge

	Knowledge	Scale
Bridging Organization	GBC: Connect actors across structural information holes IAS: Bring together a variety of knowledge nodes, especially when knowledge is distributed across space, i.e., landscapes	IAS: Link to higher level policy CA: Can link global science to local implementation, i.e., science-management partnerships
Knowledge network	CA: Lack of network diversity can inhibit knowledge translation to implementation All case studies: Connect different bits of knowledge and link knowledge systems	IAS: Conservation knowledge can navigate scale through a diversity of organizational types in a network CA: Formal bridging strategies, such as science-management partnerships, can be expanded to include a broader range of stakeholders

Fig. 6.8 Summary of lessons learned from the Crown of the Continent case studies (IAS—Invasive species management, GBC—Grizzly Bear Conservation, CA—Climate Adaptation)

the knowledge-implementation gap in conservation science. As summarized in Fig. 6.8, some key characteristics that emerge from this work include the role of bridging organizations to cross structural holes (i.e., grizzly bear conservation) and link cross-scale policy levels (i.e., invasive species management), as well as the important role of network diversity in navigating scale (i.e., invasive species management) or lack of diversity in creating barriers (i.e., climate adaptation). All cases demonstrate that knowledge networks are a good tool to connect different bits of knowledge and link knowledge systems across the scale.

6.4 Conclusion

Scientific assessments increasingly indicate human activities are pushing ecological systems outside of a normal range of variability (Sample et al. 2016), which signals that conservation practitioners will be implementing solutions in a context somewhat unfamiliar to them and facing problems never before seen. This will require new strategies for knowledge transfer and leverage networks to navigate scales and knowledge systems.

The majority of this chapter dealt primarily with three case studies in one geographic region in North America. The author's in-depth knowledge and extensive research were used to flesh out the concepts of knowledge, scale, networks, and bridging organizations and their use to bridge the knowledge-implementation gap in conservation science. However, these concepts are ubiquitous worldwide in both terrestrial and marine systems. For example, an extensive multicountry collaborative fisheries comanagement research effort found that cross-scale knowledge networks, that included NGOs and government units at different levels of organization, were critical for the conservation success of the cases studied because the cross-scale institutional linkages make adaptive management possible by bringing together groups with broad local foci and ones with narrow trans-local mandates (Wilson et al. 2006). Relatedly, research in the Solomon Islands examined a locally Managed Marine Area Network (Cohen et al. 2012) and identified specific network structures and features such as density of network ties and bridging ties across scales and to nonmembers that were critical for knowledge exchange. In Kenya, research demonstrated the network included subgroups of diverse and complementary local ecological knowledge that was shared across scale (Bodin and Crona 2009). Similar dynamics were also found in Mexico (Ramirez-Sanchez and Pinkerton 2009) and Chile (Marin et al. 2012) and networks were found to be pervasive in linking local conservation efforts, which were a part of the United Nations Development Programme Equator Initiative, to knowledge being generated at higher scale institutions (Berkes 2007).

Research on bridging organizations is also prominently found in contexts outside of North America. For example, in Vietnam forest management, bridging organizations can make inroads to increase knowledge sharing and local empowerment against a backdrop of state administrative control (KimDung et al. 2016). In Japan, research on the role of the Kyoto Model Forest Association (KMFA) as a bridging organization illustrated how a multi-stakeholder approach can provide a pathway to recouple people-forest relationships and enhance management outcomes. Research in Bali, Indonesia, showed how bridging organizations helped to navigate the "messiness" inherent in conservation settings by compensating for sparse linkages and improving knowledge sharing and ultimately conservation outcomes (Berdej and Armitage 2016). Moreover, Rathwell et al. (2015) conduct an extensive review of literature emphasizing the prominent role of bridging organizations in bridging indigenous and western knowledge systems (Rathwell et al. 2015).

These examples, as well as the case studies presented in this chapter, provide a better understanding of the barriers, opportunities, and strategies for transferring conservation knowledge across spatiotemporal scales and stakeholders. From local environments to the global earth system, identifying strategies that facilitate the practice of conservation science will be key for future social–ecological sustainability and resilience.

Acknowledgement This research was supported by a grant from USDA Forest Service Rocky Mountain Research Station and the Center for Collaborative Conservation at Colorado State University.

References

Roberge J-M, Angelstam P (2004) Usefulness of the umbrella species concept as a conservation tool. Conserv Biol 18(1):76–85

Ansell C, Gash A (2008) Collaborative governance in theory and practice. J Public Adm Res Theory 18(4):543–571

Armitage DR, Plummer R, Berkes F, Arthur RI, Charles AT, Davidson-Hunt IJ, Diduck AP et al (2009) Adaptive co-management for social–ecological complexity. Front Ecol Environ 7(2):95–102

Baird J, Plummer R, Schultz L, Armitage D, Bodin Ö (2019) How does socio-institutional diversity affect collaborative governance of social–ecological systems in practice? Environ Manag 63(2):200–214

Berdej SM, Armitage DR (2016) Bridging organizations drive effective governance outcomes for conservation of Indonesia's marine systems. PLoS One 11(1):e0147142

Berkes F (2007) Community-based conservation in a globalized world. PNAS 104(39):15188–15193

Berkes F (2009) Evolution of co-management: role of knowledge generation, bridging organizations and social learning. J Environ Manag 90(5):1692–1702

Biggs R, Westley FR, Carpenter SR (2010) Navigating the back loop: fostering social innovation and transformation in ecosystem management. Ecol Soc 15(2):9.

Bixler RP (2014) From community forest management to polycentric governance: assessing evidence from the bottom up. Soc Nat Resour 27(2):155–169

Bixler RP, Johnson S, Emerson K et al (2016) Networks and landscapes: a framework for setting goals and evaluating performance at the large landscape scale. Front Ecol Environ 14(3):145–153

Bixler RP, Reuling M, Johnson S, Higgins S, Williams S, Tabor G (2018) The crown of the continent: a case study of collaborative climate adaptation. In: Dellasala DA, Goldstein MI (eds) Encyclopedia of the anthropocene. Elsevier, Oxford, pp 307–315

Bodin Ö, Crona BI (2009) The role of social networks in natural resource governance: what relational patterns make a difference? Glob Environ Change 19(3):366–374

Borgatti SP, Halgin DS (2011) On network theory. Organ Sci 22(5):1168–1181

Burt RS (1992) Structural holes: the social structure of competition. Harvard University Press, Cambridge

Burt RS (2004) Structural holes and good ideas. Am J Sociol 110(2):349–399

Burt RS (2005) Brokerage and closure: an introduction to social capital. Oxford University Press, New York

Carlsson L, Sandström A (2008) Network governance of the commons. Int J Commons 2(1):33–54

Carolan MS (2006) Science, expertise, and the democratization of the decision-making process. Soc Nat Resour 19(7):661–668

Cash DW, Moser SC (2000) Linking global and local scales: designing dynamic assessment and management processes. Glob Environ Change 10(2):109–120

Cohen PJ, Evans LS, Mills M (2012) Social networks supporting governance of coastal ecosystems in Solomon Islands. Conserv Lett 5(5):376–386

Crona B, Hubacek K (2010) The right connections: how do social networks lubricate the machinery of natural resource governance? Ecol Soc 15(4):18

Crona BI, Parker JN (2012) Learning in support of governance: theories, methods, and a framework to assess how bridging organizations contribute to adaptive resource governance. Ecol Soc 17(1):32

Cross MS, McCarthy PD, Garfin G et al (2013) Accelerating adaptation of natural resource management to address climate change. Conserv Biol 27(1):4–13

Cumming GS, Cumming DHM, Redman CL (2006) Scale mismatches in social-ecological systems: causes, consequences, and solutions. Ecol Soc 11(1):14

Cundill G, Rodela R (2012) A review of assertions about the processes and outcomes of social learning in natural resource management. J Environ Manag 113:7–14
Duraiappah AK, Asah ST, Brondizio ES et al (2014) Managing the mismatches to provide ecosystem services for human well-being: a conceptual framework for understanding the new commons. Curr Opin Environ Sustain 7:94–100
Ekstrom JA, Young OR (2009) Evaluating functional fit between a set of institutions and an ecosystem. Ecol Soc 14(2):16
Fischer J, Hartel T, Kuemmerle T (2012) Conservation policy in traditional farming landscapes. Conserv Lett 5(3):167–175
Folke C, Hahn T, Olsson P, Norberg J (2005) Adaptive governance of social-ecological systems. Annu Rev Environ Resour 30:441–473
Funtowicz SO, Ravetz JR (1993) Science for the post-normal age. Futures 25(7):739–755
Garshelis DL, Gibeau ML, Herrero S (2005) Grizzly bear demographics in and around Banff National Park and Kanaskis country, Alberta. J Wildl Manag 69:277–297
Granovetter MS (1973) The strength of weak ties. Am J Sociol 78(6):1360–1380
Granovetter M (1983) The strength of weak ties: a network theory revisited. Sociol Theory 1:201–233
Hahn T, Olsson P, Folke C, Johansson K (2006) Trust-building, knowledge generation and organizational innovations: the role of a bridging organization for adaptive comanagement of a wetland landscape around Kristianstad, Sweden. Hum Ecol 34(4):573–592
Halofsky JE, Peterson DL, Prendeville HR (2018) Assessing vulnerabilities and adapting to climate change in northwestern U.S. forests. Clim Chang 146(1):89–102
Holley C, Gunningham N, Shearing C (2013) The new environmental governance. Routledge, London
Jasanoff S (2004) States of knowledge: the co-production of science and social order. Routledge Taylor & Francis Group, London
Jasanoff S (2010) A new climate for society. Theory Cult Soc 27(2-3):233–253
KimDung N, Bush SR, Mol APJ (2016) NGOs as bridging organizations in managing nature protection in Vietnam. J Environ Dev 25(2):191–218
Latour B (1987) Science in action. Open University Press, Milton Keynes
Lemos MC, Agrawal A (2006) Environmental governance. Annu Rev Environ Resour 31:297–325
Littell JS, Peterson DL, Millar CI, O'Halloran KA (2011) U.S. National forests adapt to climate change through science-management partnerships. Clim Change. 110:269–296
Lodge DM, Williams S, MacIsaac HJ et al (2006) Biological invasions: recommendations for U.S. policy and management. Ecol Appl 16(6):2035–2054
Marin A, Gelcich S, Castilla JC, Berkes F (2012) Exploring social capital in Chile's coastal benthic comanagement system using a network approach. Ecol Soc 17(1):13.
McLellan BN, Hovey FW, Mace RD, Woods JG (1999) Rates and causes of grizzly bear mortality in the Interior Mountains of British Columbia, Alberta, Montana, Washington, and Idaho. J Wildl Manag 63(3):911–920
McPherson M, Smith-Lovin L, Cook JM (2001) Birds of a feather: homophily in social networks. Annu Rev Sociol 27(1):415–444
Ojha HR, Hall A, Sulaiman RV (2013) Adaptive collaborative approaches in natural resource governance. Routledge, London
Olsson P, Folke C, Berkes F (2004) Adaptive comanagement for building resilience in social–ecological systems. Environ Manag 34(1):75–90
Ostrom E (1990) Governing the commons. Cambridge University Press, Cambridge
Pahl-Wostl C (2009) A conceptual framework for analysing adaptive capacity and multi-level learning processes in resource governance regimes. Glob Environ Change 19(3):354–365
Pahl-Wostl C, Lebel L, Knieper C, Nikitina E (2012) From applying panaceas to mastering complexity: toward adaptive water governance in river basins. Environ Sci Pol 23:24–34
Phelps C, Heidl R, Wadhwa A (2012) Knowledge, networks, and knowledge networks: a review and research agenda. J Manag 38(4):1115–1166

Pielke RA Jr (2004) When scientists politicize science: making sense of controversy over the skeptical environmentalist. Environ Sci Pol 7(5):405–417
Powell WVV (1990) Neither market nor hierarchy: network forms of organization. Res Organ Behav 12:295–336
Prell C, Hubacek K, Reed M (2009) Stakeholder analysis and social network analysis in natural resource management. Soc Nat Resour 22(6):501–518
Proctor MF, Paetkau D, McLellan BN, Stenhouse GB et al (2012) Population fragmentation and inter-ecosystem movements of grizzly bears in Western Canada and the Northern United States. Wildl Monogr 180(1):1–46
Ramirez-Sanchez S, Pinkerton E (2009) The impact of resource scarcity on bonding and bridging social capital: the case of fishers' information-sharing networks in Loreto, BCS. Mexico. Ecol Soc 14(1):22
Rathwell K, Armitage D, Berkes F (2015) Bridging knowledge systems to enhance governance of environmental commons: a typology of settings. Int J Commons 9(2):851–880
Raymond CL, Peterson DL, Rochefort RM (2013) The North Cascadia adaptation partnership: a science-management collaboration for responding to climate change. Sustainability 5(1):136–159
Sample VA, Bixler RP, Miller C (eds) (2016) Forest conservation in the anthropocene: science, policy, and practice, 1st edn. University Press of Colorado, Boulder
Sarewitz D (2004) How science makes environmental controversies worse. Environ Sci Pol 7(5):385–403
Sayre NF (2005) Ecological and geographical scale: parallels and potential for integration. Prog Hum Geogr 29(3):276–290
Schwartz MW, Fleishman E, Williamson MA, Williams JN, Morelli TL (2021) The use of boundary-spanning organizations to bridge the knowledge-action gap in North America. In: Ferreira CC, Klütsch CFC (eds) Closing the knowledge-implementation gap in conservation science – evidence transfer across spatiotemporal scales and different stakeholders. Springer, New York, pp 229–254
Sternlieb F, Bixler RP, Huber-Stearns H, Huayhuaca C (2013) A question of fit: reflections on boundaries, organizations and social–ecological systems. J Environ Manag 130:117–125
Urry J (2011) Climate change and society. Polity Press
Wasserman S, Faust K (1995) Social network analysis: methods and applications. Cambridge University Press, Cambridge
Weinberg AM (1972) Science and Trans-Science. Science 177(4045):211
White DD, Corley EA, White MS (2008) Water managers' perceptions of the science-policy interface in phoenix, Arizona: implications for an emerging boundary organization. Soc Nat Resour 21(3):230–243
Wilson DC, Ahmed M, Siar SV, Kanagaratnam U (2006) Cross-scale linkages and adaptive management: fisheries co-management in Asia. Mar Policy 30(5):523–533
Wilson SM (2011) Managing livestock carcasses to reduce conflicts with Grizzly bears in Montana's Blackfoot Valley. Int Bear News 20(2):21–22
Wilson SM, Neudecker GA, Jonkel JJ (2014) Human–Grizzly bear coexistence in the Blackfoot River Watershed, Montana: getting ahead of the conflict curve. In: Clark SG, Rutherford MB (eds) Large Carnivore conservation: integrating science and policy in the North American West. University of Chicago Press, pp 177–214
Wyborn C, Bixler RP (2013) Collaboration and nested environmental governance: scale dependency, scale framing, and cross-scale interactions in collaborative conservation. J Environ Manag 123:58–67
Young O (2006) Vertical interplay among scale-dependent environmental and resource regimes. Ecol Soc 11(1):27

Chapter 7
Communication: The Bridge Between Knowledge and Implementation

Christine O'Connell and Merryn McKinnon

Contents

C. O'Connell (✉)
Stony Brook University, Stony Brook, NY, USA
e-mail: christine.oconnell@stonybrook.edu

M. McKinnon
Centre for the Public Awareness of Science at The Australian National University, Canberra, ACT, Australia
e-mail: merryn.mckinnon@anu.edu.au

C. C. Ferreira, C. F. C. Klütsch (eds.), *Closing the Knowledge-Implementation Gap in Conservation Science*, Wildlife Research Monographs 4,
https://doi.org/10.1007/978-3-030-81085-6_7

7.1 Introduction

This chapter discusses different approaches to science communication and outreach to achieve conservation outcomes. Before delving into the *how*, it is useful to discuss *why* conservation scientists should be actively involved in communicating conservation issues to general publics. The use of the plural—publics—here is intentional. Within each audience, there are many different levels of expertise, values, and interest. Treating them all as one amorphous blob does them, and the communicator, a disservice. More on that later. For now, let us talk about why conservation scientists communicating with other audiences matters.

7.2 If Not You, Then Who?

At its simplest—if not conservation scientists, then who should be talking to publics about conservation issues? Who else is better qualified or could or should be trusted to present the issues, and the science, accurately?

Scientists feel positive about communicating their work either from a sense of enjoyment and accomplishment (Besley et al. 2013) or a perception of personal benefit or moral duty (Bubela et al. 2009; Carlsen and Riese 2016). Yet scientists—from all disciplines—can feel reluctant to engage with the media or general publics. This can be due to discomfort with the idea, a past negative experience, fear, or an ingrained sense of there being no value in engaging. But at what cost?

If conservation scientists are not speaking about their areas of expertise, then this leaves a gap to be filled by others. Typical alternative sources could include activists, companies, special-interest groups, scientists from other disciplines, government officials, and politicians. Previous studies have shown that audiences will use an evaluation of the perceived credibility of the messenger as a heuristic—an information shortcut—to decide if they will accept the message or not (Brewer and Ley 2013). Some of these messengers seek to misinform or misrepresent the science for their own purposes, as was seen in the tobacco industry and currently with climate change (Oreskes and Conway 2010). Others, despite being well-intentioned, may do more harm than good. Activists, which can be members of the scientific community or interested and engaged members of various publics, are often visible spokespeople for conservation and environmental issues. However, the "typical" activist may actually alienate the audience and reduce their motivation for change (Bashir et al. 2013). Government representatives and politicians will have varying degrees of credibility with different audiences. The political ideology of the audience will influence their perception of a government or politician messenger's credibility; however, audience attitudes toward government and corporations are also important predictors of whether the messenger can be trusted or not (Pechar et al. 2018).

Trust is a key component of effective communication; if you are not well represented in public areas, you risk losing that trust (Jucan and Jucan 2014). The Edelman Trust Barometer (Edelman 2018) is a global report released annually, based on surveys of over 33,000 people from 28 countries. It examines the most trusted sources of information and people's level of trust in government, media, businesses, and nongovernmental organizations. In 2018, the most trusted sources of information were technical and academic experts. These two groups have consistently been in the topmost trusted sources for over 15 years of Trust Barometer reports, and in 2018 in particular the "voices of expertise" have regained credibility as other sources declined.

The role of the media in providing information about conservation is vital to gaining public support, which in turn greatly increases the likelihood of conservation measures being successful (Rosalino and Rosalino 2012). The way media presents—or frames—an issue can influence people's interest (Nghiem et al. 2016), attitudes, and beliefs toward conservation issues (Munro 1997; Kruse 2001). Framing is not telling the audience what to think; rather it is attempting to influence what they should be thinking about. This can work both for and against the achievement of conservation goals. For example in the early 1990s in Australia, duck hunting was framed by the media as an "anti-social, cruel sport" which lead to a decline in the number of shooters from 90,000 in 1986 to 21,000 in 1994 (Munro 1997, p. 151). In comparison, an examination of frames used by media in Australia and the United States used to discuss shark attacks found that most media coverages emphasized the risk that sharks pose to people and ignored the risk of extinction being faced by many shark species because of people's actions (Muter et al. 2013). This framing of risk to people creates a public perception of shark attacks occurring frequently (Sabatier and Huveneers 2018), generating fear which does not lend itself to creating public support for their conservation (Liordos et al. 2017).

Media coverage and subsequent public perception of scientific issues, including conservation, should be shaped by the scientific community (Heupel and Simpfendorfer 2010), but this does not appear to be the case. Hurlimann and Dolnicar (2012) found that the scientific community was conspicuous in their absence from media coverage on a recycled water referendum, with politicians and non-specialists given far more coverage. Similar results were found by Muter et al. (2013) in shark attack-related stories, and Crook (2014) who found stories about mountain lions frequently cited by community members. All of the studies cited here call for conservation scientists, biologists, and environmental scientists to actively engage with publics and the media to communicate evidence-based information in an accessible way in order to increase public understanding of conservation issues and thereby bridge the knowledge-action gap.

7.3 The First Two Steps to Communicating

There are two fundamental questions that must be answered before any communication activity: Who is your audience and what is your goal?

7.3.1 Who Is Your Audience?

Knowing your audience is one of the most important tenets of science communication training (Bray et al. 2012) requiring the development of empathy and a shift of focus to your audience's needs. Communication is not about you or what you say. It is *always* about your audience. It is about whether the other person "gets it," otherwise you are just talking or writing *at* them, not *with* them. Identify where your audience is coming from (in terms of knowledge, beliefs, and/or attitude) and meet them where they are. This goes beyond demographics and education level. Think about the many publics they belong to—dig deeper: What do they care about? (Table 7.1).

For example, why are they coming to be a part of the audience in the first place? Are they part of a local community group that has extensive practical knowledge of the area? Are they students who are studying conservation science and may have a more in-depth theoretical understanding (but potentially less of the practical)? Are they a government representative who is looking for potential policy answers or projects to fund, who may not have a lot of conservation experience but also does not have a lot of time? The way to communicate with each group will be subtly different—requiring informed choices about your message, mode, and place of communication. Each of these different audiences also has a different set of motivating factors and values.

Reflecting the values of the audience serves a number of purposes. First, it shows the audience that the person talking to them has actively considered their interests and needs. This helps the communicator appropriately tailor the information provided and build rapport with the audience, which in turn more positively predisposes the audience to listen to the message. Second, it will help the communicator appear as someone more credible and trustworthy, rather than an "arrogant outsider." Third, it may also help the communicator better understand the issues in the local area by

Table 7.1 Ask yourself…

What are their interests?
What do they value or care about? How might your message conflict with, enhance, or affect these?
What do they worry about?
Where do they get their information?
How will they react to your information? Why might it be important to them?
How will it affect their life or work?

acknowledging and listening to the local experience and expertise. This can sometimes offer more accurate understanding and insight than the scientific knowledge of "outsiders" (Wynne 1992) as exemplified after the Chernobyl disaster.

In the aftermath of Chernobyl, scientists provided extensive information and advice regarding the environmental hazards from the radioactive fallout being carried from the Chernobyl site to the Lake District of Northern England. The scientists based their advice on empirical understandings of the decay of cesium isotopes in clay soils, restricting sales and movements of sheep from these hill-farm areas which created extreme financial difficulties for farmers (Wynne 1992). Bans that were initially promised for a few weeks became indefinite. Scientists were confident their predictions were accurate, but testing repeatedly showed they were not. It was eventually revealed that the clay soils upon which they were basing their advice were not actually present in the hill-farm areas. They contained peat soil that reacts entirely differently to cesium. The scientists did not acknowledge their mistake. Nor did they acknowledge the farmers' in-depth knowledge of their own lands and fields, information which actually would have assisted their experiments and formulation of advice. The refusal of the scientists to listen to the advice of the farmers led to the abandonment of experiments, undermining of the scientists' credibility, and a sense of denigration for the farmers (Wynne 1992). What this case study shows is that to create and maintain trust and credibility, scientists must also acknowledge and respect the explicit knowledge of their audience of their local area.

7.3.2 *Know Your Goal*

Once you have identified your audience, what do you want them to think, feel or do after the communication activity? This is your goal that is generally supported by a key message. If people only remember one thing from your talk/interview/conversation, what would you want that to be? This is your key message; craft it ahead of time and develop it to suit your audience. It is possible you may have different key messages for different audiences, with different intentions, i.e., what you say to a community group may be very different than to an elected official. Irrespective of whom you are talking to and what your key message is, do not fall into the "deficit model" way of thinking.

The deficit model assumes that the audience has a deficit of knowledge about a particular topic, and simply providing them with more facts and information will cause them to (a) understand, (b) care, (c) change behavior, or (d) all of the above. This is categorically and empirically untrue, yet it has proven very difficult to steer scientists away from following this approach (Baker 2019). To illustrate, think about how many people you know (or see) who still smoke cigarettes. Or those who have unhealthy diets and do not exercise. How many of those people do you think have never heard that smoking, poor diet, and no exercise are bad for their health? Simply giving people facts is in and of itself not enough to motivate, convince, or change behaviors. In some cases, it can actually have the opposite effect (Kahan et al. 2011).

Confirmation bias is a process where people have the tendency to seek information that confirms what they already believe and discard information that does not support their beliefs. Attempting to get people to accept a scientific idea by simply providing more facts can actually cause people to become more entrenched in their original positions (Cook and Lewandowsky 2016). This is a challenge for scientists and science communicators alike (e.g., anthropogenic climate change). Conflicts and controversies about scientific issues are not necessarily because of the facts; the true source of argument may be based on the values of the different parties. Attempting to argue against someone's values using facts will rarely, if ever, be successful (Dietz 2013). Knowing your audience and their values, and basing your communication there, is a more promising place to start working toward desired conservation outcomes.

What does all this mean for bridging the knowledge–implementation gap?

Simply giving people the information is not enough to take them from knowledge to action. Many desired conservation outcomes require people to "do" something, not just "know". This is a difficult step and takes time.

To get people to "do" they need to be persuaded and motivated, which means that you need to know who your audience is and what they value. Use this knowledge to shape your key message and plan your communication activities.

7.4 Difficult Audiences

A common question asked by scientists working in contentious areas, especially those which require people to change behavior, is how to engage those who are disinterested, dismissive, or antagonistic. There is no easy solution—if there were, we would not be facing increasing rates of vaccine-preventable disease and prolonged inaction on climate change!

The first thing to acknowledge is that there will never be 100% audience acceptance. There will always be those who, for whatever reason, will never accept scientific facts and conclusions. The flip side of this is that there will always be a similarly sized group of the "converted," those who will accept and support scientific findings and adopt desired behaviors. The core group of any audience that communicators should be trying to reach is in between these two poles (Fig. 7.1). These are the people who may not have engaged yet, do not feel a topic is relevant to them, avoid engaging due to fear or felt hopelessness, or who are unsure of what information source to trust and so have not yet made up their minds. This is the group where

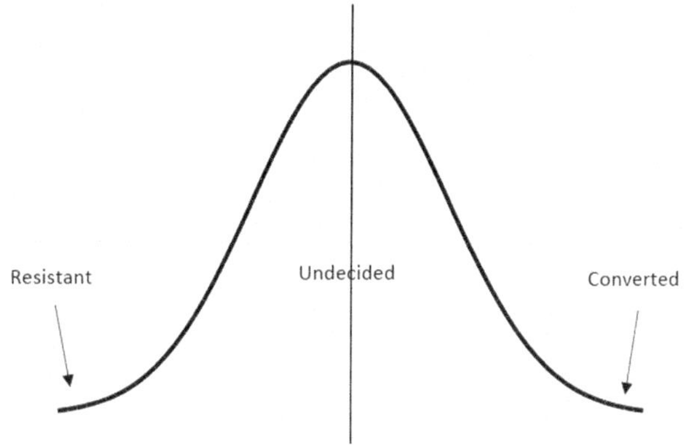

Fig. 7.1 One way to think of an audience—communicators should focus their efforts on the undecided group in the middle

communicators can have the greatest influence to achieve results and bridge the knowledge–implementation gap.

Set realistic expectations for the effectiveness of communication. Within that large group of undecided listeners, it is very likely that some will remain toward the resistant end irrespective of communication efforts. People are not rational (Baker 2019). Rather than aiming for the unattainable, instead, think of moving people along that spectrum of resistant to converted. Imagine a scale along the bottom of that curve where resistant equals one and converted equals 10. Someone who is sitting in the undecided group at around a three is unlikely to be moved straight to 10. But perhaps they can be shifted to four or five. Even incremental gains are a positive result.

You may not be able to move someone on an issue, but you may get them to trust science and the scientific process a bit more—this is still a win. You may get them to start listening, *versus* yelling or being adversarial, which is the first step in communicating effectively.

7.5 Stop Talking, Learn to Listen

Communication is often conceived of as speaking or writing, an active form of information transmission. However, communication is a two-way street; it is often more about listening than talking. Listening is the most crucial part of communication and essential to know the audience. Baram-Tsabari and Lewenstein (2017) assert that open listening and responding honestly to your audience is the basis of true communication, creating more positive interactions between scientists and the many publics they interact with. Regular formal and informal interactions between publics and scientists increase empathy for both parties, leading to mutually positive changes on how each views the other and the overall process of science communication (Besley and Tanner 2011). Learning to listen can also help you to see your science through a different lens or gain new insight, making your science stronger.

Particularly useful for audiences who may be resistant, actively listening to concerns will allow a communicator to gain vital insight into what the audience values. Practice empathy, put yourself in the other person's shoes, and see things from their point of view. This does not mean you have to agree with them, but you should make an effort to understand where they are coming from, what they really care about, and why. If someone feels heard, even if you do not agree with each other, defenses fall and they are more willing to listen.

Consider the example of Health in Harmony (2019), a nongovernment organization providing support to the Alam Sehat Lestari (ASRI) project in West Kalimantan, Indonesia. This area is one of the last remnants of lowland rainforest and an important refuge for orangutans. It is also an area where the local population lives hand to mouth, and illegal logging of the surrounding national park for agricultural land is threatening to destroy the remaining rainforest. The project began because the founder of Health in Harmony had previously been to the national park as an orangutan researcher and saw the environmental destruction, as well as the dire health needs of the community. In developing a solution to both issues, they listened to the concerns of individual community members for over 400 hours! Through listening they were able to identify that the lack of access to high-quality health care was a problem, as was a lack of economic alternatives to logging. Through this "radical listening" process, Health in Harmony were able to identify the drivers of illegal logging and the needs and values of the villagers. This shaped the structure of the program, leading to the building of a hospital for the community that provides affordable health care for the villagers via a "green credit" system. Those villagers who do not participate in illegal logging (which is verified by local logging monitoring staff) receive discounted health care, and the poorest patients are able to "pay" with seedlings for reforestation, organic manure to support sustainable farming practices or handicrafts (Health in Harmony 2019). Once the program was implemented, logging rates fell. Through using "radical listening" with village leaders, a new program (Forest Guardians) was implemented in 2011 which appoints a community member to work one on one with their neighbors who still conduct illegal logging to help them find alternative livelihoods, further reducing the rate of deforestation.

In this example, conservation goals were achieved when other needs of the villagers were met. Not all audiences and communication scenarios will fit this model. But there may be other opportunities for finding "common ground," areas where the values of communicator and audience overlap.

Conservation sits at the intersection of humans and their relationship to the environment, meaning it spans the social and natural sciences (Bickford et al. 2012). Communicating in this space then requires skills from both domains, encompassing not only the clear conveyance of scientific fact but also packaging it in such a way to inspire action. The challenge then becomes selling an idea to an audience, which is where research has begun to focus on the idea of "conservation marketing" (Wright et al. 2015). In essence, conservation marketing involves using the techniques and tools of traditional marketing, which aim to identify and influence audience members' preferences and behavior. Marketing begins with identifying a target audience and then conducting comprehensive research on their values, interests, and decision-making processes (Wright et al. 2015), much as this chapter has discussed so far. Use Activity 1 to understand your audience, and how else you may be able to reach them.

Activity 1: Power Mapping Your Audience

Other resources: https://movetoamend.org/toolkit/guide-power-mapping

Power mapping is a method to map out a strategy to successfully influence a person or issue and is commonly used by advocacy groups or political campaigns. It can also be very helpful when mapping your communication strategy with specific audiences, especially if your goal is to get your audience to "do" something, i.e., change a behavior, influence policy, or fund your research.

Power mapping allows you to visually plot influence, power, and relationships. It helps you to understand knowledge or acceptance of your topic from different points of view and whether you might be approaching a friendly or hostile audience. This exercise also allows you to map out who or what has influence over your audience, which is helpful in understanding what they care about and finding common ground.

There are two elements to creating communication power maps: graphing knowledge or power (Fig. 7.2), and then mapping circles of influence (Fig. 7.3).

First, get a marker and a big piece of paper and write your communication goal at the top. Then make a list of the different audiences you are trying to reach or might find yourself interacting with. The more specific you can be about your audience the better (i.e., do not just say the public): *2nd graders* or *Mr. X's class* is better than just *kids*, parents of 2nd graders with younger siblings is better than just parents; *Drs. O'Connell and McKinnon who are experts in science communication* is stronger than just communication researchers. The more specific you can be, the more accurately you can map your audiences.

Draw two axes: the *X*-axis is whether the person is *Likely to Agree* or *Disagree* with what you have to say, and the *Y*-axis is either *Level of Influence* or *Level of Knowledge*—pick one depending on your goal. For example, if your goal is to influence someone to do something use *Influence*; if it is to get them to think or feel something, go with *Knowledge*. Then start to plot your individual audiences on your graph (Fig. 7.2).

Once you have your audiences mapped, you can start to think about different strategies to reach each group, depending on their level of knowledge or influence, and whether or not they are likely to agree with your topic. You may have to readjust your goal for more hostile audiences or employ different communication strategies.

(continued)

Activity 1: Power Mapping Your Audience (continued)

The second part of this exercise is mapping circles of influence (Fig. 7.3). Start on another piece of paper and pick a specific audience or target from your first audience map—this is level 1. Write that target in the center of the page and draw a circle around them. Then write down all of the people or things that directly influence that person, placing them equal distance around your target—this is level 2. The next step is to go out one or two more levels, repeating the exercise for each person or place you draw on your paper. Take a step back and look at where you may have influence over something/someone on the map or where you have a connection, even if it is four levels out. This can help you find common ground with your audience, which is always a good place to start with your communication strategy. You may decide to focus your science outreach on presentations to local community groups because they have great influence over your target or choose to focus on how your issue impacts public health, as it is one of the issues your target cares about. For example, if your goal was to get local Policy Maker X to support a law that would limit single-use plastic bags, you could see from Fig. 7.2 that your policymaker is not very supportive of this legislation, but has the most influence over whether it will happen. However, Fig. 7.3 shows you that there are other people/groups that have influence over your policymaker. If you can find one that you have common ground with or that you have the ability to influence through effective communication, e.g., coastal clean-up groups, then you may have a shot at moving your policymaker to take action. Of course, it is not usually this simple and requires time, effort, and strategy; but a power map can help you make the best use of your time and energy, as well as clarify your strategy when it comes to your communication efforts.

***Goal** (What do you want your audience to think, feel or do?):* e.g., *support legislation to limit single-use plastic bags*

Audiences: e.g., *Parents at the beach, kids at the beach, Mrs. X's 9th grade class, coastal homeowners, businesses who manufacture plastic, local policymaker X, PTA President, environmental science college majors, coastal clean-up group.*

7.6 Best Practices in Science Communication

To be an effective science communicator you need to do more than just follow a playbook or read a list of tips. It is a skill that needs to be developed and honed, just like mastering any new hobby or discipline. Effective communication requires effort and practice—whether it is being a more effective speaker, writer, teacher, or being more effective with visuals, social media, or the press. Creating an authentic connection with someone else is a choice and paying dynamic attention takes work. However, the payoff can be huge for your career, science, and research.

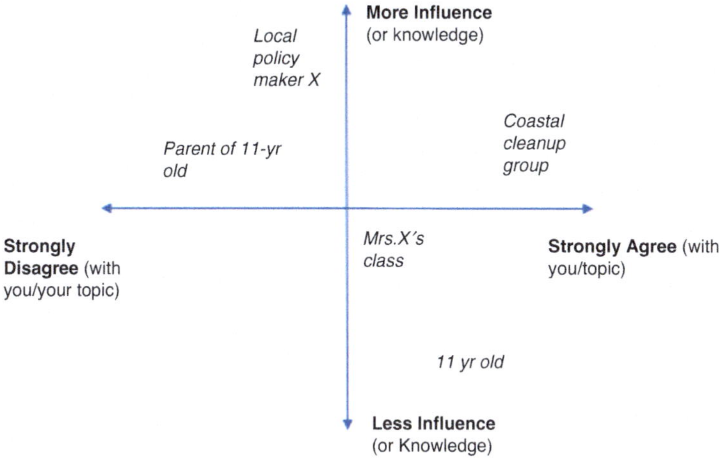

Fig. 7.2 Power mapping

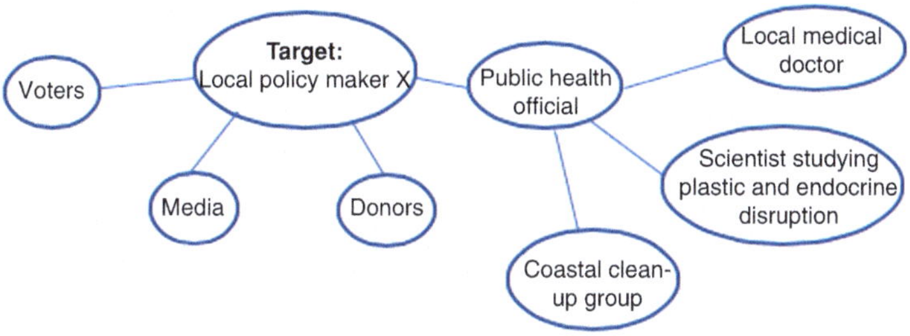

Fig. 7.3 Circles of influence

Scientists who learn to communicate effectively with different audiences are better at raising money through government grants and private donors, are better teachers, can inspire curiosity or engagement, help policy to be based on sound science, bring attention to science topics including via the media, and can even become a stronger scientist themselves. Science communication training has become a growing element in the education and professional development of scientists (Baram-Tsabari and Lewenstein 2017; Bray et al. 2012) and there are many organizations that provide different types of communication training around the world.

There are recognized best practices in science communication that will help you be a better science communicator: listening, empathy, having a clear goal, and knowing your audience. These principles hold true for written, spoken, or visual communication. Below are some science communication best practices from the literature and from our experience as science communication experts and practitioners. One of the most common mistakes when communicating science to the public is getting stuck in your own thoughts instead of focusing on and connecting with the audience. Luckily, you can learn how to focus and connect—here are some techniques that can help.

7.6.1 *Developing Listening and Empathy Through Improvisation and Connection*

There are some organizations, like the Alan Alda Center for Communicating Science and ImprovScience in the United States, that use theater and improvisation exercises to help teach scientists to be better communicators by building skills in connection, focus, audience, and presence. Improvisation—or improv—helps participants fully engage in the moment and develop an authentic and personal connection with their audience and a deeper understanding of people's behaviors and motivations (Bernstein 2014; Toivanen et al. 2011). The core tenets of improvisation involve the principles of "Yes, and" and "Making your partner look good": saying yes to the situation presented to you and adding your point of view, listening, paying deep attention to and supporting your partner (Bernstein 2014; Rossing and Hoffmann-Longtin 2016). These tenets involve the building of empathetic relationships and a shift in your focus from yourself to the audience (Kaplan-Liss et al. 2018; Bernstein 2014). When your focus is completely on your audience, your anxiety lessens and you can be more reactive and present. Improvisation facilitates participants to be in the moment, and to be mentally, emotionally, and physically engaged in listening and responding to their audience. You know when they are getting what you are saying, and you know when you have lost them and need to try something different. Improv is a great way to build skills that help you focus on your audience and help you move a conversation forward to achieve your goal.

7.6.2 *Shaping Your Goal*

Picture the best science talk you attended this year—how many things do you actually remember from it? If you are lucky, it is one or two things. Make your goal one of those things! This is incredibly important to remember for any talk, figure, or piece of writing. You should know the one or two things you want someone to remember when they leave that room or put down your article—be perfectly clear about it and repeat it a few times throughout the talk/article so it is not missed. Do not be shy about highlighting key phrases. For example, if something is important, point it out with phrases like: "the most important thing to think about is ..." or "what I am most excited about is...".

It is also important to know the goal of your audience because it may be different from your's. Imagine you want your boss to invest more money in a new program. However, your boss is responsible to a board that may be cutting her budget, so her goal is to spend less money. If you just ask for money it will be in direct conflict with her goal, but if you frame it first as a long-term way to save, or even make, the company money, she might be more receptive. You should always address your audience's goal first; otherwise, they will be thinking about their bottom line and not fully listening to what you have to say.

7.6.3 Engage Your Audience

Given the identified importance of listening (Baram-Tsabari and Lewenstein 2017) and audience focus (Evia et al. 2018; Besley and Tanner 2011) as key component skills of effective science communication, it is important to actively engage your audience during a talk or presentation: make eye contact, be present and authentic. We advise against trying to memorize a talk, as it often makes one disengaged from the audience and focused on what comes next rather than whether the audience is actually "getting it". Have a strong start to your talk and a strong ending to wrap it up. Do not be afraid to invite audience participation through questions or anecdotes (what they have seen or experienced related to your topic or issue), just make sure you can retain control of the conversation. Be cautious of asking questions that assume prior knowledge; people do not like to admit not knowing or they may be starting from a different baseline or perspective. Allow people time to think and reflect before moving on. Listen with more than just your ears. Someone might say they understood what you said, but their body language and tone might tell you something very different. Your clearly articulated goal will help you take your audience to your intended destination but get comfortable with the idea that how you get there might be different every time.

7.6.4 Build Connections

If you can link a piece of information to something someone already knows then it is easier to remember and learn. As the communicator, it is your responsibility to do this for your audience. Ask yourself: Where might they have heard of the term before? Where might you have common ground? Research shows that if you can link new information to prior knowledge then the same area of the brain that is related to learning and education becomes active (Van Kesteren et al. 2014). This is especially important when dealing with an audience who might not agree with you. Finding a connection helps people move away from being defensive and can set them up to listen to what you have to say. Always try and start from a place of connection, even if it has nothing to do with what you came to communicate (see Activity 2). Build common ground and make your topic relevant to them. Ask a question and acknowledge their experience.

Activity 2: Building Connections: The many whos

Get out a piece of paper and pen. Draw a line down the center and set your timer for 2 minutes. On the left-hand side of the sheet of paper write a list of the many whos you are. Try and write out as many things as you can.

Start each line with I am...

For example:

I am a marine scientist

I am a communicator

I am a dog lover

I am passionate about education

I am a mother

I am an ocean lover

I am a terrible cook.....

Now, put your focus on the other side of the page. Write down the name of the person you are trying to talk with at the top and be specific. It may help to pick someone you know who might exemplify the type of audience you will be communicating to. Repeat the same exercise but from their point of view.

They are a conservative

They do not trust science

They are good businessmen

They care about their kids

They love fishing

They love their boat

They value being out on the ocean

They are not trusting of regulations....

Do this for the full 2 minutes. Try and think beyond the surface of how you see this person. Instead go deeper—think about what they care about, what they value.

Now, look at the two lists and circle anything you might have in common. This is often the place you should start. You may not be able to agree on current fishery regulation numbers, but you can both agree that you love boating and being out on the ocean and want to make sure there are fish to catch. Start there.

7.6.5 Get Them to Care

The public generally cares about a science topic because it either (1) affects them (2) taps into their sense of wonder and awe (i.e., it is cool), or (3) because they care about *you*, the messenger. If something is practical or relevant to someone's life, they are more likely to care. For example, I care about cancer research because I know someone who is suffering from cancer. We also care because something is cool, exciting, or sparks our interest—science can be crazy, unexpected, and awe-inspiring. Big questions like "Why are we here and how does the world work" intrigue people. Remember the wonder, intrigue, and inspiration in your work and do not underestimate its ability to get people interested.

One of the best ways to get people to care about your science is to get them to care about you. Getting your audience to feel like they know you, even just a bit, is an extremely important tool in building trust, connection, and listening. Try and figure out a way to make it personal and put the "I" back in your science. Simple changes, like saying *I* or *Me* instead of *We* can make a big difference. However, often concerns arise from scientists regarding using *I* instead of *We*, fearing appearing arrogant or not giving credit to their team. In scientific culture this is pervasive: the "*We*" is used in writing papers (which are often written in third person) and speaking to show the research was a team/lab effort. However, people are there to listen to you; they want to hear what you care about, the piece you are most excited about, and what you are frustrated with. Instead, to give credit, try something like: "I work with an amazing team who were invaluable in conducting this research, and here is the part I am most excited to tell you about...".

7.6.6 Get to the Point

Paint the big picture, set the scene, and let us know why we should care upfront—what is the "so what"? Think about the structure of a scientific paper: there is the background, methods, results, and discussion. The discussion is the "so what" of the paper, telling us why we should care, what is next, and putting the research in context for something bigger. We are trained to save this until the very end for scientific manuscripts. This format may be appropriate for a scientific paper (although many scientists read papers backward, starting with the discussion!), but it does not work when you are trying to communicate scientific information to the public. You have to flip the order. Start with the "so what" and then build in complexity as your audience is ready for it. If you are having trouble distilling your message down to the "so what" of your work, try an improv exercise called *Half-life* described in Activity 3.

Activity 3: Half-Life Exercise

Recruit someone who is not an expert in your area of science. Set the timer for 2 minutes and talk about your research for that duration. You can pick a project or your broader research interest. When you are done, make sure to ask your friend for constructive criticism: What did they think the point was? What was most memorable to them? Where were they lost, confused, or bored?

Listen to this feedback, restructure your talk accordingly but do this quickly because you are now going to try it again, in a shorter timeframe. Now you

(continued)

Activity 3: Half-Life Exercise (continued)
only have 1 minute to talk about your research, so you have to distill it down even further. What are the most important points to get across? What is the "so what"? Start your timer and begin.

Did you get to your "so what" this time? If not, then you have to make sure you put it first and not save it for the end of your talk.

Try it again (I hope you are starting to see the pattern in this exercise. . .what do you think comes next?). Repeat and set the clock for 30 seconds (were you right?) and remember to ask for feedback after each round.

Repeat again at 15 seconds and finally at 7 seconds (yes, 7 seconds!). Your 15-second version usually ends up being your "so what?" because it is all you have time to say. Think about the 7-second version as your movie or book title—the bottom line. What can you say in the first 7 seconds that gets your audience to want to come to your talk or read the rest of the article?

Ask your partner which version they liked the best and why. Often it is the shorter versions because they got to the point clearly and concisely. Sometimes people prefer the longer version and this is usually because the speaker does a great job of engaging them using stories, examples, metaphors, or analogies.

7.6.7 Keep It Simple

I am sure you have had that experience—talking about a topic you know very well and watching the other person's eyes glass over, or look down at their phone, or worse, make an excuse and walk away. Perhaps even you have done this after meeting someone and having them launch into a technical and dense explanation of what they are working on. This is the danger of jargon.

While jargon has its purpose and can be efficient, accurate, and familiar, it can often be a barrier for public audiences leading to their disengagement and disinterest. Instead, when communicating with publics, avoid jargon and use everyday language. You can simplify your language and still be accurate.

If you have a term you want your audience to know, then define the term first before you introduce it. For example, if you want them to understand hypoxia, do not start with the term itself, but describe the concept. Once they get it, then introduce the term: *Similar to how people need oxygen to breathe, so do fish and other organisms in our bays and harbors. Sometimes that oxygen gets used up in our waterways and animals get stressed or die. When there is little oxygen left in the water, we call that hypoxia.* If you mention the term first and people have to stop and think about what it means, you have already lost them—they have stopped listening.

7.6.8 *Beware of the Curse of Knowledge*

We all suffer from the curse of knowledge, no matter our field of work. When you study and work on something with resolve, you forget what it is like not to know (Carmer et al. 1989). This is where empathy is essential. You must put yourself in the shoes of your audience using the techniques described in this chapter. Often, as an expert studying and interacting with something every day, the curse of knowledge prevents us from remembering what drew us to do this work in the first place. We forget our passion and often forget what is exciting, novel, and surprising about our own research. For example, you might have forgotten what it was like to see plankton under the microscope, or seen a particular species in the wild, for the first time. These are the moments that inspire people, do not forget to share them.

7.6.9 *Make it Memorable*

Visual imagery, language, or details are great ways to make your science memorable, i.e., make it stick. How can you use words and details to show your audience a picture of your research or get them to care about a conservation topic? How can you use body language and tone of voice to show you care, *versus* just telling your audience you do? People are visual. Showing your audience something carries more weight than simply telling them. If someone sees something, even if it is just a picture in their head, it makes it easier for them to remember. Give an example whenever you can. Try using words like "imagine" or "picture this"—these can help engage the audience and get them actively involved, even empathize with you!

Do not shy away from emotion in your science communication. Emotion can be a charged word for scientists. However, evoking emotion when communicating science is crucial. If there is an emotion attached to something, it is more memorable (Kuriyama et al. 2010; McGaugh 2013). Emotion comes in many forms: wonder, awe, excitement, frustration, passion, fear, curiosity, humor. Show your passion for your work in your communication. Try and get your audience to leave your talk feeling something instead of just thinking something—it is a much stronger goal.

All data have a story to tell, or sometimes many stories. A good visual can help you focus on the right story, the one you want your audience to see. Visuals follow the same rules as speaking or writing—know your audience, have one or two clear goals, and highlight the data that supports your goal. All too often scientists will try and display all of their data or have multiple points they are trying to make on the same graph. This can be overwhelming and confusing for the audience. Pick your goal first, decide what story you want to tell with the data, and then design your presentation or visuals around these. This applies to the type of graph you use, too. Sometimes you may decide not to use a graph at all. Instead, an image might better suit your goal and audience. Pictures can be much more memorable because they elicit emotion; your audience will focus on you and what you are saying for the content, instead of being distracted trying to decipher a graph on the screen behind you.

Metaphors and analogies are extremely helpful tools in getting your science to stick and paint a picture for your audience. They can be especially helpful in explaining complex conservation topics. But, do not try to make every aspect of your science concept fit within the metaphor or analogy. This is where people get into the weeds. You only have to compare one thing, one aspect of your topic. For example, you can say, "Climate change is kind of like a coffee mug in this way, but here is how it is different..." Coming up with a metaphor or analogy is easier than you might think. Try it out in Activity 4.

Activity 4

Pick a conservation topic you want to explain and then pick three random objects from the room you are in (e.g., an orange, a mug, and a pen).

Give yourself the challenge of using one of those objects to come up with a way to describe the concept. It will not be perfect but we bet you can make one thing fit! Try this with colleagues. Everyone sees things differently and it's a fun way of exploring ideas.

What does all this mean for bridging the knowledge–implementation gap?

To get people to take action, you need them to understand and to care. The best way you can get people "on your side" is to actively demonstrate that you care about them! By taking the time to listen, to understand their pre-existing knowledge, beliefs, values, and concerns, you can then make sure that what you say—and how you say it—is appropriately tailored for their needs.

How you interact with your audience will influence their perception of the issue. Sometimes the most powerful tool in communication is not the message, but the messenger.

7.7 The Power of Stories

Stories are the secret weapon of science communication. They evoke emotion, visuals, and personal connection—all ingredients to engage an audience and make your science stick (see Case Study 1). Civilizations from thousands of years ago passed down important information from generation to generation in the form of legends, myths, or stories and we still do this today. Stories resonate and stick with audiences over just facts, and our brains are wired to receive them—they help us to connect. Studies show that similar areas in our brains light up in both listeners and storytellers (Stephens et al. 2010).

> Stories are meaning-making devices. They help us collectively process ambiguous events and decide who our heroes and villains are, what were the consequences of an action, and

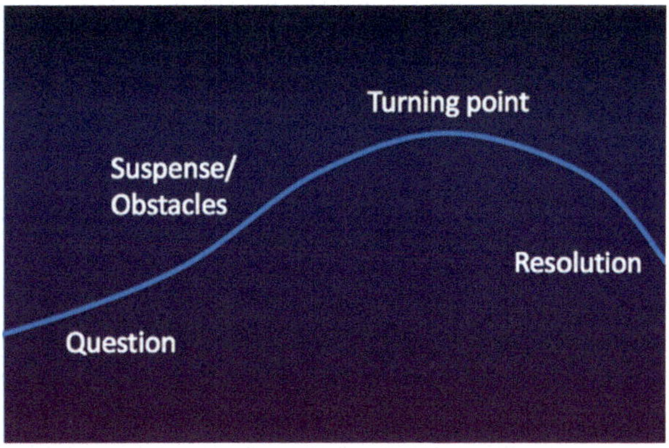

Fig. 7.4 Story curve

how we will behave in similar situations in the future. When you compare them to evidence-based argumentation, the best research we have on hand suggests that narratives are more engaging, more comprehendible, and more persuasive. Given the enormity and heart-breaking scope of all the environmental and social issues we face right now, I firmly believe we need to use every tool at our disposal to inspire, engage, and take collective action.—Liz Neeley, Principal, Liminal Creations.

Science and storytelling are natural partners. Science is filled with stories: someone wants something, struggles to get it, or faces obstacles along the way, then there is a turning point and they either achieve what they were hoping to or they do not, but either way, something has changed. They never end up in the same place as they left off—they have gained new insight, discovered a new species, method or theory. Every time a scientist steps into a lab or starts a new experiment, the elements of a story come into play. However, it is important that science stories are grounded in robust data and sound science (Leslie et al. 2013).

Narratives can increase the comprehension and appreciation of complex scientific findings and research (Dahlstrom 2014). Stories are a great way to show *versus* tell—show the motivations or strengths of a character, the importance of a lesson, the relevance of a point. But you have to learn how to craft it; storytelling is an art. Think about good storytelling as a left (or negative)-skewed parabola (Fig. 7.4). A good story, just like a left-skewed parabola, builds up slowly onto a high point and then drops off more quickly afterward to a resolution.

Think about what makes a good story for a second. What do your favorite authors do in the beginning to get you engaged in the book? Most likely, they set the scene: they get you to empathize or connect with the character and see their emotions and state of mind. They paint a vivid picture in your mind with their words, and then they build suspense or anticipation. They may get you to ask yourself a question: Will they make it? Will they find what they are looking for? Good stories, just like good communication, have a goal or a point. This is where you want to spend some time setting the scene and getting your audience engaged. Then focus on the build, let us know what prompted your journey of investigation, the struggles you overcame along the way, and the moment you realized you would (or would not) achieve your goal, then what happened and what is next (O'Connell 2016).

Stories can include both robust science and the not-so-glamorous descriptions of fieldwork, challenges you faced along the way of collecting the data, or successes and triumphs (Gross et al. 2018). Often the obstacles in your research and how you dealt with them are the most interesting part. How you overcame the barriers will often show your resilience, passion, and leadership qualities, something that might be important for you to communicate to the public or policymakers for instance. Anyone can say "I'm passionate about this" or "I was meant to do this job." But again, you want to show as opposed to tell—tell me a story that shows your passion for your research, show me why you care about the science. All too often when dealing with communicating data we start with the question and jump to what we found, leaving out the good part—the suspense or struggles and the obstacles, the part that makes it a story instead of just a report.

Stories always have an element of time attached—it is what makes them different from textbooks or lectures. It is why most fairytales start with "Once upon a time..." Something always changes in a story (even if it is just the main character's point of view), and for something to change, there has to be an element of time associated with it. If you are ever struggling with how to tell a story, a good place to start is with the word "when," as it automatically introduces the element of time.

Case Study 1: Storytelling and Making Conservation Movies

Interview with Carl LoBue, PhD, New York Ocean Programs Director at The Nature Conservancy, New York

Many conservation outcomes require that policymakers get involved and make decisions. If policymakers do not have backing from their constituents (even if the individual policymaker supports it) they may decide not to take action or worse—they could make decisions that are in opposition to the science.

I was working with some conservation scientists on issues related to water quality and nitrogen on Long Island (LI) in New York. Much of the water quality problems on LI have to do with the biogeochemistry of the nitrogen cycle. So, we funded a series of lectures—if people understood the science, they would understand the urgency.

We had an expert scientist give a 1.5 h technical seminar on the Nitrogen Cycle with over 80 people in attendance, including local politicians, members of our board, and the public. It became obvious that this might not be working the way we wanted from the questions and looks on people's faces. People told us afterward that they felt like they knew less coming out of the meeting than they did coming in.

We canceled most of the other lectures. We realized that people do not need that much information—what they need is to put their trust in experts like

(continued)

Case Study 1: Storytelling and Making Conservation Movies (continued)
us. For example, when my electricity does not work, I hire an electrician. I do not need to know all of the details for how they will perform the work, I just need to know that I trust them and that they can fix it.

We developed a series of water quality videos and distilled the message to: (1) it is a serious problem, but (2) there are solutions and it is solvable, but (3) we need to make the investments in it now. Then we used real stories of actual residents to amplify our message, each 3-minute narratives pieced together from hour-long interviews. Our movies have been very effective for people who see them.

Each story will not resonate with everyone so we filmed enough to have a broad reach. When I go talk to a specific audience, I think about which one of these dozen stories I should use. The latest one is with an orthodox Christian priest. It ruffled some feathers because people have preconceived notions of Christianity—however, it really resonated with other audiences. We targeted it to people whose faith has an impact on how they make decisions about life.

About the Films

The animated films were about getting a few facts out that we were unable to do in the personal stories. This was how simplified we needed to get after the failed scientific seminar events. WAY less detail: Where does it go when we flush? (https://www.youtube.com/watch?v=YtP6VfeZsws&feature=youtu.be), What is nitrogen pollution? (https://www.youtube.com/watch?v=Gv5kfXRARN0&feature=youtu.be), and Where does our water come from? (https://www.youtube.com/watch?v=pGgb0EvO6NA&feature=youtu.be).

The 11 personal stories are packaged as a series and preceded by a trailer: https://vimeo.com/165524793. All of the films can be accessed at www.nature.org/longislandwater. Last year, we won Telly awards for *We're Oyster Farmers* (https://vimeo.com/129951378), and *Foraging the High Seas* (https://vimeo.com/237809165) about five Long Islanders' perspectives on why we need to change the way we manage bunker fish.

7.8 Listen and Practice

7.8.1 Listen

It is a great practice to watch, listen, and learn from others. Take note of where and why you were most engaged with what they were saying or writing. The more you listen to or read other science communicators, the more you learn what works and what does not. Next time you listen to a science podcast or great talk, think about why the speaker was engaging (or not), and what exactly about their content, voice, or visuals made it so compelling? Do not underestimate the value of one-on-one interactions you have in the field, at the market, or with family members at a holiday celebration. These can often be the most informative because you can directly build trust with your audience and also get direct feedback to improve your communication by asking questions about what was interesting, boring, or confusing for them as listeners.

7.8.2 Practice

Explore the many fun and creative forms of science communication. Be inquisitive, experiment, learn and revise—in other words, be a scientist! Collaborate with others, including non-scientists. For example, scientists can partner with artists or historians to communicate their work. There are centers, fellowships, and festivals that help promote such collaborations such as Beakerhead in Alberta, Canada (see Case Study 2). Other suggested outlets include:

- *Social media*—applications like Twitter and Instagram force you to distill your message into a limited number of characters and/or with impactful images.
- *Podcasts*—perfect your storytelling either about your research or work, your personal journey in conservation, or something else entirely!
- Writing personal or professional *blogs* to communicate to various audiences and practice being clear, concise, and engaging. There are many online platforms that will take guest blogs or commentary pieces including Scientific American, Nature, MassiveSci.org, or The Conversation.
- Engage local *media*; it is often easier to get published and they can have a large following. Local radio, TV, or newspapers will often send reporters to cover local science stories. Start to develop and foster relationships with individual reporters and media outlets. Offer your expertise on various topics and invite reporters to the lab or out in the field.
- Write *press releases* when you have exciting or novel research findings or want to get information on a science topic or event out to the public. Communication and marketing staff at your organization are also excellent sources of assistance with this, and in preparing for interviews.
- *Opinion Editorials or Letters to the Editor* are a great way to learn to write clearly, concisely (most are only 150–200 words) and get across a point of view. They are also relatively easy to get published and are widely read. As a scientist, you bring an expert opinion to the conversation and your letters can carry more weight.
- *Kids and teenagers* are great audiences to practice with. Can you explain your science to an 11-year-old child in an engaging way? Many newspapers aim to write for a 7th or 8th grade audience, so it is good practice to aim your talks for a general audience at a 7th-grade comprehension level. If you can clearly and engagingly explain it to an 11-year-old, then you know your stuff (O'Connell 2016).

Case Study 2: Art and Science

Interview with Carolyn Hall, Historical Marine Ecologist and Professional Dancer (also, artist, adjunct professor, communication coach). (www.carolynjhall.com)

Communicating Science Through Art

Science communication is crucial because there are many really important issues (e.g., health of our waterways, impending climate change impacts) that affect people daily that don't get communicated, get misrepresented, or get ignored because people do not understand. The best way for people to understand how something in science is affecting them is to *make it local and make it something they can touch, see, hear or feel.* Science communication needs to be personal; people need to understand a way that they can relate to it. That is where I come in as an Artist.

Art evokes emotion and connection. Art is personal whether it is performative or visual. It is a sensory and embodied way of learning. Science, when it is presented, can often be very cerebral and static. Art provides another lens/way of learning—it gives people another way in.

Art has helped me to be a better science communicator and scientist. When I start a new research project, I begin thinking about ways to translate it to people who are not in my research community. I imagine how I can represent it in a more physical or visual way for someone to better understand it. From my experience, scientists who collaborate with artists find the experience helps them to ask questions differently and look at their research through a broader lens.

The Sunk Shore Project (https://clarindamaclow.com/projects/tryst-overview/tryst-sunk-shore/)

One project I did in lower Manhattan (2017) and Governors Island (2018) is called Sunk Shore. It incorporates art and experiential learning to communicate local facts and speculative futures about the effects of climate change on our shores—while walking along an actual shoreline. Partnering with Clarinda Mac Low and Paul Benney, we base the walks on factual data about the past and present of a specific site, and explore how the weather, flora, fauna, and terrestrial and aquatic ecosystems of a place have changed over time and how they have been and may be affected by our human activities. We *make the data come to life* by putting people into a scenario where they move forward in time, or time travel, and include them in the narrative requiring imagination, props, physical movement, and engagement. With sensory cues and on-site interaction, we guide our travelers to move from the past to the future by re-envisioning existing structures or landmarks of the site, by utilizing props we have designed, and by executing simple choreography all the while delivering data about how the ecosystem has or will be changing—they get to see, hear, and experience the effects of climate change in a tangible way

(continued)

Case Study 2: Art and Science (continued)
(Fig. 7.5). By *asking them to also create, they invest more personally and deeply in the outcome* which inspires ownership and care of the site and of possible solutions.

Participants may walk past a building their whole life but never think about what was there before or how it will be affected in the future. For example, what once was a sandy beach was built into a facility that houses electrical equipment, but, given rising sea levels in New York City, it will continue to get flooded and the equipment will be ruined. Now, what can or should it be? Bringing people into the discussion with the site in front of them makes climate change very tangible and present. And, *people have fun: learning facts is not work, their curiosity is sparked, and they have more questions*: "What can we do to avoid the worst-case scenario?" "What can *I* do?" Instead of being overwhelmed by data and consequences, they see things they value and want to protect.

Artists and scientists can make great partners, especially for people who have even the smallest sense of activism in them and want their work to serve a bigger purpose. Many artists work on social, political, and environmental issues. When they are inspired by an environmental issue or natural resource, artists want to know more. Art and science can be very similar—they both require creative states of mind, trial and error, and experimentation.

Best Practices Using Art and Science to Communicate

My most successful science talks are when people are more immersed in the environment and actually doing the collection or being able to see and touch equipment or subjects being used in the fieldwork or laboratory. An artistic and teaching moment was successful when it was more tangible. For example, instead of just telling people about the natural beaches covered in shells that used to exist where they could only see chain link fences and broken down docks in front of them, we covered a small area with oyster shells and had people walk over them toward the water. They had a tactile experience and change of sense of place that involved hearing, feeling, and seeing—they were transported even though it was very simple. *People desire a physical reference and experience to be grounded in an area; otherwise, it can still feel like a lecture, just in a nice place.* When people are physically doing and feeling things, it helps them relate better to the information they are receiving.

7.9 Bridging the Knowledge–Implementation Gap

Communication is the bridge between knowledge and implementation. Facts alone will not drive change. Make people care, connect with them as humans and have conversations based on a common ground of shared values and beliefs. Our planet is a literal and metaphorical common ground—we all have a vested interest in its

Fig. 7.5 Participants at *The Sunk Shore Project*

survival. Connect, listen, empathize, and persevere. Communication is like a muscle—the more you use it, the stronger and more effective it will become.

References

Baker B (2019) Biologists as Communicators: Lessons from social science can help researchers get their message across. Bioscience 69(2):101–107

Baram-Tsabari A, Lewenstein BV (2017) Science communication training: what are we trying to teach? Int J Sci Educ Part B 7(3):285–300

Bashir NY, Lockwood P, Chasteen AL, Nadolny D, Noyes I (2013) The ironic impact of activists: negative stereotypes reduce social change influence. Eur J Soc Psychol 43(7):614–626
Bernstein R (2014) Communication: spontaneous scientists. Nature 505(7481):121–123
Besley JC, Tanner AH (2011) What science communication scholars think about training scientists to communicate. Sci Commun 33(2):239–263
Besley JC, Oh SH, Nisbet M (2013) Predicting scientists' participation in public life. Public Underst Sci 22(8):971–987
Bickford D, Posa MRC, Qie L, Campos-Arceiz A, Kudavidanage EP (2012) Science communication for biodiversity conservation. Biol Conserv 151(1):74–76
Bray B, France B, Gilbert JK (2012) Identifying the essential elements of effective science communication: what do the experts say? Int J Sci Educ Part B 2(1):23–41
Brewer PR, Ley BL (2013) Whose science do you believe? Explaining trust in sources of scientific information about the environment. Sci Commun 35(1):115–137
Bubela T, Nisbet MC, Borchelt R, Brunger F, Critchley C, Einsiedel E, Geller G et al (2009) Science communication reconsidered. Nat Biotechnol 27(6):514–518
Carlsen B, Riese H (2016) High stakes. Nordicom Rev 37(1):85–99
Carmerer C, Loewenstein G, Weber M (1989) The curse of knowledge in economic settings: an experimental analysis. J Polit Econ 97(5):1232–1254
Cook J, Lewandowsky S (2016) Rational irrationality: modeling climate change belief polarization using Bayesian networks. Top Cogn Sci 8(1):160–179
Crook SES (2014) Information spread in a region of human-mountain lion coexistence. Hum Dimens Wildl 19(6):555–558
Dahlstrom MF (2014) Using narratives and storytelling to communicate science with nonexpert audiences. PNAS 111(Suppl 4):13614–13620
Dietz T (2013) Bringing values and deliberation to science communication. PNAS 110(3):14081–14087
Edelman (2018) 2018 Edelman Trust Barometer – Global Report. Retrieved from http://www.edelman.com/global-results/
Evia JR, Peterman K, Cloyd E, Besley J (2018) Validating a scale that measures scientists' self-efficacy for public engagement with science. Int J Sci Educ Part B 8(1):40–52
Gross L, Hettinger A, Moore JW, Neeley L (2018) Conservation stories from the front lines. PLoS Biol 16(2):e2005226
Health in Harmony (2019) What is "radical listening"? Online. https://healthinharmonyorg/approach/radical-listening/. Accessed 10 Jan 2019
Heupel MR, Simpfendorfer CA (2010) Science or slaughter: need for lethal sampling of sharks. Conserv Biol 24(5):1212–1218
Hurlimann A, Dolnicar S (2012) Newspaper coverage of water issues in Australia. Water Res 46(19):6497–6507
Jucan MS, Jucan CN (2014) The power of science communication. Procedia Soc Behav Sci 149: 461–466
Kahan DM, Jenkins-Smith H, Braman D (2011) Cultural cognition of scientific consensus. J Risk Res 14(2):147–174
Kaplan-Liss E, Lantz-Gefroh V, Bass E, Killebrew D, Ponzio NM, Savi C, O'Connell C (2018) Teaching medical students to communicate with empathy and clarity using improvisation. Acad Med 93(3):440–443
Kruse CR (2001) The movement and the media: framing the debate over animal experimentation. Polit Commun 18(1):67–87
Kuriyama K, Soshi T, Fujii T, Kim Y (2010) Emotional memory persists longer than event memory. Learn Mem 17(3):130–133
Leslie HM, Goldman E, McLeod KL, Sievanen L, Balasubramanian H, Cudney-Bueno R, Fuerstein A, Knowlton N, Lee K, Pollnac R, Samhouri JF (2013) How good science and good stories can go hand-in-hand. Conserv Biol 27(5):1126–1129

Liordos V, Kontsiotis VJ, Anastasiadou M, Karavasias E (2017) Effects of attitudes and demography on public support for endangered species conservation. Sci Total Environ 595:25–34
McGaugh JL (2013) Making lasting memories: Remembering the significant. PNAS 110 (Suppl 2):10402–10407
Munro L (1997) Framing cruelty: the construction of duck shooting as a social problem. Soc Anim 5(2):137–154
Muter BA, Gore ML, Gledhill KS, Lamont C, Huveneers C (2013) Australian and U.S. news media portrayal of sharks and their conservation. Conserv Biol 27(1):187–196
Nghiem LTP, Papworth SK, Lim FKS, Carrasco LR (2016) Analysis of the capacity of Google trends to measure interest in conservation topics and the role of online news. PLoS One 11(3): e0152802
O'Connell C (2016) Use the proven power of stories to explain data. The water cooler: a blog of new ideas and findings from organizational Leaders and researchers. Google's re:Work. May 30, 2017. Permanent address: https://rework.withgoogle.com/blog/use-stories-to-explain-science-and-data/
Oreskes N, Conway EM (2010) Merchants of doubt. Bloombury Press, New York
Pechar E, Bernauer T, Mayer F (2018) Beyond political ideology: the impact of attitudes towards government and corporations on trust in science. Sci Commun 40(3):291–313
Rosalino LM, Rosalino C (2012) Nature conservation from a junior high perspective. J Nat Conserv 20(3):153–161
Rossing JP, Hoffmann-Longtin K (2016) Improv(ing) the academy: applied improvisation as a strategy for educational development. Improve Acad 35(2):303–325
Sabatier E, Huveneers C (2018) Changes in media portrayal of human-wildlife conflict during successive fatal shark bites. Conserv Soc 16(3):338–350
Stephens GJ, Silbert LJ, Hasson U (2010) Speaker-listener neural coupling underlies successful communication. PNAS 107(32):14425–14430
Toivanen T, Komulainen K, Ruismäki H (2011) Drama education and improvisation as a resource of teacher student's creativity. Procedia Soc Behav Sci 12:60–69
Van Kesteren MTR, Rijpkema M, Ruiter DJ, Morris RGM, Fernández G (2014) Building on prior knowledge: schema-dependent encoding processes relate to academic performance. J Cogn Neurosci 26(10):2250–2261
Wright AJ, Veríssimo D, Pilfold K, Parsons ECM, Ventre K, Cousins J, Jefferson R, Koldewey H, Llewellyn F, McKinley E (2015) Competitive outreach in the 21st century: why we need conservation marketing. Ocean Coastal Manage 115:41–48
Wynne B (1992) Misunderstood misunderstandings: social identities and public uptake of science. Public Underst Sci 1(3):281–304

Chapter 8
Making an Impact: How to Design Relevant and Usable Decision Support Systems for Conservation

David Christian Rose, Katherine Despot-Belmonte, James A. Pollard, Olivia Shears, and Rebecca J. Robertson

Contents

D. C. Rose (✉)
School of Agriculture, Policy and Development, University of Reading, Reading, UK
e-mail: d.c.rose@reading.ac.uk

K. Despot-Belmonte
UN Environment World Conservation Monitoring Centre, Cambridge, UK

J. A. Pollard · O. Shears
Department of Geography, University of Cambridge, Cambridge, UK

R. J. Robertson
School of Biology, Irene Manton Building, University of Leeds, Leeds, UK

C. C. Ferreira, C. F. C. Klütsch (eds.), *Closing the Knowledge-Implementation Gap in Conservation Science*, Wildlife Research Monographs 4,
https://doi.org/10.1007/978-3-030-81085-6_8

8.1 Introduction

The interfaces between evidence, policy, and practice have been the subject of much research in the fields of environmental management and conservation (e.g., Marshall et al. 2017; Rose et al. 2018a; Young et al. 2014). In briefly summarising this large body of literature, there are several common themes which have led scholars to characterise these interfaces as complex and messy (Lawton 2007). The lack of linearity between evidence and decision has been commonly identified (see Evans et al. 2017a, b) and studies have widely described a 'gap' between the different worlds of science, policy, and practice (see Rose et al. 2018a). Such a gap makes it difficult for evidence to be communicated to decision-makers, while the needs of practitioners and policy-makers struggle to shape scientific agendas (Arlettaz et al. 2010). Decision-making, therefore, is usually never based on evidence alone, particularly in controversial or 'wicked' issue contexts, nor in fact should it be in a functioning democracy where values, beliefs, and interests matter (Owens 2015; Rose 2018). Furthermore, research has illustrated that evidence can take many forms, including knowledge that may be considered scientific within academic communities, but also lay or indigenous knowledges based on experience, observation, and a close place-based connection with the environment (Montana 2017; see also Part I of this book).

In this chapter, we adopt a normative position which sees scientific evidence as important to robust decision-making. It is fair to take such a position in the light of continuing calls for 'evidence-based' or 'evidence-informed' decision-making in conservation (e.g., Gardner et al. 2018; Sutherland and Wordley 2017). In the simplest form, decision-makers desire evidence so that adopted policies have the best chance of succeeding in practice. Or put another way, they use evidence to minimise the chances of an incorrect decision and to increase decision transparency. In many fields, such as economics (OECD 2015), medicine (Sackett et al. 1996), and increasingly conservation (Sutherland et al. 2004), it is generally accepted that decisions should be informed by robust evidence. In medicine, for example, few, if any, patients would want their doctor to make a diagnosis without consulting the evidence. Patients would then expect their doctor to adopt an evidence-informed treatment plan based on clinical trials.

In nature conservation, a plea for evidence-informed decision-making has been made by a number of scholars (Gardner et al. 2018; Sutherland et al. 2004; Walsh et al. 2015; Sutherland and Wordley 2017). It is argued that we should expect conservation decision-makers, such as reserve managers, to base actions on evidence. If decisions are not evidence-informed, then actions may be undertaken that do not work, which wastes time and money without providing any tangible benefit to the target species or habitat, or indeed to people. There are, of course, differences between medical and conservation decision-making; for example, in medicine it is usually the case that a specific drug will cure a specific illness whether a patient has the same illness in Australia or Canada. Although conservation actions are somewhat generalizable (Roughgarden et al. 1994), there is more uncertainty associated

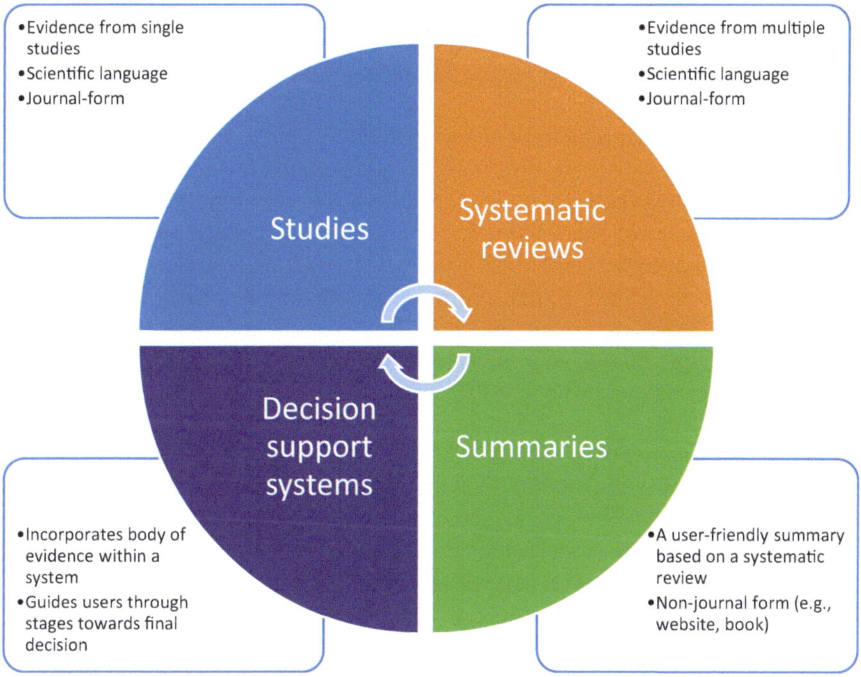

Fig. 8.1 Four routes through which evidence may be communicated to decision-makers, with a description, and the format in which they are usually presented (based on the '4S hierarchy', Dicks et al. 2014)

with comparing outcomes between different places, as there are far more factors to control for (Sutherland et al. 2017). For example, a successful strategy to conserve coastal saltmarsh in East Anglia, UK, may be inappropriate in different parts of the UK, let alone in a different international context because a number of factors vary (such as tidal range, climate, level of development; see also Chap. 7; O'Connell and McKinnon 2021).

Despite the challenges of comparability, it is still logical to argue that conservation actions would be more successful if they were informed by evidence of what works (Gardner et al. 2018; Sutherland et al. 2017). In the midst of a so-called communication gap between science, policy, and practice, Dicks et al. (2014) highlight four formats in which scientific evidence can be presented to decision-makers—studies, systematic reviews, summaries, and decision support systems (Fig. 8.1). Figure 8.1 provides a summary of these different formats and notes the predominant style in which they are generally (but not exclusively) presented. Decision-makers could be government policy officials at different levels, reserve managers from NGOs, or other people who make environmental management decisions, such as farmers, fishers, or businesses.

Studies are single pieces of research that may describe the results of an intervention aimed at one or more species. Systematic reviews collate lots of single studies together to give a broad overview of the body of literature (similar in many ways to meta-analysis that seek to combine data from different sources), while summaries take the results of a systematic review and offer a precis in simple, non-academic language.

There are, however, problems with trying to deliver evidence in these formats to practitioners. Firstly, single studies may provide selective evidence, an issue that may be overcome by systematic reviews/meta-analyses, which can give an overview of the body of evidence. This overview is likely to be more robust, and studies have found that decision-makers welcome syntheses of evidence, rather than individual studies (Rose et al. 2018a). Systematic reviews though, like studies, are generally inaccessible to decision-maker communities (Rose et al. 2018a), or may be written in jargonistic, complicated language which is difficult to interpret. Summaries attempt to overcome this problem by presenting a clear precis in relevant language, but again this relies upon decision-makers finding this information, interpreting it, and applying it to a decision context.

Decision support systems (DSS), however, offer a further layer of sophistication, providing a route through which evidence can be delivered in a usable form (Dicks et al. 2014). They tend to be computer-based, either in software- or app-based formats (although they can be paper-based), and incorporate evidence within the inner workings of the tool. In so doing, the tangible tool is able to take users through various decision stages towards a final decision. Systems can be dynamic in nature, in other words manipulating inputs provided by the decision-maker before suggesting evidence-based outputs, or they can act as information sources, offering further evidence in a cumulative decision process (Rose et al. 2016). DSS aim to integrate complex process-based models in an accessible interface, helping decision-makers use data to solve unstructured problems (Addison et al. 2013; McIntosh et al. 2011; Schwartz et al. 2018), therefore acting as 'boundary objects', bridging the gap between scientific evidence and the decision-making process.

In this chapter, we will give an overview of research on DSS in conservation and related disciplines up to 2019, with the express aim of identifying lessons to guide the design of good, impactful systems. As the discussion below will show, many DSS have been designed at great expense, but have sometimes failed to make an impact in practice as their intended audience did not use them. Since DSS have great potential to deliver evidence in a usable format to conservation decision-makers (and indeed to others outside of conservation), we thus need to ensure that they are well-designed so that they are actually used in practice. To this end, we draw out key principles for good system design and delivery with a focus on participatory user-centred design.

8.2 The Use of Decision Support Systems Beyond Conservation

A proliferation of formal decision support approaches has developed to assist with evidence challenges in decision-making (Ascough et al. 2008). DSS are increasingly considered, by both policy-makers and experts, to be productive routes to support complex decision-making structures (Van Kouwen et al. 2007). As previously stated, they can help to deliver evidence in a usable form for decision-makers and may help them to overcome complex challenges associated with biological, socio-economic, and political trade-offs. But, we make it clear at this early stage of the chapter that we do not present an argument that sees DSS as the only way to make conservation decisions, nor do we say that the advice of such systems should be blindly followed. Systems will only ever contribute to decisions, since there are other forms of knowledge (e.g., place-based knowledge) available, which can sometimes be just as valuable as the information behind technical algorithms (Rose 2018).

Before looking specifically at conservation DSS, it is worth reflecting on the considerable research on system design and uptake which has been conducted in fields with similar characteristics to conservation—medicine, agriculture, and coastal management are good examples because all should involve practitioners (e.g., doctor, farmer, coastal manager) making evidence-informed decisions. Much of this research draws on behavioural models which identify the factors affecting uptake of decision support systems or technology in general. The most well-known of these models is the Unified Theory of the Acceptance and Use of Technology by Venkatesh et al. (2012), which predicts that various factors determine technology uptake—these include whether the system performs well (performance expectancy), whether it is easy to use (effort expectancy), as well as a range of social (e.g., habits) and personal characteristics (IT education, age), and facilitating conditions (e.g., IT infrastructure).

In medicine, DSS have been designed to help medical practitioners use evidence to support their decisions (Rawson et al. 2017; Shibl et al. 2013; Thursky and Mahemoff 2006). Thursky and Mahemoff (2006), for example, report on the successful introduction of an antibiotic prescribing system for Intensive Care Unit use. This system reduced the time taken to perform prescribing tasks and it was readily used in practice. Furthermore, Shibl et al. (2013) discuss systems aimed at improving the use of evidence by General Practitioners. Their study concluded that various factors influenced system use, which built on the early work of Venkatesh and others—namely, performance expectancy, ease of use, existing decision habits, facilitating conditions, and age, experience, and gender (Shibl et al. 2013).

In agriculture, there has been much research on the subject of decision support system adoption over at least two decades (see Rose et al. 2016). Much research, however, has found limited uptake of systems by their intended end audience, usually farmers (Gent et al. 2013; Hochman and Carberry 2011; McCown 2002). In response to the problem of implementation, studies have set out to identify successful characteristics of systems that are actually used in practice; for example, Rose et al. (2016) listed 15 factors that influence whether a system is used (Box 8.1).

Box 8.1 Fifteen Design Features of Effective Decision Support Systems (from Rose et al. 2016)

1. ***Performance expectancy***—how useful a system is and whether it works well
2. ***Ease of use***—how easy a system is to use
3. ***Peer recommendation***—a system that is recommended by peers has a greater chance of widespread uptake
4. ***Trust***—how far end users trust the evidence underpinning the system or the manufacturer themselves
5. ***Cost***—whether a system is free, cheap, or expensive to buy is a key factor
6. ***Habit***—whether using a system matches existing decision-making habits or not, flagged by Rose et al. (2016) as a key factor
7. ***Relevance to user***—a system which gives information relevant to the user is important
8. ***Farmer-adviser compatibility***—whether a system was used by linked advisors
9. ***Age***—younger farmers tended to use computer-based systems more
10. ***Business scale***—bigger farmers used more decision support
11. ***Farming type***—different farming enterprises (e.g., arable *vs.* livestock) used systems more or less often
12. ***IT education***—farmers with higher IT education used computer-based decision support
13. ***Facilitating conditions***—farms with good internet or broadband connectivity were more likely to use DSS
14. ***Compliance***—whether a system helps farmers satisfy legislative or market requirements was important
15. ***Marketing***—the user had to know about the system in order to use it

There are thus a number of important considerations for system designers, which move beyond well-known criteria such as performance and ease of use. In addition, designers need to understand who the users are, including their decision-making habits, age, level of IT education, workflows, and individual circumstances, as well as assess the necessary infrastructure (e.g., connectivity) for system use, and adopt strategies for marketing, delivery, and implementation. Thus, it is inadequate merely to design a sophisticated system which is easy to use.

Other studies have shown the importance of involving users in the design of agricultural decision support systems (Allen et al. 2017; Evans et al. 2017a, b; Nelson et al. 2002; Lindblom et al. 2017) so that products are relevant, usable, trusted, well-known, sustainable, and easy to use. Indeed, participatory user-centred design in which users are involved in the conception, design, and implementation phases is now widely considered to be vitally important (McIntosh et al. 2011; Parker and Sinclair 2001; Rose et al. 2018b; Santoro et al. 2013). We discuss user-centred design in more detail later in this chapter. Important also to note is the tendency to focus on changing the behaviour of users in relation to technology, rather than focusing on the design of the product itself. This has resulted in users being blamed for non-adoption, rather than the technology, which may have been poorly designed (De Oca Munguia and Llewellyn 2020).

8.3 Using Decision Support Systems in Conservation

As of 2019, there are many examples of DSS being used to make evidence-informed decisions in conservation (e.g., in strategic land conservation planning) though there are likely to have been advances since then (Gibson et al. 2017; Anderson and Rex 2019). While decision-making in conservation sometimes involves conducting interventions in 'the dark' without good data (Cook et al. 2010; Regan et al. 2005), there is often at least some evidence of what is likely to work (Sutherland et al. 2017). The strength of this evidence will vary by taxa or location, but there are generally some studies that will help to guide the intervention. This evidence of what works has, for example, been usefully synthesised by the 'Conservation Evidence' platform, which is described in more detail below. DSS can take this synthesised evidence and use it to underpin risk-based conservation decision-making, increasing the chances that policies or interventions will be effective, thereby saving time, money, and effort, and helping to achieve objectives (Addison et al. 2013; Dicks et al. 2014; Sutherland et al. 2017).

In their analysis of system use, Gibson et al. (2017) provide many examples of systems, including the Ecosystem Management Decision Support System, which is used to guide landscape analysis in the USA, and Marxan, which is a tool designed for cost effectiveness analysis in relation to the selection of conservation areas. It is claimed that the latter system has over 6000 users across 182 countries. A further suite of systems was the subject of a user testing workshop by Rose et al. (2017), and McIntosh et al. (2011) identified a number of different systems for environmental management. Furthermore, a team at the University of Queensland have built a decision support system to help policy-makers with biodiversity offsetting, and there is evidence that this collaboration between researchers and government has been successful (see http://www.uq.edu.au/research/impact/stories/a-calculated-approach/).

Interestingly, a study in Pennsylvania by Rittenhouse et al. (2018) on the use of the 'SILVAH-Oak' decision support tool, which provides forest management alternatives based on ecological and decision thresholds, found that managers used it as a key part of decision-making. They found that a large percentage of forest managers (69%) were following recommendations made by the tool, although there was sometimes disagreement based on the threshold data.

A systematic literature review would likely identify a plethora of systems that could be used to inform conservation. It is speculated, however, that returned papers would describe what systems do (e.g., Bottero et al. 2013), rather than exploring how, why, and if they are used (Rose et al. 2016).

8.3.1 Examples of Successful Decision Support Systems

Below we provide six examples of DSS that are being used to guide decision-making in conservation at various scales, choosing to focus on a range of decision-maker audiences: policy-makers, practitioners, and business users. Although these systems have tended to be developed by Western conservationists (although a detailed review might challenge this assertion), many are being applied globally. It is worth noting here that some scholars have argued that decision support system development for environmental management needs to be better encouraged in developing countries (see, e.g., Mackay et al. 2018—context of Pacific Small Island Developing States). All information was correct at the time of drafting in 2019.

8.3.1.1 Tool: Conservation Evidence (https://www.conservationevidence.com/)

Purpose Conservation Evidence, a project led from a group at the University of Cambridge (UK) collates evidence on conservation interventions (see Sutherland et al. 2019). Scientific literature that is usually locked behind paywalls or difficult to find is summarised in plain English and made available for free on Conservationevidence.com. Alongside the website, there is an offline pdf and hard copies of synopses and 'What Works in Conservation'. This project also identifies where there is no evidence and can inform future research and conservation efforts. The search function helps searching through the 7662 individual studies (3119 interventions) using keywords and filters. This helps decision-makers find studies on similar topics in similar systems and countries. End users can also download a bespoken summary of evidence by selecting the interventions that they are interested in, this creates their own offline reference document unique to their questions.

The evidence for an intervention is then assessed using the Delphi technique (Mukherjee et al. 2015) giving a score for effectiveness, certainty of evidence and potential harms of the interventions for the target group (i.e., 'set longlines at night to reduce seabird bycatch' in some cases can increase bycatch of white-chinned petrels but decrease other types of bycatch). This tool also tries to integrate both grey literature (evidence from unpublished sources such as government agency documents or organisations reports) and non-English evidence into the tool by searching and summarising the literature. For non-English studies, the title is displayed in the original language along with the English title. It is also possible to search the grey literature and non-English literature in a similar way, for example, the subject (birds) and language (Japanese) to refine the evidence further. Furthermore, an associated journal, Conservation Evidence, allows practitioners to send in evidence of successful or failed interventions, which is then automatically integrated into the decision support system.

End users Decision-makers ranging from a nature reserve manager to a policymaker can easily find the available evidence summarised in short paragraphs organised under groups such as amphibians, control of freshwater invasive species or Mediterranean farmland.

Format Web application.

Evidence of use/outcomes It has been used by conservation NGOs from small (Echo) to large (Society for the Protection of Birds Netherlands). For example, the People's Trust for Endangered Species check the system before they decide on what interventions to perform and ask those seeking funding to reference 'Conservation Evidence' materials in applications and to write up the effectiveness of interventions for the associated journal. Furthermore, the funding body, the Whitley Fund for Nature require applicants to demonstrate the effectiveness of interventions by referring the 'Conservation Evidence' materials and also encourage authors to write up their findings for the associated journal. Organisations that formally agree to integrate Conservation Evidence into their processes are part of the Evidence Champions initiative.

8.3.1.2 Tool: Toolkit for Ecosystem Services Site-Based Assessment (TESSA)

Purpose TESSA is a decision support tool for carrying out ecosystem services assessments. It provides practical stepwise guidance to producing baseline estimations of ecosystem services (provisioning, regulating, cultural, and supporting services) and their value at the site scale. TESSA guides the user through a selection of relatively accessible, low-cost, and simple methods. The methods allow the user to identify which ecosystems services may be important at a site and to evaluate the magnitude of benefits that people currently obtain from them, compared with those expected under an alternative state (e.g., changing land use, restoration, degradation of the site). The toolkit is designed to overcome obstacles such as costs and complexity by providing practical guidance and methodologies to assess ecosystem services. The interactive PDF takes the user through steps to identify (1) which services may be significant at a site of interest; (2) which data are needed to measure them; (3) which methods or sources can be used to obtain these data; and (4) how to effectively communicate the results (Fig. 8.2).

The toolkit has attempted to find a balance between simplicity of inputs and usability of outputs (Fig. 8.3) and therefore excludes consideration of some of the more advanced ecosystem service science. It can be applied by non-experts within a limited time, using limited resources and at a relatively low cost, yet still provides

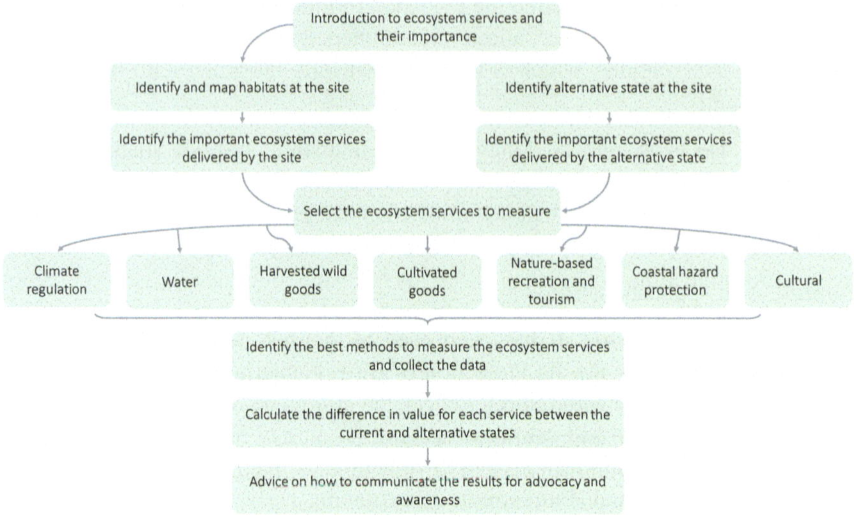

Fig. 8.2 Diagram showing the steps within the TESSA toolkit. Within each step, there are a series of structured flow charts and decision-trees which guide the user through the methods required to collect data and the processes by which to estimate ecosystem services using these data. Adapted from 'Measuring and monitoring ecosystem services at the site scale' BirdLife International. Available at: http://www.birdlife.org/worldwide/science/tessa-publicationsand http://datazone.birdlife.org/sowb/sowbpubs#Ecoservices2011http://tessa.tools/

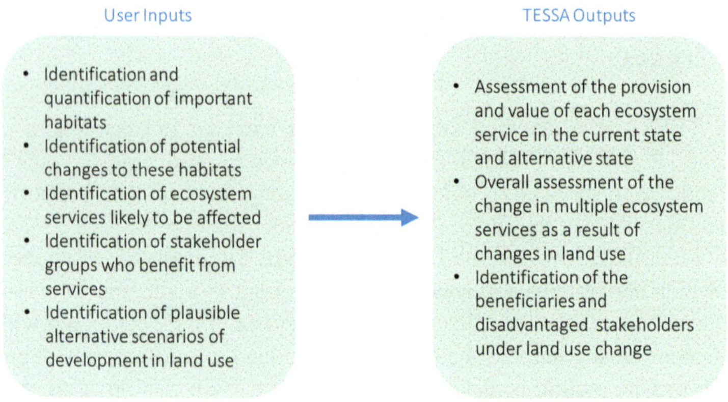

Fig. 8.3 User inputs and outputs of TESSA Toolkit

scientifically robust information (Peh et al. 2013). The toolkit recommends using existing data where appropriate and places emphasis collecting accessible field data.

End users It is aimed at supporting non-specialist conservation practitioners and decision-makers at the local scale.

Format Downloadable and interactive PDF.

Table 8.1 Some examples of TESSA usage in academic literature

Use location	Project description	References
Nepal, Phulchoki Mountain Forest Important Bird and Biodiversity Area (IBA)	Compared multiple ecosystem service values (including carbon storage, greenhouse gas sequestration, water provision, water quality, harvested wild goods, and nature-based recreation) provided by the site in current state and a state where community forestry practices had not been implemented	Birch et al. (2014)
Centre Hills, Montserrat	Estimate the effect of feral livestock control on ecosystem services—global climate regulation, nature-based tourism, harvested goods, and water provisioning. TESSA was employed to measure and compare ecosystem service provision in the presence and absence of feral livestock	Peh et al. (2015)
Nepal—across network of 27 Important Bird and Biodiversity Areas (IBAs)	Participatory rapid appraisal approach used to assess ecosystem services—developed as part of a more comprehensive methodology to measure services at individual sites using TESSA	Thapa et al. (2016)
UK, East England	Study quantifies the differences in ecosystem service (climate change regulation, cultivated goods, nature-based recreation, and flood-risk mitigation) provision under two common mineral site after-use—nature conservation and agriculture	Blaen et al. (2015)
Wanglang National Nature Reserve, China	Study quantified the differences in ecosystem services (global climate change regulation, water related services, grazing and harvested wild goods, nature-based recreation) provision of two alternative conservation approaches: (1) existing strict regulation and (2) local community use of natural resources	Liu et al. (2017)
Bwindi Impenetrable National Park, Uganda	Pilot study to identify and assess the diversity of ecosystem services in the park and benefits to local stakeholders. Comparison between Bwindi, which is managed by the Uganda Wildlife Authority and Echuya Central Forest Reserve which has a Collaborative Forest Management agreement	Nature Uganda, 2018
Sierra de Bahoruco National Park, Dominican Republic	Study to generate information about the benefits that people in the reserve receive from the ecosystem services (global climate regulation, water services, harvested wild goods, cultivated goods, nature-	Angarita-Martinez et al. LINK

(continued)

Table 8.1 (continued)

Use location	Project description	References
	based recreation). Compared a well-conserved vision to a highly degraded state	
Natewa Tunuloa, Fiji	TESSA was applied at three forest sites, including Natewa Tunuloa, to identify and highlight the ecological and socio-economic values of forests and therefore sustainable forest management. Compared the current state with two alternatives—one featuring more logging and grassland and the other more plantation forest	Valu et al. LINK
Khe Nuoc Trong, Vietnam	Part of a wider initiative to explore the potential to develop a sustainable management model for conserving the forest—TESSA was used to compare the global climate regulation, harvested wild goods, water provision and flood protection services under a 'business as usual' scenario of extraction and exploitation and a 'forest of hope' scenario of restoration and management	Merriman et al. LINK
Copal Community Forest, Cameroon	Three sites in the COPAL community forest were investigated using TESSA—(1) the current community forest; (2) certified cocoa plantations; (3) non-certified cocoa plantation. The benefits in terms of global climate regulation, water services, harvested wild goods and cultivated goods were valued	Mbosoo LINK
Yala Swamp Complex, Kenya	TESSA study assessed the value of harvested wild goods and cultivated goods through surveys in 16 villages within the area. The alternative state was a better managed Yala wetland, which used the Lake Kanyaboli National Reserve as a comparison site	Akwany LINK

Evidence of use/outcomes TESSA has been used to assess varying ecosystem services in a wide range of locations (Table 8.1), including Kenya, Uganda, Cameroon, Nepal, Cambodia, Vietnam, China, Dominican Republic, Fiji, Romania, and the UK. See links here: http://www.birdlife.org/assessing-ecosystem-services-tessa/case-studies and here: http://www.birdlife.org/worldwide/science/tessa-publications. The methods are designed as templates and allowing users to adapt them to local conditions at a particular site.

8.3.1.3 Tool: Ape Seizure Database

Purpose The Ape Seizure Database was developed for recording instances of seized apes (i.e., chimpanzees, gorillas, orangutans and bonobos) in order to tackle the illegal trade of great apes and ensure their long-term survival. Data are uploaded on the ground via smartphones, and the records are then validated by a panel of great ape experts from around the world. The system is fully responsive and caters to users with poor and unstable internet connections.

End user This tool is used on the ground by The Great Apes Survival Partnership (GRASP), an alliance of nearly 100 *national governments, conservation organisations, research institutions, UN agencies,* and *private companies.*

Format Web Application and Database.

Evidence of use/outcomes Data gathered through the Apes Seizure Database enable users to quantify displaced apes and improve accuracy in terms of scale and scope of illegal trade to better inform decisions and efforts for tackling illegal activity. The tool also helps identifying key geographic areas of concern where law enforcement efforts need to be strengthened.[1]

8.3.1.4 Tool: Protected Planet https://protectedplanet.net/

Purpose This is a publicly available online platform that provides up-to-date spatial data and site information on the World's 237,000 protected areas (see Fig. 8.4). Data on protected areas are updated monthly with submissions from governments, non-governmental organisations, landowners, and communities. Protected Planet is managed by the UN Environment World Conservation Monitoring Centre (UNEP-WCMC) with support from IUCN and its World Commission on Protected Areas (WCPA). Users can access information on protected areas, statistics, and download data from the World Database on Protected Areas (WDPA). The database is updated on a monthly basis and the website has the most up-to-date information.

End users National governments, academics/scientists, businesses.

Format Web Application.

Evidence of use/outcomes Protected Planet provides up-to-date protected area data that inform decision-making, policy development, and conservation planning. A range of businesses (including finance, mining, and oil/gas) use the information for identifying biodiversity risks and opportunities. The WDPA is a key resource for tracking progress towards the achievement of global targets. For example, the WDPA data are used for five official indicators for Aichi Biodiversity Target

[1]For more examples of use and outcomes, see http://www.bbc.co.uk/news/science-environment-37513707

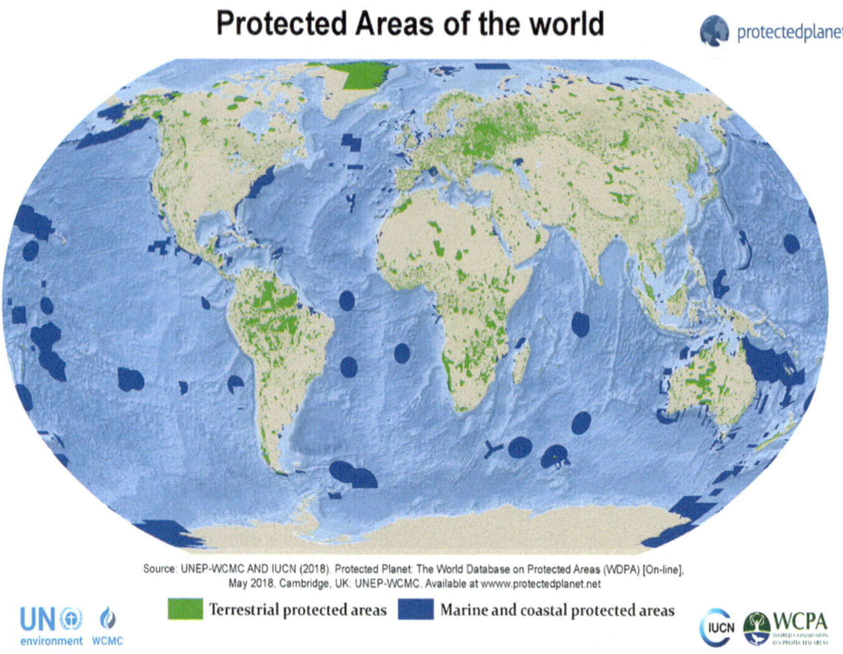

Fig. 8.4 Protected planet image (source embedded in image)

11 (regarding Protected Areas) of the CBD Strategic Plan for Biodiversity 2011–2020,[2] and official indicators for three targets within the Sustainable Development Goals—namely, Targets 14.5 of Goal 14 (Life below Water) and 15.1 and 15.4 of Goal 15 (Life on Land).[3] The Millennium Challenge Corporation (MCC), a US government aid agency, uses data from the WDPA to measure the effectiveness of policies related to Natural Resource Protection in order to assign funds to recipient countries.[4]

---

[2]The list of Aichi Biodiversity Targets and official indicators for the CBD Strategic Plan for Biodiversity 2011–2020 is available here: https://www.cbd.int/doc/decisions/cop-13/cop-13-dec-28-en.pdf

[3]The list of SDG targets and official indicators is available here: https://undocs.org/A/RES/71/313

[4]More information on users and outcomes soon to be published in Heather C Bingham, Diego Juffe Bignoli, Edward Lewis, Brian MacSharry, Neil D Burgess, Piero Visconti, Marine Deguignet, Murielle Misrachi, Matt Walpole, Jessica L Stewart and Naomi Kingston. *The World Database on Protected Areas: the past, present and future of a major conservation database* (in review).

8.3.1.5 Tool: Online Reporting System http://ors.ngo/

Purpose To streamline national reporting for Multilateral Environmental Agreements (MEAs) and support countries with meeting their reporting obligations to MEAs.

End user Secretariats of MEAs, and country officials reporting on MEAs.

Format Web Application.

Evidence of use/outcomes This tool streamlines the reporting obligations contracting parties have to the various MEA secretariats and makes data available to inform decisions on biodiversity. The tool is being used by eight MEAs, including the Convention on International Trade in Endangered Species (CITES), the Ramsar Convention on Wetlands of International Importance (Ramsar), and the Convention on the Conservation of Migratory Species (CMS).

8.3.1.6 Tool: The Cool Farm Tool (https://coolfarmtool.org/coolfarmtool/)

Purpose An online greenhouse gas, water, and biodiversity calculator for farming helping farmers/growers, food manufacturers, and retailers to improve environmental management. The mission of the Cool Farm Alliance, which is comprised of a network of industry groups, supermarkets, universities, and others (see below), is to enable millions of growers around the world to make more informed on-farm decisions that reduce their environmental impact. The Cool Farm Tool enables on-farm greenhouse gas calculations for all major crops globally; biodiversity assessments for farms in temperate forest biomes; and soon, water footprinting for 25 crops globally' (from Cool Farm Alliance website, https://coolfarmtool.org/coolfarmtool/). The biodiversity component of the tool allows farmers and buyers to see which species are benefiting from management practices, suggest different strategies, and monitor impacts on biodiversity. It is free for farmers.

End user Across the supply chain, including farmers, food manufacturers, and retailers.

Format Online application.

Evidence of use/outcomes The Cool Farm Alliance is now comprised of well over 30 members, including agricultural industry groups (e.g., Yara, Syngenta), supermarkets (e.g., Tesco, M & S), food manufacturers (e.g., Kellogg's, Nestle, McCain), other food retailers (e.g., McDonalds), universities (e.g., Wageningen, Aberdeen), and environmental initiatives (e.g., European Initiative for Sustainable Development in Agriculture). Various service providers support tool implementation and training.

8.4 Barriers for Uptake of Decision Support Systems in Conservation

In a similar way to the studies outlined in the previous section, DSS in conservation are sometimes underutilised, or not used at all by their intended audiences (Addison et al. 2013; Gibson et al. 2017). Contrastingly with fields such as agriculture, however, there is much less critical social science research that has looked at the problem of lack of uptake in conservation (Dick et al. 2017; Gibson et al. 2017; Rodela et al. 2017). This includes limited work on what practitioners think about systems that have been created for them (Dick et al. 2017), such as GIS-based spatial tools (Bottero et al. 2013; Rodela et al. 2017). There is certainly a gap in the literature for further research of this nature. Of those few studies that have addressed the problem of implementation, the explanations are not dissimilar to those found to explain lack of uptake in fields such as agriculture or medicine. Prominent barriers to uptake include:

- *Lack of system relevance for decision-makers*—for example, a system does not help policy-makers address key policy objectives (Addison et al. 2013; Gibson et al. 2017; Weatherdon et al. 2017).
- *Limited trust between designer and user*—noted, for example, in studies by McIntosh et al. (2011), Addison et al. (2013), and Gibson et al. (2017). The lack of a user-centred approach, where intended end users are involved in the design process to ensure that systems are relevant and easy to use, may be a contributory factor here (Addison et al. 2013; Rose et al. 2018b). Poor communication between designers and stakeholders is also a problem (Addison et al. 2013; Schwartz et al. 2018).
- *Unstructured decision procedures don't fit with the use of systems*—mentioned, for example, in a study by Johnson et al. (2015). The authors describe how decision-makers, including conservation practitioners, rarely use systematic and transparent procedures through which to make decisions. In other words, decisions are not made in a step-by-step fashion with detailed consideration of the evidence, and transparency with respect to how the final decision was taken. Systematic DSS may, therefore, not fit in well with such 'messy' decision-making processes (Johnson et al. 2015).
- *Poorly designed or maintained systems*—systems can be difficult to use, or may quickly become obsolete if they are not maintained after funding ends (Rose et al. 2018b). Rittenhouse et al. (2018) found that early versions of the 'SILVAH-Oak' tool were not as user-friendly as possible, leading to some mistakes in its use.
- *Inflexibility when dealing with uncertain information*—some systems are perceived to be poor at working with uncertain or missing information (Gibson et al. 2017), which is commonplace in the complex problem of conservation.
- No evidence champions in organisations—there is some evidence to suggest that systems will be used if they are championed first by particular individuals, who then recommend them to peers and colleagues (Gibson et al. (2017).

[see Gibson et al. (2017) and McIntosh et al. (2011) for more barriers]

Based on research in other fields, it is likely that factors such as poor delivery and lack of marketing are also significant barriers to uptake (Rose et al. 2016). The fact that similar design and delivery flaws are being noted in the conservation literature suggests that lessons have not been widely learned from other fields. Put simply, therefore, although there are examples of DSS being used in conservation, there are still barriers to uptake which need to be overcome. The next section provides tips on how to conduct good user-centred design of systems. The aim of this exercise is to ensure that we design systems that users want to use in the first instance and then that they continue using them once adopted.

8.5 Designing Usable, Impactful Systems: Tips for Good Participatory Design

With reference to the prominent barriers to uptake listed above, it makes logical sense that systems would be more impactful if they did not suffer from common design and delivery flaws. Although it is sometimes difficult to define what success looks like for DSS (McIntosh et al. 2011), we argue that widespread use by the intended end user is a suitable measure. To overcome the problem of implementation, several protocols have been suggested, including by McIntosh et al. (2011)[5] and Rose et al. (2018b). Focusing on the latter protocol here, Rose et al. (2018b) constructed a multi-stage approach to guide the user-centred design of DSS. Shown in Fig. 8.5, this approach attempts to reconfigure the dominant top-down knowledge transfer approach associated with existing decision support system projects. This process depends on involving the user at all stages (Addison et al. 2013; Cerf et al. 2012; Parker and Sinclair 2001), embracing the end user 'throughout the design and development process' (McIntosh et al. 2011, 1389).

Following this process should prevent the design of flawed systems that do not adequately consider their end user. We will briefly discuss each stage:

1. *Think* ***user***—identifying the user is key to understanding what their questions of interest are and their workflows. Understanding user's problems is important so that systems are relevant (Addison et al. 2013); this will include identifying the needs of policy-makers, for example, how a system can help them to satisfy reporting requirements (French and Geldermann 2005; Weatherdon et al. 2017).
2. *Think* ***value***—the system has to have value for the use. If we want to make a difference in practice, the system has to be useful for the end user, and not just be scientifically sophisticated enough to result in an academic publication. The designer of the tool should be able to provide some metrics for the potential

[5]Five suggested stages of success were: (1) Design for ease of use, (2) Design for usefulness, (3) Establish trust, (4) Promote plan for longevity, (5) Start simple, develop incrementally (McIntosh et al. 2011)

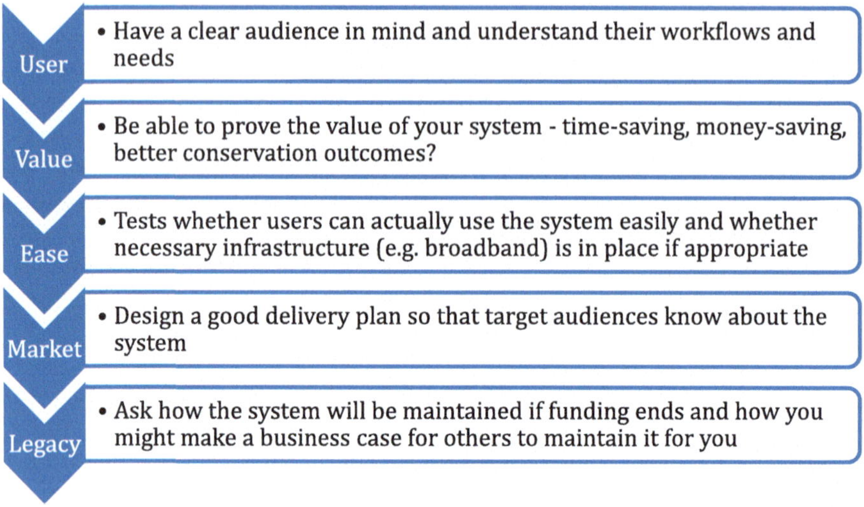

Fig. 8.5 Five-stage process for designing an impactful decision support system (based on Rose et al. 2018b)

performance of the system; these may include, for example, the amount of time saved in making a decision or the amount of money saved in making a more effective, efficient decision.

3. *Think **ease** of use*—this is a key consideration, but this should be from the perspective of users. Systems must be easy to use, but testing must be conducted with the intended end user, rather than on like-minded colleagues. Furthermore, an assessment of the site of implementation is needed to check that the system can physically be used in a given location (e.g., internet access, IT knowledge). This is likely to vary by location. For example, remote rural locations, particularly in developing countries, are likely to suffer from poor broadband access, making it difficult to use internet-based systems. Different audiences are also likely to vary with respect to IT competency. There are many examples in the literature where innovations designed in a scientific 'laboratory' are unsuitable for application on the ground (see, e.g., Lash et al. 1996) and thus the context of implementation always needs to be considered before developing a system. It may be that paper-based decision frameworks are required in areas of poor connectivity. In all cases, designers might also consider the language of their systems, and whether it can be available in multiple languages (see Amano et al. 2016).
4. *Think **market***—all businesses must market products in order to increase awareness. Why should it be any different for conservation DSS? Conservation policy-makers, practitioners, and business cannot use a system if they do not know that it exists.
5. *Think **legacy***—DSS, including in conservation, often need to be maintained for accuracy. The business model should be considered at an early stage so that maintenance is guaranteed once, for example, academic funding ends. Designers

may ask how they can convince third parties (e.g., businesses or NGOs) to maintain the system for them if it cannot be self-maintained in the long-term?

[all based on Rose et al. 2018b]

To stress the point again, you cannot adequately address any of the stages above without considering and involving the user throughout. Research on co-innovation processes in agriculture, for example, encourages designers to ensure that a range of relevant actors are brought together allowing shared priorities to be identified and mutual trust to be built (Fielke et al. 2018).

While it has been claimed that the design of decision support systems has become more participatory (see, e.g., Dick et al. 2017), there is limited evidence that user-centred practice is widespread. One example may be the QUICKSCAN software tool for ecosystem services decision-making (Dick et al. 2017), which used stakeholder workshops in Scotland (representatives from farming, fishing, bird protection, tourism, Cairngorms National Park Authority, etc.) to test the tool and provide feedback on its relevance and usability. Yet, it is unclear whether this process was truly participatory in the sense that users were involved at the conception phase. All too often, researchers or other tool designers have an idea to build a system and then initiate a participatory exercise to validate the idea (Chilvers and Kearnes 2016). This fact was noted by Mann and Schäfer (2018) who reported on a so-called trans-disciplinary water and land management in Germany in which a decision support system was originally intended to be user-centred. However, designers seldom involved the end users in the development process, and a system was produced with limited relevance. Hence, pro-innovation, top-down bias still often predominates.

Some previous projects have unwisely involved intended users at a late stage, trying to identify ways of incentivising uptake and perhaps even changing behaviour. Yet, if intended end users were involved at an early, upstream stage, then the ability to design a relevant and usable tool, which users trust and have knowledge of, is much enhanced (Fielke et al. 2018; McIntosh et al. 2011; Parker and Sinclair 2001; Rose et al. 2018b). Users would be more likely to adopt it in the first instance, and then continue to use it as it would be relevant and user-friendly.

Consequently, as a research community, we need to make progress in two areas in order to build the capacity for participatory research. Firstly, we need to understand better how to engage end users better so that we can establish successful two-way dialogue, and we then secondly require a clear methodology for involving users in system design which does not currently exist (Rodela et al. 2017). This will require a change in research and design cultures to move away from top-down knowledge transfer, which builds a product and then tries to influence or change user behaviour to adopt it. We need to change our own behaviour so that we can build tools that match the workflows of end users, fit their tasks, and understand their needs and constraints (Gibson et al. 2017).

We may need help to do so. If developers of systems, including researchers, are going to invest time and money into a trans-disciplinary mode of participatory development, then encouragement is needed. In academia, we need better incentives to focus on impact, rather than scientific publication, and much greater emphasis on impact from those who fund research (see Rose et al. 2019; Tyler 2017). One simple idea is to encourage funders of research to require applicants, and subsequently successful bidders, to report against the five-step criteria above when applying for, or carrying out, the project. Such a reporting protocol would ask developers to show that they have (1) considered their audience, (2) identified a system that would be useful and relevant, (3) assessed the site/s of implementation and tested ease of use from a user perspective, (4) considered how to market the product, and (5) developed a long-term sustainability plan. Satisfactory reporting against such criteria would limit the chances of a system being designed that was useless, irrelevant, poorly designed, and poorly maintained.

Thus far, we have provided tips about how designers can change their behaviour to develop better systems. However, it is worth noting that decision-makers need to play their part too if decision support systems are to be better utilised (Johnson et al. 2015). Johnson et al. (2015) describe how conservation practitioners may require better training to use decision support systems, although presumably this would not have to be too onerous if systems were easy to use in the first instance. Furthermore, they describe how messy decision-making processes, which are rarely transparent and step-based, do not lend themselves to the systematic use of decision support systems. Addison et al. (2013) would concur as they argue that unstructured decision-making might lead to subjective judgements that rely on hidden assumptions or individual interests. In response to this problem, Addison et al. (2013) suggest that conservation decision-makers should adopt structured decision-making frameworks which encourage a transparent step-based approach. Ultimately, the adoption of structured decision-making frameworks creates the right conditions for DSS to be used; to this end, it is argued that problems must be clearly formulated between those designing systems and end user decision-makers, communication should be effective between all stakeholders, and system designers should ensure that their product is relevant to decision contexts.

However, the ideal of structured decision-making is not easy to achieve. While we may wish that conservation decision-making was systematic, evidence-informed, and transparent (Sutherland and Wordley 2017; Gardner et al. 2018), in reality it is usually complex and multi-faceted with several 'decision-makers' (including stakeholders) involved in the process (Evans et al. 2017a, b). With this in mind, therefore, it is perhaps worth remembering a point that we stressed at the start of this chapter. Decision support systems are useful tools which can make a contribution to decision-making; however, they will only ever be a contributory tool and not the only factor in that decision-making. We should not expect the unstructured nature of most conservation decision-making to be replaced easily with a structured process where DSS tell users what to do. Thus, they may be used as a decision aid within a messy process, but designers should try as hard as possible to ensure that systems are flexible enough to work in such scenarios (see section below).

8.6 Using Decision Support Systems for Uncertain, 'Wicked' Problems

As an additional consideration to the above steps, it is worth mentioning how DSS may be used to address uncertain problems. Environmental decision-making is characterised by situations in which some factors or outcomes are not known (Hulme 2005; Regan et al. 2005) because predictions of environmental change can be highly uncertain (Ascough et al. 2008; Newbold et al. 2016; Polasky et al. 2011). This uncertainty in environmental decision-making arises from (1) the non-linear nature of the bio-physical processes which underpin the system, (2) the variable impacts of the socio-political processes which surround the system and (3) difficulties, imprecision, and inaccuracy in collecting empirical information about these processes and their impacts (Morgan and Henrion 1990). Effective environmental management can thus be considered a function of the ability to make good decisions under uncertainty and limited knowledge of all parameters (Polasky et al. 2015; Wynne 1992).

In complex, uncertain conservation problems, it is difficult to interpret recommendations made by DSS (Gibson et al. 2017). A user might not necessarily trust the evidence underpinning the system, or the context-specific nature of conservation may mean that DSS work better in some places than others. Yet, it is unrealistic to expect that DSS can only be used in situations where there is little or no uncertainty. Such uncertainty-free scenarios rarely exist, and thus we need to help users understand how these tools can be used to address complex problems.

Coastal zone management can provide a useful case study example of DSS use under uncertainty. The coastal zone arguably represents one of the most complex systems for management, featuring interactions between natural hydrology, geomorphic and bio-physical processes, and socio-cultural and political influences (Arkema et al. 2013; French 2004; Nicholls et al. 2007). Coastal management is characterised by many of the challenges identified by Maier et al. (2008): (1) it is concerned with complex systems, many of which are not well-understood; (2) it tends to involve large numbers of stakeholders, with competing objectives; and (3) there are multiple potential management options. In many cases, nature conservation is a key component of coastal zone management.

Consequently, coastal decision-makers are placed in a situation of high political stakes, substantial uncertainties, and numerous potential solutions (Sarewitz 2004), a situation not atypical of most conservation management scenarios. DSS have been used to offer guidance across varying areas of coastal decision-making, including aquaculture site-selection (e.g., Halide et al. 2009; Nath et al. 2000) and fisheries optimisation (e.g., Rice and Rochet 2005), flood warning systems (e.g., Alfieri et al. 2012; Billa et al. 2006), and marine spatial planning (e.g., Duarte et al. 2016; Villa et al. 2014) (see Table 8.2).

Coastal DSS use various methods to account for and communicate uncertainty, as illustrated by Table 8.2. For example, scenario modelling allows tool users to assess outcomes under varying conditions. The use of scenarios is often supported through

Table 8.2 Examples and brief description of emerging generation of integrated models for decision support at the coast (based on Van Kouwen et al. 2007; van Dongeren et al. 2014; FAST 2015; Peh et al. 2013)

Decision support system	Description	Areas of uncertainty	Communicating uncertainty
FAST project MI-SAFE Tool	User-friendly tool showing coastal profiles, flood risk and attenuation from vegetation. Uses satellite imagery alongside, where available, in-situ vegetation properties, elevation and sediment stabilisation data	Resolution of input data, applicability of model to different areas and coastline types	Colour coded bands which describe the level of confidence attached to each data series for a particular area
RISC-KIT	Suite of tools for assessing risk and vulnerability to coastal storms and flooding, including risk assessment frameworks, a storm impact database, a high-resolution quantitative evaluation hotspot risk reduction analysis tool, a multi-criteria analysis tool, and a web-based management guide	Input data uncertainty, nested scales with greater detail at smaller scales	Hotspot analysis allows user to zoom in on area. Use of detailed descriptions and data from past storm events
ARCoES	Series of animations and interactive map-viewer which illustrate the potential sea level rise and storm surge risk for populated coastal areas	Input data uncertainty (e.g., DEM resolution) and model uncertainty	Scenarios variable by sea level and storm surge, disclaimer describing uncertainty
TESSA Coastal Module	Interactive PDF document which guides users through a suite of practical methods to assess the ecosystem service provision—coastal hazard regulation—of a particular site	Resolution of collected data, accuracy of data using simple methods, uncertain future boundary conditions	Assess against two plausible 'alternative states' which restricts the outcomes. Disclaimer describing uncertainty and visual aids to show confidence in methods

GIS and mapping interfaces which integrate and spatially resolve varying social, environmental, and economic information into a common interactive interface. Almost all coastal tools aim to communicate spatial uncertainty using maps as part of their outputs. For example, ARCoES provides an interactive map-viewer to display sea level and storm surge risks (Knight et al. 2015) and the RISC-KIT tool allows the user to zoom into 'hotspots' (van Dongeren et al. 2014). The MI-SAFE tool attempts to provide a simple visualisation of the various scenarios for the user by colour-coding results (i.e., green for more confident, red for least confident). In the TESSA Coastal Module, the uncertain nature of coastal hazards is clearly explained, thus, being transparent about uncertainties could be useful. Outputs, such as the maps described, are used alongside sketches (Milligan et al. 2006), animations (Lieske et al. 2014) and even some 3D visualisations (Jude et al. 2006; Jude 2008) to communicate uncertainty at the coast to stakeholders.

Moving forwards, we could learn lessons about communicating uncertainty in conservation DSS from these coastal management examples. Firstly, we should be transparent about the uncertainties present in using the system to guide management. Secondly, we could find ways of presenting uncertainty in a clear fashion, for example by presenting different colour-coded scenarios showing the level of confidence of each recommendation. Thirdly, we could aim to ensure that systems use engaging visualisations to enable the user to understand uncertainties. Ultimately, these steps will improve the usability of systems in uncertain situation and increase trust from users.

8.7 Concluding Remarks

This chapter has shown the enormous potential for decision support systems to make a difference in conservation, improving the chances of evidence-informed decision-making. We should, therefore, all be interested in ensuring that systems are designed in such a way as to make them impactful on the ground. It serves no one in the conservation community to support the design of systems that will just 'sit on the shelf'. To ensure impact, systems must be relevant, useful, easy to use, sustainable, and well-marketed, so that they are used by their intended audience.

We suggest that researchers make use of the five-stage design protocol outlined above, crucially involving the user at every stage in a participatory user-centred approach. We also argue that funders and other supporters of system design, which can include research councils, government agencies, technology companies, and conservation NGOs, should use the outlined protocol (or something similar) to judge the strength of research proposals that seek to build a decision support system. If applicants are required to make it clear how they intend to: identify and characterise a clear audience (stage 1), determine a useful purpose (stage 2), assess existing infrastructure for the system (stage 3), ensure ease of use (stage 4), establish a clear delivery plan (stage 5), and (stage 6) guarantee long-term sustainability, then the chances of an obsolete system being produced will be limited.

Applicants who are able to show that their methodological approach will, over the course of time, involve the user to satisfy each stage should be supported and required to report on their progress against each milestone throughout the project. Applicants who are not able to show convincingly that users will be involved to determine such things as relevance and ease of use should not be funded, or at least not prioritised if the aim of the funder is to support activities that are going to make an impact on the ground, rather than simply be published in a high impact academic journal. This will require a simultaneous recalibration of academic reward systems to prioritise and reward policy relevant impact work (see Tyler 2017).

Acknowledgments DCR acknowledges the support of the School of Environmental Sciences at the University of East Anglia for funding to hold various meetings associated with the book chapter.

References

Addison PFE, Rumpff L, Bau SS, Carey JM, Chee YE et al (2013) Practical solutions for making models indispensable in conservation decision-making. Divers Distrib 19(5-6):490–502

Alfieri L, Salamon P, Pappenberger F, Wetterhall F, Thielen J (2012) Operational early warning systems for water-related hazards in Europe. Environ Sci Policy 21:35–49

Allen W, Cruz J, Warburton B (2017) How decision support systems can benefit from a theory of change approach. Environ Manag 59(6):956–965

Amano T, González-Varo JP, Sutherland WJ (2016) Languages are still a major barrier to global science. PLoS Biol 14(12):e2000933

Anderson CC, Rex A (2019) Preserving the scenic views from North Carolina's Blue Ridge Parkway: a decision support system for strategic land conservation planning. Appl Geogr 104:75–82

Arkema KK, Guannel G, Verutes G, Wood SP, Guerry A et al (2013) Coastal habitats shield people and property from sea-level rise and storms. Nat Clim Chang 3(10):913–918

Arlettaz R, Schaub M, Fournier J, Reichlin TS, Sierro A et al (2010) From publications to public actions: when conservation biologists bridge the gap between research and implementation. BioScience 60(10):835–842

Ascough JC II, Maier HR, Ravalico JK, Strudley MW (2008) Future research challenges for incorporation of uncertainty in environmental and ecological decision-making. Ecol Model 219(3–4):383–399

Billa L, Shattri M, Mahmud AR, Ghazali AH (2006) Comprehensive planning and the role of SDSS in flood disaster management in Malaysia. Disaster Prev Manag 15(2):233–240

Birch JC, Thapa I, Balmford A, Bradbury RB, Brown C et al (2014) What benefits do community forests provide, and to whom? A rapid assessment of ecosystem services from a Himalayan forest, Nepal. Ecosyst Serv 8:118–127

Blaen PJ et al (2015) Rapid assessment of ecosystem services provided by two mineral extraction sites restored for nature conservation in an agricultural landscape in Eastern England. PLoS One 10(4):p.e0121010

Bottero M, Comino E, Duriavig M, Ferretti V, Pomarico S (2013) The application of a multicriteria decision support system (MCSDSS) for the assessment of biodiversity conservation in the Province of Varese (Italy). Land Use Policy 30(1):730–738

Cerf M, Jeuffroy M-H, Prost L, Meynard J-M (2012) Participatory design of agricultural decision support tool: taking account of the use situations. Agron Sustain Dev 32:899–910

Chilvers J, Kearnes M (2016) Remaking participation. Science, environment and emerging publics. Routledge, Abingdon

Cook CN, Hockings M, Carter RW (2010) Conservation in the dark? The information used to support management decisions. Front Ecol Environ 8(4):181–186

De Oca Munguia OM, Llewellyn R (2020) The adopter versus the technology: which matters more when predicting or explaining adoption? Appl Econ Perspect Policy 42(1):80–91

Dick J, Verweij P, Carmen E, Rodela R, Andrews C (2017) Testing the ecosystem service cascade framework and QUICKScan software tool in the context of land use planning in Glenlivet Estate Scotland. Int J Biodivers Sci Ecosyst Services Manage 13(2):12–25

Dicks LV, Walsh JC, Sutherland WJ (2014) Organising evidence for environmental management decisions: a '4S' hierarchy. Trends Ecol Evol 29(11):607–613
Duarte GT, Ribeiro MC, Paglia AP (2016) Ecosystem services modeling as a tool for defining priority areas for conservation. PLoS One 11(5):e0154573
Evans MC, Davilla F, Toomey A, Wyborn C (2017a) Embrace complexity to improve conservation decision making. Nat Ecol Evol. 1(11):1588
Evans K, Terhorst A, Ho Kang B (2017b) From data to decisions: helping crop producers build their actionable knowledge. Crit Rev Plant Sci 36(2):71–88
Fielke SJ, Botha N, Reid J, Gray D, Blackett P, Park N, Williams T (2018) Lessons for co-innovation in agricultural innovation systems: a multiple case study analysis and a conceptual model. J Agric Educ Ext 24(1):9–27
Foreshore Assessment using Space Technology (FAST) EU FP7 Project-607131 (2015). Available at: http://www.fast-space-project.eu/. Accessed 01 July 2017
French PW (2004) The changing nature of, and approaches to, UK coastal management at the start of the twenty-first century. Geogr J 170(2):116–125
French S, Geldermann J (2005) The varied contexts of environmental decision problems and their implications for decision support. Environ Sci Pol 8(4):378–391
Gardner CJ, Waeber PO, Razafindratsima OH, Wilmé L (2018) Decision complacency and conservation planning. Conserv Biol 32(6):1469–1472
Gent DH, Mahaffee WF, McRoberts N, Pfender WF (2013) The use and role of predictive systems in disease management. Annu Rev Phytopathol 51:267–289
Gibson FL, Rogers AA, Smith ADM, Roberts A, Possingham H et al (2017) Factors influencing the use of decision support tools in the development and design of conservation policy. Environ Sci Pol 70:1–8
Halide H, Stigebrandt A, Rehbein M, McKinnon AD (2009) Developing a decision support system for sustainable cage aquaculture. Environ Model Softw 24(6):694–702
Hochman Z, Carberry P (2011) Emerging consensus on desirable characteristics of tools to support farmers' management of climate risk in Australia. Agric Syst 104(6):441–450
Hulme PE (2005) Adapting to climate change: is there scope for ecological management in the face of a global threat? J Appl Ecol 42(5):784–794
Johnson FA, Eaton MJ, Williams JH, Jensen GH, Madsen J (2015) Training conservation practitioners to be better decision makers. Sustainability 7(7):8354–8373
Jude S (2008) Investigating the potential role of visualization techniques in participatory coastal management. Coastal Manage 36:331–349
Jude S, Jones AP, Andrews JE, Bateman IJ (2006) Visualisation for participatory coastal zone management: a case study of the Norfolk Coast, England. J Coast Res 6(226):1527–1538
Knight PJ, Prime T, Brown JM, Morrissey K, Plater AJ (2015) Application of flood risk modelling in a web-based geospatial decision support tool for coastal adaptation to climate change. Nat Hazards Earth Syst Sci 15(7):1457–1471
Lash S, Szerszynski B, Wynne B (1996) Risk, environment & modernity: towards a new ecology. Sage, London
Lawton JH (2007) Ecology, policy and politics. J Appl Ecol 44(3):465–474
Lieske DJ, Wade T, Roness KA (2014) Climate change awareness and strategies for communicating the risk of coastal flooding: a Canadian Maritime case example. Estuarine Coastal Shelf Sci 140:83–94
Lindblom J, Lundström C, Ljung M, Jonsson A (2017) Promoting sustainable intensification in precision agriculture: review of decision support systems development and strategies. Precis Agric 18(3):309–331
Liu P, Jiang S, Zhao L, Li Y, Zhang P, Zhang L (2017) What are the benefits of strictly protected nature reserves? Rapid assessment of ecosystem service values in Wanglang Nature Reserve, China. Ecosyst Serv 26(Part A):70–78

Mackay S, Brown R, Gonelevu M, Pelesikoti N, Kocovanua T et al (2018) Overcoming barriers to climate change information management in small island developing states: lessons from pacific SIDS. Clim Pol 19(1):125–138

Maier HR, Ascough JC, Wattenbach M, Renschler CS, Labiosa WB, Ravalico JK (2008) Uncertainty in environmental decision making: issues, challenges and future directions. In: Jakeman AJ, Voinov AA, Rizzoli AE, Chen SH (eds) Environmental Modelling, Software and Decision Support: State of the Art and newPerspectives. Elsevirer, Amsterdam, pp. 69–85

Mann C, Schäfer P (2018) Developing sustainable water and land management options: reflections on a transdisciplinary research process. Sustain Sci 13(1):205–217

Marshall N, Adger N, Attwood S, Brown K, Crissman C et al (2017) Empirically derived guidance for social scientists to influence environmental policy. PLoS One 12(3):e0171950

McCown RL (2002) Changing systems for supporting farmers' decisions: problems, paradigms, and prospects. Agric Syst 74(1):179–220

McIntosh BS, Ascough JC II, Twery M, Chew J, Elmahdi A et al (2011) Environmental decision support systems (EDSS) development – challenges and best practices. Environ Model Softw 26(12):1389–1402

Milligan J, O'Riordan T, Watkinson A (2006) Designing coastlines fit for the future. English Nature Research Reports, No 702. Retrieved from http://publications.naturalengland.org.uk/publication/61017?category=56001

Montana J (2017) Accommodating consensus and diversity in environmental knowledge production: achieving closure through typologies in IPBES. Environ Sci Policy 68:20–27

Morgan M, Henrion M (1990) Uncertainty: a guide to dealing with uncertainty in quantitative risk and policy analysis. Cambridge University Press, Cambridge

Mukherjee N, Hugé J, Sutherland WJ, McNeill J, Van Opstal M et al (2015) The Delphi technique in ecology and biological conservation: applications and guidelines. Methods Ecol Evol 6(9):1097–1109

Nath SS, Bolte JP, Ross LG, Aguilar-Manjarrez J (2000) Applications of geographical information systems (GIS) for spatial decision support in aquaculture. Aquac Eng 23(1–3):233–278

Nelson RA, Holzworth DP, Hammer GL, Hayman PT (2002) Infusing the use of seasonal climate forecasting into crop management practice in North East Australia using discussion support software. Agric Syst 74(3):393–414

Newbold T, Hudson LN, Arnell AP, Contu S, De Palma A et al (2016) Has land use pushed terrestrial biodiversity beyond the planetary boundary? A global assessment. Science 353(6296):288–291

Nicholls RJ, Wong PP, Burkett VR, Codignotto JO, Hay JE et al (2007) Coastal systems and low-lying areas. In: Parry ML, Canziani OF, Palutikof JP, van der Linden PJ, Hanson CE (eds) Impacts, adaptation and vulnerability. Contribution of working group II to the fourth assessment report of the intergovernmental panel on climate change. Cambridge University Press, Cambridge, pp 315–356

O'Connell C, McKinnon M (2021) Communication: the bridge between knowledge and implementation. In: Ferreira CC, Klütsch CFC (eds) Closing the knowledge-implementation gap in conservation science – evidence transfer across spatiotemporal scales and different stakeholders. Springer, New York, pp. 171–197

OECD (2015) Scientific advice for policy making: the role and responsibility of scientists. OECD, Paris

Owens S (2015) Knowledge, policy, and expertise: the UK Royal Commission on environmental pollution 1970–2015. Oxford University Press, Oxford

Parker CG, Sinclair M (2001) User-centred design does make a difference. The case of decision support systems in crop production. Behav Inf Technol 20(6):449–460

Peh KS-H, Balmford A, Bradbury RB, Brown C, Butchart SHM et al (2013) TESSA: a toolkit for rapid assessment of ecosystem services at sites of biodiversity conservation importance. Ecosyst Serv 5:51–57

Peh KS-H, Balmford A, Birch JC, Brown C, Butchart SHM et al (2015) Potential impact of invasive alien species on ecosystem services provided by a tropical forested ecosystem: a case study from Montserrat. Biol Invasions 17(1):461–475

Polasky S, Carpenter SR, Folke C, Keeler B (2011) Decision-making under great uncertainty: environmental management in an era of global change. Trends Ecol Evol 26(8):398–404

Polasky S, Tallis H, Reyers B (2015) Setting the bar: standards for ecosystem services. PNAS 112(24):7356–7361

Rawson TM, Moore LSP, Hernandez B, Charani E, Castro-Sanchez E et al (2017) A systematic review of clinical decision support systems for antimicrobial management: are we failing to investigate these interventions appropriately? Clin Microbiol Infect 23(8):524–532

Regan HM, Ben-Haim Y, Langford B, Wilson WG, Lundberg P et al (2005) Robust decision-making under severe uncertainty for conservation management. Ecol Appl 15(4):1471–1477

Rice JC, Rochet MJ (2005) A framework for selecting a suite of indicators for fisheries management. In ICES Journal of Marine Science. United Nations Commission on Sustainable Development, Washington, DC, pp 516–527

Rittenhouse J, Leites LP, Derham K, Miller S (2018) Insights on the use of decision-support tools to sustain Forest ecosystems from a case study in Pennsylvania, USA. J For 116(4):391–395

Rodela R, Bregt AK, Ligtenberg A, Pérez-Soba M, Verweij P (2017) The social side of spatial decision support systems: investigating knowledge integration and learning. Environ Sci Pol 76: 177–184

Rose DC (2018) Avoiding a post-truth world: embracing post-normal conservation science. Conserv Soc 16(4):518–525

Rose DC, Sutherland WJ, Parker C, Lobley M, Winter M, et al (2016) Decision support tools in agriculture: towards effective design and delivery. Agric Syst 149:165–174

Rose DC, Addison P, Ausden M, Bennun L, Mills C, O'Donnell SAL, Parker C, Ryan M, Weatherdon L, Despot-Belmonte K, Sutherland WJ, Robertson RJ (2017) Decision support tools in conservation: a workshop to improve user-centred design. Res Ideas Outcomes 3: e21074

Rose DC, Sutherland WJ, Amano T, González-Varo JP, Robertson RJ et al (2018a) The major barriers and their solutions for evidence-informed conservation policy. Conserv Lett. 11(5): e12564

Rose DC, Parker C, Park C, Fodey J, Sutherland WJ, Dicks LV (2018b) Involving stakeholders in agricultural decision support systems: improving user-centred design. Int J Agric Manage 6(3–4):80–89

Rose DC, Amano T, González-Varo JP, Mukherjee N, Robertson RJ et al (2019) Calling for a new agenda for conservation science to create evidence-informed policy. Biol Conserv. 238:108222

Roughgarden J, Pennington T, Alexander S (1994) Dynamics of the rocky intertidal zone with remarks on generalization in ecology. Philos Trans R Soc B 343(1303):79–85

Sackett DL, Rosenberg WM, Gray JM, Haynes RB, Richardson WS (1996) Evidence based medicine: what it is and what it isn't. BMJ 312:71

Santoro F, Tonino M, Torresan S, Critto A, Marcomini A (2013) Involve to improve: a participatory approach for a decision support system for coastal climate change impacts assessment. Ocean Coast Manage 78:101–111

Sarewitz D (2004) How science makes environmental controversies worse. Environ Sci Pol 7(5):385–403

Schwartz MW, Cook CN, Pressey RL, Pulling AS, Runge MC et al (2018) Decision support frameworks and tools for conservation. Conserv Lett 11(2):e12385

Shibl R, Lawley M, Debuse J (2013) Factors influencing decision support system acceptance. Decis Support Syst 54(2):953–961

Sutherland WJ, Wordley CFR (2017) Evidence complacency hampers conservation. Nat Ecol Evol. 1:1215–1216

Sutherland WJ, Pullin A, Dolman P, Knight T (2004) The need for evidence-based conservation. Trends Ecol Evol 19(6):305–308

Sutherland WJ, Dicks LV, Ockendon N, Smith RK (2017) What works in conservation. Openbook Publishers, Cambridge
Sutherland WJ, Taylor NG, MacFarlane D, Amano T, Christie AP et al (2019) Building a tool to overcome barriers in research-implementation spaces: the conservation evidence database. Biol Conserv 238:108199
Thapa I, Butchart SHM, Gurung H, Stattersfield AJ, Thomas DHL, Birch JC (2016) Using information on ecosystem services in Nepal to inform biodiversity conservation and local to national decision-making. Oryx 50(1):147–155
Thursky KA, Mahemoff M (2006) User-centered design techniques for a computerised antibiotic decision support system in an intensive care unit. Int J Med Inform 76(10):760–768
Tyler C (2017) Wanted: academics wise to the needs of government. Nature 552(7683):7
van Dongeren A, Ciavola P, Viavattene C, de Kleermaker S, Martinez G et al (2014) RISC-KIT: resilience-increasing strategies for coasts – toolKIT. J Coast Res 70(66):366–371
Van Kouwen F, Dieperink C, Schot P, Wassen M (2007) Applicability of decision support systems for integrated coastal zone management. Coast Manag 36(1):19–34
Venkatesh V, Thong JYL, Xu X (2012) Consumer acceptance and use of information technology: extending the unified theory of acceptance and use of technology. MIS Q 36(1):157–178
Villa F, Bagstad KJ, Voigt B, Johnson GW, Portela R et al (2014) A methodology for adaptable and robust ecosystem services assessment. PLoS One 9(3):e91001
Walsh JC, Dicks LV, Sutherland WJ (2015) The effect of scientific evidence on conservation practitioners' management decisions. Conserv Biol 29(1):88–98
Weatherdon LV, Appeltans W, Bowles-Newark N, Brooks TM, Davis FE et al (2017) Blueprints of effective biodiversity and conservation knowledge products that support marine policy. Front Mar Sci 4(96):1–16
Wynne B (1992) Uncertainty and environmental learning. Reconceiving science and policy in the preventive paradigm. Glob Environ Chang 2(2):111–127
Young JC, Waylen KA, Sarkki S, Albon S, Bainbridge I et al (2014) Improving the science-policy dialogue to meet the challenges of biodiversity conservation: having conversations rather than talking at one-another. Biodivers Conserv 23:387–404

Part IV
The Knowledge-Action Outcome(s)

Chapter 9
The Use of Boundary-Spanning Organizations to Bridge the Knowledge-Action Gap in North America

Mark W. Schwartz, Erica Fleishman, Matthew A. Williamson, John N. Williams, and Toni Lyn Morelli

Contents

M. W. Schwartz (✉)
Department of Environmental Science and Policy, University of California, Davis, CA, USA
e-mail: mwschwartz@ucdavis.edu

E. Fleishman
Department of Environmental Science and Policy, University of California, Davis, CA, USA

College of Earth, Ocean, and Atmospheric Sciences, Oregon State University, Corvallis, OR, USA
e-mail: Erica.fleishman@oregonstate.edu

M. A. Williamson
Department of Environmental Science and Policy, University of California, Davis, CA, USA

Human Environment Systems, Boise State University, Boise, ID, USA
e-mail: mattwilliamson@boisestate.edu

J. N. Williams
Department of Environmental Science and Policy, University of California, Davis, CA, USA

Centro Interdisciplinario de Investigación para el Desarrollo Integral Regional, Unidad Oaxaca, Instituto Politécnico Nacional, Oaxaca, Mexico
e-mail: jnwill@ucdavis.edu

T. L. Morelli
USGS Northeast Climate Adaptation Science Center, Amherst, MA, USA
e-mail: morelli@umass.edu

C. C. Ferreira, C. F. C. Klütsch (eds.), *Closing the Knowledge-Implementation Gap in Conservation Science*, Wildlife Research Monographs 4,
https://doi.org/10.1007/978-3-030-81085-6_9

9.1 Introduction

The 24.7 million km^2 continent of North America extends from above the Arctic Circle (83° N) to the Isthmus of Panama (9° N) (Fig. 9.1). The majority of its land (88%) is within the contemporary boundaries of Canada, the United States, and Mexico. The remainder falls within the six smaller countries of southern Central America (Belize, Costa Rica, Guatemala, Honduras, Nicaragua, and Panama). A chain of mountains extends along the western edge of North America, and minimum and maximum continental elevations range from 85 m below sea level (Death Valley, California, USA) to 6190 m (Mt. Denali, Alaska, United States). Application of different sets of values and criteria to identify global conservation priority areas in North America has recognized most of the continent (e.g., Brooks et al. 2006). For example, the northern latitudes were singled out as a part of the "Last of the Wild" (Sanderson et al. 2002), the eastern United States, central plains, and southern Central America were identified as "Crisis Ecoregions" (Hoekstra et al. 2005), and Central America and California were identified as hotspots of species richness and endemism (Myers et al. 2000).

Threats to biological diversity are diverse across this continent. The loss of species' habitats is a leading threat in numerous ecoregions. Non-native invasive species and climate change also threaten ecosystems and native species. Iconic examples of these threats include non-native Burmese pythons (*Python bivittatus*), which have reduced the density of native birds and mammals in the Everglades (Florida, United States) by more than 60% (Dorcas et al. 2012); and one of the first extinctions driven in part by climate change, that of the golden toad (*Incilius periglenes*) in Costa Rica (Pounds et al. 2006). Conservation challenges include managing species with large territories, such as brown bear (*Ursus arctos*) and gray wolf (*Canis lupus*), and diverse migratory species such as woodland caribou (*Rangifer tarandus caribou*), Neotropical migratory songbirds, monarch butterflies (*Danaus plexippus*), and salmonids. Endangered species legislation in Canada, the USA, and Mexico protects more than 2000 taxa, but the number of taxa at risk of extinction likely exceeds 10,000. As an example, as of January 2019, NatureServe (www.natureserve.org) listed 6713 full species of plants and animals as imperiled or critically imperiled in Canada and the USA, excluding Hawaii. Full lists of

Fig. 9.1 Approximate boundaries of three sets of global conservation priority areas in North America. The *Last of the Wild* (Sanderson et al. 2002) are regions with low human influence. *Crisis Ecoregions* (Hoekstra et al. 2005) are ecoregions under particularly heavy human threat. Hotspots of biodiversity (Myers et al. 2000) are regions with high species endemism and species richness

threatened and endangered species in Central America are not readily available but are likely to exceed the number in Canada and the USA given the strong gradient of increasing species richness toward the tropics in the western hemisphere (Hillebrand 2004).

Laws and policies that compel conservation action vary widely among Canada, the USA, and Mexico. A robust suite of laws foster the conservation of nature in the USA (e.g., Endangered Species Act, Clean Water Act, Wilderness Act, Wild and Scenic Rivers Act) (Doremus et al. 2008). Among these, the Endangered Species Act (ESA) is one of the world's strongest pieces of legislation that aims to prevent extinction of species (Bean 2009). The ESA requires planning, action, and reporting on listed species. Application of the ESA is not without serious concerns, but few species have gone extinct while listed under the act (Evans et al. 2016). Similarly, Canada has a set of laws that compel conservation. Foremost among these is the Species At Risk Act (SARA). The SARA empowers Canada's Committee on the

Status of Endangered Wildlife in Canada (COSEWIC) to list species at risk of extinction and provides for some protective measures (Waples et al. 2013). Mexico operates under the guidance of the General Law of Ecological Balance and Environmental Protection (LGEEPA) (https://www.wipo.int/edocs/lexdocs/laws/en/mx/mx028en.pdf), which is intended to preserve, restore, and improve the environment in addition to protecting at—risk species.

Complex public and private conservation efforts across the continent span all levels of ecological organization. Approximately 89% of the area of Canada and 28% of the area of the USA (Vincent et al. 2017) is public land. By contrast, Mexico and Central America contain relatively little public land. Instead, local communities and municipalities often drive land use in these countries. For example, much of the public land in Mexico is owned communally, whether as *ejidos* (communal land slated for agrarian use that is registered with the National Agrarian Registry; as of 1992, *ejidos* may be privatized) or as *bienes comunales* (communal land that may be partitioned to community members for approved personal uses such as crop cultivation or forestry but may not be sold to private parties). Despite the communal ownership of these lands, the federal government restricts their use. National regulations apply to activities such as water diversion, timber harvesting, and the take of rare or threatened plant and animal species. In contrast, countries with large portfolios of public lands, such as the USA and Canada, have an extensive bureaucracy dedicated to managing lands in the public portfolio, although only a portion of these public lands is dedicated to conservation.

9.2 Boundary-Spanning Organizations as Knowledge Action Actors

Numerous efforts have sought to improve the integration of science into the process of making decisions about resource management and conservation (Meadow et al. 2015; Wall et al. 2017). Making decisions with respect to protecting natural resources and conservation often is socially contentious. Therefore, the process of decision-making can be as important as the information that bears on those decisions (Cash et al. 2003; Cook et al. 2013). Efforts to improve decision-making include processes that foster the creation of science that informs specific actions (Meadow et al. 2015; Wall et al. 2017) and methods to support transparent, logic- and information-driven decisions (e.g., structured decision-making, systematic conservation planning) (Schwartz et al. 2018). Efforts also include experimenting with organizational structures that maximize the use of knowledge [e.g., Climate Adaptation Science Centers (Morisette et al. 2017; DeCrappeo et al. 2018)] and facilitate inclusive decisions [e.g., Joint Ventures (Giocomo et al. 2012; Doherty et al. 2015; Behnken et al. 2016)]. These efforts have spawned newly defined subdisciplines, such as translational ecology, which focus on improving the process of collaboration to produce actionable science and inform decision-making (Enquist et al. 2017).

Two fundamental hurdles face resource management. First, despite widespread recognition of their value, objectives rarely are clear and measurable (Sanchirico et al. 2014; Redford et al. 2018). Given social complexity and the potential for societal resistance to some management goals, some managers may believe, not without justification, that intentionally vague objectives provide a means of avoiding conflict (Weible and Sabatier 2009; Weible et al. 2009). Second, because natural systems are complex, there is a high level of uncertainty that an action will result in the desired outcome (Regan et al. 2002; Martin et al. 2012). This uncertainty pertains both to actions that directly affect people (e.g., restricting certain behaviors) and actions that directly or indirectly affect nature (e.g., managing water flows). Furthermore, actions may not lead to desired outcomes because both people and nature respond unpredictably. The consequence is that setting vague objectives makes it more difficult to assert management failure, and hence may seem like a safer choice for resource managers (Wilson 1989). Nevertheless, vague objectives also make it difficult to document and evaluate the success of actions, learn from those actions, or even contend that the proposed actions emerge from a logical process intended to accomplish explicit goals.

Box 9.1 Seven terms commonly used to classify conservation actors. The terms typically are defined loosely, and most resource management professionals fit in multiple classes. The classifications are more relevant to individuals in the context of a given action than to a given individual in all cases

Decision-maker. A person who has sole or shared decision-making authority over a resource management issue. Decisions may range from authorization of comprehensive policies to on-the-ground implementation of tasks specified in management plans.

Policy-maker. A person representing a public or private organization who has decision-making authority over a policy or a set of policies that guide the actions of that organization.

Practitioner. A person who implements policies or practices.

Researcher. A person with advanced training in any discipline related to resource management (e.g., biology, economics, sociology) whose principal responsibility is to conduct research.

Scientist. A person with advanced training in science who is employed to conduct research or apply science to management.

Stakeholder. Any person with a substantial interest in the outcome of a decision or action.

Boundary-spanning organizations, which support communication between information producers and users (Cook et al. 2013), often are launched with the core assumption that a platform to convey applicable knowledge to decision-makers, practitioners, and stakeholders (Box 9.1) will increase the likelihood that the objectives of natural resource decisions are met. For example, Climate Adaptation Science Centers (CASCs), joint US Geological Survey—university endeavors, were launched with the goal of improving climate adaptation through coproduction of knowledge and direct delivery of relevant science to decision-makers and practitioners (Morisette et al. 2017; DeCrappeo et al. 2018). Taking a broad view of boundary-spanning organizations as entities that strive to link knowledge to action through activities such as information sharing and stakeholder convening, we readily identified over 100 boundary-spanning organizations within Canada, the United States, and Mexico. We acknowledge that many additional philanthropic foundations, local government entities, non-governmental organizations, and citizen action groups act within this sphere (Table 9.1). We cannot hope to assess all boundary-spanning efforts comprehensively. Our aim is to use representative examples to elucidate the gradient of success of these organizations. We seek to understand accomplishments and challenges that characterize boundary-spanning efforts.

Testing whether actions have their desired effects deeply challenges conservation science in general (Miteva et al. 2012; Baylis et al. 2016). Evaluating the core assumption of boundary-spanning organizations, that they foster better decision-making, also remains a challenge. Metrics that could be used to assess the effectiveness of these organizations include frequency of communication, the number of meetings or the position the organization holds in a social network (see Chap. 6; Bixler 2021). These metrics would assess the degree to which organizations have moved out of academia and increased the flow of information but do not assure better outcomes for nature. A separate, but important, metric is whether the increase in communication affected decisions. A fundamental assumption of boundary spanning is that failure to use available science in decision-making is associated with low rates of communication amongst researchers and decision-makers. A corollary assumption is that increased use of science in decisions indicates that greater breadth of knowledge, including non-scientific knowledge, was applied to making decisions.

Evaluating whether boundary-spanning organizations increase incorporation of science into decision-making and whether scientifically informed decisions are more effective is complicated by the fact that major environmental plans often are subject to many layers of both process and policy. For instance, revision by the US Forest Service of a Forest Management Plan, which defines acceptable actions, is a deliberative process that is guided and bounded by specific policies (Schultz et al. 2013). Making the decisions reflected in the revision takes years, and the process is structured to document the use of science in evaluating the potential outcomes of sanctioned alternatives. Other decision processes are less constrained (e.g., voluntary conservation agreements in conservation easements). Thus, the breadth of conservation decisions and actions makes it difficult to generalize about the use of knowledge to inform actions.

Furthermore, the unique context of individual actions makes it difficult to gauge what would have happened in the absence of a boundary-spanning organization or in the absence of any particular knowledge. Additionally, outcomes of actions often are poorly monitored (Lyons et al. 2008). As a result, it may be impossible to accurately discern where, when, and how knowledge was used, who was responsible for

Table 9.1 A representative selection of over 100 boundary-spanning organizations within Canada, the USA, and Mexico

Organization (country)	Year established	Purpose	Website
A. Federally funded knowledge exchanges			
Climate Adaptation Science Centers. Geological Survey (USA)	2008	Deliver science to address stakeholder-driven, high-priority needs for climate knowledge. Organized into eight regional centers with coordination through a National Center	https://casc.usgs.gov/
Landscape Conservation Cooperatives. Department of the Interior (USA)	2010	Provide science capacity and technical expertise to address shared natural and cultural resource priorities. Organized into 22 biogeographically defined units. Note: These were defunded from late 2017 through 2018; some have continued under other funding and names, whereas others no longer exist	https://lccnetwork.org/
Regional Integrated Science Assessments. National Oceanographic and Atmospheric Administration (USA)	2000	Support research teams that help expand and build the nation's capacity to prepare for and adapt to climate variability and change. 11 networks currently are active	https://cpo.noaa.gov/Meet-the-Divisions/Climate-and-Societal-Interactions/RISA
Fire Science Exchange Network. Department of Agriculture and Department of the Interior (USA)	2010	Provide the most relevant, current wildland fire science information to federal, state, local, tribal, and private stakeholders within ecologically similar regions. 15 exchanges currently are active	https://www.firescience.gov/jfsp_exchanges.cfm
B. Federally funded, regionally focused partnerships among public and private actors			
Collaborative Forest Landscape Restoration Program. Forest Service (USA)	2009	Encourage the collaborative, science-based restoration of priority forest ecosystems. Approximately 40 projects are in progress, with up to ten new projects initiated per year	https://www.fs.fed.us/restoration/CFLRP/
Puget Sound Partnership (USA)	2009	Accelerate the collective effort to recover and sustain the Puget Sound ecosystem	http://www.psp.wa.gov/

(continued)

Table 9.1 (continued)

Organization (country)	Year established	Purpose	Website
Chesapeake Bay Program (USA)	1983	Improve the condition of the Chesapeake Bay ecosystem by meeting science-based goals	https://www.chesapeakebay.net/
Migratory Bird Joint Ventures (Multi-country)	1987	Conserve habitat for the benefit of priority bird species, other wildlife, and people. There currently are 22 such Joint Ventures, which are collaborative, regional partnerships among government agencies, non-profit organizations, corporations, tribes, and individuals	https://mbjv.org/
Crown Managers Partnership (multi-country)	2001	Build common awareness of Crown interests and issues, shape relationships, and identify collaborative and complementary tasks that the participating jurisdictions can pursue	http://www.crownroundtable.net/partnerships-in-the-crown.html
Chicago Wilderness (USA)	1990s	Preserve, improve, and expand nature and quality of life by connecting leaders in conservation, health, business, science, and beyond	https://www.chicagowilderness.org/
Corredor Biológico Mesoamericano (Mexico)	1997	Promote connectivity between protected areas from Mexico to Panama to facilitate movements of migratory species, gene flow, and increases in effective population size	https://www.biodiversidad.gob.mx/corredor/corredorbiomeso.html
Áreas Voluntariamente Destinadas a la Conservación (Mexico)	2002	Assist communities and private landowners in creating, managing, and monitoring voluntarily protected areas. An initiative of the Mexican National Commission of Protected Natural Areas (CONANP)	https://www.gob.mx/conanp/acciones-y-programas/areas-destinadas-voluntariamente-a-la-conservacion

(continued)

Table 9.1 (continued)

Organization (country)	Year established	Purpose	Website
C. Philanthropic foundations and non-governmental organizations			
COMPASS (USA)	1999	Effectively engage more scientists in the public discourse about the environment	https://www.compassscicomm.org/
Point Blue (USA)	1965	Reduce the impacts of climate change, habitat loss, and other environmental threats while developing nature-based solutions to benefit both wildlife and people. Partners include land and water managers, fishermen, ranchers, farmers, cities, and counties	https://www.pointblue.org/
Wilburforce Foundation (USA)	1991	Support land, water, and wildlife conservation efforts in western Canada, the USA, and Mexico	http://www.wilburforce.org/
Conservation Biology Institute (USA)	1997	Provide scientific expertise to support the conservation and recovery of biological diversity in its natural state through the application of research, education, planning, and community service	https://consbio.org/
D. Privately organized partnerships that aim to inform decisions with science			
HCP/NCCP science advisory panels (USA)		Conservation planning for protected species by private landowners at both the California (NCCP) and federal (HCP) levels often engages scientists who provide expert advice and may facilitate communication among project proponents and agencies	e.g., http://www.co.contra-costa.ca.us/depart/cd/water/HCP/overview.html
Cooperative Weed Management Areas (USA)	various	Manage undesirable invasive species in a defined area. These areas are managed by partnerships among federal, state, and local government agencies, tribes, individuals, and other groups	https://www.invasive.org/cismas/

(continued)

Table 9.1 (continued)

Organization (country)	Year established	Purpose	Website
Ecosystem management partnerships (multi-country)	1991	As an example, the Malpai Borderlands Group aims to restore and maintain the natural processes that create and support a diverse community of human, plant, and animal life along the border between the USA and Mexico	http://www.malpaiborderlandsgroup.org/
Watershed action groups (multi-country)	various	As an example, the Watershed Action Alliance of Southeastern Massachusetts aims to promote the sustainable use of water and the implementation of growth policies that protect the natural environment and keep water local	http://www.watershedaction.org/

The four organizations or classes of organizations in each cluster vary with respect to the source of funding and focal resource issues

making the knowledge available and relevant, or whether the knowledge and its transfer led to actions that achieved objectives. At best, one might assess subjectively whether boundary-spanning organizations appear to convey relevant information to decision-makers that otherwise might not be accessible, whether decision-makers find the information useful, and whether decision-makers indicate that the information affects their actions.

We present examples of boundary-spanning organizations that strive to assist stewards of private or public lands in achieving conservation or other management objectives. We qualitatively interpret their success in terms of four criteria: (1) whether the boundary-spanning organization affected what knowledge was made accessible to decision-makers, (2) whether decision-makers indicated that the boundary-spanning organization's efforts were worthwhile, (3) whether the evidence suggests that science was used to inform the decision, and (4) whether the action resulting from the decision was effective.

9.3 Meeting the Challenges

Conservation action generally requires social negotiation (Sabatier 1987). For example, the capacity for actions such as purchasing land or conservation easements by local land trusts is affected strongly by the willingness of landowners. Similarly, public agencies must negotiate the local, regional, and national social environment.

Considerable social challenges emerge from federal management actions on public lands, enforcement of regulatory compliance on private lands (e.g., permitting the incidental take of endangered species), and actions by willing landowners. In each case, actions can be affected by advocacy coalitions (Sabatier 1987) that may selectively emphasize science that reinforces their values or goals. In this chapter, we address how knowledge informs socially contentious conservation decisions.

Knowledge is mostly likely to inform actions if it generally is accepted as relevant, salient, and legitimate (Cash et al. 2003). Boundary-spanning organizations may seek to negotiate contentious social environments such that all forms of knowledge are applied effectively to decision-making (Kirchhoff et al. 2013). The literature describing the knowledge to action gap (Knight et al. 2008) is consistent with the notion that decisions should be science-driven. This vision of science-based decision-making is reasonable from the perspective of natural or social scientists. However, the social-ecological systems literature suggests that knowledge has many forms [natural science, social science (including economics), law, cultural knowledge, and cultural practices] and that these bodies of knowledge are integrated and acted on by some decision-makers (Ostrom 2000, 2009; Dietz et al. 2003). Therefore, we recognize that knowledge gaps may result from the failure to use available, relevant science in making a decision.

An effort to identify high-priority research questions that, if answered, will increase the effectiveness of policies related to conservation and management of natural resources found that a small set of themes are common among policymakers and researchers (Fleishman et al. 2011; Rudd and Fleishman 2014). For example, policymakers and researchers indicated that research on water availability, evaluation of trade-offs among benefits humans receive from ecosystems, and hydrological effects of climate change are highly applicable to policy. Translating inferences from these research topics into policy requires coordination and cooperation among scientists, government agencies, and private landowners. Boundary-spanning organizations have been launched to facilitate such coordination and cooperation, especially for cases in which the development of policy is controversial.

The recent academic specification of boundary-spanning organizations provides the opportunity to consider organizations, both new and old, that operate under the assumption that knowledge increases the likelihood of achieving conservation objectives (Safford et al. 2017). Some organizations are focused on particular threats or regions, whereas many others work on multiple levels with multiple objectives. Boundary-spanning organizations further may be classified on the basis of their genesis, whether a top-down government mandate, a grassroots private initiative, or both.

The following case studies reflect our collective experience and are broadly representative of the types of boundary-spanning efforts that characterize conservation in the USA and, to a lesser extent, Canada and Mexico. We do not mean to imply that other organizations have faced or met the same challenges or have experienced the same failures. Our choice of boundary-spanning organizations is opportunistic. We cannot speak to the vast diversity of motivations and goals among boundary-spanning organizations. However, on the basis of our experiences, we offer subjective insights into the gradient of success encountered by boundary-spanning organizations.

9.4 Case Studies

9.4.1 Migratory Species Joint Ventures

Arguably the most well-recognized and effective set of boundary-spanning organizations in Canada, the USA, and Mexico are the Migratory Bird Joint Ventures (MBJVs). The MBJVs are public–private partnerships dedicated to improving the management of habitat for migratory birds. Generally defined biogeographically, more than 24 MBJVs currently are active: three in Canada, 15 within the United States, three that span national boundaries (Sonora and Rio Grande [Mexico and United States], Pacific Bird [Canada and United States]), and three that are defined as Species Joint Ventures (Arctic Goose, Black Duck, and Sea Duck) (www.mbjv.org). The MBJVs were launched in 1986 to garner funding and implement actions to meet the objectives of the North American Waterfowl Management Plan (United States, Canada, and Mexico) (Giocomo et al. 2012).

The MBJVs are overseen by management boards that include representatives from the main partner organizations. These organizations are expected to bring resources to the MBJVs. Technical teams within the MBJVs coordinate communication among the partners. The technical teams also coordinate research and provide decision support (Giocomo et al. 2012). Therefore, MBJVs join not only multiple actors in resource management but research with management. As a consequence of their development of sound monitoring protocols for a well-defined suite of migratory bird species, MBJVs have been described as an exemplar of successful implementation of adaptive management (Giocomo et al. 2012; Fig. 9.2). In their first 25 years, MBJVs reportedly raised more than US$4.7 billion and protected more than 7 million hectares of habitat for migratory birds in the USA alone.

The MBJVs have served as a model for the conservation of monarch butterflies, eastern brook trout (*Salvelinus fontinalis*) (Giocomo et al. 2012), and ecosystems. For example, the Regional Alliance for the Conservation of Chihuahuan Desert Grasslands is a joint venture led by Mexican federal agencies (http://www.cec.org/content/spreading-joint-venture-model-regional-alliance-conservation-chihuahuan-desert-grasslands).

Although not formally a joint venture, the Mesoamerican Biological Corridor, established in 1997, is an effort among seven countries (Belize, Costa Rica, El Salvador, Guatemala, Honduras, Mexico, Nicaragua, and Panama) to conserve species that migrate from southern Mexico to Panama. The initiative, which was implemented in Mexico in 2002, coordinates transnational policies that promote sustainable uses of forests and other lands in corridors between protected areas. The Mesoamerican Biological Corridor assumes that such policies will promote connectivity, gene flow, and increases in effective population size (Godoy 2003).

Fig. 9.2 Three species of migratory waterfowl—Green-winged Teal (*Anas carolinensis*), American Wigeon (*Mareca americana*), and Northern Pintail (*Anas acuta*)—that benefit from management informed by Migratory Bird Joint Ventures. Photograph by M. Schwartz

9.4.2 Consultation on Areas Voluntarily Designated for Conservation

In 2002, the Mexican National Commission of Protected Natural Areas (CONANP), which is part of the Secretariat of Environment and Natural Resources (SEMARNAT), initiated a voluntary community protected areas program, *Áreas Destinadas Voluntariamente a la Conservación*. The program is a collaboration between communities or private landholders and the federal government in which one of the former entities formally sets aside some fraction of their lands for conservation. CONANP helps the community or landholder to establish legal protection for the land and assists them in developing a land-use strategy, protection criteria, and management actions. The government also may assist the community or landholder in defending the conservation status of the land if local, state, or other external parties seek to conduct activities incompatible with conservation. As of November 2018, more than 400 areas in 23 states were certified through this mechanism, protecting more than 512,000 hectares and representing the interests of 11 ethnic groups and nearly 93,000 people (https://www.gob.mx/conanp/acciones-y-programas/areas-destinadas-voluntariamente-a-la-conservacion).

9.4.3 Climate Adaptation Science Centers

Communicating climate science to inform adaptation is a recognized challenge in environmental decision-making (Kirchhoff et al. 2013). The US Congress authorized the establishment of the National Climate Change and Wildlife Science Center (now the National Climate Adaptation Science Center [NCASC]) within the US Department of the Interior in 2008. Further direction was promulgated by Secretarial Order 3289 (Salazar 2009). The NCASC and eight regional Climate Adaptation Science Centers (CASCs), each of which is led by USGS scientists and a university consortium selected through a competitive process, partner with natural and cultural resource managers to provide science that addresses adaptation of species, ecosystems, and the human communities they support to climate change. The scope of the CASCs encompasses myriad issues, from silviculture to landscape design, and their areas of emphasis include multiple states and large regions. For example, the Northeast CASC covers 22 states; the Alaska CASC covers over 1.7 million km^2. The CASCs' focus ranges from the delivery of climate data (Oakley and Daudert 2016) to examining the cumulative effects of climate change and other stressors (e.g., wildfire).

The CASCs have both bottom-up and top-down organizational attributes. From the bottom up, federal and academic leaders of the CASCs interact with state, federal, and tribal resource managers within their regions to set priorities for research. Modes of interaction include workshops, conferences and regional meetings, and one-on-one or small-group meetings. These research needs are matched with existing expertise. When relevant expertise is not available among the federal and university members of the CASC, needs can sometimes be filled through awards to external partners. From the top down, the USGS competitively awards funds for research.

The university consortium model allows the federal government to tap into a flexible and diverse set of experts, including highly respected climate scientists. Translation and coproduction of science are integral to the CASCs (Enquist et al. 2017). Nevertheless, eliciting research needs from management partners has been quite challenging. Researchers expect managers to state explicit needs, whereas managers expect researchers to explain what managers need to know. Although such exchanges are an improvement over well-intentioned guesses at management interests by researchers, their inefficiencies may discourage participation by the university and federal researchers who are evaluated on the basis of peer-reviewed publications (Hallett et al. 2017).

Successes of the CASC network include stakeholder-driven designs for regional conservation, such as the Southeast Conservation Adaptation Strategy (http://secassoutheast.org/) and Nature's Network (http://naturesnetwork.org/), inputs to management documents such as State Wildlife Action Plans (https://necsc.umass.edu/projects/integrating-climate-change-state-wildlife-action-plans, https://www.tandfonline.com/doi/abs/10.1080/08941920.2017.1290178) and tribal drought management plans (McNeeley and Beeton 2017), syntheses on major environmental topics such as drought (Crausbay et al. 2017), and training of early-career scientists in translational science (Schwartz et al. 2017).

9.4.4 Fire Science Exchange Network

The Joint Fire Science Program (JFSP), which is funded by the US Department of the Interior and USFS, launched the Fire Science Exchange Network in 2010. The network of 15 science exchanges supports the JFSP's mission of funding and delivering science associated with the management of wildland fire, fuels, and ecosystems in response to the needs of practitioners, managers, and policymakers. The network is intended to increase awareness, understanding, and use of scientific information on wildland fire.

The exchanges, which largely are delineated on the basis of major ecosystems, also aim to increase communication among fire managers, practitioners, and researchers and to deliver information not only to government employees but to tribal and private stakeholders. Moreover, their objectives include the development of methods to assess the quality and applicability of research, provision of support for adaptive management, and identification of research and related needs. Methods of communication and knowledge transfer include field trips, demonstration projects, workshops, and conferences, networks of experts, training programs, research syntheses and briefs, electronic and social media, and regional databases.

The JFSP has evaluated the network annually since 2011 via methods including surveys of fire managers and practitioners, fire researchers and scientists, and the general public; metrics and experiences related to the use of each exchange's website; and interviews with exchange personnel. The evaluation examines whether JFSP's logic model is consistent with the outcomes of the network. Results are intended to assist the JFSP's governing board and exchanges in improving and supporting the individual exchanges and the network.

On the whole, evaluations suggest that both fire managers and practitioners and fire researchers and other scientists perceive that the network improves communication between these two groups (Hunter 2016; Maletsky et al. 2018). Across years, fire managers and practitioners generally indicated that the use of fire science increased their effectiveness on the job, but was equivocal about whether fire science readily was applicable to their work. Evaluations also suggested that fire managers and practitioners believe that a mechanism to facilitate sharing of fire science in their regions is warranted, but not necessarily that the network has led to improvements in environmental conditions, possibly because there are lags between communication or action and outcomes. In addition, two independent assessments of whether science sponsored by the JFSP has been used in decision-making concluded that the network had increased considerably the use of fire science in management (Hunter 2016; Maletsky et al. 2018).

There is great respect for the Fire Science Exchange Network, and the JFSP, among the management and research communities. Moreover, wildfire seasons are becoming longer, leading to increased losses of life and property and creating substantial economic losses. Nevertheless, funding for all JFSP activities has declined steadily in recent years. In spring 2018, managers and researchers appealed to both chambers of the US Congress to maintain funding at historical levels. Signatories to a letter sent to congressional appropriations committees noted that the JFSP is distinct in its ability to support research in the physical, natural, and social sciences in direct response to the priorities of practitioners. At the time of this writing, it is unclear whether the JFSP and the network will be sustained.

9.4.5 The Puget Sound Partnership

The 31,440 km^2 Puget Sound watershed encompasses most of the west coast of Washington and its border with British Columbia, Canada, including flows from the Olympic Mountains and Cascade Range (Dunagan 2015). Puget Sound is a major regional fishery, shipping zone, and tourist attraction. Primary environmental stressors to Puget Sound include land-use change, non-point source pollution and toxicants, shoreline hardening, in-channel barriers to water and sediment transport, and direct threats to species (Georgiadis 2015). The Puget Sound watershed is home to nearly 4.8 million people (Birkeland et al. 2015).

The Puget Sound Partnership (PSP; http://www.psp.wa.gov/) organizes myriad government agencies, non-profit organizations, and citizen groups that are responsible for or interested in the environmental management of the watershed. The Partnership, a state agency, was launched in 2010 to establish region-wide action priorities and foster a collaborative, science-based approach to ecosystem management (Koontz and Thomas 2018). The PSP requires its partners to use the Open Standards for the Practice of Conservation as a planning framework (Schwartz et al. 2012; Redford et al. 2018) within the context of their action agenda (Koontz and Thomas 2018). The structure of the PSP includes nine geographically based Local Integrating Organizations (LIOs) that create priorities and action plans that specify targets, stressors to those targets, mechanisms to alleviate threats, and metrics of progress toward explicit goals (PSP 2017).

The PSP streamlined planning for environmental management in the Puget Sound watershed. Although not all participants appear enthusiastic about the process, the effort has resulted in uniform reporting of progress toward well-defined goals (PSP 2017). Each LIO reports on near-term actions and effectiveness in a report-card format that is linked to 50 region-wide vital signs. The Partnership notes that progress is constrained by resource limitations and insufficient commitment to the project, leading to insufficient funding of actions (PSP 2017). Ten of the defined vital signs are reported to be improving. The status of nine is equivocal, six have not changed significantly, and four are declining. Information is insufficient to gage the status of 21 other vital signs. Targets that are improving include restoration of riparian lands, floodplains, and estuaries. Populations of killer whales (*Orcinus orca*) and Pacific herring (*Clupea pallasii*) are declining, as is the condition of marine waters (PSP 2017).

We (MWS, MAW) tried to integrate structured decision-making (Martin et al. 2009) into the operations of the PSP to make collaborative prioritization of projects more transparent. We discovered that the organization had produced a formidable set of documents that justifies management priorities and describes the integrative, collaborative approach used to identify these priorities. An analysis of the use of science in executive and technical committee meetings of the LIOs found little evidence that the LIOs were strongly science-driven (Koontz and Thomas 2018). Although technical committees discussed scientific issues, they rarely referenced peer-reviewed scientific documents to support their decisions (Koontz and Thomas 2018).

Nevertheless, nearly a decade of collaborative effort has standardized discussion and understanding of management issues by institutions in the Puget Sound watershed. A large number of independent organizations now use a common language to establish management goals, describe pressures to their targets, and measure success. Although it is impossible to evaluate what the condition of the Puget Sound watershed would be without the PSP, it appears that collaborating agencies and private stakeholders find value in the organization.

9.4.6 Chicago Wilderness

Chicago Wilderness, established in 1996 (Moskovits et al. 2004), is a regional alliance of local, state, and federal agencies, universities, and businesses working to protect and restore the biological diversity of the greater Chicago region. The alliance now has over 360 member organizations and operates in the region from southeastern Wisconsin to southwestern Michigan along the shore of Lake Michigan. Chicago Wilderness also includes the nine counties of the greater Chicago region (Watkins et al. 2015). Strategies, work plans, and decisions about focal areas are made by an executive council. A number of member-initiated committees meet periodically to address issues of regional interest that extend beyond the focus areas. In 2002, Chicago Wilderness created a corporate council to leverage the resources, capabilities, and influence of the region's corporations to meet the organization's mission.

Since its inception, Chicago Wilderness has received over $15 million dollars to support conservation projects. These funds have supported the removal of non-native invasive species, tree planting, prairie restoration, and educational programs of partner organizations throughout the region. Chicago Wilderness developed the first atlas of biodiversity for the region, a biodiversity recovery plan to help coordinate management activities throughout the region, a children's outdoor bill of rights to encourage participation in outdoor activities, a green-infrastructure vision document to guide open space acquisition, and a climate action plan (https://www.chicagowilderness.org/page/publicationsnew).

Chicago Wilderness has been instrumental in providing a venue for the region's natural resource managers, researchers, consultants, and government officials to develop shared strategies for solving the region's environmental challenges. Examinations of language used to define ecological restoration by Chicago Wilderness suggest a strong, common understanding of goals and objectives (Watkins et al. 2015). Implementation of these strategies, however, can be challenging because governments—particularly at the local and county level—must reconcile Chicago Wilderness objectives with their political concerns. Provision of financial resources to these governments for the implementation of Chicago Wilderness's vision helps to meet these challenges, but money is increasingly scarce. Ebb and flow in Chicago Wilderness membership also makes it difficult to maintain continuity. The plans and documents developed by Chicago Wilderness motivated the members and inspired higher degrees of participation. In 2017, the Chicago Wilderness Trust (the organization formed to support Chicago Wilderness) announced that it was ceasing operations. Currently, the executive council is managing operations with a fiscal sponsor, and the future of the alliance is unclear.

9.4.7 Scientific Advisors to Habitat Conservation Plans

The US Fish and Wildlife Service (FWS) encourages applicants for incidental take permits under section 10(a) of the US Endangered Species Act to engage independent scientists in the development of habitat conservation plans (HCP). In California, some HCPs are coupled with natural community conservation plans (NCCPs), which also require the engagement of independent scientists. California's Natural Community Conservation Planning Act aims to protect the state's biological diversity while reducing conflicts between the protection of natural resources and economic development. Incidental take permits allow non-federal entities to take (loosely, kill or harm) species listed as endangered or threatened in the course of otherwise lawful activities provided that the take is minimized and mitigated (California Environmental Quality Act, Title 14, Section 15357). Independent scientific input is intended to guide conservation strategies, design of reserves, development of goals for monitoring and other aspects of adaptive management, and reduction of data gaps and uncertainties. The scientists or groups of scientists who serve as advisors to the development of HCPs and NCCPs sometimes function as boundary-spanning organizations.

In many cases, FWS personnel are concerned that direct engagement between science advisors and permit applicants will compromise the advisors' independence. However, in our experience, direct and sustained interaction among advisors, applicants, and regulators increase the applicability of scientific input and the likelihood that the HCP (or HCP/NCCP) not only will be informed by science but will achieve conservation objectives. In our experience, the integrity of the science is not compromised in the process. The East Contra Costa County, California HCP/NCCP (approved in 2007) and the town of Apple Valley, California's multiple-species

HCP/NCCP (planning in progress as of this writing) illustrate the ability of science advisors to serve as boundary-spanning groups.

The East Contra Costa County HCP Association (HCPA), which represents Contra Costa County, the Contra Costa Water District, the cities of Brentwood, Clayton, Oakley, and Pittsburg, and the East Bay Regional Park District, felt strongly that independent scientific input would help ensure that their HCP/NCCP was based on relevant and valid scientific principles and techniques. The HCPA also agreed that including scientific input from the earliest phases of development of the HCP/NCCP would help to identify and resolve scientific challenges before they threatened the schedule and budget for the plan. All meetings of the science advisors were open to the public, and there was an opportunity for public comment at each meeting. Issues considered by the science advisors included adequacy of data for the development of the HCP/NCCP, identification of data gaps and sources of uncertainty, formulation of biological goals and objectives for conserving species and natural communities, and development of adaptive management guidelines. Representatives of the HCPA and the consultant team were present at each meeting of the science advisors to explain all relevant components of the HCP/NCCP, receive comments and request clarification from the science advisors.

At the conclusion of their work, the science advisors offered a number of comments and suggestions about the science advisory process. Some science advisors believed it would have been useful for their group to communicate directly with political leaders involved in the development of the HCP/NCCP. They also felt that more interaction with members of the public who attended their meetings might have been useful. Despite continued discomfort with the process by FWS, the scientists felt that the presence of HCPA members and the public did not hinder the science advisory process. There was no perceived need for confidentiality during the science advisory process for this HCP.

Similarly, as part of its development of an HCP/NCCP, the town of Apple Valley, California, convened a science advisory committee. The town, science advisory committee members, and consultants to the town agreed on four goals for the engagement of the science advisory committee in the planning process. First, understand the town's biological and social goals and objectives for the HCP/NCCP and the process that the town is following in developing the plans. Second, provide realistic scientific guidance in the context of the town's constraints and opportunities, such as existing levels of regional urbanization and fragmentation. Third, address not only species and natural communities but maintenance of ecological processes via open space, both within the planning area and regionally. Fourth, emphasize the incorporation of science not only into the development of the HCP / NCCP but into decision-making during the implementation of the HCP/NCCP. Again, all parties felt that open, direct communication greatly increased information transfer and applicability among applicants, regulators, and scientists.

9.5 Discussion

In this chapter, we presented examples of boundary-spanning organizations that strive to assist stewards of private or public lands in achieving conservation or other management objectives. We then qualitatively assessed four aspects of the success of boundary-spanning organizations: first, whether the boundary-spanning organization affected what knowledge was made accessible to decision-makers; second, whether decision-makers indicated that the boundary-spanning organization's efforts were worthwhile; third, whether the evidence suggests that science was used to inform the decision; and fourth, whether the action resulting from the decision was effective. We acknowledge that our evidence is largely anecdotal and often subjective and complex. There are few external evaluations of the effectiveness of boundary-spanning organizations (Pitt et al. 2018).

Our evidence suggests that in most cases, boundary-spanning organizations affect the knowledge that is conveyed to decision-makers. Each of the boundary-spanning organizations in our case studies convenes researchers and practitioners, providing access to and, in some cases coproducing scientific knowledge through some combination of meetings, webinars, websites, synthetic papers, reports, and popular writing (Fig. 9.3). These activities are intended to provide decision-relevant information, relieving practitioners of the need to discover or navigate sometimes limited access to the primary literature. Both Chicago Wilderness and the Puget Sound Partnership created groups of decision-makers that meet regularly to identify priorities collectively. Migratory Bird Joint Ventures have technical committees that integrate science into decision-making. The Fire Science Exchanges and the Climate

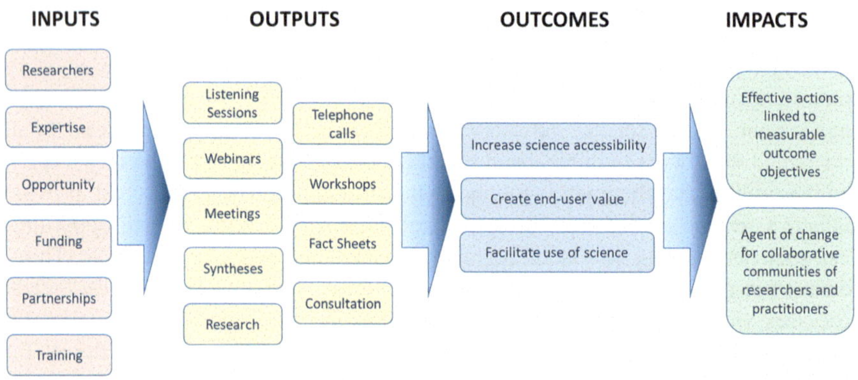

Fig. 9.3 A theory of change for boundary-spanning organizations in conservation, as interpreted through the authors' collective experience. Boundary-spanning organizations contribute to change by providing researchers and practitioners with relevant expertise to work collaboratively, often supporting the interaction with funding. Outputs generated through two-way dialog to increase accessibility and use of science and the value end users attribute to the organization. The activities of these organizations increase the effectiveness of actions and appreciation of the process by both researchers and practitioners

Adaptation Science Centers host meetings, webinars, or other interactions to deliver information to decision-makers. The science advisors to habitat conservation plans are tasked by a subset of the decision-makers to collate, interpret, and deliver scientific information to planners and regulators.

In some cases, it is clear whether decision-makers indicated that the boundary-spanning organization's efforts were worthwhile. For example, that Chicago Wilderness has existed and grown for over 20 years is a powerful testament that decision-makers realize a value in its existence. When top-down government sponsorship creates the boundary-spanning organization (e.g., Puget Sound Partnership, CASCs, Fire Science Exchanges), it is much more difficult to assess the value that decision-makers attribute to the organization. The Fire Science Exchange Network has systematically evaluated whether it increases the use of science in decisions (Hunter 2016; Maletsky et al. 2018). Two observations are particularly salient. First, researchers and practitioners felt that the most effective way for science to inform the decision process was direct personal communication between end users of the science and the researchers who provided the science (Hunter 2016). Second, practitioners reported that the value of working with researchers increased over time (Maletsky et al. 2018). These two observations suggest that a primary role of the boundary-spanning organization is to foster connections between individual researchers and practitioners. It is unclear whether this represents sufficient value in the eyes of practitioners. Researchers, staff of the boundary-spanning organizations, and practitioners all recognize there is lack of awareness that the boundary-spanning organizations exist (Hunter 2016), which may reflect the limited budgets of these organizations relative to the breadth of the decisions that they are trying to inform.

The data on whether the science conveyed by boundary-spanning organizations informs decisions is equivocal. Each of the authors has worked with, or for, boundary-spanning organizations. We have done this, in part, because we believe that working with boundary-spanning organizations increases our capacity to connect research and practice. The engagement of researchers with practitioners is largely a personal choice. Doing so may reduce scientific productivity as measured by number of publications. Nevertheless, many researchers who engage with practitioners feel that their science gains credibility, relevance, and legitimacy (Wall et al. 2017), and thus is more likely to have a tangible effect. However, the effects of researcher-practitioner communications appear to be inconsistent. An assessment of the Puget Sound Partnership found that although science was discussed at technical committee meetings, there is little documentation that science-informed decisions were made by the constituent organizations (Koontz and Thomas 2018). In contrast, analyses of the Fire Science Exchanges demonstrated that science disseminated by the exchanges is integrated into planning (Hunter 2016; Maletsky et al. 2018).

It is quite difficult to assess whether decisions informed by science that was produced or communicated by boundary-spanning organizations were effective. Understanding whether a decision met objectives may require hindsight that is not yet available. Chicago Wilderness, the oldest of the organizations that we described, may offer the best opportunity to evaluate decisions. However, the evidence is mostly anecdotal and indirect. For example, language and actions related to ecological restoration across the region are consistent, suggesting that spanning boundaries helps to develop a shared understanding (Watkins et al. 2015). Chicago Wilderness is currently struggling with its future. The reasons are unclear, but Chicago Wilderness may be failing now simply because of its previous successes. The organizations convened by Chicago Wilderness are working as a collective, and Chicago Wilderness per se may no longer be necessary.

Our assessment highlighted several other challenges and opportunities for boundary-spanning organizations. For example, navigating multi-jurisdictional resource management over large areas, an aim of most boundary-spanning organizations is difficult. Furthermore, many boundary-spanning organizations work on resource management topics that are subject to social conflict (e.g., Fire Science Exchanges, CASCs). As a result, these organizations may function as knowledge brokers that strive to evaluate the weight of scientific evidence that could be used to support alternative actions.

It is a challenge to deliver management-relevant science to decision-makers in a timely fashion. There are many constituents for this information and few staff in boundary-spanning organizations. The result is limited capacity for one-on-one or small-group communication, the form of communication that arguably is most effective and trusted by practitioners. Also, it is a challenge for researchers to build and sustain personal connections with practitioners. Even researchers that recognize such connections as a critical step in actionable science may find their attention, funds, and performance metrics more focused on the research itself. Moreover, practitioners rarely reach out to the research community; learning about new research also takes time.

We believe that the effectiveness of communicating science may vary predictably among organizations. Most top-down governmental organizations (e.g., CASCs) have insufficient resources to achieve their broad scope fully. The consequence is a strong need for prioritization of issues and partners. Organizations with a narrower scope (e.g., Joint Ventures) may be more effective with limited resources. Organizations that adopt explicit frameworks for planning (e.g., the Open Standards) must create a community of practitioners willing to use those methods to describe and prioritize targets and actions. The focus on the Open Standards for the Puget Sound Partnership, in particular, may come at the expense of examining how emerging research might inform decisions. Bottom-up organizations that are driven by the interests of practitioners rather than an external body (e.g., Chicago Wilderness) appear to have the capacity to communicate a common vision. Small and highly focused groups, such as science advisors to an HCP/NCCP process, effectively can synthesize information on the potential effects of alternative actions on rare species.

Generally, it seems that the more focused the target, the more likely it is that consensus can be reached on a narrow set of objectives and priorities. When an objective is focused, the body of relevant science likely will be bounded and readily conveyed. In these cases, the boundary-spanning organization likely exists because the action alternatives are socially contested. Therefore, these groups' primary challenge may be to provide salient science while remaining impartial and therefore credible. Moreover, transparency and inclusiveness bolster perceived legitimacy or fairness, which can increase trust and the use of science. In contrast, the main challenge for organizations with a broader scope is to focus limited resources on reaching a defined set of practitioners with relevant and compelling information.

9.6 Conclusions

Understanding patterns in science delivery, use, and integration can help boundary-spanning organizations such as CASCs set and realize objectives. Researchers and practitioners are in the early phases of determining which approaches to boundary spanning are effective. Therefore, there also is considerable investment in the evaluation of process (e.g., Watkins et al. 2015; Wall et al. 2017; Maletsky et al. 2018). These assessments, and our own experience, suggest that boundary-spanning organizations have the potential to be highly valuable, but the path to value is not smooth. Creating an organization that spans boundaries does not, in and of itself guarantee success. Success requires a skilled group of people who define explicit priorities, develop ways for researchers and practitioners to interact productively, and assure lasting and meaningful personal connections. Furthermore, political support for boundary-spanning efforts can be difficult to maintain. As an example, the Landscape Conservation Cooperative network within the USA was modeled after the Joint Ventures but focused more broadly on how changing climate may affect the protection of biological diversity (Table 9.1). The network was disbanded after the 2016 national election and the subsequent administrative reorganization of agency priorities. Although the network itself has disbanded, some are still active by virtue of the former partners choosing to remain engaged and take on a leadership role to maintain the cooperative endeavor effectively.

Nevertheless, we believe that boundary-spanning organizations in North America will continue to be prominent in fostering the use of knowledge in conservation actions. We acknowledge that convening groups to negotiate everything from what constitutes salient information to how decisions are made is slow and laborious process. However, there appears to be an unwavering consensus that integrating people who will be affected by decisions into the decision-making process increases the success and acceptance of those decisions. At present, such integration includes structures that foster the application of science to decisions. Boundary-spanning organizations continue to play a substantial and visible role in bringing knowledge exchange to bear on major environmental decisions in North America.

Acknowledgements The authors thank the USGS Southwest and Northeast Climate Adaptation Science Centers and California Agriculture and Natural Resources project CA-D-ESP-2094-H for support.

References

Baylis K, Honey-Rosés J, Börner J et al (2016) Mainstreaming impact evaluation in nature conservation. Conserv Lett 9(1):58–64

Bean MJ (2009) The endangered species act science, policy, and politics. Ann NY Acad Sci 1162(1):369–391

Behnken JA, Groninger JW, Akamani K (2016) Institutional constraints to collaborative ecosystem management within a wetlands conservation partnership. J Contemp Water Res Ed 158(1):19–33

Birkeland C, Biedenweg K, Christie P (2015) Human dimensions. In: 2015 Puget Sound fact book. Puget Sound institute, University of Washington, Tacoma, Washington

Bixler RP (2021) The knowledge network: identifying actors and structural dimensions of knowledge transfer. In: Ferreira CC, Klütsch CFC (eds) Closing the knowledge-implementation gap in conservation science – evidence transfer across spatiotemporal scales and different stakeholders. Springer, New York, pp 143–170

Brooks TM, Mittermeier RA, da Fonseca GAB et al (2006) Global biodiversity conservation priorities. Science 313(5783):58–61

Cash DW, Clark WC, Alcock F (2003) Knowledge systems for sustainable development. PNAS 100:8086–8091

Cook CN, Mascia MB, Schwartz MW et al (2013) Achieving conservation science that bridges the knowledge-action boundary. Conserv Biol 27(4):669–678

Crausbay SD, Ramirez AR, Carter SL et al (2017) Defining ecological drought for the twenty-first century. Bull Am Meteorol Soc 98(12):2543–2550

DeCrappeo NM, Bisbal GA, Meadow AM (2018) A path to actionable climate science: perspectives from the field. Environ Manag 61:181–187

Dietz T, Ostrom E, Stern PC (2003) The struggle to govern the commons. Science 302(5652):1907–1912

Doherty KE, Evans JS, Walker J et al (2015) Building the foundation for international conservation planning for breeding ducks across the US and Canadian Border. PLoS One 10(2):e0116735

Dorcas ME, Wilson JD, Redd RN et al (2012) Severe mammal declines coincide with proliferation of invasive Burmese python in Everglades National Park. PNAS 109(7):2418–2422

Doremus H, Lin AC, Rosenberg RH et al (2008) Environmental law policy: problems, cases, and readings, 5th edn. Foundation Press, New York

Dunagan C (2015) Overview: Puget Sound as an estuary. In: Puget Sound fact book. Puget Sound Institute, University of Washington, Tacoma, Washington, pp 8–15

Enquist CAF, Jackson ST, Garfin GM et al (2017) Foundations of translational ecology. Front Ecol Environ 15(10):541–550

Evans DM, Che-Castaldo JP, Crouse D et al (2016) Species recovery in the United States: increasing the effectiveness of the endangered species act. Issues in Ecology, Report Number 20. Ecological Society of America. Washington, DC. 27 pp.

Fleishman E, Blockstein DE, Hall JA et al (2011) Top 40 priorities for science to inform US conservation and management policy. Bioscience 61(4):290–300

Georgiadis N (2015) Threats. In: Puget Sound fact book. Puget Sound Institute, University of Washington, Tacoma, Washington, pp 97–103

Giocomo JJ, Gustafson M, Duberstein JN et al (2012) The role of joint ventures in bridging the gap between research and management. In: Sands JP, DeMaso SJ, Schnupp MJ, Brennan LA (eds) Wildlife science: connecting research with management. CRC Press, Boca Raton, pp 239–251

Godoy JC (2003) Corredor Biológico Mesoamericano: iniciativa de integración regional para promover la conservación del bosque. http://www.fao.org/docrep/article/wfc/xii/ms15-s.htm

Hallett LM, Morelli TL, Gerber LR et al (2017) Navigating translational ecology: creating opportunities for scientist participation. Front Ecol Environ 15(10):578–586

Hillebrand H (2004) On the generality of the latitudinal diversity gradient. Am Nat 163(2):192–211

Hoekstra JM, Boucher TM, Ricketts TH et al (2005) Confronting a biome crisis: global disparities of habitat loss and protection. Ecol Lett 8(1):23–29
Hunter ME (2016) Outcomes of fire research: is science used? Int J Wildland Fire 25(5):495–504
Kirchhoff CJ, Lemos MC, Dessai S (2013) Actionable knowledge for environmental decision making: broadening the usability of climate science. Ann Rev Environ Resour 38:393–414
Knight AT, Cowling RM, Rouget M et al (2008) Knowing but not doing: selecting priority conservation areas and the research-implementation gap. Conserv Biol 22(3):610–617
Koontz TM, Thomas CW (2018) Use of science in collaborative environmental management: evidence from local watershed partnerships in the Puget Sound. Environ Sci Pol 88:17–23
Lyons JE, Runge MC, Laskowski HP et al (2008) Monitoring in the context of structured decision-making and adaptive management. J Wildlife Manage 72(8):1683–1692
Maletsky LD, Evans WP, Singletary L et al (2018) Joint fire science program (JFSP) fire science exchange network: a national evaluation of initiative impacts. J Forest 116(4):328–335
Martin J, Runge MC, Nichols JD et al (2009) Structured decision making as a conceptual framework to identify thresholds for conservation and management. Ecol Appl 19(5):1079–1090
Martin TG, Burgman MA, Fidler F et al (2012) Eliciting expert knowledge in conservation science. Conserv Biol 26(1):29–38
McNeeley SM, Beeton TA (2017) Wind River reservation: drought risk and adaptation in the interior (DRAI) report. North Central climate science center. Fort Collins, Colorado. https://www.sciencebase.gov/catalog/item/59b7041ae4b08b1644ddf9b4
Meadow AM, Ferguson DB, Guido Z et al (2015) Moving toward the deliberate coproduction of climate science knowledge. Weather Clim Soc 7(2):179–191
Miteva DA, Pattanayak SK, Ferraro PJ (2012) Evaluation of biodiversity policy instruments: what works and what doesn't? Oxf Rev Econ Pol 28(1):69–92
Morisette JT, Cravens AE, Miller BW et al (2017) Crossing boundaries in a collaborative modeling workspace. Soc Nat Resour 30(9):1158–1167
Moskovits DK, Fialkowski C, Mueller GM et al (2004) Chicago wilderness - a new force in urban conservation. Ann N Y Acad Sci 1023:215–236
Myers N, Mittermeier RA, Mittermeier CG et al (2000) Biodiversity hotspots for conservation priorities. Nature 403:853–858
Oakley NS, Daudert B (2016) Establishing best practices to improve usefulness and usability of web interfaces providing atmospheric data. Bull Am Meteorol Soc 97(2):263–274
Ostrom E (2000) Collective action and the evolution of social norms. J Econ Perspect 14(3):137–158
Ostrom E (2009) A general framework for analyzing sustainability of social-sociological systems. Science 325(5939):419–422
Pitt R, Wyborn C, Page G et al (2018) Wrestling with the complexity of evaluation for organizations at the boundary of science, policy, and practice. Conserv Biol 32(5):998–1006
Pounds JA, Bustamante MR, Coloma LA et al (2006) Widespread amphibian extinctions from epidemic disease driven by global warming. Nature 439:161–167
Puget Sound Partnership (2017) 2017 State of the sound. Olympia, Washington. November 2017, 84 pp. www.psp.wa.gov/so
Redford KH, Hulvey KB, Williamson MA et al (2018) Assessment of the conservation measures partnership's effort to improve conservation outcomes through adaptive management. Conserv Biol 32(4):926–937
Regan HM, Colyvan M, Burgman MA (2002) A taxonomy and treatment of uncertainty for ecology and conservation biology. Ecol Appl 12(2):618–628
Rudd MA, Fleishman E (2014) Policymakers' and scientists' ranks of research priorities for resource-management policy. Bioscience 64(3):219–228
Sabatier PA (1987) Knowledge, policy-oriented learning, and policy change: an advocacy coalition framework. Knowl 8(4):649–692

Safford HD, Sawyer SC, Kocher SD et al (2017) Linking knowledge to action: the role of boundary spanners in translating ecology. Front Ecol Environ 15(10):560–568

Salazar K (2009) Secretarial Order 3289. Department of the Interior, Washington, DC

Sanchirico JN, Springborn MR, Schwartz MW et al (2014) Investment and the policy process in conservation monitoring. Conserv Biol 28(2):361–371

Sanderson EW, Jaiteh M, Levy MA et al (2002) The Human Footprint and the Last of the Wild: The human footprint is a global map of human influence on the land surface, which suggests that human beings are stewards of nature, whether we like it or not. Bioscience 52(10):891–904

Schultz CA, Sisk TA, Noon BR et al (2013) Wildlife conservation planning under the United States Forest Service's 2012 planning rule. J Wildlife Manage 77(3):428–444

Schwartz MW, Deiner K, Forrester T et al (2012) Perspectives on the open standards for the practice of conservation. Biol Conserv 155:169–177

Schwartz MW, Hiers JK, Davis FW et al (2017) Developing a translational ecology workforce. Front Ecol Environ 15(10):587–596

Schwartz MW, Cook CN, Pressey RL (2018) Decision support frameworks and tools for conservation. Conserv Lett 11(2):e12385

Vincent CH, Hanson LA, Argueta CN (2017) Federal land ownership: overview and data. Congressional Research Service, Washington, DC

Wall TU, Meadow AM, Horganic A (2017) Developing evaluation indicators to improve the process of coproducing usable climate science. Weather Clim Soc 9(1):95–107

Waples RS, Nammack M, Cochrane JF et al (2013) A tale of two acts: endangered species listing practices in Canada and the United States. Bioscience 63(9):723–734

Watkins C, Westphal LM, Gobster PH et al (2015) Shared principles of restoration practice in the Chicago Wilderness region. Hum Ecol Rev 21(1):155–177

Weible CM, Sabatier PA (2009) Coalitions, science, and belief change: comparing adversarial and collaborative policy subsystems. Policy Stud J 37(2):195–212

Weible CM, Sabatier PA, McQueen K (2009) Themes and variations: taking stock of the advocacy coalition framework. Policy Stud J 37(1):121–140

Wilson JQ (1989) Bureaucracy: what government agencies do and why they do it. Basic Books, New York

Chapter 10
Progress and Gaps in Biodiversity Data Mainstreaming and Knowledge Transfer for Conservation in South America

Carmen E. Josse and Miguel Fernandez

Contents

C. E. Josse (✉)
Fundación EcoCiencia, Quito, Ecuador
e-mail: carmenjosse@ecociencia.org

M. Fernandez
NatureServe, Arlington, VA, USA

Universidad Mayor de San Andrés, Instituto de Ecología, La Paz, Bolivia

Smithsonian-Mason School of Conservation and Department of Environmental Science and Policy, George Mason University, Fairfax, VA, USA
e-mail: miguel_fernandez@natureserve.org

C. C. Ferreira, C. F. C. Klütsch (eds.), *Closing the Knowledge-Implementation Gap in Conservation Science*, Wildlife Research Monographs 4,
https://doi.org/10.1007/978-3-030-81085-6_10

10.1 Introduction

Five out of the 12 South American countries are on the list of the world's 17 megadiverse countries, with an additional one close to qualifying as such (Mittermeier et al. 1997). At the same time, the latest Living Planet Report (WWF 2018) ranked South America—along with Central America and the Caribbean—as the biogeographic realm that has suffered the most dramatic decline in the population size of vertebrate species, with an overall loss of 89% in 1040 populations of 689 species compared to 1970.

At this rate, and with most of the region's national economies still relying on extractive industries, agro-industry, and trade of natural resources (such as timber, hydropower, and marine resources), it is a challenge for the region to sustain the wealth of ecosystem services it provides within and beyond its borders into the future. Governments, the private sector, multilateral financial agencies, local communities, scientists, and conservation NGOs, are all stakeholders in need to promote sustainable development that fulfills health, food and water security, and human well-being objectives. At the core of this is the generation and transfer of the necessary knowledge and its mainstreaming into policy development and implementation.

10.1.1 The Knowledge–Implementation Continuum in South America

Although considerable financial resources have been dedicated to biodiversity research in South America (Brooks et al. 2006), relatively little attention has been paid to ensuring that the results of scientific research are used effectively by natural resource managers or policymakers to inform decisions at different levels of governance. Effective and timely public policies that integrate biodiversity conservation and development depend not only on the generation of high-quality information, but also on the ability to integrate, synthesize, and deliver relevant information to the right users at the right time (Han et al. 2017).

The existing gap between data producers and information users is exacerbated by many factors including:

1. Large taxonomic, spatial, and temporal biases that limit the usability of information

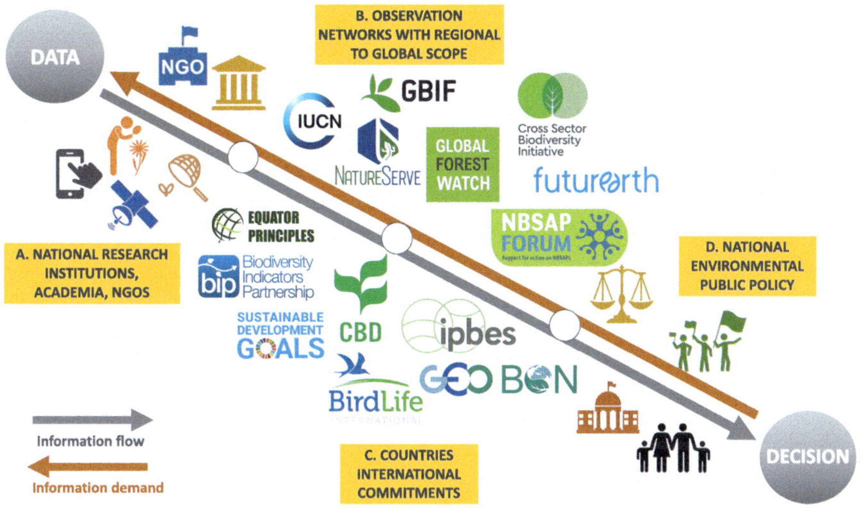

Fig. 10.1 The biodiversity data mainstreaming continuum, from data to decisions. Note that the institutions indicated here are mere examples of each level and do not represent the myriad of institutions that can play a role at these different levels

2. The lack of official channels that maintain a bidirectional flow of communication regarding information provision and demand
3. The absence of adequate information transfer platforms

These shortcomings create silos where new information tends to remain in academic contexts, limiting its access to policy- and decision-makers (e.g., Fernández et al. 2015). In order to facilitate communication, biodiversity data mainstreaming can be conceptualized as data generation that, once harmonized and standardized, can be integrated into variables (e.g., Essential Biodiversity Variables—EBVs; Pereira et al. 2013) and potentially feed into different indicators (e.g., Biodiversity Indicators Partnership, Global Biodiversity Change Indicators). These indicators, measured over time, become valuable tools to support decision-making processes in biodiversity conservation (Han et al. 2017).

In this chapter, we focus on four fundamental steps of the biodiversity data mainstreaming process and how they are being implemented in South America. These steps refer to the levels of complexity of the information as it moves along the biodiversity data mainstreaming continuum, from data generation to decision-making. Figure 10.1 associates these steps with a representative sample of the most relevant players involved.

On one end (Fig. 10.1a) there are agents responsible for the production of the raw data and for collecting biological and physical environmental data, such as academic institutions, national research institutes or programs, nongovernmental research organizations, and independent researchers. This group contributes to the production and complementation of baseline information that feeds the subsequent levels. In

South America, this group is often quite technical and limited in relation to the ability to re-contextualize and re-purpose the information they produce.

Next in the chain are a group of actors that feed of the information produced by the previous level (Fig. 10.1b). These are the national, regional, and global biodiversity observation networks that may or may not have a thematic focus. Their work focuses primarily on the development of conceptual frameworks, data sockets, and information gateways that enable data integration and harmonization (e.g., Element Occurrence, Darwin Core Standard, EBVs). Analogous to the climate change community, this group has a similar role to that of the regional and global climate networks that support data integration, storage, and exchange. In South America, the development of this level is in progress and there are successful examples that often result from collaborating with international organizations (e.g., Bosques Andinos Program; Red GLORIA-Andes; Latin American Seasonally Dry Tropical Forest Floristic Network—DryFlor; Amazon Forest Inventory Network—Rainfor; NatureServe; Global Biodiversity Information Facility—GBIF; Group on Earth Observations Biodiversity Observation Network—GEO BON).

Further along in the continuum are the various international conventions, agreements, and multilateral forums that countries ratify (Fig. 10.1c; e.g., Convention on Biological Diversity—CBD; Intergovernmental Science-Policy Platform on Biodiversity and Ecosystem Services—IPBES), and their different implementation mechanisms (e.g., Sustainable Development Goals—SDGs, Aichi Targets, IPBES Regional and Thematic Assessments, National Biodiversity Strategies and Action Plans). This level is particularly data-hungry, as it requires a constant flow of information from the other two levels to be operational and ensure countries comply with international commitments.

The final and possibly most relevant level for the purposes of this chapter is that of national public policy (Fig. 10.1d). It is precisely at this level that the transfer of knowledge needed to generate biodiversity indicators, which are produced on a global and regional scale, tends to fail to ensure that they are applied at the subnational or local level. The indicators that are informed by data generated at the national level may be in a better position to meet local needs, to be replicated over time, and to form the basis of local policy-making. Nonetheless, evidence shows that the availability of nationally produced indicators and the capacity and willingness to generate such data is uneven across countries of the region. Reasons for these differences could be related to uneven access to technology and financial support for data collection and management, availability of conservation expertise and analysis, as well as limited information accessibility and interoperability (Han et al. 2014).

10.1.2 Structure of the Chapter

Our chapter is divided into four sections that tackle different aspects of the biodiversity data mainstreaming continuum as described in Fig. 10.1. The first section discusses the interactions, synergies, and gaps in the flow of information between data production and science-informed decision-making. We draw on the results and conclusions of a recent study which involved one of the authors of the present chapter, conducted in the Andean-Amazon countries of Bolivia, Colombia, Ecuador, and Peru that focused on the demand for and supply of biodiversity information (Josse and Vásconez 2016). The demand for information is discussed from the perspective of international commitments and information needs to inform national public policy. We also assess the extent to which the public and private sectors provide relevant information on biodiversity. Finally, after analyzing the efficiency in the use of existing information, we provide recommendations that can fill in identified gaps.

In the following section, to expand our scope to all countries of South America, we analyze National Biodiversity Strategies and Action Plans (NBSAPs). Here, we evaluate the progress made and challenges faced by different countries in the region in relation to the production and mobilization of knowledge and the outcomes in terms of conservation action. We carried out a brief analysis of the institutional framework required for the successful transfer of knowledge and finally, evaluated how this translates into policy and actions in different countries.

Subsequently, we focus on two case studies relevant to the topic of the chapter. The first case study describes the national efforts made in Bolivia to collect and integrate observational data to support a global network of biodiversity observations, promoted by GEO BON. In this context, we explore the challenges, but also the opportunities, experienced at the national level to set up such a network. The second case study refers to the IUCN RLTS and the RLE. We contrasted both standards and assessed their potential to contribute in a concerted manner to supporting conservation efforts in information-deficient regions.

Finally, in the last section, we return to a more objective and regional vision. Here, we highlight how little is known about the cost and effectiveness of efforts aimed at reducing the gap between information production and demand and how critical it is for local initiatives to be fully integrated into regional and global efforts for biodiversity mainstreaming and biodiversity knowledge transfer in South America to become successful.

10.2 Interactions, Synergies, and Gaps in the Information Flow Between Data and Decisions

Bridging the gap between knowledge and action for advancing biodiversity conservation and securing the long-term sustainability of natural resources is, to some extent, linked to the availability of the required information and to the political environment which demands or enables the use of that information to implement conservation actions.

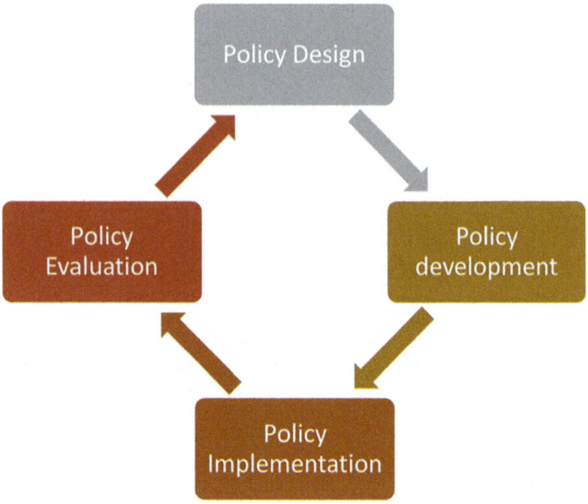

Fig. 10.2 Public policy cycle

Public policies are driven by the changing nature of existing problems and are designed through an iterative cycle (Fig. 10.2). To illustrate this in terms of environmental and biodiversity public policy, we draw on the results and conclusions of a recent study, published as a technical report, conducted in the Andean-Amazonian countries of Bolivia, Colombia, Ecuador, and Peru, that focused on both how and to what extent such policies affect the generation, type, and use of biodiversity information by different actors (Josse and Vásconez 2016). While these four countries share similar levels of biodiversity, their political contexts are quite distinct, making them representative examples for the South American context. For this reason, the comparative analysis of policy and knowledge generation served as an opportunity to identify common challenges and solutions for the effective use of biodiversity evidence in policy design in this region of the world.

After reviewing approx. 60 relevant policies, a questionnaire was designed and used to interview *in-loco* a total of 52 stakeholders during 2015. The respondents were identified by their country, their category as an information user or generator (or both), the type of organization they represented (public institution, NGO, university, public research institute), and their decision-making level (director or mid-level specialists). The interviews were transcribed and subsequently systematized. Below, we present the most relevant results of said published technical report.

The study analyzed both *information demand*—the way by which conservation policies stimulate the generation of biodiversity data needed to inform the implementation and evaluation of the policy, and *information supply*—the influence of documented evidence about biodiversity on environmental public policy in all four countries.

The results show that scientific evidence is now being more widely used in the design of public policies which, to some extent, then serve to catalyze the production of more biodiversity information. However, the response to this demand remains disorganized and/or insufficiently institutionalized. Moreover, the growing complexity of environmental and socioeconomic problems that public policies are striving to resolve in the region demands new forms of knowledge that need to provide responses incorporating multidisciplinary information. The latter then needs to be translated and delivered through user-friendly information systems that offer multiple consultation options. It is increasingly difficult to generate and mobilize this kind of knowledge, and the supply of information lags in comparison to the demand.

10.2.1 The Knowledge Demand Side

There is a demand for knowledge to inform national environmental public policy and to support compliance with country-specific commitments to international conventions and platforms (e.g., CBD, IPBES, United Nations SDGs, Convention on International Trade in Endangered Species—CITES, and the United Nations Framework Convention on Climate Change—UNFCCC). This requires further elaboration to understand how these two sources of demand for biodiversity and environmental information interact—or not—as triggers of knowledge production in South America.

The CBD (https://www.cbd.int/convention/) came into effect toward the end of 1993 and was adopted by 168 signatory countries, setting a milestone in the consideration of biological resources as vital for humankind's economic and social development. The initial implementation of the CBD was quickly followed by the construction of environmental legal frameworks in countries aligned with the Convention. For the aforementioned four Andean countries, environmental protection gained prominence in public policy even prior to the CBD, in the early 1990s, as follows:

- In Peru, the 1990 Environment and Natural Resources Code paved the road for a pivotal national environmental policy.
- In 1992, Bolivia enacted its first Environmental Law.
- In Colombia, the first Ministry of Environment was created in 1993 following the enactment of this country's Environmental Law.
- In Ecuador, the Environmental Advisory Commission to the Presidency, a predecessor of the Ministry of Environment, was created in 1992.

In the late 1990s and early 2000s, as government institutions and their regulatory bodies were created at unprecedented rates to develop and regulate environmental policies, NGOs and scientists in these countries played an important role in providing conceptual frameworks, methodologies, and benchmarks for advancing the production of knowledge for policy-making and policy implementation.

As a more recent example of increased knowledge demand, the Ministry of Environment and Sustainable Development in Colombia has implemented offsetting measures for biodiversity loss or for the impact caused by development projects since 2012. These measures follow the principle of "no net-loss" and apply a hierarchical scheme aimed at limiting environmental losses through three main actions, namely avoidance, mitigation, and compensation (Ministerio de Ambiente y Desarrollo Sostenible de la República de Colombia 2012). This new regulatory framework is a paradigm shift from reforestation/restoration as compensatory measures to biodiversity conservation, and its application depends on the quality, relevance, and availability of key information at each step of the process. In addition, this policy reflects the scope of the National Policy for the Integral Management of Biodiversity and Ecosystem Services (Mendoza et al. 2012).

In Peru, in the last 10 years, several institutions were created or modernized with the following objectives: to protect biodiversity, to regulate the use of natural resources, and to mitigate environmental impacts. These institutions include the National Natural Protected Areas Service created in 2008, the Environmental Evaluation and Oversight Organism (OEFA) in 2008, the National Environmental Certification Service (SENACE) in 2012, and the National Forest and Wildlife Service renovated in 2017. These examples reflect the public sector's dynamism in adjusting the institutional and regulatory framework to meet the challenges of intersectorial public policies that focus on biodiversity and that guarantee the provision of essential environmental services.

Bolivia and Ecuador also made special efforts incorporating environmental rights in their constitutions (Ecuador-2008, Bolivia-2009), recognizing the importance of ensuring biodiversity conservation and sustainable development at the highest legislative and institutional level.

These four countries have also shown significant regulatory innovation in relation to the protection of ecosystem services. Specifically, this was achieved by their adherence to the UNFCCC in relation to REDD+ (Reducing Emissions from Deforestation and Degradation + foster conservation, sustainable management of forests, and enhancement of forest carbon stocks in developing countries). Bolivia was the exception in this process proposing an alternative mechanism for climate change mitigation and adaptation. Country commitments to the UNFCCC require forest inventories and regular monitoring and reporting on deforestation, carbon stocks, and emissions, not to mention the wealth of relevant knowledge required to implement the *plus* in REDD+, as well as for adaptation measures.

Decentralization of public functions (i.e., responsibilities and jurisdictions) has additionally become a repeated strategy toward more effective implementation of biodiversity and natural resource management and environmental regulations in each of the four countries (Table 10.1).

These decentralized and participatory processes for natural resource management have increased the need for relevant local to subnational information on biodiversity and, at the same time, the need for capacity building. From the implementation standpoint, decentralization is a complex process that has advantages as well as disadvantages, and in practice needs a series of preconditions to work as suggested by theory (Carrillo and Casellas 2016). Being a political process concerned with the distribution of power and resources across a nation, decentralization outcomes will depend, among others, on the relationships established between local and central governments, power relations, and the existing capacities at the local level, the

Table 10.1 Institutional and legal framework to implement national decentralization policies in four Andean-Amazon countries

Country	Decentralization policies that assign responsibilities to different multilevel authorities in natural resource management
Colombia	National Environmental Information System (SIAC) is composed of all the national, regional, departmental, and municipal institutions that deal with environmental issues. Regional Autonomous Corporations (CAR) have special jurisdiction over parts of the territory and the mission to support the implementation of public policies and the control and monitoring of natural resources
Ecuador	Autonomous and Decentralized Territorial Planning Organic Code (COOTAD), approved in 2010, establishes decentralized autonomous governments (GAD) in provinces and municipalities with increasing functions in the planning and management of local natural resources
Peru	Regional governments (i.e., departmental or state-level governments) have the mandate to manage natural resources through the creation of Regional Environmental Authorities
Bolivia	Co-management is provided for territories where protected areas have been superimposed with original indigenous communities to promote the participation of locals in biodiversity management and conservation decision-making

implementation of a system of check and balances, and the consideration of existing local structures for resource governance (Larson and Ribot 2004).

This policy framework snapshot is not intended to be an exhaustive review of the situation of a continent spanning ~17.8 million square kilometers, but to rather use these four countries as a means of providing evidence that the demand for biodiversity knowledge exists in South America and that it is increasingly challenging to meet, given the diverse technical areas, as well as the different scales of implementation. Similarly, it demonstrates the lack of knowledge mobilization tools required for an effective implementation of the policy framework.

When analyzing the demand from environmental public policies for biodiversity knowledge production, findings in Bolivia, Colombia, Ecuador, and Peru confirmed that domestic policies act as the major catalysts of the development of information by different means and at different stages of the policy cycle and not the opposite, i.e., that information precedes policy (Fig. 10.3; Josse and Vásconez 2016). The results further suggest that top-down pressures, such as international commitments and international cooperation, when combined, also play a major role in the generation of environmental public policy in the region (Fig. 10.3). In these countries, international cooperation—mainly at the governmental level—has played a fundamental role in catalyzing national environmental and biodiversity policies.

Eventually, the need to organize all the biodiversity-related information, critical for policy-making and implementation, has demanded the development of environmental/biodiversity information systems. The main differences between countries are the age of existing systems and their level of consolidation. This consolidation depends on numerous factors such as institutional stability, available human and financial resources, and a degree of communication that ensures integration between

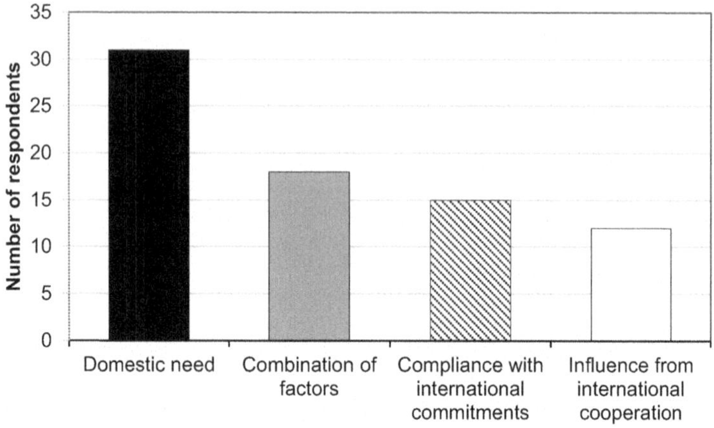

Fig. 10.3 Number of information users and generators responding to the question "How does environmental public policy arise?" in Bolivia, Colombia, Ecuador, and Peru. Replies were grouped into four categories ("Domestic need" refers to policies formulated to address issues identified at the national level, "Combination of factors" includes the other three categories)

knowledge producers and users across levels of organization as well as between public, private, and academic sectors (Josse and Vásconez 2016). Variations in the advancements of these factors contribute to the effective fulfillment—or lack thereof—of public policy objectives in terms of knowledge production and mobilization.

In addition to responding to national priorities, the generation and transfer of knowledge have been institutionalized, in part as a result of regular requests from the international agreements mentioned earlier. This institutionalization has had positive results, but at the same time, it promotes a separation between what is considered official information and what is not. While the latter can be of high quality, it is less likely to be used in public policy cycles as found in the study by Josse and Vásconez (2016). Yet, key information that cannot be generated by government institutions (e.g., the state of species conservation, their distribution, population status, ecological integrity indexes, etc.), is often supplied by national and international academic researchers and NGOs. In more advanced examples of institutionalization of biodiversity data, such as Colombia's SIB (Biodiversity Information System; https://www.gbif.org/country/CO/summary), the system serves as the hub for the country in which biodiversity observations are integrated, mobilized, harmonized, and finally mapped to the Global Biodiversity Information Facility (GBIF).

Policy evaluation (Fig. 10.2) aims to determine whether the adopted policy has had the expected effects on solving the problem, and what obstacles were encountered during the process. It is difficult to find documentation on this step of the cycle for any of the countries investigated; however, an internal document evaluating the *Guidelines for the implementation of the environmental offset measures for biodiversity loss from development projects in Colombia* (Colombia MdAyDSd 2012), points out several deficiencies in policy implementation due to:

- Lack of clarity on conceptual issues
- The adequacy and scale of the available technical information
- Lack of use of supplementary social information included in the methodology
- Lack of actual economic evaluation of (lost) biodiversity and/or ecosystem services
- Protocols for the traceability of the measures taken

This analysis highlights the magnitude of the challenge when it comes to supporting the implementation of conservation policy (i.e., conservation action) with the right knowledge.

Finally, another gap between knowledge and action is of political rather than technical nature. The economic emphasis on extractive industries (Marques et al. 2019) presents an additional challenge for biodiversity conservation and management in the region. Activities related to gas, petroleum, and mining, along with the enabling infrastructure, are expanding into well-identified biodiversity hotspots and protected areas across the continent, despite the fact that biodiversity has been recognized as a strategic asset in national development plans, and its conservation as a key objective for the well-being of the people of these countries.

10.2.2 The Knowledge Supply Agents

A critical component to address the knowledge–implementation gap, related to the knowledge supply end of the pipeline, is the identification of the key public and private stakeholders that generate and use biodiversity and environmental information.

The findings of Josse and Vásconez (2016) suggest that, in all four countries, NGOs see themselves as key information suppliers for the first and intermediate stages of the policy cycle (i.e., policy design, development, and implementation), providing continuous technical support. Figure 10.4 presents the results of a question intended to understand the origin of the research that influences public policy.

The most frequent answer of the "user" category was that national and subnational universities are the entities that produce the most research relevant to public policy, followed by national research institutes. The responses from user-generators, who tend to be researchers from institutes or public agency specialists, suggested that key research generators are NGOs, research centers, and international universities, with national universities being the least common option. Meanwhile, information generators frequently identified NGOs and national universities as the most important generators, followed by research institutes.

Most of the NGO agents interviewed confirmed that the strategic planning of their organization seeks to support the policy cycle up to the implementation of relevant conservation policies at the national level. The information they produce consists of comprehensive or multidisciplinary, applied, and geospatial analyses, policy briefs, and geoportals, in addition to outreach materials designed for local communities.

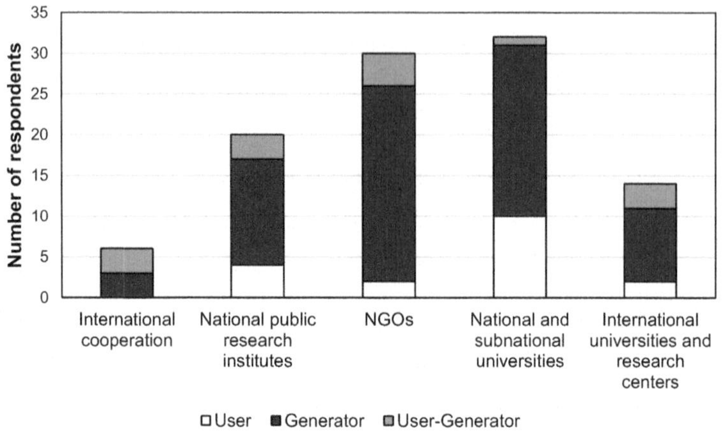

Fig. 10.4 Number of respondents, classified as information users, generators, or both, on the origin of research used in public policy in Bolivia, Colombia, Ecuador, and Peru

Because of their institutional frameworks, only a few NGOs generate basic biodiversity information, therefore, species inventories, geo-referenced mapping, and assessments of conservation actions usually are conducted by academic institutions.

It is important to note that in general in South America, as in many other parts of the world, there is a clear detachment between academic research and public-sector needs (Finch and Patton-Mallory 1993). While information users in the public sector see universities as the primary source of biodiversity information, academics do not see themselves as information providers for public policy (Josse and Vásconez 2016). The challenge is to bridge this gap. On the one hand, the fundamental role of universities is acknowledged, and information generated by this sector is commonly deposited in the public domain, but at the same time, users feel that academic researchers distance themselves from the political process and express a lack of willingness to share and make the results of their research available to the wider public by publishing technical reports, sharing data over information platforms or systems, or being more efficient in the publication of peer-reviewed papers. This is paradoxical, because while most universities receive public funding, academics are not responding to the public sector's information needs.

The role of national research institutes as knowledge generators is significant in the public sector. Again, this is very tangible in the case of Colombia, where they fulfill a primary need for information generation and research, providing hard data for public policy-making, through the periodical updating of biodiversity or natural resources information, the monitoring of indicators, and other specific or tailored information produced in response to a demand. The pathway to ensure greater involvement of researchers in this process includes developing methodological standards and protocols for monitoring data and incentives for following them, creating data management tools, and establishing communication channels that promote interaction with public sectorial entities as end users (see also Chap. 8;

Rose et al. 2021). The successful model of knowledge production and mobilization in Colombia highlights the importance of institutional autonomy of these research entities and their public funding, two conditions that need to be met to assure stability across different governmental cycles.

10.2.3 Efficiency and Gaps of Current Knowledge Mobilization Structures in South America

Biodiversity information is frequently used by policymakers and practitioners to improve natural resource management and regulate biodiversity use. This type of data includes georeferenced species observations and inventories, knowledge on conservation priorities and threats, status of species populations, and trends in land use change over time. Interviewees from Colombia and Peru indicated that subnational authorities require regular access to this type of information more than officials at the national level (Josse and Vásconez 2016). This might be related to the earlier implementation in the two countries of decentralization policies where the management of natural resources and the enforcement of environmental regulations is the responsibility of subnational governments.

As for the issue of resolving knowledge gaps to implement policies, most interviewees, from different decision-making levels and countries, responded: "*partially, information exists, but there are gaps in the applied research*" (Josse and Vásconez 2016). This deficiency was mentioned earlier, and the authors speculate that it is because biodiversity conservation and adequate natural resource management are increasingly complex, multi-sectorial, and multidisciplinary, requiring information on pressures, cumulative effects, cause-effect relationships, trends, biodiversity responses, cycles, thresholds, resilience, and vulnerability. This type of information is not readily available in existing information systems. It will require integration, harmonization, and a degree of abstraction to make it available which will be difficult to achieve for a subnational governmental organization or even a country-level biodiversity observing infrastructure because of insufficient financial support for data collection, synthesis and management, and limited coordination and networking among researchers.

It should be noted, however, that the second most common response to the latter question about the adequacy of knowledge was "*yes, existing information meets internal needs*," and it came mainly from forest sector officials, responsible for managing forestry concessions and controlling deforestation. Evidence for this sector is available from UNFCCC-related commitments and funds made available for its implementation.

The results of this study demonstrated that most knowledge users have the flexibility to use the information and constantly incorporate it into their work area. The main constraints they mentioned were the timelines and the official nature of the information they are expected to use. Authorities often need to make urgent decisions in a short period of time and the information they need is not always available or is easily accessible.

The progress and the shortcomings in the use of scientific evidence in the design and implementation of public policies described in the previous paragraphs of this section, in relation to the four countries encompassed in Josse and Vásconez (2016), certainly illustrate the challenges of the whole South American continent for the use of evidence in conservation action, as mediated by policy demand.

While policies serve as catalysts at certain levels to produce information on biodiversity, its supply through institutional channels is insufficient, thematically and timewise. The information demanded may exist or may be feasible to produce by research institutes or academic researchers; however, absence of communication and coordination among actors, coupled with insufficient financial and political support to build integrated information systems, results in disconnected and costly consulting services whose products are not incorporated into public, accessible, and institutionalized information systems.

The solution to this will require the participation of both information providers and users to determine what biodiversity information needs to be produced and how to channel it according to the needs of public users. Improving interactions between actors can be achieved through the following means:

- Bringing together primary knowledge producers, such as universities, and end users to maintain dialog and seek coordination (see Chap. 8; Rose et al. (2021))
- Developing long-term agreements between the parties to incentivize targeted research
- Easing procedures for research permits and having more efficient schemes for handing back research results to the environmental authority or the public domain
- Increasing funding for biodiversity research and the generation of information in formats relevant to decision-makers
- Funding structures that incentivize sharing and open access to data

The communication and distribution of information can be improved by the following measures:

- Improved definition and communication of public-sector strategic information needs
- Development of relevant information in a timely manner in line with policy cycles
- Institutional autonomy of research entities and their public funding
- Establishment of more effective dissemination channels

10.3 An Overview of the Region as Captured by an Analysis of the NBSAPs to the Convention on Biological Diversity

NBSAPs are the primary instruments for implementing the Convention at the national level (following Article 6 on General Measures for Conservation and Sustainable Use). The Convention requires countries to prepare a national biodiversity strategy (or equivalent instrument) to ensure that this strategy becomes national policy and thus is mainstreamed into the planning and activities of all those sectors that can have an impact on biodiversity.

For this section, we reviewed the latest national submissions of the NBSAPs to the CBD by signatory countries in the region (in 2018), to provide a comprehensive overview of progress and the current challenges that these countries face regarding knowledge production and mobilization, ultimately leading to obstacles to conservation action.

The latest submitted NBSAPs comply with the commitment of the countries to align their national biodiversity strategies with the 20 global targets, known as the Aichi Targets, set out as targets of the 2011–2020 CBD Strategic Plan for Biodiversity during COP 10 held in 2010 in Nagoya, Japan.

The NBSAPs' structure usually includes the following components:

(a) The legal and institutional frameworks, and the national policy context that provides a vision of how the country internalizes the concept of biodiversity conservation in its various policy tools and relevant sectors
(b) A description of the state of the biodiversity components and the pressures they endure
(c) A description of the access to current biodiversity knowledge and the systems devised to mobilize it to achieve knowledge transfer to relevant actors and decision-makers in a timely manner, with the right contents, and in the right formats

With this background as the foundation for the Strategy, NBSAPs then describe in detail the strategic planning process with the identification of goals and indicators to be measured.

Therefore, the NBSAPs submitted by South American countries served as the main source of information for this section. We extracted key data on three components of the knowledge management cycle (knowledge production, knowledge transfer, and conservation action) from these documents that served as indicators of the nations' progress toward bridging the knowledge–implementation gap in biodiversity conservation and natural resource management. Each component was given a score from 0 to 2, depending on their stage of development as per the national reports' descriptions (Table 10.2). The following sections will examine in more detail the different components of Table 10.2.

10.3.1 Knowledge Production

The IUCN RLTS is one of the best-established protocols to measure the extinction risk of species. Red list assessments help determine which species need immediate attention and specific management and/or conservations strategies. Also, because of the requirements of the assessment protocol, RLTS demand significant expert knowledge not only on species distribution, species population trends, and

Table 10.2 Scoring of South American countries for each component of the knowledge management cycle (Knowledge production, Knowledge transfer, and Conservation action) supported by an analysis of the latest National Biodiversity Strategy and Action Plan (NBSAPs) or National Reports submitted to the CBD

	KNOWLEDGE PRODUCTION					KNOWLEDGE TRANSFER		CONSERVATION ACTION				SUM PER COUNTRY /22
COUNTRY	National Red Lists			Monitoring Programs (evidence-based changes in biodiversity and forest cover)	Planning and funding of priority research in biodiversity and environment	Participatory institutional framework	National environmental and biodiversity information systems	Management and conservation plans: threatened species and protected areas network	Payment for ecosystem services	Mainstreaming of biodiversity through cross-sectorial policies (transport, oil mining, energy, agriculture, timber)	Traditional knowledge (access and benefit sharing)	
	RLTS	RLE	RLI									
Argentina	2	0	1	1	1	2	1	1	1	1	0	11
Bolivia	2	0	1	1	0	1	0	1	0	0	2	8
Brazil	2	0	1	2	2	2	2	2	2	1	2	18
Chile	1	2	0	2	2	1	1	2	1	1	1	14
Colombia	1	2	1	2	2	1	2	2	2	1	2	18
Ecuador	2	1	1	1	2	1	1	2	2	1	2	16
Guyana	1	0	0	1	0	1	0	1	0	1	1	6
Paraguay	1	0	0	0	0	1	0	1	1	0	1	5
Peru	1	0	0	1	2	2	2	2	2	2	2	16
Suriname	1	0	0	0	0	1	0	1	0	0	1	4
Uruguay	1	0	0	1	1	2	1	1	0	1	0	8
Venezuela	2	2	1	1	1	1	1	1	0	0	1	11
SUM PER PARAMETER /24	17	7	6	13	13	16	11	17	11	9	15	

Each component is defined by different parameters, described in-depth in the main text. *RLTS* Red List of Threatened Species, *RLE* Red List of Ecosystems, *RLI* Red List Index. Score: 0 = does not exist; 1 = exists, partially developed; 2 = exists, well developed. Sum of scores per country in the region and per parameter of the knowledge management cycle are also presented

ecological requirements, but also data on threats. These temporal and spatial attributes, when combined with other datasets, have proven useful to determine the most urgent priority species and areas for conserving biodiversity (Hoffmann et al. 2008). RLTS are developed to assess the species' threat category at different geographical scales. In order to be relevant for status monitoring over time, species need to be regularly re-assessed following the same protocol, which enables the calculation of the Red List Index (*RLI*), a critical metric for monitoring the effect of conservation action (Bubb et al. 2009). Here, we focused on the national RLTS because this assessment process demands the involvement of local scientists and nationally developed input data. Therefore, a value of 1 in Table 10.2 was assigned to countries where the national assessments did not attempt to comprehensively assess all species in a taxonomic group (i.e., birds, amphibians, and mammals), and a score of 2 means that the assessments included all species within the taxonomic group.

The RLI has been calculated only for subsets of species with national assessments conducted on at least two occasions, or it has been calculated using the global re-assessments for the species occurring in the country, instead of conducting the re-assessment at the national level. Therefore, no country scored 2. Despite the wide use of threat categories to convey the urgency of action to prevent species from extinction, the development of the countries' species red lists is generally lagging behind, with less than half of the countries having completed taxonomic group assessments and very few generating RLI (scoring 1, Table 10.2).

The more recent IUCN RLE applies the same principle of the RLTS to ecosystems and habitats to assess the immediate risk of ecosystem collapse (Keith et al. 2013). The RLE could be an alternative tool to produce biodiversity knowledge quite useful for policy-making at the national level, yet very few countries in the region have generated it. Some countries, namely Chile, Colombia, and Venezuela, have already implemented this standard, and have therefore scored 2, with the exception of Ecuador which has done an RLE assessment but only for the ecosystems of its Amazon region (hence, scoring 1). Countries with no RLE assessment scored 0 for this parameter.

A key issue underlying the efforts in the kind of periodical reporting that the CBD requires is that countries should begin to generate consistent data that effectively detect changes in biodiversity, i.e., biodiversity indicators. A report from Bubb et al. (2011) documented that 66% of the National Reports presented in the CBD lack evidence-based information on biodiversity changes. For this reason, we looked for references within the NBSAPs regarding environmental and biodiversity monitoring programs (summarized in column '*Monitoring Programs*,' Table 10.2). Countries with a score of 2 have institutionalized monitoring programs of different sorts, all relevant to environmental issues, comparatively more sophisticated in the number and type of indicators monitored. Still, a score of 2 does not mean that there are not challenges yet to be addressed. NBSAPs themselves state that there is no adequate information for negotiations with the production sectors (i.e., industrial, agriculture, extractive, etc.). Evidence on correlation, causality, thresholds of ecosystems and their services, economic trade-offs, are critical areas of research that are not sufficiently developed.

Finally, the foundation of a sound knowledge production infrastructure lies in its funding mechanisms and in how targeted the research investment is to obtain strategic knowledge that addresses the demands of policy implementation. The results for this parameter (column '*Planning and funding of priority research in biodiversity and environment*,' Table 10.2) are encouraging with 8 out of the 12 countries having a priority research agenda and funding in place, even if at different stages of development.

10.3.2 Knowledge Transfer

Participatory institutional framework refers to the set of interacting entities of different hierarchies that each country assembles to respond to the needs and challenges of biodiversity conservation and environmental regulation. Taken together, these are in charge of designing and implementing policies and assessing their impacts. The "participatory" qualifier implies that different actors of the civil society, especially those from academia, conservation NGOs, and sectors regulated or impacted by policy, make part of these entities. Given that a key characteristic of (good) knowledge transfer is how organic it is (e.g., there is feedback), we gave a higher score to countries that showed such kind of participatory framework, based on the countries' accounts of the way their institutional framework operates. Only four countries' reports explicitly describe this kind of participation (scoring 2, Table 10.2), with the rest reporting only governmental participation in their environmental policy institutional frameworks (scoring 1, Table 10.2).

National environmental and biodiversity information systems refer to the means developed to transfer existing knowledge to all targeted and potential users to obtain effective and positive conservation outcomes, especially important when the investment in knowledge production is made by the public sector. Most South American countries have developed national environmental and biodiversity information systems to ensure knowledge or evidence is available for the institutional framework to do its job. Therefore, the scoring criteria were based on how institutionalized the systems are (which brings stability, relevance, and timeliness), their versatility, and the efforts put into training the target groups for their use. As a result, Brazil, Colombia, and Peru stand out as having fully institutionalized systems (scoring 2, Table 10.2), whereas the rest of the countries in the region present a mix of either embryonic information systems or none, at least not recognized or led by the governmental sector.

10.3.3 Conservation Action

For this component in the knowledge management cycle, we selected actions that are reported in the NBSAPs that also correspond to the targets of the CBD Strategic Plan for Biodiversity 2011–2020:

1. Management and conservation plans
2. Payment for ecosystem services and other forms of economic valuation
3. Mainstreaming of biodiversity through cross-sectorial policies
4. Traditional knowledge (access and benefit sharing)

These four were selected to illustrate what has been the countries' progress on conservation actions that depend on existing relevant knowledge to be successful.

Management and conservation plans for threatened species and in-situ conservation in protected areas are the most common actions adopted. South American countries have designated significant portions of their geography for biodiversity protection. For example, the most extensive protection coverage achieved at a continental level is for Latin America and the Caribbean, where 4.85 million km^2 (24%) of land is protected (UNEP-WCMC and IUCN 2016). Half (2.47 million km^2) of the entire region's protected land is in Brazil, making it the largest national terrestrial protected area network in the world (UNEP-WCMC and IUCN 2016). It is precisely the enormity of the designated protected area networks in most of the South American countries that economically and logistically challenges their adequate management. For this assessment, a score of 2 means that besides the amount and representability of the area designated, the country reports having performed assessments of the management quality and state of its protected area network, sustainability analyses, or evaluations of the ecosystem services provided by protected areas. In sum, countries receiving a score of 2 (Table 10.2) have taken steps to improve the impact and integrity of their protected area network based on solid evidence. However, less than half (5 out of 12) of the evaluated countries received the maximum score of 2 whereas the other countries received a score of 1, indicating that while counting with a protected area network, as per their NBSAPs' reports, these countries have not carried out assessments of its management quality and conservation effectiveness.

Payment for Ecosystem Services After the first studies attempting to estimate the value of ecosystem services (Costanza et al. 1997), the economic evaluation of the services and mechanisms to compensate their stewards have multiplied and are being implemented around the world (Grima et al. 2016). In terms of conservation action, these *payment schemes for ecosystem services* (PES) aim to raise awareness of nature conservation and fundraise for it, but they can go beyond to internalize environmental externalities in the market prices (Grima et al. 2016). In any case, the implementation of PES must be supported by knowledge to value the services, monitor the provision of the service, and assess whether the implemented scheme is working. This poses several challenges but there are many successful examples to encourage their implementation. The scoring for this parameter is based on the level of efforts reported by the countries to implement PES and other ways of valuing nature's derivatives, such as the water supply provided by the national system of protected areas or by enacting the regulatory framework that facilitates legal implementation of PES schemes (i.e., who pays and who benefits from the payment under which conditions). A handful of countries in the region have successfully introduced PES schemes in the systems of water supply for large urban centers, and more recently, Ecuador and Peru put in place direct payments to private or communal landowners for the conservation of forests and other ecosystems which deliver important services. Here too, the PES schemes require good baseline information and monitoring systems.

Mainstreaming of Biodiversity Through Cross-sectorial Policies As the NBSAPs identify, biodiversity mainstreaming requires innovative and attractive incentive mechanisms for the national production and planning sectors and the subsequent policies and corresponding regulatory framework. From the knowledge standpoint, the implementation of biodiversity mainstreaming policies within NBSAPs requires evidence to convince the private sector about the beneficial outcomes of potential tradeoffs, and implementation mechanisms should be accessible to all stakeholders. Several countries reflect in their NBSAPs the growing realization of the need to mainstream biodiversity conservation and protection of ecosystem services as part of the contributions of all economic sectors to national development. Peru's NBSAP stands out detailing important progress with concrete, institutionalized mechanisms that tackle the participation of all involved sectors, from planning and production ministries in the central government to a body of trained officers in the subnational governments to enforce compliance.

Traditional Knowledge (Access and Benefit Sharing—ABS) Integrating countries with high biological and cultural diversity, South America is a region immensely rich in genetic resources and *traditional knowledge* (Aswani et al. 2018; Saylor et al. 2017—for further information on Traditional Knowledge, see Chap. 5; Ens et al. (2021), for an overview about the state of genetic diversity protection in South America, see Chap. 3; Klütsch and Laikre (2021)). The Nagoya Protocol on Access to Genetic Resources and the Fair and Equitable Sharing of Benefits Arising from their Utilization is a supplementary agreement to the CBD that entered into force in October 2014, aimed at providing the parties with a standard and transparent legal framework for the effective implementation of one of the three objectives of the CBD: the fair and equitable sharing of benefits arising from the utilization of genetic resources. The Nagoya Protocol also considers Traditional Knowledge associated with genetic resources and the benefits arising from its utilization. Issues related to the regime of ABS have been discussed in the countries of the region since the 1990s. At the regional level, the Andean Community countries, encompassing Bolivia, Colombia, Ecuador, and Peru, agreed on ABS measures, with Decision 391, a Common Regime on Access to Genetic Resources (ten Kate 1997). Beyond the ratification/acceptance of the Protocol (7 out of 12 countries), the progress in the legal frameworks directly connected to ABS, and in the research and development investments that ultimately deliver on the realization of benefits from genetic resources and Traditional Knowledge, we reviewed the national NBSAPs considering how explicit they are regarding the level of recognition given to the rights of indigenous peoples, their territories, and Traditional Knowledge through different sets of national legislations. Only five countries in the region scored 2 for this parameter, which explicitly recognizes indigenous people's rights to their land. This is important because the cultural diversity and Traditional Knowledge in

South America are linked to concrete territories, where people have developed knowledge to use biodiversity for their livelihoods (see also Chap. 5; Ens et al. (2021)). This fact constitutes a widely unrecognized contribution to date to the conservation of much of the natural ecosystems, services, and wildlife that we still enjoy in the region.

10.3.4 What and Where Are the Gaps from Knowledge to Conservation Action?

A comparison of how countries in South America are doing in relation to the different components of the knowledge management cycle is made in this section. A simple sum of the scores from Table 10.2 by country across all the factors considered, shows that Brazil, Colombia, Ecuador, Peru, and Chile are ahead of the other countries of the region in matters related to the effective use of knowledge for advancing conservation and maintaining long-term sustainability of natural resources (Table 10.2).

The performance across countries and across the three groups of factors (i.e., knowledge production, knowledge transfer, and conservation action; Fig. 10.5 and Table 10.2) was compared by calculating the percentage based on the sum of the

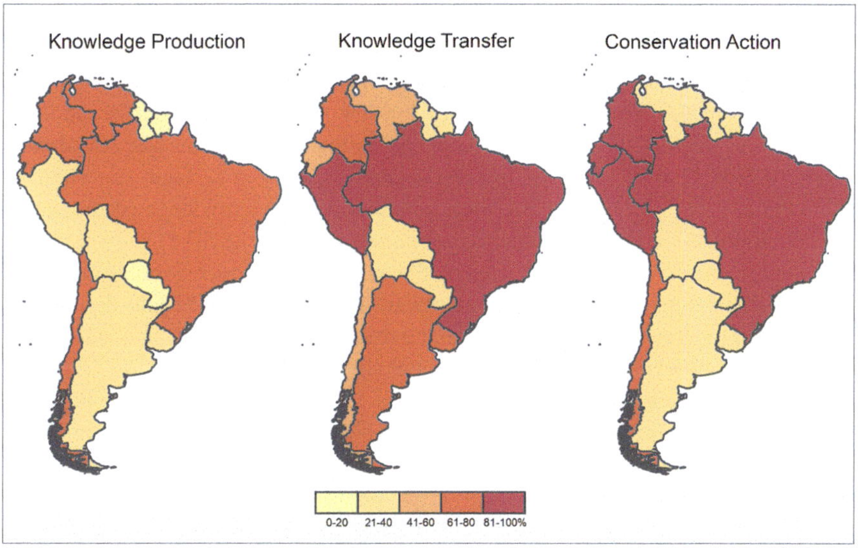

Fig. 10.5 South American nations' performance for the three components of the knowledge management cycle (knowledge production, transfer, and conservation action) based on the latest National Biodiversity Strategy and Action Plan or National Report countries in the region submitted to the CBD. Higher % and warmer colors reflect higher performance

countries' scores for parameters within each of the three groupings depicted in Table 10.2. For example, for the parameters grouped under knowledge transfer, Ecuador scored a total of 2 for two different parameters. The highest possible value was 4 and therefore, the calculated percentage was 50% for knowledge transfer. Thus, Fig. 10.5 shows the relative performance of countries in % for the three factor groups, respectively.

The first key message that arises from the analysis of Fig. 10.5 is that the degree to which knowledge is being produced deserves more attention, as the majority of the countries in the region are still underperforming, lacking assessments of biodiversity status, monitoring programs based on indicators that provide information critical for a more focused biodiversity conservation action, and/or investment in biodiversity research. Furthermore, it may seem paradoxical that the subsequent components of the knowledge management cycle, i.e., transfer and conservation action, show overall more development in the continent. This could reflect a larger investment in recent years in some countries in their institutional framework because ultimately governments can create public institutions and legislation, whereas the production of purposeful knowledge relies on a combination of enabling conditions and coordination among various stakeholders which are more difficult to fulfill. Previous sections of this chapter also discussed the weaknesses in knowledge production in the region. This pattern could also reflect the fact that action is taken without appropriate knowledge support. In recent years more international funding has been available to support enhanced governance and public policy-making than for baseline biodiversity research. However, in the absence of baseline knowledge, it is impossible to track the effectiveness and success of targeted conservation action.

Additionally, countries' commitments to the UNFCCC were characterized by improving the protection of significant proportions of native forests to reduce emissions from the land use change and forestry sector through conservation of forest carbon stocks. The level of investments in this sector in the past decade has surpassed biodiversity conservation investments, resulting in de facto conservation action but without necessarily contributing to biodiversity knowledge production.

The parameters used to evaluate *conservation action* (Fig. 10.5) show a strong divide within the continent as indicated by the presence of high-performing countries in dark red and weaker performing countries in yellow in Fig. 10.5. This divide may be explained by different actions taken in the respective regions. Some countries are supporting biodiversity conservation through improved protected areas management, economic incentives, PES schemes, and the recognition of indigenous land rights, another important form of in-situ conservation; leading to higher performance in conservation action taken. Other countries, however, have directed their policies toward relaxing environmental regulation pursuing investments in several production sectors or simply have not invested comparatively similar resources in the last several years to ensure the conservation of biodiversity while also failing to create institutionalized mechanisms to mainstream biodiversity through cross-sectorial policies.

10.4 Initiatives Contributing to Narrow the Knowledge–Implementation Gap in the Region

The following case studies from South America are about the production of knowledge on which conservation action is based. The first example shows how, with commitment and openness to share knowledge and work together, researchers linked to the production of raw data (i.e., species inventories, distribution, status assessments, etc.) could deliver enhanced products for national consumption while improving the uptake and use of this knowledge by global or regional biodiversity information networks. The second example discusses the efficiency of resource investment in species threat assessments, and that of other components and variables of biodiversity, and how they can be optimized to serve as more comprehensive indicators of the conservation status of biodiversity in the region.

10.4.1 Biodiversity Observation Networks: The Global Initiative

Biodiversity research has witnessed massive growth worldwide both in the collection of observational data and model outputs, with many new efforts underway. These increasingly large and complex data sets, uneven in space and time, require urgent ontological alignment, harmonization, and a degree of abstraction (e.g., models) that meet the expectations and requirements of end users (e.g., indicators for Aichi Biodiversity Targets, national decision-making, conservation planning, environmental impact assessment). However, it is difficult for individual scientists, research groups, or even national research infrastructures to develop and implement the platforms required to retrieve, share, and leverage data. Hence, investments to facilitate and improve collaboration, integration, and harmonization of data sets to inform end users on pressing issues such as global environmental change, are needed.

To date, very few harmonized monitoring systems deliver regular and timely data on biodiversity change to support all levels of governance, management, and decision-making at different spatial scales in South America. Despite progress in biodiversity data production and mobilization, there is still no consensus or clear guidelines on what and how to monitor with the limited human and financial resources available. The GEO BON is working on the implementation and progression of the EBVs which could serve as a foundation for interoperable subnational, national, regional, and global monitoring initiatives. However, the multidimensional and multi-scale nature of biodiversity makes it extremely difficult to determine what is essential in order to assess the pulse of the planet's biodiversity.

10.4.1.1 The Bolivian Biodiversity Observation Network

Effective sustainable development policy and actions that can reverse biodiversity loss fundamentally rely on consistently generated, high-quality biodiversity information. However, the capacity to generate this information is inadequate in biodiversity-rich regions such as South America (Fig. 10.5). Existing biodiversity observation efforts suffer from insufficient integration and lack connectivity to policies. Despite deficiencies in the mobilization and integration of knowledge, advances in biodiversity data collection methodologies and technologies are rapidly emerging and, combined with an emergent culture of collaboration, integration, and sharing, offer a major opportunity to improve the efficiency and impact of biodiversity observing systems. The grand challenge is to capitalize on these advances and opportunities to establish a harmonized and efficient network of national observatories that can help us not only to understand the biosphere as a system, but also how fast, where, and why it is changing in order to inform targeted, effective, and timely conservation actions.

Bolivia is a country that is performing particularly poorly in all three components of the knowledge management cycle (Fig. 10.5). In response to this challenge, in 2014, a team of more than 30 Bolivian scientists made a fundamental first step toward the creation of a Bolivian Biodiversity Observation Network (B-BON). Working closely with experts from national research institutions, academia, NGOs, and observation networks with regional to the global scope (Fig. 10.6), an assessment of the adequacy of Bolivian biodiversity observations was conducted to inform effective conservation decisions and contribute to the Global Earth Observation System of Systems (GEOSS).

Fig. 10.6 Network of organizations involved in the establishment of the Bolivian Biodiversity Observation Network

This assessment provided insights into current biodiversity observation capacity in Bolivia and the extent of existing biodiversity observations data holdings (Fig. 10.7).

This first assessment played a key role in the establishment of a Bolivian Biodiversity Observatory that will deliver relevant biodiversity data for decision-makers in user-friendly formats to facilitate sound conservation action. Our initial assessment utilized biodiversity observations, assessment, and networking tools found in GEO BON's BON-in-a-Box (a state-of-the-art, customizable global online toolkit for improving biodiversity observation capacity and harmonization). These tools facilitated the collection and integration of a number of existing, multi-scale,

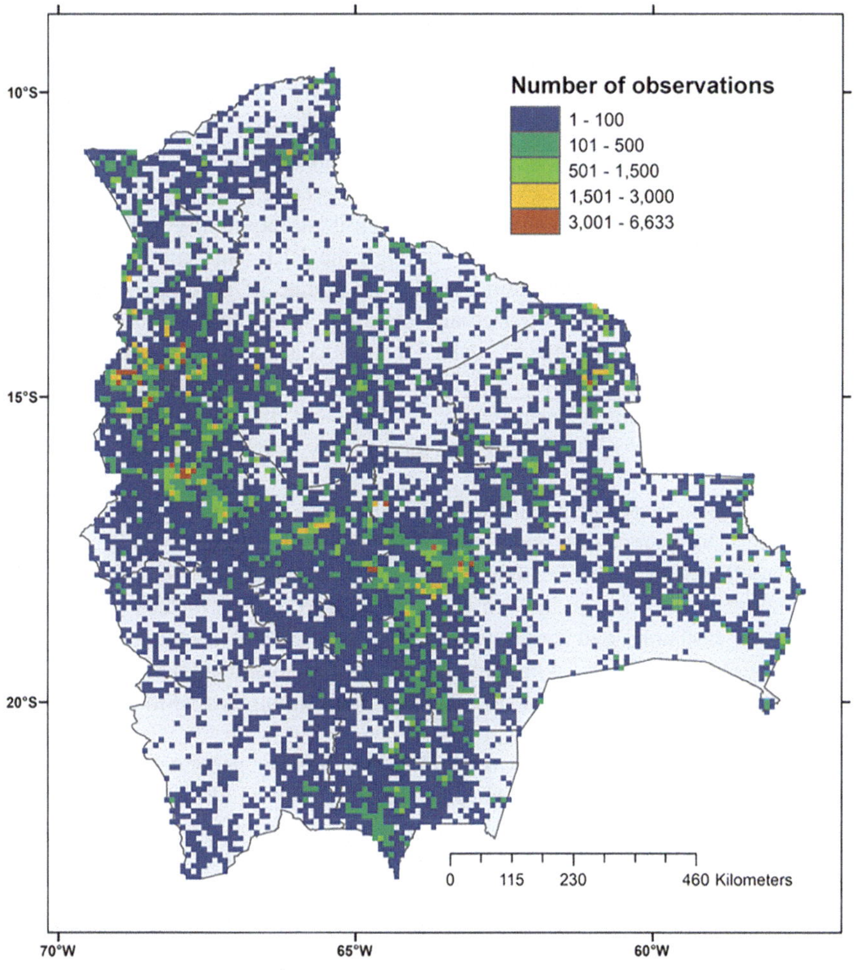

Fig. 10.7 Number of unique biodiversity observations on a 10 × 10 km grid compiled by the Bolivian Biodiversity Observation Network (adapted from Fernández et al. 2015)

and multi-temporal biodiversity datasets from major institutions and individuals within and outside Bolivia. This effort resulted in the most comprehensive biodiversity database ever compiled for Bolivia (Fernández et al. 2015). This biodiversity baseline served as the basis for a national biodiversity monitoring strategy and contains more than 600,000 records involving more than 27,000 species referenced in time and space. Not only did this effort assess existing data holdings but served to identify priority taxonomic and spatio-temporal gaps in Bolivia's biodiversity record, thereby informing the development of a targeted and efficient Bolivian biodiversity observation effort.

10.4.2 Red Listing of Species and Ecosystems: Advances and Challenges

One of the most important tools that helps bridge biodiversity science and policy is the RLTS developed by the IUCN. Its primary objective is to estimate species' immediate extinction risk at relevant scales (Rodrigues et al. 2006). As previously stated in this chapter, a key feature of the RLTS is the possibility to re-evaluate a species' conservation status over time, through the RLI. This entails not only a re-analysis of available information but also an update of information regarding temporal changes in population sizes and distribution of each species (Rondinini et al. 2014). Despite the potential of the RLTS and RLI in narrowing the knowledge–implementation gap in conservation, it remains of limited use to monitor biodiversity trends at different scales in a sustainable and efficient manner (Possingham et al. 2002; Stuart et al. 2010) especially in biodiversity-rich regions such as South America.

This statement may seem counterintuitive at first given the apparent current abundance of local, regional, and global red listing exercises (Hobbie et al. 2003; Edwards 2004; Scholes et al. 2008; Ruggles et al. 2015; Table 10.2). However, a large amount of information has not been equating to more informed biodiversity conservation decisions and more sustainable use of natural resources worldwide (Fernández et al. 2015). In South America, two reasons can explain this phenomenon: lack of integration (Walls et al. 2014) and information bias (Hortal et al. 2008; Trimble and van Aarde 2012).

Information is the new currency (Castells 2003; Soberon and Peterson 2004) and the speed at which it is produced far exceeds our ability to integrate it. However, integration is a key requirement that allows us to analyze patterns and trends over time (Jetz et al. 2012). Simultaneously, our current knowledge about the earth's biodiversity is still very limited. While we have managed to catalog ~1.5 million species (Costello et al. 2013), we do not know much about various groups, such as Neotropical insects (Régnier et al. 2015). This bias in information is even more evident in regions with high biological diversity such as the Amazon, where 43% of the area has never been surveyed by botanists (Schulman et al. 2007). To contextualize the importance of these issues, recent estimates suggest that the current rate of species extinction is much higher than the background rate of extinction, using the last 65 million years as a reference (Ceballos et al. 2015). As scientists, it is difficult to disregard these estimates. When we are only certain of the identity of one and a half million species and this represents 14% (±2) of the planet's total biodiversity (Mora et al. 2011), we need to be cautious when providing recommendations given the high levels of uncertainty in these estimates.

Considering the limitations and uncertainties of the knowledge associated with the species-based approach (such as that produced under the RLTS), then it needs to be supplemented with additional levels of organization. To this end, the adoption of the IUCN RLE (Rodríguez et al. 2011, 2015; Keith et al. 2013) has gained traction in the region as a complementary analysis that provides an integrated solution for biodiversity monitoring. The RLE provides a framework for evaluating threats to ecosystems by using the concept of ecosystem collapse as an analog to species extinction (Keith et al. 2013) and assesses degradation and loss of habitat as well as land conversion. Advantages of the ecosystem assessment approach include the identification and protection of species interactions as well as interactions of species with their abiotic environment, aspects that are missing from the RLTS. Also, common species that are rarely considered in RTLS can be key species in ecosystem functions and essential for ecosystem assessments (Keith et al. 2013). Hence, the RLE has a stronger focus on the assessment and protection of ecosystem functions and services. For the estimation of extinction risks, the RLTS requires specific information that can only be collected in the field (e.g., species population size, but see Santini et al. 2019). Contrastingly, the assessment criteria for the RLE enables the use and integration of ecological and environmental data collected both in situ and remotely. It is precisely this flexibility that facilitates repeated multi-scale assessments (Skidmore et al. 2015) in information-deficient regions such as South America.

If we look at the existing priorities of South American governments, it is important to consider the economic cost of the tools that bridge conservation science with policy-making. In 2013, US$~4.7 million was allocated to the IUCN RLTS to evaluate ~76,000 species globally, which is equivalent to 2.4% of the total number of species described until now (Stuart et al. 2010; Rondinini et al. 2014; Juffe-Bignoli et al. 2016). In contrast, the first implementation of the RLE in the Americas approximately cost US$2 million (including all ecosystems, training of scientific staff, analysis time, and coordination meetings), suggesting that the RLE might be a more cost-effective solution to monitor and assess the status of natural environments especially in data-deficient and species-rich regions like South America. This economic sustainability has led Chile, as one of the first countries in South America, to include the results of the RLE assessment as an element to guide its latest National Biodiversity Strategy and Action Plan (CBD 2018; Table 10.2). However, challenges remain for the wide implementation of the RLE (Keith 2015; Bland et al. 2018). These include appropriate ecosystem classifications, assessment of ecosystem dynamics, and standardized definitions of ecosystem collapse (Keith 2015; Bland et al. 2018). Thus, these complementary approaches have advantages and disadvantages but combined, RLTS and RLE have the potential to provide an efficient toolkit that supports strategic biodiversity conservation in information-deficient regions to promote decisions that tilt toward the conservation of important biodiversity areas.

In data-deficient regions such as South America, where reliance on the RLTS alone may lead to missing too many potentially threatened species, the RLE may provide a valuable additional tool as the large-scale protection of ecosystems potentially protects currently unknown biodiversity as a byproduct. Finally, it is urgent to implement mainstreaming mechanisms that better support conservation planning and sustainable development policies and strategies, and that build on and integrate regionally important conservation tools.

10.5 Challenges and Recommendations to Bridging the Biodiversity Data Mainstreaming Gaps in South America

The findings of this chapter suggest that the knowledge transfer and conservation action components in the biodiversity data mainstreaming continuum have improved in South America in the past decades, largely as a consequence of the pressure imposed by international commitments. However, it is striking how little is still known about the cost and effectiveness of interventions that aim to narrow the gap between knowledge production and demand. The shortcomings found in knowledge production could also reflect a critical lack of organization and integration of the information produced rather than limited research capacity. More intensive and strategic efforts seem critical to target research and knowledge mobilization and achieve more successful conservation practices. This involves improving working conditions and securing tenure for scientists in the academic sector, and research funding structures that incentivize sharing and open access to data while rewarding research that responds to prioritized needs identified for policy implementation and evaluation.

This implies not only dedicated resources but also the provision of leadership to create an institutional framework that encourages information generators to participate and integrate their data for practical purposes. In the absence of basic information that allows us to measure the speed, magnitude, and direction of change caused by human activities in the essential components of biodiversity and ecosystem services, the scientific community in South America and the world is limited to educated opinions.

The results of basic and applied research should be systematically reviewed to identify promising implementation techniques and areas where more research is required. Among them, there is the question of integrated management of data, which means approaching the understanding of ecosystems functioning using biotic and abiotic data to explain cause-effect relationships, for example. Undertaking reviews in these areas is difficult because of the inherent complexity, the variability in the methodological and analytical approaches used, and the difficulty of generalizing study findings across spatio-temporal scales and taxa.

Several regional and global Earth observation networks are joining forces to help meet the need for systematic reviews of the current best evidence on the effects of global change on the essential dimensions of biodiversity (Lindenmayer et al. 2018). Therefore, we must seek to ensure that each effort, no matter how local, is seamlessly integrated into national, regional, and global monitoring initiatives, contributing to the larger goal of conservation and sustainable development in the countries of South America.

Finally, the development of public policy in South America is extremely variable and complex, with periods of stability followed by abrupt periods of rapid change. Thus, more effort needs to be placed in the development, empowering, and stabilization of national biodiversity monitoring networks. While the tools that could support biodiversity conservation in South America already exist, the country-level institutional support that can provide the resources for the maintenance of a national biodiversity observation network and guarantee the continuity of these efforts is still limited in the region. Without the long-term financial and institutional stability that shields these institutions against abrupt changes in government priorities, they will not be able to collect the basic but vital information that allows us to detect changes in the essential components of the biodiversity and environmental services, and consequently steer public policy in the right direction toward conservation action.

Acknowledgements The study by Josse C, and Vãsconez S about the relationship between public environmental policy and the development and use of biodiversity information in Bolivia, Colombia, Ecuador, and Peru, was generously funded by JRS Biodiversity Foundation.

References

Aswani S, Lemahieu A, Sauer WHH (2018) Global trends of local ecological knowledge and future implications. PLoS One 13(4):e0195440

Bland LM, Rowland JA, Regan TJ, Keith DA, Murray NJ, Lester RE, Linn M, Rodriguez JP, Nicholson E (2018) Developing a standardized definition of ecosystem collapse for risk assessment. Front Ecol Environ 16(1):29–36

Brooks TM, Mittermeier RA, da Fonseca GAB, Gerlach J, Hoffmann M, Lamoreux JF, Mittermeier CG, Pilgrim JD, Rodrigues ASL (2006) Global biodiversity conservation priorities. Science 313(5783):58–61

Bubb P, Butchart S, Collen B, Dublin H, Kapos V, Pollock C, Stuart S, Vié JC (2009) IUCN Red List Index: guidance for national and regional use

Bubb PP, Chenery A, Herkenrath P, Kapos V, Mapendembe A, Walpole M (2011) National Indicators, Monitoring and Reporting for the Strategic Plan for Biodiversity 2011–2020, A review of experience and recommendations in support of the CBD AD HOC Technical Expert Group (AHTEG) on Indicators for the Strategic Plan 2011-2020

Carrillo P, Casellas A (2016) Descentralización y gestión ambiental en América Latina: un análisis de las publicaciones académicas Cuadernos de Desarrollo Rural 13

Castells M (2003) The power of identity: the information Age: economy, society and culture, Volume II (The information age), 2nd edn. Wiley-Blackwell, Oxford

Ceballos G, Ehrlich PR, Barnosky AD, Garcia A, Pringle RM, Palmer TM (2015) Accelerated modern human-induced species losses: entering the sixth mass extinction. Sci Adv 1(5): e1400253

Colombia MdAyDSd (2012) Manual para la Asignación de Compensaciones por Pérdida de Biodiversidad

Costanza R, d'Arge R, De Groot R, Farber S, Grasso M, Hannon B, Limburg K, Naeem S, O'Neill RV, Paruelo J, Raskin RG, Sutton P, van den Belt M (1997) The value of the world's ecosystem services and natural capital. Nature 387:253–260

Costello MJ, May RM, Stork NE (2013) Can we name Earth's species before they go extinct? Science 339(6118):413–416

Edwards JL (2004) Research and societal benefits of the global biodiversity information facility. Bioscience 54(6):485–486

Ens E, Reyes-García V, Asselin H, Hsu M, Reimerson E, Reihana K, Sithole B, Shen X, Cavanagh V, Adams M (2021) Recognition of indigenous ecological knowledge systems in conservation and their role to narrow the knowledge-implementation gap. In: Ferreira CC, Klütsch CFC (eds) Closing the knowledge-implementation gap in conservation science – evidence transfer across spatiotemporal scales and different stakeholders. Springer, New York, pp 109–139

Fernández M, Navarro LM, Apaza-Quevedo A, Gallegos SC, Marques A, Zambrana-Torrelio C, Wolf F, Hamilton H, Aguilar-Kirigin AJ, Aguirre LF, Alvear M, Aparicio J, Apaza-Vargas L, Arellano G, Armijo E, Ascarrunz N, Barrera S, Beck SG, Cabrera-Condarco H, Campos-Villanueva C, Cayola L, Flores-Saldana NP, Fuentes AF, García-Lino MC, Gómez MI, Higueras YS, Kessler M, Ledezma JC, Limachi JM, López RP, Loza MI, Macía MJ, Meneses RI, Miranda TB, Miranda-Calle AB, Molina-Rodriguez RF, Moya-Diaz MMRMI, Ocampo M, Perotto-Baldivieso HL, Plata O, Reichle S, Rivero K, Seidel R, Soria L, Terán MF, Toledo M, Zenteno-Ruiz FS, Pereira HM (2015) Challenges and opportunities for the Bolivian Biodiversity Observation Network. Biodiversity 16:86–98

Finch DM, Patton-Mallory M (1993) Closing the gap between research and management. In: Finch DM, Stangel PW (eds) Status and management of neotropical migratory birds: September 21–25, 1992, Estes Park, Colorado. Gen. Tech. Rep. RM-229. Fort Collins, Colo.: Rocky Mountain Forest and Range Experiment Station, US Dept. of Agriculture, Forest Service: 12-16, 229

Grima N, Singh SJ, Smetschka B, Ringhofer L (2016) Payment for Ecosystem Services (PES) in Latin America: analysing the performance of 40 case studies. Ecosyst Serv 17:24–32

Han X, Smyth RL, Young BE, Brooks TM, de Lozada AS, Bubb P, Butchart SHM, Larsen FW, Hamilton H, Hansen MC, Turner WR (2014) A biodiversity indicators dashboard: addressing challenges to monitoring progress towards the Aichi biodiversity targets using disaggregated global data. PLoS One 9(11):e112046

Han X, Josse C, Young BE, Smyth RL, Hamilton HH, Bowles-Newark N (2017) Monitoring national conservation progress with indicators derived from global and national datasets. Biol Conserv 213:325–334

Hobbie JE, Carpenter SR, Grimm NB, Gosz JR, Seastedt TR (2003) The US long term ecological research program. BioScience 53(1):21–32

Hoffmann M, Brooks TM, Da Fonseca GAB, Gascon C, Hawkins AFA, James RE, Langhammer P, Mittermeier RA, Pilgrim JD, Rodrigues ASL (2008) Conservation planning and the IUCN Red List. Endang Species Res 6(2):113–125

Hortal J, Jiménez-Valverde A, Gómez JF, Lobo JM, Baselga A (2008) Historical bias in biodiversity inventories affects the observed environmental niche of the species. Oikos 117(6):847–858

Jetz W, McPherson JM, Guralnick RP (2012) Integrating biodiversity distribution knowledge: toward a global map of life. Trends Ecol Evol 27:151–159

Josse C, Vásconez S (2016) The relationship between public environmental policy and the development and use of biodiversity information in Bolivia, Colombia, Ecuador, and Peru Technical Report NatureServe

Juffe-Bignoli D, Brooks TM, Butchart SHM, Jenkins RB, Boe K, Hoffmann M, Angulo A, Bachman S, Böhm M, Brummitt N, Carpenter KE et al (2016) Assessing the cost of global biodiversity and conservation knowledge. PLoS One 11:e0160640

Keith DA (2015) Assessing and managing risks to ecosystem biodiversity. Austral Ecol 40(4):337–346

Keith DA, Rodríguez JP, Rodríguez-Clark KM, Nicholson E, Aapala K, Alonso A, Asmussen M, Bachman S, Basset A, Barrow EG, Benson JS, Bishop MJ, Bonifacio R, Brooks TM, Burgman MA, Comer P, Comin FA, Essl F, Faber-Langendoen D, Fairweather PG, Holdaway RJ, Jennings M, Kingsford RT, Lester RE, Nally RM, McCarthy MA, Moat J, Oliveira-Miranda MA, Pisanu P, Poulin B, Regan TJ, Riecken U, Spalding MD, Zambrano-Martinez S (2013) Scientific foundations for an IUCN Red List of ecosystems. PLoS One 8(5):e62111

Klütsch CFC, Laikre L (2021) Closing the conservation genetics gap: integrating genetic knowledge in conservation management to ensure evolutionary potential. In: Ferreira CC, Klütsch CFC (eds) Closing the knowledge-implementation gap in conservation science – evidence transfer across spatiotemporal scales and different stakeholders. Springer, New York, pp 51–82

Larson AM, Ribot JC (2004) Democratic decentralisation through a natural resource lens: an introduction. Eur J Dev Res 16(1):1–25

Lindenmayer DB, Likens GE, Franklin JF (2018) Earth observation networks (EONs): finding the right balance. Trends Ecol Evol 33(1):1–3

Marques A, Martins IS, Kastner T, Plutzar C, Theurl MC, Eisenmenger N, Huijbregts MAJ, Wood R, Stadler K, Bruckner M, Canelas J, Hilbers JP, Tukker A, Erb K, Pereira HM (2019) Increasing impacts of land use on biodiversity and carbon sequestration driven by population and economic growth. Nat Ecol Evol 3:628–637

Mendoza, J, Amaya J, Terán P, Ramos A, Vargas N, Cediel M, Ramírez A, Sanabria A, Higuera D, Calderón M (2012) Política Nacional para la gestión integral de la biodiversidad y sus servicios ecosistémicos–PNGIBSE Ministerio de Ambiente y desarrollo sostenible 1–134

Ministerio de Ambiente y Desarrollo Sostenible de la República de Colombia (2012) Manual para la Asignación de Compensaciones por Pérdida de Biodiversidad. Ministerio de Ambiente y Desarrollo Sostenible (MADS), Autoridad Nacional de Licencias Ambientales (ANLA), The Nature Conservancy (TNC), World Wildlife Fund (WWF) y Conservación Internacional (CI), Colombia

Mittermeier RA, Robles-Gil P, Mittermeier CG (eds) (1997) Megadiversity. Earth's Biologically Wealthiest Nations CEMEX/Agrupación Sierra Madre, Mexico City

Mora C, Tittensor DP, Adl S, Simpson AGB, Worm B (2011) How many species are there on Earth and in the Ocean? PLoS Biol 9(8):e1001127

Pereira HM, Ferrier S, Walters M, Geller GN, Jongman RHG, Scholes RJ, Bruford MW, Brummitt N, Butchart SHM, Cardoso AC, Coops NC et al (2013) Essential biodiversity variables. Science 339(6117):277–278

Possingham HP, Andelman SJ, Burgman MA, Medellín RA, Master LL, Keith DA (2002) Limits to the use of threatened species lists. Trends Ecol Evol 17(11):503–507

Régnier C, Achaz G, Lambert A, Cowie RH, Bouchet P, Fontaine B (2015) Mass extinction in poorly known taxa. Proc Natl Acad Sci 112(25):7761–7766

Rodrigues AS, Pilgrim JD, Lamoreux JF, Hoffmann M, Brooks TM (2006) The value of the IUCN Red List for conservation. Trends Ecol Evol 21(2):71–76

Rodríguez JP, Rodríguez-Clark KM, Baillie JE, Ash N, Benson J, Boucher T, Brown C, Burgess ND, Collen BE, Jennings M, Keith DA (2011) Establishing IUCN Red List criteria for threatened ecosystems. Conserv Biol 25:21–29

Rodríguez JP, Keith DA, Rodríguez-Clark KM, Murray NJ, Nicholson E, Regan TJ, Miller RM, Barrow EG, Bland LM, Boe K, Brooks TM (2015) A practical guide to the application of the IUCN Red List of Ecosystems criteria. Philos Trans R Soc Lond B Biol Sci 370:20140003

Rondinini C, Di Marco M, Visconti P, Butchart SH, Boitani L (2014) Update or outdate: long-term viability of the IUCN Red List. Conserv Lett 7(2):126–130

Rose DC, Despot-Belmonte K, Pollard JA, Shears O, Robertson RJ (2021) Making an impact: how to design relevant and usable decision support systems for conservation. In: Ferreira CC, Klütsch CFC (eds) Closing the knowledge-implementation gap in conservation science –

evidence transfer across spatiotemporal scales and different stakeholders. Springer, New York, pp 199–226

Ruggles K, Bokor J, Lundgren L (2015) Taking learning to the field: exploring ecology with technology. In: Society for information technology & teacher education international conference, vol 1, pp 10538–10542

Santini L, Butchart SHM, Rondinini C, Benítez-López A, Hilbers JP, Schipper AM, Cengic M, Tobias JA, Huijbregts MA (2019) Applying habitat and population-density models to land-cover time series to inform IUCN Red List assessments. Conserv Biol 33(5):1084-1093

Saylor CR, Alsharif KA, Torres H (2017) The importance of traditional ecological knowledge in agroecological systems in Peru. Int J Biodivers Sci Ecosyst Serv Manage 13(1):150–161

Scholes RJ, Mace GM, Turner W, Geller GN, Jürgens N, Larigauderie A, Muchoney D, Walther BA, Mooney HA (2008) Toward a global biodiversity observing system. Science 321(5892):1044–1045

Schulman L, Toivonen T, Ruokolainen K (2007) Analysing botanical collecting effort in Amazonia and correcting for it in species range estimation. J Biogeogr 34(8):1388–1399

Secretariat CBD (2018) National Biodiversity Strategies and Action Plans (NBSAPs)

Skidmore AK et al (2015) Environmental science: agree on biodiversity metrics to track from space. Nature 523:403–405

Soberón J, Peterson AT (2004) Biodiversity informatics: managing and applying primary biodiversity data. Philos Trans R Soc Lond B Biol Sci 359:689–698

Stuart SN, Wilson EO, McNeely JA, Mittermeier RA, Rodríguez JP (2010) The barometer of life. Science 328(5975):177

ten Kate K (1997) The common regime on access to genetic resources in the Andean. Pact Biopolicy 2:Paper 6 (PY97006). http://hdl.handle.net/1807/98

Trimble MJ, van Aarde RJ (2012) Geographical and taxonomic biases in research on biodiversity in human-modified landscapes. Ecosphere 3(12):1-16

UNEP-WCMC and IUCN (2016) Protected planet report 2016. UNEP-WCMC/IUCN, Cambridge/Gland

Walls RL, Deck J, Guralnick R, Baskauf S, Beaman R, Blum S, Bowers S, Buttigieg PL, Davies N, Endresen D, Gandolfo MA et al (2014) Semantics in support of biodiversity knowledge discovery: an introduction to the biological collections ontology and related ontologies. PLoS One 9(3):e89606

WWF (2018) Living planet report – 2018: aiming higher. In: Grooten M, Almond REA (eds). WWF, Gland

Chapter 11
Conservation Science in Africa: Mainstreaming Biodiversity Information into Policy and Decision-Making

P. J. Stephenson, Mohamed Bakarr, Nadine Bowles-Newark, Fritz Kleinschroth, Abisha Mapendembe, Yaa Ntiamoa-Baidu, David Obura, Nanie Ratsifandrihamanana, John Simaika, Noah Sitati, Peter Sumbi, and Taye Teferi

---

P. J. Stephenson (✉)
IUCN SSC Species Monitoring Specialist Group, c/o IUCN, Gland and Ecosystem Management Group, ETH Zürich, Zürich, Switzerland
e-mail: stephensonpj@gmail.com

M. Bakarr
Global Environment Facility, Washington, DC, USA
e-mail: mbakarr@thegef.org

N. Bowles-Newark · A. Mapendembe
UN Environment World Conservation Monitoring Centre, Cambridge, UK
e-mail: Nadine.Bowles-Newark@unep-wcmc.org; abisha.mapendembe@unep-wcmc.org

F. Kleinschroth
Ecosystem Management Group, Department of Environmental Systems Science, ETH Zürich, Zürich, Switzerland
e-mail: fritz.kleinschroth@usys.ethz.ch

Y. Ntiamoa-Baidu
Department of Animal Biology and Conservation Science and Centre for African Wetlands, University of Ghana, Legon, Accra, Ghana
e-mail: ynbaidu@ug.edu.gh

D. Obura
CORDIO East Africa, Mombasa, Kenya
e-mail: Kenya.dobura@cordioea.net

N. Ratsifandrihamanana
WWF-Madagascar, Antananarivo, Madagascar
e-mail: nratsifandrihamanana@wwf.mg

J. Simaika
Department of Water Science and Engineering, IHE Delft, Delft, The Netherlands

Department of Soil Science, Stellenbosch University, Stellenbosch, South Africa
e-mail: simaikaj@sun.ac.za

C. C. Ferreira, C. F. C. Klütsch (eds.), *Closing the Knowledge-Implementation Gap in Conservation Science*, Wildlife Research Monographs 4,
https://doi.org/10.1007/978-3-030-81085-6_11

Contents

11.1 Introduction: The Need for Biodiversity Data in Africa

Species and habitats are being lost worldwide and, as human pressures on the environment increase, ecosystems become degraded with negative implications for human well-being and environmental sustainability (Cardinale et al. 2012; Bernstein 2014; Secretariat of the Convention on Biological Diversity 2014). The situation in Africa is a particular cause for concern as a high proportion of the population depends on natural resources directly for their food and livelihoods yet many of these resources are in decline (Craigie et al. 2010; Cardinale et al. 2012; WWF 2014). Furthermore, growing demand for agricultural land (Secretariat of the Convention on Biological Diversity 2014) and expansion of urban areas (Seto et al. 2012) will likely exacerbate loss of critical habitats and areas of global importance for biodiversity. While ecosystem services are a foundation for many economic sectors (e.g., tourism, agriculture), natural assets are rarely recognized and quantified (Brown et al. 2014).

N. Sitati
WWF Tanzania Country Office, Dar es Salaam, Tanzania
e-mail: n_wasilwa@yahoo.com

P. Sumbi
Tanzania Forest Conservation Group, Dar es Salaam, Tanzania

T. Teferi
TRAFFIC International, c/o IUCN-EARO, Nairobi, Kenya
e-mail: tteferi1@gmail.com

In order to address biodiversity conservation issues and ensure sustainable livelihoods, decisions at multiple levels across multiple sectors need to be guided by information on the state of the environment. Relevant information will include species (presence, abundance, range), offtake, trade and threat status; habitat cover and distribution; protected area (PA) coverage and management effectiveness.

Building on a recent assessment (Stephenson et al. 2017a), we acknowledge a diversity of government decisions requiring biodiversity information in Africa across ministries such as:

- The development of environmental resource policies and legislation.
- National planning and budgeting for resource management across sectors (e.g., PAs, forestry, fisheries, agriculture, infrastructure, mining, urban planning, water management), including internationally recognized biodiversity sites (World Heritage Sites, Ramsar Sites, Biosphere Reserves, etc.) and delivery of multilateral environmental agreements (MEAs) such as the Convention on Biological Diversity (CBD), the Ramsar Convention, the Convention on Migratory Species (CMS), and the Convention on International Trade in Endangered Species of Wild Fauna and Flora (CITES).
- Planning at landscape and seascape levels, including "ridge to reef" and marine spatial planning approaches that match natural system scales and dynamics.
- Transboundary and global planning and collaboration, when managing shared resources and contributing to global goals such as the Aichi Biodiversity Targets and the UN Sustainable Development Goals (SDGs).
- Access and benefit sharing and the control and licensing of resource use (e.g., mining, hunting quotas, forest concessions, water allocation, etc.)
- Defining and managing nature-based tourism, especially in and around PAs.
- The measurement and mitigation of human impacts on the environment, such as legal and illegal exploitation of resources, infrastructure development, threats from invasive species, and health-related issues such as those around Ebola.
- Mitigation of resource-related conflicts and human-wildlife conflict.
- Calculating and managing the values of ecosystem services in national accounting.

In addition to governments, other stakeholders require biodiversity data for decision-making. Conservation NGOs require data for monitoring the status of their target species and habitats and the impacts of their projects. Local communities require information on the location and status of resources to manage either individually or collectively their natural capital. Donors need to monitor their return on investment and the performance of funded projects. Science and education institutions base their research and teaching on biodiversity statistics. Companies from a range of business sectors, especially infrastructure, energy and extractives, require, or are legally obliged to collect, data on the state of biodiversity for environmental impact assessments (EIAs) and ongoing monitoring of their impacts. Investors in these companies also need to know how natural resources (e.g., soils, forestry, fisheries) support their business and what costs and losses will be caused through bad publicity when investing in destructive projects in areas of high conservation value. Urban planners and municipal governments need spatially explicit data on biodiversity patterns to identify areas and targets for protection and for achieving habitat connectivity in urban landscapes.

There is growing demand for more evidence-based conservation, with data informing decisions and evaluating performance (e.g., Segan et al. 2011; Stephenson et al. 2015a). In addition, governments need to report on progress towards global goals such as the Aichi Targets and the SDGs (Brooks et al. 2015). Therefore, there is an increasing demand for data, but that demand is not always being met.

11.2 Challenges Blocking Data Collection and Use

Even though many stakeholders require information for decision-making, numerous challenges block access to, and use of, biodiversity data, including gaps or other inadequacies in indicators, data sets and capacity (e.g., Secades et al. 2014; Stephenson et al. 2015a, 2017b). Barriers to using biodiversity information in decision-making in Africa can be clustered into four main categories (*sensu* Stephenson et al. 2017a).

11.2.1 Availability of Data

A strong geographical bias persists in conservation science, with 40% of recent studies carried out in Australia, the UK, or the USA, compared with only 10% in Africa, and research is poorly aligned with biodiversity distribution and conservation priorities (DiMarco et al. 2017). It is therefore not surprising that many global data sets have taxonomic, temporal, and geographic gaps in coverage (Butchart et al. 2010; Stephenson et al. 2015a). For example, much of the available plant species-occurrence data for Africa has large gaps in central Africa and "pronounced temporal biases towards older specimens" (Stropp et al. 2016). Most existing monitoring programs have been designed at localized scales and often produce information that is disaggregated, heterogeneous, and non-standardized when considered at national or regional scales (Han et al. 2014). Across Africa, existing data are housed in a range of different institutions and government departments. These data are often not shared because of political tensions and poor institutional connections, the poor links between science and biodiversity policies, and the limited interaction between the data gatherers (such as academic institutions and NGOs) and the data users within ministries (Stephenson et al. 2017a). Therefore, national and global data sets are often inadequate or not accessible to decision makers. Data accessibility can also be blocked by lack of capacity to identify, collate, and use it (see below), as well as lack of clear national policies and legislation on data sharing.

11.2.2 Usability and Quality of Data

Available data are frequently of inadequate quality or timeliness and rarely presented in a way that is conducive to use in decision-making (Stephenson et al. 2017a), due to, for example, the use of overly technical jargon or a lack of adequate interpretation (Segan et al. 2011; Roe and Mapendembe 2013). Many African states conduct regular wildlife surveys, yet the resultant data are rarely analyzed and presented in a form that could be of direct use to decision makers (Bubb et al. 2011). In some cases, data presentation and use are influenced by a donors placing conditions on sharing.

11.2.3 Willingness to Collect, Use, and Share Data

There is circumstantial evidence that some government, NGO, and business leaders may not always be keen to use data if those data do not show the trends they want or reflect badly on their management decisions. The reluctance to use biodiversity data is often compounded by the low level of importance given to environmental issues. For example, in countries such as Angola, South Africa, and Uganda, environmental strategies are often seen as lower political priorities than social and development ones (health, education, housing, security, etc.) and sometimes are even perceived as conflicting (Crouch and Smith 2011; Wilhelm-Rechmann and Cowling 2011; Ministério do Ambiente 2014). Political interests may therefore drive managers to disregard biodiversity information (National Environment Management Authority 2015).

11.2.4 Capacity

There are ongoing challenges with biological research and monitoring capacity in Africa (e.g., Yevide et al. 2016; Cresswell 2018). In a review of national reports submitted to CBD by African governments (Stephenson et al. 2017a), inadequate technical and financial capacity were highlighted as key obstacles to implementing National Biodiversity Strategy and Action Plans (NBSAPs). This reflects a broader lack of funding and capacity for conservation monitoring (Martin et al. 2012). Many monitoring tools are expensive, difficult to implement, and poorly adapted to local needs (Thapa et al. 2016). New technologies are often tested first in low-biodiversity countries. Even with satellite-based remote sensing, where data sharing has increased in recent years, numerous challenges remain in data use, most of them relating to national-level capacity (Secades et al. 2014). In Africa, the problems of data sharing and use are often compounded not only by more limited resources to pay for raw images and/or data processing, but also by limited internet capacity (Roy et al. 2010). Many of the recent assessments of African biodiversity data have been conducted by scientists who are predominantly based outside Africa (see, e.g., Beresford et al. 2013). For example, of the 3942 publications produced on Madagascar biodiversity from 1960 to 2015, only 8.9% had a lead author based at a Malagasy institution (Waeber et al. 2016). These trends reflect the fact that most of the global data sets, and most of the scientists with access and capacity to analyze them, are housed in Europe or North America, as are many of the museum collections that often present the best available taxonomic data and serve as historic references for biodiversity. Many local communities use indigenous knowledge on a range of variables, including biological indicators, for local decisions on farming and resource use (e.g., Mapfumo et al. 2016), yet this capacity is rarely tapped for more formal decision-making processes in Africa.

11.3 Case Studies

We present here a series of case studies to explore the collection and use of biodiversity data by different stakeholders in Africa and Madagascar. We then assess what the case studies and the literature tell us of challenges and potential solutions for enhancing the uptake of data for decision-making and conservation action across the continent.

11.3.1 Planning for Protected Areas and Conservation Action

Case Study 1 Biodiversity Knowledge Accelerates Protection of the Eastern Arc Mountains, Tanzania

The Eastern Arc Mountains (EAM) of Tanzania cover 2.3 million hectares of the Eastern Afromontane biodiversity hotspot, one of the world's most biodiversity-rich places (Plate 11.1). A series of research projects led to the accumulation of biodiversity and socio-economic data on the values and issues in the region, including the high levels of species richness, endemism, and threat (e.g., Burgess et al. 2007; Rovero et al. 2014) and the importance of timber and non-timber forest products to community livelihoods (Schaafsma et al. 2014). Such research started in the 1980s and 1990s (e.g., Rodgers and Homewood 1982; Lovett and Wasser 1993) and the resulting data have significantly influenced the government of Tanzania to continually upgrade the protection offered to the EAM forests.

In 1985, a logging ban was imposed across all national forest reserves within the EAM and the management emphasis was shifted towards catchment and biodiversity protection. In the 1990s a number of forest reserves were upgraded. For example, in 1992 the Udzungwa Mountains National Park (UMNP; 199,000 ha) was created from three national forest reserves; in 1999, the Amani Forest Nature Reserve was established in the East Usambara Mountains by amalgamating forest reserves and

Plate 11.1 (**a**) East Usambara Mountain Forests, part of the Eastern Arc Mountains in Tanzania that have been protected due to the availability of biodiversity data. © Peter Sumbi. Flagship species for the Eastern Arc Montane Forests include: (**b**) the gray-faced sengi *Rhynchocyon udzungwensis* © Michele Menegon; (**c**) Eastern double-collared sunbird *Cinnyris mediocris* © Michele Menegon.; (**d**) Parker's Tree Frog *Leptopelis parkeri* © Michele Menegon

unprotected land. As part of a larger strategy for the EAM region (www.easternarc.or.tz), the government's Forestry and Beekeeping Division in 2005 proposed the creation of 11 new nature reserves to safeguard high biodiversity sites. Since 2008, the government upgraded and merged 23 forest reserves to form 12 nature forest reserves which, as IUCN category Ia strict nature reserves, ensure improved protection for 312,677 ha of forest.

Support for biological studies and analysis of available data were provided by donors such as the Critical Ecosystem Partnership Fund (CEPF) and the Global Environment Facility (GEF) through the Project *Conservation and Management of Eastern Arc Mountains Forests* in 2004–2007. In 2009, UNESCO provided a World Heritage preparatory grant to the Division of the Antiquities of the Ministry of Natural Resources and Tourism for technical assistance in the preparation of the World Heritage Site nomination dossier for EAM forests. The government of Tanzania and its partners are currently revising the application in response to comments received from UNESCO World Heritage Centre, and it is hoped that up to nine forest blocks (451,000 ha, including 50% of remaining forest cover), such as

Amani Nature Reserve, Mkingu Nature Reserve, UMNP and Magamba Nature Reserve, will soon be designated World Heritage Sites.

Today, more than 150 reserves exist in the EAM and are managed for different purposes, including nature conservation (national parks and nature reserves), catchment protection (national forest reserves), and production (production forest reserves). The creation of new nature reserves was driven by the availability of data on the distribution and abundance of endemic and threatened species. These key data arose from field surveys of the region (Burgess et al. 2007) and of poorly known reserves (funded by CEPF and the GEF), as well as the Valuing the Arc Project, which mapped and quantified ecosystem services between 2003 and 2013 (Rovero et al. 2014). The Eastern Arc Mountains therefore demonstrate how biodiversity data can lead to conservation outcomes when made available to decision makers.

Case Study 2 Biodiversity Data: Uses and Deficiencies in Kenya's National Parks

Kenya has a well-developed PA system that covers 12.5% of its national territory. Management of PAs is guided by plans that are developed by the Kenya Wildlife Service (KWS; http://www.kws.go.ke/), the agency mandated for wildlife conservation in the country. With about 4000 personnel and an annual budget of US$50 million, KWS is relatively well-developed, well-staffed, and well-funded compared to most similar organizations in Africa and has its own Training Institute to build capacity for natural resource management. The development of PA management plans is led by a unit headed at a Deputy Director level with several ecologists and biologists working in headquarters and field stations. Kenya has additional advantages over most other African countries in that many international conservation organizations (e.g., Conservation International, International Fund for Animal Welfare, International Union for Conservation of Nature, The Nature Conservancy, TRAFFIC, Wildlife Conservation Society, WWF) have a strong presence, providing funding as well as technical support and an additional source of field data.

In 2006, KWS developed a Protected Area Planning Framework to standardize the process and structure of PA management plans and make them more practical day-to-day guides for management. Many observers and PA managers (e.g., Y. Wato and M. Said, personal communication) note that there is effective use of biodiversity data in the development of management plans in Kenya. Species-focused plans, especially for high profile wildlife such as elephants and rhinos that are also widely supported by international NGOs, have an even richer supply of data for monitoring and reporting from aerial and ground censuses (Litoro et al. 2012). However, when management plans are implemented on the ground, managers face serious difficulties in monitoring the change in status of biodiversity due to a lack of sufficient resources and a paucity of skilled staff to collect data. In an environment where funding is short and poaching rife (Weru 2016), PA managers are forced to prioritize park security, anti-poaching, and the monitoring of illegal activity over the monitoring of species status so, even where well-designed, evidence-based PA management plans are in place, often the post-collection of biological monitoring data, crucial to assess measure performance, falls by the wayside.

The revised Wildlife Act of 2013 requires KWS and the relevant ministry to produce a State of Wildlife report every 3 years. The first report was produced in 2015. This is an encouraging development and will likely motivate KWS and its partners to enhance the collection and use of data on the status of wildlife species and habitat conditions in the future.

Case Study 3 Water Bird's Data and the Designation of Ramsar Sites in Ghana Up until 1992, the wildlife PA system in Ghana included only forest and savanna sites, comprising one strict nature reserve, seven national parks, six resource reserves, and four wildlife sanctuaries, each with different levels of protection and permitted utilization (Grainger 1994). Strict nature reserves enjoy maximum protection and minimal use for the preservation of flora, fauna, and physical features, while resource reserves target selected species with the aim of ensuring sustained production of wildlife products.

In the mid-1980s, concern for the declining populations of the roseate tern (*Sterna dougallii*) and the identification of the Ghana coast as a hotspot for tern trapping by children (Everett et al. 1987) led to an inventory of wetlands and water birds along the coast (Ntiamoa-Baidu and Hepburn 1988). This survey covered around 50 lagoons, estuaries, marshlands, flood plains, and salt pans, ranging in size from a few hectares to several tens of thousands of hectares. Subsequent monthly surveys on selected sites identified eight that were important for biodiversity, mainly for water birds, marine turtles, and fish. The survey also identified threats, including establishment of settlements and the associated demand for land, farming and industrial developments, as well as coastal erosion and domestic, industrial, and agricultural pollution. Clearly, some level of protection was essential to secure the wetland habitats and their biodiversity. However, all the wetlands were situated in areas with dense human populations which depended on them for diverse socio-economic and cultural activities (Ntiamoa-Baidu 1991). Lagoon fisheries provided an important source of protein and livelihoods for the coastal communities (Koranteng et al. 2000), while other wetland resources such as salt, grasses, and mangroves were exploited heavily. It was obvious that none of the existing PA categories, which excluded people and limited access and use of natural resources, would be acceptable to the local communities.

The Ramsar Convention's concept of "wise use," where a designated site is allowed multiple uses as long as the ecological character of the site is maintained, offered the best approach for protecting the wetlands. The water bird counts provided strong justification for designating Ramsar sites (internationally important wetlands) on the Ghana coast (Ntiamoa-Baidu and Gordon 1991), since one of the designation criteria was that a wetland supported a total bird population of 20,000 or 1% of the flyway population of any one species. In 1992, Ghana designated five coastal wetland Ramsar sites (Keta, Songor, Sakumo, Densu Delta and Muni-Pomadze) based on water bird populations ranging from 23,000 to 111,000, with 3–12 species occurring in internationally important numbers. The Ramsar designation requires monitoring of the changes in the ecological character of the site. Hence a very comprehensive monitoring protocol was designed and advocated for the sites. However, this protocol could not be implemented due to a lack of resources.

In spite of the capacity challenges, counts of water birds have continued for over three decades, initially conducted by the local NGO, Ghana Wildlife Society, and subsequently by the Centre for African Wetlands, University of Ghana. The counts show declining populations of certain species and a reduction in the international importance of some sites. Evidence of habitat loss from encroachment by human settlement, industrial development and erosion, and decreases in habitat quality from pollution, have been documented through single-site studies (e.g., Osei et al. 2010; Addo and Adeyemi 2013). However, there are no data to feed into management interventions addressing the declining ecological status of the sites. The lack of a comprehensive long-term monitoring program, compounded by the lack of resources to manage the sites effectively and the ongoing negative impacts of climate change on coastal ecosystems, poses tangible threats to the existence of Ghana's coastal Ramsar sites.

Case Study 4 Planning for Conservation in Urban Areas: The Case of South Africa

South Africa is a megadiverse country, with a vast array of plant species found nowhere else on Earth. Securing this rich natural heritage requires a systematic approach to conservation planning, for which the country is known as a leader (Balmford 2003). In addition to identifying targets for establishing PAs, the systematic approach to conservation planning has also played a critical role in mainstreaming biodiversity across sectors to help inform decision-making on conservation actions and investments by diverse stakeholders (Holmes et al. 2012). The expansion and development of cities is a major threat to biodiversity, so the major municipalities are critical stakeholders in conservation. In Cape Town, Durban, and Johannesburg, biodiversity conservation has been integrated into urban development planning.

South Africa's approach to systematic conservation planning (Margules and Pressey 2000) extends from the national level (e.g., Reyers et al. 2001) to specific biomes such as the Succulent Karoo (e.g., Desmet et al. 2002) and the Cape Floristic Region (e.g., Cowling et al. 2003). This approach has been driven largely by the availability of data for the biomes, ecosystems, habitats, and species assemblages across the country. The South African National Biodiversity Institute (SANBI; www.sanbi.org) plays a major role in providing the data, which enables diverse stakeholders to make informed decisions about conservation. The overall framework has helped to align scientific and policy processes for biodiversity conservation and supports decision-making at the lowest jurisdiction where investments and actions are needed. As a result, the city of Cape Town, which is located in the globally important Cape Floristic Region, has embraced biodiversity conservation as a major priority in its urban planning.

Cape Town's approach to systematic conservation planning in an urban context utilized high quality data on species and habitats to establish targets for conservation, identify critical areas for improved management and restoration, and inform stakeholders' policies and actions (Holmes et al. 2012). For example, approximately 90% of the remaining natural remnants, representing 35% of the city area, have been considered as existing conservation areas or planned critical biodiversity areas and critical ecological support areas. The challenge, however, remains in translating the priorities into action, including the need to balance tradeoffs in meeting the mainstream urban development needs such as space for housing (Holmes et al. 2012).

11.3.2 Monitoring Projects, Programs, and Sites

Case Study 5 Using SMART for Community-Based Forest Management in Madagascar

Since 1996, a legal framework for community-based natural resource management has been in place in Madagascar. By 2017, the framework was being implemented across 2.45 million ha of mostly natural forest areas through 1248 contracts signed between the government and local communities which devolve management authority to the latter. Community-based management schemes often form a buffer zone around PAs. Madagascar expanded its PA coverage from 1.7 million ha to 7.2 million ha in 2015, with new governance models such as community PAs and co-managed areas.

The Spatial Monitoring and Reporting Tool (SMART; http://smartconservationtools.org/) was established to measure, evaluate, and improve the effectiveness of wildlife law enforcement patrols and site-based conservation activities. Based on standardized data collection protocols, it consists of a software application that enables the collection, storage, communication, and evaluation of ranger-based data on patrol efforts, patrol results, and threat levels. The use of SMART in PAs and community-managed areas in Madagascar started in 2014 and has since been expanded to a network of 46 national parks and reserves managed by Madagascar National Parks and 23 community-managed conservation areas supported by WWF, WCS, and the Durrell Wildlife Conservation Trust. SMART is integrated into surveillance and patrolling plans and conducted by community members. One key challenge in data collection is the community members' limited capacity to use equipment such as GPS units, due to the high level of illiteracy in rural areas. To counter this, a system for capturing data in graphic form has been developed using CyberTracker and a program of regular training and assessment is conducted by supporting NGOs (Plate 11.2). Nonetheless, data collection continues to be hampered by other factors, such as the distance to cover for monitoring and patrolling large areas, the remoteness of the sites, and poor or absent internet connections.

Since 2016, a roadmap has been implemented to promote adoption of SMART for the entire PA system, and a national SMART+ technical protocol has been adopted by the Ministry of Environment, Ecology and Forests (MEEF) and PA managers. This technical protocol includes a shared data configuration and standards for patrolling and surveillance within PAs (e.g., size and frequency of patrols,

Plate 11.2 Local Malagasy villagers being trained in SMART do not need to be literate since Cybertracker software allows them to collect data using icons. © Rabenandrasana Clark/ WWF MDCO

composition, level of patrolling). The process for national upscaling will rely on four main components:

1. adequate institutional and policy frameworks, through the establishment of a dedicated SMART unit and capacity within the MEEF, as well as a decree and a manual of procedures;
2. implementation structures from local to regional and national levels, with clear definitions of roles and responsibilities between the entities involved (government agencies, NGOs, community-based organizations) and capacity building at all levels;
3. a charter and a system for data sharing agreed between stakeholders; and
4. an early warning system to signal the level of intervention needed (from the local community to the national enforcement departments) to respond to a detected threat.

The implementation of these pillars has started, and it is hoped they will help overcome the earlier blockages experienced in collecting and using data for community-based forest management.

Case Study 6 Biodiversity Data from the Private Sector: Challenges and Opportunities

Many companies require biodiversity data for decision-making either because their operations have a potential impact on the environment or because their operations are based on the exploitation of ecosystems. Yet, the quality and usefulness of these data depend on the consistency and transparency of collection methods and the level of availability to different stakeholders.

Assessing the Environmental Impacts of Infrastructure Projects

Major improvements in transport and hydropower infrastructure are currently underway in Africa, leading to an unprecedented wave of road (Laurance et al. 2015) and dam (Zarfl et al. 2015) construction. Such projects are among the principal drivers of habitat destruction and fragmentation, especially when they are located in wilderness or frontier areas and serve as a conduit for human pressures (Laurance et al. 2009). Private companies are legally required to collect data on biodiversity for environmental impact assessments (EIAs), which are a policy mechanism to reduce the impacts of infrastructure projects. Commonly, EIAs are carried out by private consulting firms hired by the project investors. Unfortunately, there is no globally accepted standard for how biodiversity data should be used as baseline information for EIAs. Consultants generally use broad international standards as an orientation to reduce social and environmental risks such as the Equator Principles (http://equator-principles.com/) or the policies and standards of the International Finance Corporation of the World Bank Group (www.ifc.org/PerformanceStandards). However, due to widespread limitations in data availability, accessibility, and financial constraints, biodiversity assessments for EIAs have strong limitations in terms of quality and the coverage of taxonomic groups. While EIA reports are often extensive, there is no consistency in form, transparency of methods, and publication of the results. While for some projects funded by the African Development Bank, studies are available online (see https://www.afdb.org/en/documents/environmental-and-social-assessments), many EIAs are still only distributed to a narrow range of stakeholders. New data collected by consultancies are privately owned and barely accessible to the public. The entanglement of EIAs in a dependency relationship between project funders and consultants is a key issue with potential negative effects on quality (Spiegel 2017). Lucrative contracts are often issued to consultants whose past work provided assurance of basic acceptance of a project (Carr 2017). While many of the EIA issues are not specific to Africa and have been reported elsewhere (e.g., Alamgir et al. 2018), they are compounded by weak governance. Overcoming these shortcomings will require better involvement of independent NGOs and more efficient data sharing across sectors. Private companies should be involved as partners rather than opponents in this process.

Inventories Conducted due to Commercial Interests

Private companies dependent on natural resources are often more efficient than public bodies in creating databases. Although companies have the only reliable information on biodiversity in large parts of Africa outside national parks, they often have no interest in sharing their data. For example, logging companies collect biodiversity data for commercial purposes. They need high quality, accurate, spatially explicit information on tree species in their concessions to plan infrastructure and the volumes of wood to be harvested. Pre-logging inventories provide the most extensive source of information on tree species species diversity for the Congo Basin (Réjou-Méchain et al. 2011). Extremely high numbers of sampling replications and a consistent methodology allow novel analyses of species traits and assemblages (Réjou-Méchain et al. 2014), as well as forest types (Gond et al. 2013), and historic disturbances (Morin-Rivat et al. 2017). However, whenever commercial interests are involved, accessibility of data is affected. Companies fear that making their inventory data public exposes commercially valuable information to competitors and risks opening them up for criticism from conservation NGOs. Many companies therefore only agree to make their data public if it is anonymized and bundled.

A potential framework for sharing commercial data in Central Africa is the Congo Basin Forest Partnership, which brings together stakeholders from the private sector (mostly logging companies and their collective bodies) with organizations of civil society, international NGOs, donors, research institutes, and multilateral bodies. The Central African Forests Commission (COMIFAC), composed of the administrations of ten countries, is leading this initiative. They make data available through their scientific and technical branch OFAC (Observatoire COMIFAC) run in collaboration between different international research institutes. Raw data on forest exploitation can be found online (www.observatoire-comifac.net), while analyses and summaries are published regularly to present the state of the forest (de Wasseige et al. 2014) and the state of PAs (Doumenge et al. 2015). While data quality and completeness are still far from satisfactory, such a platform is an important first step in building trust and illustrating the mutual interests in data sharing between the private sector and other stakeholders. It might be a model worth replicating elsewhere.

Case Study 7 Human–Wildlife Conflict Monitoring

Human-wildlife conflict (HWC) is a major threat to conservation and a driver of species extinctions (Conover 2001). Across Africa, where wildlife habitats and migration corridors are being destroyed, animals such as large carnivores, primates, ungulates, and elephants cause considerable damage to human property, including crop damage, livestock predation, and human injuries and deaths (Conover 2001; Sitati and Walpole 2006). People develop negative attitudes towards the wildlife responsible for the conflicts (Okello et al. 2014) leading to retaliatory attacks, habitat destruction, and support for poachers (Wakoli and Sitati 2012).

An example of an HWC study in Africa was the long-term monitoring of livestock predation in Amboseli (Gichohi et al. 2014). This research measured the levels of livestock losses (4272 cattle, sheep, goats, and donkeys over 4 years) and identified the six wild carnivores responsible: lion (*Panthera leo*), cheetah (*Acinonyx jubatus*), spotted hyena (*Crocuta crocuta*), leopard (*Panthera pardus*), jackal (*Canis* spp), and caracal (*Caracal caracal*). The economic cost of predation was estimated at between US$697,880 and US$728,941 (Gichohi et al. 2014). The calculated costs derived from effective and efficient monitoring informed national policy formulation around compensation. However, although Kenya's Wildlife Conservation and Management Act 2013 made provision for compensating farmers for livestock losses and crop damage, the government has been unable to pay due to the high costs and the complex nature of verification. Insurance schemes are now being proposed by some countries, including Namibia and Kenya, as an alternative to cushion pastoralists from predation. Improved livestock enclosures in Amboseli have also reduced predation incidences from homesteads by over 80% (Okello et al. 2014).

However, the appropriate choice and implementation of policies to tackle HWC can only be facilitated through long-term monitoring using standardized and locally adapted protocols (Sitati et al. 2005). HWC is often misunderstood and politicized (Walpole et al. 2006) so data on the causes and costs of HWC are needed to help inform politicians and other decision makers. Future efforts will need to collect data on the types of mitigation measures employed and the socio-economic and ecological factors affecting HWC. More effort is also needed to tap into indigenous knowledge (Sitati and Ipara 2012). Capacity and funding will be required and the final users of the data (especially government agencies and local people) consulted in the design of monitoring systems.

Case Study 8 Biodiversity Monitoring and Citizen Science in Africa

Citizen science fosters communication and collaboration among local residents and scientists and is increasingly cited as an important means of acquiring biodiversity data (e.g., Bonney et al. 2014; Chap. 4; Phillips et al. (2021)). Here, we summarize key lessons from efforts to test citizen science approaches in Africa.

The Event Book system of monitoring communal conservancies in Namibia is one of the earliest successes of citizen science in Africa. Communities monitor various aspects of their conservancy, from wildlife numbers to economic returns to patrolling records (Stuart-Hill et al. 2005). It differs from conventional monitoring in that communities dictate what needs to be monitored, and scientists only facilitate the design process. Conservancy members undertake all data analysis. Communities needed training and technical support on the interpretation and use of the monitoring information and there was need for a national support body. But the system went on to be adopted widely and replicated in many countries. Key lessons learned included the need to build on small successes rather than be too ambitious, and "be patient and pay attention to building sustainable monitoring systems rather than obtaining data at all costs" (Stuart-Hill et al. 2005).

Using citizen science methods developed in Europe, Wotton et al. (2020) collected data on bird populations in Botswana, Kenya, and Uganda and demonstrated that, with technical support and modest investment (approx. US$30,000 per scheme per year), meaningful biodiversity indicators could be measured. In most instances, observations increased after the first year, suggesting it took time for the volunteers to learn and improve their species identification skills. In all three schemes, staff from government wildlife agencies, who were often highly skilled observers and required to accompany those visiting national parks, were also involved.

In South Africa, Stellenbosch University encouraged high school students to participate in ant monitoring schemes in the Cape Floristic Kingdom (Braschler et al. 2010). Regular school visits by scientists helped ensure data quality and analysis, and the use of the data in research; throughout the project students learned about biodiversity in the local context. The approach was felt to be replicable because ants are relatively easy to identify, are abundant almost everywhere, and ant sampling methods are accepted and standardized.

Stephenson et al. (2007) reported on schemes to monitor and mitigate human-elephant conflict in Tanzania. Local enumerators were trained to capture georeferenced data using standardized data collection and analysis protocols. The enumerators each cost about US$1000 per year and were a cost-effective way of obtaining good quality data. If recruited locally, they were more readily accepted by the villagers.

Larson et al. (2016) surveyed villagers around a PA in Sierra Leone to quantify levels of human-wildlife conflict. As part of the study they investigated the possibility of collecting longer-term data through citizen science. The potential benefits identified for local involvement in research and monitoring included increased researcher access to community lands and to indigenous knowledge. However, some participants believed researchers did not listen to advice, and these cultural misunderstandings could result in distrust and negative relations.

There were multiple benefits and positive elements across citizen science schemes reviewed and growing evidence that the approach can work in Africa. Education and capacity building were a common motivation for local people to participate. Common challenges included recruiting, training, and retaining volunteers. Financial rewards were the greatest source of motivation in some places (e.g., Larson et al. 2016) and some projects paid data collectors (e.g., Stephenson et al. 2007). However, this strategy can lead to jealousy among other local residents who then fail to collaborate. Sustainability is often an issue, with local data collectors leaving their roles either to find full-time jobs or because the project was timebound and approaching its end. In several instances, low literacy rates or language barriers were a hindrance. In every case, the absence of outside support (technical and/or financial) was likely to lead to an end to data collection. For long-term success in citizen science, more effort needs to be made to ensure data users' needs are assessed and prioritized (Amano et al. 2016; Pocock et al. 2019). In addition, data generated by citizen science need to be curated in secure, neutrally governed institutional homes—the function fulfilled in South Africa by SANBI—and converted into formats of use for decision makers, such as reports, graphs, and maps (Barnard et al. 2017). Bonney et al. (2014) encourage the creation of citizen science centers that create and manage centralized repositories of volunteer-collected data on biodiversity and also answer queries on tools and methods. Although most monitoring schemes focus on larger vertebrates, "scientists should advertise less charismatic species and develop societal initiatives that specifically target neglected organisms" (Troudet et al. 2017).

11.3.3 Overlooked Species: Taxonomic and Geographic Data Gaps

Case Study 9 Data Gaps for Invertebrates: Causes, Consequences, and Solutions

Our well-being is intricately linked to the survival of invertebrates (Samways 2015). For example, without pollinating insects, on which an estimated 75% of crop plants depend globally (IPBES 2016), the world would face a reduction in food security, reducing the number of crops available for consumption and increasing the cost to the consumer. Similar arguments can be made for the role of invertebrates in soil formation and fertility, maintenance of trophic structures, and the cycling of nutrients (Schowalter et al. 2018). Of the two million species described, more than 95% are invertebrates (IUCN 2020). However, it is only those invertebrates which are either directly useful (pollinators, food species), indirectly useful (aquatic macroinvertebrates for biomonitoring) or harmful (pests, vectors of disease) that are relatively well studied. The vast majority—an estimated 7.8 million—are not even described yet. With biodiversity in decline (e.g., WWF 2014), time is running out for the unknown majority of species, some of which will go extinct before description. With a few exceptions (e.g., butterflies), it is likely that it is their utility that will save most invertebrates rather than their inherent appeal.

Although invertebrates are a universal umbrella taxon and offer numerous opportunities for biodiversity monitoring (Cardoso et al. 2011), in many African countries (especially in vulnerable aquatic ecosystems) they are not monitored due to lack of expertise, awareness, and capacity. The taxa commonly monitored, that are relatively well known in temperate regions and some southern African countries, are mayflies (Ephemeroptera), caddisflies (Trichoptera), stoneflies (Plecoptera), and dragonflies (Odonata). In southern Africa, a rapid bioassessment method for rivers uses aquatic invertebrates as indicators of ecological health (Dickens and Graham 2002). This works well, as the relative diversity of taxa is low, and there are taxonomic keys available and sufficient capacity to train technicians. In tropical west and east Africa this system will not work given the present lack of knowledge of invertebrates. The Dragonfly Biotic Index (Simaika and Samways 2008, 2011), based on the use of adult dragonflies and also developed in South Africa, has potential for use in rivers and wetlands throughout Africa. The success of this method is in part due to the relatively low diversity of dragonflies compared with other insect taxa, their large size, their well-resolved taxonomy, public interest in these insects, and the burgeoning numbers of freshwater assessment handbooks (Samways and Simaika 2016), field guides (Tarboton and Tarboton 2015), and taxonomic texts (Dijkstra and Clausnitzer 2014) dedicated to this taxon. The Dragonfly Biotic Index has been used to inform the ecological status of rivers (Diedericks et al. 2013), to assess the restoration of rivers (Samways et al. 2011) and succession in wetland habitats (Harabiš et al. 2013) and is therefore a model for how invertebrate data can be used for monitoring and decision-making.

Case Study 10 The Global Coral Reef Monitoring Network: Filling a Data Gap in the Western Indian Ocean

Coral reefs are among the most valued marine ecosystems, providing numerous benefits to biodiversity as well as people through fisheries, tourism, and coastal protection. Many of the marine protected areas (MPAs) in East and Southern Africa focus on coral reefs and monitoring of the reefs for conservation purposes started in some locations (e.g., Kenya, South Africa) in the 1980s. Since then monitoring expanded greatly, culminating in a regional network under the Global Coral Reef Monitoring Network (GCRMN) that has been active from the early 2000s (Obura et al. 2017). The network prepares periodic regional assessments to identify minimum standards for monitoring indicators and recommendations for coral reef policy and management. The network complements additional efforts by the Global Ocean Observation System and the Marine Biodiversity Observation Network of GEOBON (Global Earth Observations: Biodiversity Observation Network).

Nevertheless, biodiversity in marine ecosystems is generally less well known than in terrestrial biomes (Webb and Mindel 2015), and significant gaps exist in long-term coral reef monitoring programs (Obura et al. 2019). Variance in the number of sites monitored each year within countries has been high due to lack of core funding in government and NGO monitoring programs. Lack of support results in gap years without monitoring and the loss of trained staff, leading to poor and variable data quality. Inconsistent support also translates into variable methods (Obura 2014) and data quality. In some cases, short-term projects with specific objectives impose changes in methodologies to suit their purposes that undermine the integrity and value of long-term data sets.

Coral reef data feed directly into MPA management, though often this is through informal channels and the engagement of managers in monitoring, rather than formal management processes. However, IUCN standards and databases are encouraging data collection and use. Coral reef monitoring data provide the foundation for the IUCN Red List of Threatened Species for corals (Carpenter et al. 2008) and the IUCN Red List of Ecosystems for coral reefs (Keith et al. 2015). To address management needs, more focused monitoring protocols are under development, such as for coral bleaching (Gudka et al. 2020), to incorporate social and economic aspects (Obura et al. 2019, Wongbusarakum and Heenan 2019), and to assess management, such as through the IUCN Green List of Protected and Conserved Areas.

Case Study 11 Data Gaps for African Small Mammals

Among the 5850 species of mammals assessed in the IUCN Red List of Threatened Species, 872 (14.9%) remain Data Deficient, meaning there is inadequate information to assess extinction risk based on distribution and/or population data (IUCN 2020). The proportion of mammals that are Data Deficient is higher in Africa than in most other regions (Stephenson 2017). Yet, even for many species not considered Data Deficient, the information needed for Red List assessments is often incomplete. This is demonstrated well within two small mammal taxa endemic to Africa and Madagascar: the 55 species in the Afrosoricida (tenrecs, otter shrews, and golden moles) and the 20 species of Macroscelidea (sengis or elephant shrews). These taxa have a number of threatened species, yet no population estimates or trends exist. For each of the seven tenrec species considered threatened, the priority conservation action is to assess their range, populations, and threats (e.g., Stephenson et al. 2016a, b). There are no accurate population estimates for the Chrysochloridae and some golden mole species have been recorded only a handful of times. For example, only three individuals of the rough-haired golden mole (*Chrysospalax villosus*) have been found since 1980 (Bronner 2015). Each new field survey seems to bring a change in status of otter shrews (Stephenson et al. 2018). In the Macroscelidea, historical estimates are available only for the threatened golden-rumped sengi (*Rhynchocyon chrysopygus*), but these were over limited time frames and no data are available since 2009 (Fitzgibbon and Rathbun 2015). Threat monitoring is essential for the success of mammal conservation projects (Crees et al. 2016) yet, across both families, there are no quantitative data on levels or rates of habitat loss and the impact of hunting.

A recent review of the conservation status and needs of the two families and related taxa (Kennerley et al. 2018) highlighted that conservation is hindered by the lack of adequate data on species distributions, abundance, habitat needs and threats, as well as ongoing confusion over taxonomy. The same is true for rodents and bats. This lack of monitoring reflects a broader trend where small mammals are generally subject to less research and conservation attention than larger species (Entwistle and Stephenson 2000), and Africa and Asia are the most understudied regions for conservation research (Velasco et al. 2015). Therefore, dedicated research projects targeting key information gaps are essential. Since biodiversity monitoring in Africa is unlikely to ever focus primarily on small mammals, it may be prudent to integrate small mammal monitoring into schemes focused on larger, more charismatic species like primates, pachyderms, and carnivores, especially in PAs.

11.3.4 National and Regional Mainstreaming of Data

Case Study 12 National Mainstreaming of Data: Lessons from the UNEP-WCMC Connect Project

Data availability and accessibility and the capacity to use it have been highlighted as significant hurdles to biodiversity mainstreaming, especially in Africa (IIED and UNEP-WCMC 2017). Decision makers (or "end-users" of information) need to access biodiversity data that helps them understand the impacts of their decisions, and data providers need to understand how biodiversity information can be integrated into decision-making processes. To that end, the GEF-funded Connect Project (www.connectbiodiversity.com/) was launched in 2016 to help achieve sustainable development objectives by bringing biodiversity and ecosystem services information into government decision-making. The project goal is to ensure biodiversity is taken into account in decision-making across government sectors by improving end-users' access to, and use of, biodiversity information and embedding biodiversity information within national development decision-making processes. Connect has initiated a collaborative process in three pilot countries: Ghana, Mozambique, and Uganda. Data providers and data users will co-develop biodiversity information products which will be focused for use in a decision-making process selected through in-country stakeholder engagement. The national objectives are to:

- Understand in-country demands for, and the barriers to using, biodiversity information within government decision-making.
- Mobilize and repackage existing biodiversity data and information to meet identified demands.

- Strengthen the connection between government decision makers and data providers in order to provide policy-relevant, spatially explicit information that meets national needs.

Looking at Ghana as an example, there is a disconnect between providers of biodiversity information and end-users at multiple levels. In cases where biodiversity information exists, it is not easily accessible. A series of national workshops, bringing together a cross-sectoral group of data providers and data users in each country, were organized to identify capacity needs and barriers to the use of biodiversity information. Participants in the Ghanaian workshop in November 2017 proposed enhancing organizational coordination to bring together existing biodiversity data scattered across organizations and advocating to politicians the importance of biodiversity data in the planning and implementation of the national development agenda. The decision-making process in the cocoa sub-sector in Ghana was mapped as a case study. Biodiversity data on types of forest trees, fauna, and the suitability of soils are provided by the Forestry Commission and research institutes and are accessed by many organizations (e.g., Cocoa Board, Ministry of Finance, Ministry of Food and Agriculture, National Development Planning Commission, cocoa processing companies). The main barrier to the use of such information is the complex way it is presented. In addition, data are usually of poor quality, are only partly computerized and not stored in an easily accessible format. The group suggested the promotion of research and development for cocoa farming in collaboration with decision makers and conducting meetings and policy fora to better communicate biodiversity information. Standardizing the format for data collection and storage is also important.

UNEP-WCMC and its partners expect the Connect Project to generate more lessons and recommendations for mainstreaming biodiversity data into national decision-making.

Case Study 13 Multilateral Environment Agreements as Drivers of National Data Collection and Use

While many of the case studies in this chapter highlight the paucity of biodiversity data across Africa, there are more data available today than 10–15 years ago. It can be argued that many of the MEAs that have been signed by African states have had a significant influence on data availability.

The 196 countries that are Parties to the CBD are obliged to develop NBSAPs and report on their implementation and this has led to a major thrust for data collection and analysis. The international NGO community has supported CBD Parties in developing appropriate scalable indicators that can be applied at national level and rolled up to assess the global situation (see e.g., Bubb et al. 2011). The Biodiversity Indicators Partnership (BIP; www.bipindicators.net/) has been the main forum for such development. In turn, data around these common indicators are analyzed at a national level for reports to CBD (see https://www.cbd.int/reports/) and at the global level (e.g., Butchart et al. 2010; Secretariat of the Convention on Biological Diversity 2014).

Other MEAs have also encouraged data collection. For example, under CITES, monitoring systems have been developed and implemented by Parties to monitor the illegal killing of elephants, and TRAFFIC, the wildlife trade body of IUCN and WWF, was mandated to set up the Elephant Trade Information System. These two systems have greatly enhanced the data available for elephants (Burn et al. 2011; Underwood et al. 2013) to inform policy; similar systems are being developed for other species. Similarly, governments that are signatories to the Ramsar Convention and the World Heritage Convention are encouraged to assess and continually monitor sites of conservation value.

Therefore, while we still do not have the volume and quality of data we would like, we have much more information as a result of the MEAs. The SDGs are now rallying governments around a new set of global targets, and SDG 14 (Life Below Water) and SDG 15 (Life on Land) are focused on the environment. National progress in implementing these goals will need to be monitored and many of the tools and indicators developed for CBD will be applicable (Brooks et al. 2015). Therefore, we can expect the SDGs, as well as continued implementation of MEAs, to provide a strong stimulus for enhanced biodiversity monitoring in coming years.

Case Study 14 Marine Spatial Atlas for the Western Indian Ocean: Marine Data Used Regionally

A key challenge in Africa is to make available the biodiversity data collected and produced locally. Few programs have the knowledge, resources, or capacity to publish data sets digitally, let alone maintain an online platform to facilitate data sharing. However, this challenge is being addressed for coral reefs and related marine data from the east coasts of Africa, including the Indian Ocean, through the portal of the Marine Spatial Atlas for the Western Indian Ocean (MASPAWIO; https://maspawio.net/). The portal uses GeoNode software to make georeferenced data layers available through the internet, compatible with larger scale biodiversity resources such as the Regional Reference Information Systems developed by the BIOPAMA project (http://rris.biopama.org/). A second objective of MASPAWIO is to harvest available data sets from other sources and package them in ways useful to the region to make them more accessible locally. For example, environmental variables such as sea surface temperature can be extracted from global databases and combined with local data (e.g., through web-mapping tools). The goal is for data sets to be widely accessible and, ultimately, to build a broader community using and sharing biodiversity data.

MASPAWIO was started in 2015 with support from IUCN and the French development agency, AFD, and has continued through further regional projects, including the Western Indian Ocean Marine Science Association and the Inter-Governmental Agency for Development's Biodiversity Management Program. Eighty-four data layers have been built to illustrate the value of marine spatial planning in the Western Indian Ocean and of marine biodiversity surveys for seascape and county-level planning in Djibouti and Kenya. Ecological Metadata Language standards and a Creative Commons CC-BY 4.0 License are being applied, to ensure compatibility with global standards. These also allow compatibility with the Global Biodiversity Information Facility (GBIF) and the Ocean Biogeographic Information System of UNESCO, to ensure that data layers in MASPAWIO can be harvested by these and other larger scale databases. This impressive effort to link local and global marine data in the Western Indian Ocean should be replicated for other biomes in other parts of Africa.

Case Study 15 Enabling Factors and Barriers to the Use of Biodiversity Information in Decision-Making: A Review of the GEF Mainstreaming Portfolio
As part of the Connect Project (case study 12), UNEP-WCMC undertook a systematic review of GEF (Global Environment Facility) projects which aimed to support the mainstreaming of biodiversity data into decision-making. The aim was to evaluate how the success of biodiversity mainstreaming is influenced by the type of biodiversity data used, barriers to biodiversity data accessibility and use, and the activities undertaken to overcome such barriers. A search using the term "mainstreaming" yielded 284 completed projects from the "biodiversity" focal area, of which 67 contained sufficient details for analysis (21 in Latin America, 17 in Asia, 15 in Africa, 13 in Europe/Central Asia, one global). Project information was extracted from mid-term reviews and final evaluations. Findings emerging from this review included:

- Most projects used information on biodiversity status and trends (55), followed by threats (48), spatial distribution (46), conservation and management status (35), economic valuation and accounting (21), impacts of different management options (12), and ecological requirements of species and ecosystems (3).
- The principal barriers that projects addressed were availability of data (65), data accessibility (57), capability (28), and willingness to use data (20).
- The majority of projects (64) focused on providing tools and guidelines to use data, with 62 focusing on information sharing, 62 on compilation of existing biodiversity information, and 61 on institutional capacity building and training.
- The most important factors influencing project success related to the ability to use and receive guidance and support on data use and integration. Collaboration and capacity building were key.
- Projects that included spatial data tended to be more successful than those that included other types of information.

We conclude that producing more biodiversity data and improving its accessibility does not, in isolation, result in biodiversity becoming mainstreamed. Data and tools need to be coupled with collaborative and capacity building activities. While the production of biodiversity data and tools can be relatively quick, the social and economic benefits obtained from mainstreaming typically manifest themselves in the longer term (10–15 years). Therefore, sustained engagement is required to obtain support and funding that will ensure biodiversity is maintained on the political agenda. Developing the capacity to mainstream biodiversity information often takes people out of their traditional silos and alters and questions their approaches, values, and workflows, which can be met with resistance and skepticism. In part, this can be overcome via targeted and clear communications which link mainstreaming to political agendas (e.g., poverty-environment issues) and generate a shared vision.

11.3.5 Main Lessons Identified from Case Studies

While the case studies reflect the authors' own experiences, they provide a cross-section of representative examples of some of the issues associated with collecting and using biodiversity data in Africa and Madagascar. Key lessons can be extracted from across these case studies.

Data are often used to justify the choice of sites at which to establish PAs, ensuring they maximize the conservation of threatened species and habitats. This suggests governments do take on board data in decisions surrounding PA creation. However, data are then rarely used to monitor the state of biodiversity within established PAs, largely due to inadequate capacity and resources, which seem to be less of a problem before reserve creation. In some instances, this may be linked to governments and their donors being more willing to fund surveys to create PAs than fund ongoing monitoring. Environmental data are also used in urban planning although in this case the plans are not always put into action, suggesting that the lack of ongoing data use is a problem across sectors.

Global and regional efforts to monitor and share data on coral reefs demonstrate what is possible when conservation scientists and practitioners collaborate to collect and share data and make it openly available. The private sector also needs to be involved. Investors in large infrastructure projects and companies that exploit natural ecosystems for commercial purposes would benefit from more reliable and accessible public databases yet, while most companies collect their own data, other stakeholders can rarely access it. Logging companies are traditionally very secretive, but the Congo Basin Forest Partnership demonstrates how data can be shared and made more openly available, paving the way for its broader use in decision-making.

The case studies generally demonstrate the importance of governments working with civil society and academic institutions if blockages to data collection and use are to be overcome. Citizen science may offer possibilities for data collection if supported, coordinated, and resourced by government agencies or NGOs and enough literate volunteers can be found. The creation of regional citizen science centers to help provide the necessary support and coordination, store data, and make it available to decision makers might be one option for the future.

Taxonomic data bias is common, with neglected species including invertebrates and small mammals. However, simple, cost-effective, and replicable schemes targeting insects provide opportunities for a range of stakeholders to collect useful data. In addition, the monitoring of less well-known species needs to be integrated into systems for more charismatic species, like carnivores, pachyderms, and primates. Threat monitoring is also important, and SMART has potential applications in community-based natural resource management schemes as well as in and around

PAs. Human-wildlife conflict is a key threat that needs more research and monitoring to help inform policies and local decision-making.

Ultimately, as demonstrated by the Connect Project and the review of the GEF portfolio, streamlining biodiversity data into national decision-making will need more effort to build capacity for presenting good quality data in the right format at the right time, and digitizing it to make it more easily accessible. The SDGs, as well as multilateral environment agreements like CBD, are a significant stimulus for the development of capacity for data collection and use at the national level.

11.4 Solutions to Enhance Evidence-Based Decision-Making in Africa

Evidence-based management of African natural resources across sectors, and the mainstreaming of data into decision-making, can be attained in coming years if there are concentrated efforts to: enhance the application and use of the latest technologies, methods, and tools; build capacity in national institutions; and work together in partnership to improve the science–policy interface.

11.4.1 Using the Latest Technologies, Methods, and Tools to Access and Share Data

Data collection methods and tools are becoming cheaper and more easily available. There is also growing consensus on indicators of use across countries, often driven by the Aichi Targets and SDGs. National efforts to address data collection can be enhanced by harmonizing measures across scales and programs and expanding existing efforts to standardize and share data (Pereira et al. 2010; Stephenson et al. 2015a).

Several global databases are of use to national decision makers in planning and monitoring (Stephenson and Stengel 2020; see https://www.speciesmonitoring.org/data-sources.html). Examples of how global data can be used to assess local trends include analyses of factors affecting species populations in African PAs (Craigie et al. 2010; Daskin and Pringle 2018). There are also a number of biodiversity databases focused on Africa, such as the ARCOS (Albertine Rift Conservation Society) Biodiversity Management Information System (http://arbmis.arcosnetwork.org/) and FishBase for Africa (http://www.fishbaseforafrica.org/), which, as appropriate, can supplement data collated by national biodiversity centers (e.g., Egypt's National Biodiversity Unit, South Africa's SANBI, Uganda's National Biodiversity Data Bank) (Case study 14 has coral reef examples).

As well as accessing global, regional, and national databases, African managers need to use the latest monitoring techniques and technology to collect data

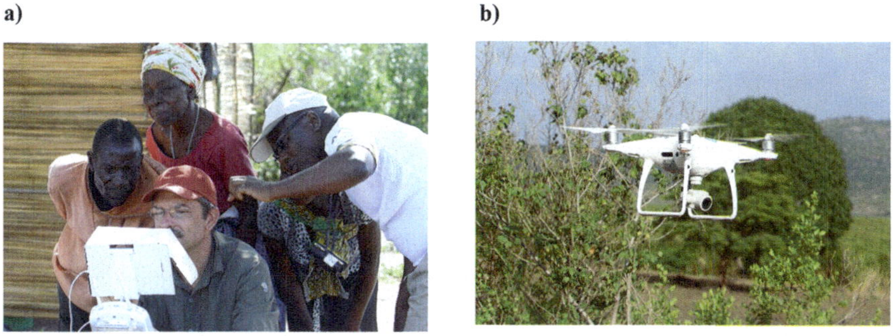

Plate 11.3 Research being conducted in the Mara River region in Tanzania by IHE Delft (**a**) involves local communities in monitoring freshwater habitats, and (**b**) modern drone technology. © IHE Delft

locally where it is relevant to local needs and capacity (Stephenson 2019). The newest generation of sensors include camera traps, which have already been used widely in Africa (Swanson et al. 2015; Rovero and Zimmermann 2016), acoustic recording devices (e.g., Blumstein et al. 2011), and drones (e.g., Wich and Koh 2018; Plate 11.3). Emerging methods such as environmental DNA monitoring (e.g., Valentini et al. 2016) also need exploring. As highlighted in case studies 9–11, more effort needs to be made to monitor lesser studied species. Recent work has expanded standardized protocols to include invertebrates and plants (e.g., Van Swaay et al. 2015; Borges et al. 2018), a trend that needs to continue.

Greater investment is required to collect data in PAs which, as we demonstrated, are rarely monitored adequately. A good starting point would be to encourage adoption of standard assessment tools such as the Management Effectiveness Tracking Tool (WWF 2007), as well as linked schemes such as the Rapid Assessment and Prioritization of Protected Areas Management Methodology (RAPPAM; Ervin 2003) and the Conservation Assured Tiger Standards (Conservation Assured 2017; soon to be adapted for rhinos and other species), which all encourage monitoring and could be used to assess gaps. They also allow assessments of relative performance between PAs and types of PA (Knights et al. 2014), helping inform decisions on management and resource allocation.

Data use will be enhanced if we can ensure synergies between national, regional, and local data to allow decision-making at the relevant levels in the appropriate context. This is particularly crucial for designing conservation strategies for species and ecosystems that extend across boundaries. Therefore, organizations hosting global and regional databases need to facilitate access to data that are disaggregated to levels relevant to local decision-making.

Data collection requires local inputs. Equitable participation of data providers and users, including local communities, can lead to better results and sustainability (Danielsen et al. 2014). Citizen science initiatives offer an opportunity to enhance data collection and efforts should be expanded. A concerted awareness campaign to encourage citizen scientists to collect data for less well-represented species might be required (Stephenson 2017). Data presentation is also key (see Chap. 8; Rose et al. (2021)). Scientists and decision makers need to produce more data-derived products in forms of relevance to people who use them, such as maps and dashboards that facilitate data interpretation and analysis and encourage its use for adaptive management (Han et al. 2014; Stephenson et al. 2015a; Stephenson 2019). The focus should be on ensuring simplicity and on open access to underlying data and methodologies to encourage transparency and easy replication.

The conservation community needs to learn and adapt and "document and share examples of monitoring, with case studies of what works well and less well" (Stephenson et al. 2015b). The case studies we present in this chapter should be built on and disseminated across Africa.

11.4.2 Building Capacity for Data Collection and Use

Capacity building in relevant national institutions is essential (Stephenson et al. 2015a, b). Workshops for CBD monitoring in eastern and southern Africa demonstrated that, in most countries, at least a few indicators of national relevance can be produced from existing data (Bubb et al. 2011). Work in Africa by the GBIF (https://www.gbif.org/) has underlined the importance of increased technical capacity and information resources to assist data mobilization. Some data sharing platforms and communities have been developed that provide access to African data, including the AfriBES social network of scientific and technical information for Africa (http://afriseb.net), the Africa Marine Atlas and African portal on the Ocean Biogeographic Information System (http://www.iobis.org/), and the ARCOS Biodiversity Information Management System for the Albertine Rift. The SANBI information system helped track biodiversity measures and stimulate biodiversity mainstreaming in South Africa (Huntley 2014); similar institutional structures may be useful elsewhere.

Capacity issues are often linked to resources, but data collection may not be overly expensive, especially if more stakeholders access shared and publicly available data sets. Initiating integrated biodiversity monitoring programs in sub-Saharan Africa could require as little as US$30,000–50,000 per country per year (Pereira et al. 2010; Wotton et al. 2020) and taxa for which monitoring capacity exists could be prioritized. As we saw in several case studies, NGO support may be needed to help build capacity in government authorities and local communities. Studies have shown that establishing monitoring systems in Africa requires more investment at the outset to support training and awareness creation and pay for equipment and materials (Bennun et al. 2005). Citizen science contributions may often be cheaper and offer opportunities to monitor less charismatic species (case study 8), but also require resources to support relevant associations, online toolkits, and network portals (Chandler et al. 2017). The global conservation community needs to find ways of sharing the costs of data acquisition to support African nations more directly and make data available more easily to governments who need them.

11.4.3 Working Together in Partnership to Improve the Science–Policy Interface

The gaps between the different data collectors and users will only be bridged through improved coordination and collaboration (Secades et al. 2014; Stephenson et al. 2017a; Vimal 2017). Local and international NGOs as well as academia have a significant role to play in supporting government agencies and it is encouraging to see that several global efforts to improve biodiversity monitoring explicitly target high biodiversity countries (Stephenson et al. 2017b; Stephenson 2018). Some of the large biodiversity databases could be useful tools for business throughout project planning and implementation (see Bennun et al. 2018), so businesses could in turn share data of use to resource-strapped governments.

Credible science-policy interfaces need to be created where scientists and decision makers from across sectors can put in place structures and incentives for interactive dialogue and work together to frame research and policy jointly (Young et al. 2014). Data collectors need to understand decision makers' needs and priorities and co-develop tools and information products that directly address those needs (Cowling et al. 2008).

There may not be one common solution across Africa. In some countries, certain government ministries may take the lead in convening and building structures for dialogue around data; in others, MEA secretariats or NGOs could facilitate national-level dialogues of science and policy actors from different sectors and help mobilize resources for their functioning (such as the Nairobi Convention and its science-policy platform for regional marine environmental issues). The Intergovernmental Science-Policy Platform on Biodiversity and Ecosystem Services (IPBES) could play a role, as could communities of practice, such as the NBSAP Forum (http://nbsapforum.net/), Biodiversity Observation Networks (Wetzel et al. 2015), and GBIF data holders' networks. Another approach is to organize meetings where scientists with data and data analysis capacity interact with government and NGO conservationists in regional planning processes, as was done for Guinean moist forest and the Congo Basin (da Fonseca et al. 2000; Brooks et al. 2001). The proposed IUCN Green List of Species (Akçakaya et al. 2018) will also provide opportunities to use biodiversity data in more positive ways, demonstrating how conservation action has helped avoid extinction and helping make a better case to decision makers for the impact of their policies and actions. In turn, policy makers should be encouraged to develop legislation only when supported by reliable data.

Stephenson et al. (2017a) recommend that African governments, NGOs, and academic bodies test different sorts of science–policy interfaces in a handful of pilot countries or regions to see what works best, building on existing methods and support systems (e.g., Dicks et al. 2014). The AfriBES network of scientific and technical information for Africa could play a brokering role as its aims revolve around information sharing and south–south collaboration.

11.5 Conclusions

Our case studies and literature review have demonstrated that, while many decision makers still struggle to access and use biodiversity data, especially those in under-resourced government departments or PAs, best practices exist, and many are already being put to use in Africa. Going forward, the SDGs should provide a stimulus for more governments to use data for monitoring across sectors and thereby encourage the necessary inter-disciplinary research and collaboration. There will be more data uptake if indicators are used that respond predictably to policy changes (Jones et al. 2011). In places where biodiversity goals and measures in Africa have been aligned to national development priorities, such as in Namibia (e.g., in communal conservancies) and in South Africa, they have received greater attention and uptake, with positive outcomes for biodiversity and development (Tallis et al. 2008; Brown et al. 2014).

Building on initial successes will require a concerted, collaborative effort. African governments will need to be open to collaboration with other states, with NGOs and with academia, within strong, open, and transparent partnerships and credible science-policy fora. Only by sharing and upscaling the solutions to data collection and use will we be able to improve the mainstreaming of biodiversity into decision-making and ultimately enhance sustainable development and stop biodiversity loss in Africa.

References

Addo KA, Adeyemi M (2013) Assessing the impact of sea-level rise on a vulnerable coastal community in Accra, Ghana. Jàmbá: J Disaster Risk Stud 5(1):1–8

Akçakaya HR, Bennett EL, Brooks TM et al (2018) Quantifying species recovery and conservation success to develop an IUCN Green List of species. Conserv Biol 32(5):1128–1138

Alamgir M, Campbell MJ, Sloan S et al (2018) Road risks and environmental impact assessments in Malaysian road infrastructure projects. Jurutera:13–16

Amano T, Lamming JDL, Sutherland WJ (2016) Spatial gaps in global biodiversity information and the role of citizen science. Bioscience 66(5):393–400

Balmford A (2003) Conservation planning in the real world: South Africa shows the way. Trends Ecol Evol 18(9):435–438

Barnard P, Altwegg R, Ebrahim I, Underhill LG (2017) Early warning systems for biodiversity in southern Africa – how much can citizen science mitigate imperfect data? Biol Conserv 208:183–188

Bennun L, Matiku P, Mulwa R et al (2005) Monitoring important bird areas in Africa: towards a sustainable and scaleable system. Biodivers Conserv 14(11):2575–2590

Bennun L, Regan EC, Bird J et al (2018) The value of the IUCN Red List for business decision-making. Conserv Lett 11(1):e12353
Beresford AE, Eshiamwata GW, Donald PF et al (2013) Protection reduces loss of natural land-cover at sites of conservation importance across Africa. PLoS One 8(5):e65370
Bernstein AS (2014) Biological diversity and public health. Annu Rev Publ Health 35:153–167
Blumstein DT, Mennill DJ, Clemins P et al (2011) Acoustic monitoring in terrestrial environments using microphone arrays: applications, technological considerations and prospectus. J Appl Ecol 48(3):758–767
Bonney R, Shirk JL, Phillips TB et al (2014) Next steps for citizen science. Science 343(6178):1436–1437
Borges PAV, Cardoso P, Kreft H et al (2018) Global Island monitoring scheme (GIMS): a proposal for the long-term coordinated survey and monitoring of native island forest biota. Biodivers Conserv 27:2567–2586
Braschler B, Mahood K, Karenyi N et al (2010) Realizing a synergy between research and education: how participation in ant monitoring helps raise biodiversity awareness in a resource-poor country. J Insect Conserv 14(1):19–30
Bronner G (2015) *Chrysospalax villosus*. The IUCN Red List of threatened species 2015: e. T4829A21290416. Downloaded on 10 May 2018
Brooks TM, Balmford A, Burgess N et al (2001) Toward a blueprint for conservation in Africa: a new database on the distribution of vertebrate species in a tropical continent allows new insights into priorities for conservation across Africa. Bioscience 51(8):613–624
Brooks TM, Butchart SHM, Cox NA et al (2015) Harnessing biodiversity and conservation knowledge products to track the Aichi targets and sustainable development goals. Biodiversity 16(2-3):157–174
Brown C, Reyers B, Ingwall-King L et al (2014) Measuring ecosystem services: guidance on developing ecosystem service indicators. UNEP-WCMC, Cambridge
Bubb P, Chenery A, Herkenrath P et al (2011) National indicators, monitoring and reporting for the strategic plan for biodiversity 2011–2020: a review of experience and recommendations in support of the CBD ad hoc technical expert group (AHTEG) on indicators for the strategic plan 2011–2020. UNEP-WCMC, Cambridge
Burgess ND, Butynski TM, Cordeiro NJ et al (2007) The biological importance of the eastern Arc Mountains of Tanzania and Kenya. Biol Conserv 134(2):209–231
Burn RW, Underwood FM, Blanc J (2011) Global trends and factors associated with the illegal killing of elephants: a hierarchical Bayesian analysis of carcass encounter data. PLoS One 6(9): e24165
Butchart SHM, Walpole M, Collen B et al (2010) Global biodiversity: indicators of recent declines. Science 328(5982):1164–1168
Cardinale BJ, Duffy JE, Gonzalez A et al (2012) Biodiversity loss and its impact on humanity. Nature 486(7401):59–67
Cardoso P, Erwin TL, Borges PAV et al (2011) The seven impediments in invertebrate conservation and how to overcome them. Biol Conserv 144(11):2647–2655
Carpenter KE, Abrar M, Aeby G et al (2008) One-third of reef-building corals face elevated extinction risk from climate change and local impacts. Science 321(5888):560–563
Carr CJ (2017) River basin development and human rights in Eastern Africa: a policy crossroads. Springer International, Cham
Chandler M, See L, Copas K et al (2017) Contribution of citizen science towards international biodiversity monitoring. Biol Conserv 213:280–294
Conover MR (2001) Resolving human-wildlife conflicts: the science of wildlife damage management. CRC Press, Boca Raton
Conservation Assured (2017) CA|TS Manual Version 1.4. March 2017, Conservation Assured, Singapore
Cowling RM, Pressey RL, Rouget M, Lombard AT (2003) A conservation plan for a global biodiversity hotspot—the Cape Floristic Region, South Africa. Biol Conserv 112:191–216

Cowling RM, Egoh B, Knight AT et al (2008) An operational model for mainstreaming ecosystem services for implementation. PNAS 105(28):9483–9488
Craigie ID, Baillie JEM, Balmford A et al (2010) Large mammal population declines in Africa's protected areas. Biol Conserv 143:2221–2228
Crees JJ, Collins AC, Stephenson PJ et al (2016) A comparative approach to assess drivers of success in mammalian conservation recovery programs. Conserv Biol 30:694–705
Cresswell W (2018) The continuing lack of ornithological research capacity in almost all of West Africa. Ostrich 89(2):1–7
Crouch NR, Smith GF (2011) Informing and influencing the interface between biodiversity science and biodiversity policy in South Africa. Bot J Linn Soc 166(3):301–309
da Fonseca GA, Balmford A, Bibby C et al (2000) . . . following Africa's lead in setting priorities. Nature 405(6785):393–394
Danielsen F, Jensen PM, Burgess ND et al (2014) A multicountry assessment of tropical resource monitoring by local communities. Bioscience 64(3):236–251
Daskin JH, Pringle RM (2018) Warfare and wildlife declines in Africa's protected areas. Nature 553(7688):328–332
de Wasseige C, Flynn J, Louppe D et al (2014) The forests of the Congo Basin – state of the Forest 2013. Publications Office of the European Union, Luxembourg
Desmet PG, Cowling RM, Ellis AG, Pressey RL (2002) Integrating biosystematic data into conservation planning: perspectives from southern Africa's succulent Karoo. Syst Biol 51(2):317–330
Dickens CWS, Graham PM (2002) The south African scoring system (SASS) version 5 rapid bioassessment method for rivers. Afr J Aquat Sci 27(1):1–10
Dicks LV, Walsh JC, Sutherland WJ (2014) Organising evidence for environmental management decisions: a '4S' hierarchy. Trends Ecol Evol 29(11):607–613
Diedericks G, Simaika J, Roux F (2013) A Survey of Adult Odonata Along the Crocodile-Inkomati River Main Stem from Source to Ocean. Report to the Mpumalanga parks and tourism Agency, 56.
Dijkstra KD, Clausnitzer V (2014) The dragonflies and damselflies of eastern Africa: handbook for all Odonata from Sudan to Zimbabwe. Royal Museum for Central Africa
DiMarco M, Chapman S, Althor G et al (2017) Changing trends and persisting biases in three decades of conservation science. Global Ecol Conserv 10:32–42
Doumenge C, Palla F, Scholte P et al (2015) Aires protégées d'Afrique Centrale - Etat 2015. OFAC, Kinshasa
Entwistle AC, Stephenson PJ (2000) Small mammals and the conservation agenda. In: Entwistle AC, Dunstone N (eds) Priorities for the conservation of mammalian biodiversity: has the panda had its day? Cambridge University Press, Cambridge, pp 119–139
Ervin J (2003) WWF: rapid assessment and prioritization of protected area management (RAPPAM) methodology. WWF, Gland
Everett MJ, Hepburn IR, Ntiamoa-Baidu Y, Thomson GJ (1987) Roseate terns in Britain and West Africa. RSPB Conserv Rev 1:56–58
FitzGibbon C, Rathbun GB (2015) *Rhynchocyon chrysopygus*. The IUCN Red List of threatened species 2015: e.T19705A21287265
Gichohi N, Warinwa F, Lenaiyasa P et al (2014) Long-term monitoring of livestock depredation in Amboseli ecosystem, Kenya. J Agric Nat Resour Sci 1(3):186–194
Gond V, Fayolle A, Pennec A et al (2013) Vegetation structure and greenness in Central Africa from Modis multi-temporal data. Philos Trans R Soc Lond Ser B Biol Sci 368:20120309
Grainger J (1994) A protected area system plan to conserve biodiversity in Ghana. Ghana Wildlife Department, Accra and IUCN, Gland
Gudka M, Obura D, Mbugua J, Ahamada S, Kloiber U, Holter T (2020) Participatory reporting of the 2016 bleaching event in the Western Indian Ocean. Coral Reefs 39(1):1–11

Han X, Smyth RL, Young BE et al (2014) A biodiversity indicators dashboard: addressing challenges to monitoring progress towards the Aichi biodiversity targets using disaggregated global data. PLoS One 9(11):e112046
Harabiš F, Tichanek F, Tropek R (2013) Dragonflies of freshwater pools in lignite spoil heaps: restoration management, habitat structure and conservation value. Ecol Eng 55:51–61
Holmes PM, Rebelo AG, Dorse C, Wood J (2012) Can Cape Town's unique biodiversity be saved? Balancing conservation imperatives and development needs. Ecol Soc 17(2):28
Huntley BJ (2014) Good news from the south: biodiversity mainstreaming – a paradigm shift in conservation? S Afr J Sci 110(9/10):01-04
IIED, UNEP-WCMC (2017) Mainstreaming biodiversity and development: guidance from African experience 2012–17. IIED, London
IPBES (2016) Summary for policymakers of the assessment report of the intergovernmental science-policy platform on biodiversity and ecosystem services on pollinators, pollination and food production. Secretariat of the Intergovernmental Science-Policy Platform on Biodiversity and Ecosystem Services, Bonn
IUCN (2020) The IUCN red list of threatened species. Version 2019-3. Downloaded on 12 March 2020
Jones JP, Collen B, Atkinson G et al (2011) The why, what, and how of global biodiversity indicators beyond the 2010 target. Conserv Biol 25(3):450–457
Keith DA, Rodríguez JP, Brooks TM et al (2015) The IUCN Red List of ecosystems: motivations, challenges, and applications. Conserv Lett 8(3):214–226
Kennerley RJ, Lacher TE, Mason VC et al (2018) Conservation priorities and actions for the Orders Cingulata, Pilosa, Afrosoricida, Macroscelidea, Eulipotyphla, Dermoptera and Scandentia. In: Wilson DE, Mittermeier PA (eds) Handbook of the mammals of the world – vol 8. Insectivores, Sloths and Colugos. Lynx Edicions, Barcelona
Knights K, Cuadros I, Zamora C et al (2014) A preliminary assessment of protected area management within the WWF 'Coastal East Africa' priority place, eastern Africa. Parks 20(2):77–88
Koranteng K, Ofori-Danson P, Entsua-Mensah M (2000) Fish and fisheries of the Muni lagoon in Ghana, West Africa. Biodivers Conserv 9(4):487–499
Larson LR, Conway AL, Hernandez SM, Carroll JP (2016) Human-wildlife conflict, conservation attitudes, and a potential role for citizen science in Sierra Leone, Africa. Conserv Soc 14(3):205–217
Laurance WF, Goosem M, Laurance SGW (2009) Impacts of roads and linear clearings on tropical forests. Trends Ecol Evol 24(12):659–669
Laurance WF, Sloan S, Weng L et al (2015) Estimating the environmental costs of Africa's massive "development corridors". Curr Biol 25(24):3202–3208
Litoro M, Omondi P, Kock R, Amin R (2012) Conservation and management strategy for the elephant in Kenya, 2012–2021. Kenya Wildlife Service, Nairobi
Lovett JC, Wasser SK (eds) (1993) Biogeography and ecology of the rain forests of Eastern Africa. Cambridge University Press, Cambridge
Mapfumo P, Mtambanengwe F, Chikowo R (2016) Building on indigenous knowledge to strengthen the capacity of smallholder farming communities to adapt to climate change and variability in southern Africa. Clim Dev 8(1):72–82
Margules CR, Pressey RL (2000) Systematic conservation planning. Nature 405(6783):243–253
Martin LJ, Blossey B, Ellis E (2012) Mapping where ecologists work: biases in the global distribution of terrestrial ecological observations. Front Ecol Environ 10(4):195–201
Ministério do Ambiente (2014) Fifth National Report on biodiversity in Angola 2007–2012. Ministério do Ambiente, Angola
Morin-Rivat J, Fayolle A, Favier C et al (2017) Present-day central African forest is a legacy of the 19th century human history. elife 6:e20343
National Environment Management Authority (2015) National Biodiversity Strategy and Action Plan (2015–2025). National Environment Management Authority, Republic of Uganda

Ntiamoa-Baidu Y (1991) Conservation of coastal lagoons in Ghana: the traditional approach. Landscape Urban Plan 20(1–3):41–46

Ntiamoa-Baidu Y, Gordon C (1991) Coastal wetlands management plans: Ghana. Report prepared for the World Bank and Environmental Protection Council, Ghana, 131 pp

Ntiamoa-Baidu Y, Hepburn I (1988) Wintering waders in Coastal Ghana. RSPB Conserv Rev 2:85–88

Obura DO (2014) Coral reef monitoring manual South-West Indian Ocean Islands. Indian Ocean Commission, Ebène

Obura DO, Gudka M, Abdou Rabi F et al (2017) Coral reef status report for the Western Indian Ocean. Global Coral Reef Monitoring Network (GCRMN)/International Coral Reef Initiative (ICRI), p 144

Obura DO, Aeby G, Amornthammarong N et al (2019) Coral reef monitoring, reef assessment technologies, and ecosystem-based management. Front Mar Sci 6:580

Okello MM, Kiringe JW, Warinwa F (2014) Human-carnivore conflicts in private conservancy lands of Elerai and Oltiyiani in Amboseli Area, Kenya. Nat Resour 5(8):375–391

Osei J, Nyame FK, Armah TK et al (2010) Application of multivariate analysis for identification of pollution sources in the Densu Delta wetland in the vicinity of a landfill site in Ghana. J Water Resour Protect 2:1020–1029

Pereira HM, Belnap J, Brummitt N et al (2010) Global biodiversity monitoring. Front Ecol Environ 8(9):459–460

Phillips TB, Parker A, Bowser A, Hakley M (2021) Publicly generated data: the role of citizen-science for knowledge production, action, and public engagement. In: Ferreira CC, Klütsch CFC (eds) Closing the knowledge-implementation gap in conservation science – evidence transfer across spatiotemporal scales and different stakeholders. Springer, New York, pp 83–107

Pocock MJ, Roy HE, August T et al (2019) Developing the global potential of citizen science: assessing opportunities that benefit people, society and the environment in East Africa. J Appl Ecol 56(2):274–281

Réjou-Méchain M, Fayolle A, Nasi R et al (2011) Detecting large-scale diversity patterns in tropical trees: can we trust commercial forest inventories? For Ecol Manag 261(2):187–194

Réjou-Méchain M, Flores O, Pélissier R et al (2014) Tropical tree assembly depends on the interactions between successional and soil filtering processes. Glob Ecol Biogeogr 23(12):1440–1449

Reyers B, Fairbanks DHK, van Jaarsveld AS, Thompson M (2001) Priority areas for the conservation of South African vegetation: a course-filter approach. Divers Distrib 7(1-2):79–95

Rodgers WA, Homewood KM (1982) Species richness and endemism in the Usambara Mountains forests, Tanzania. Biol J Linn Soc 18(3):197–242

Roe D, Mapendembe A (2013) Biodiversity and development mainstreaming: a state of knowledge review: discussion paper. International Institute for Environment and Development (IIED) and the United Nations Environment Programme (UNEP), London

Rose DC, Despot-Belmonte K, Pollard JA, Shears O, Robertson RJ (2021) Making an impact: how to design relevant and usable decision support systems for conservation. In: Ferreira CC, Klütsch CFC (eds) Closing the knowledge-implementation gap in conservation science – evidence transfer across spatiotemporal scales and different stakeholders. Springer, New York, pp 199–226

Rovero F, Menegon M, FjeldsAa J et al (2014) Targeted vertebrate surveys enhance the faunal importance and improve explanatory models within the Eastern Arc Mountains of Kenya and Tanzania. Divers Distrib 20(12):1438–1449

Rovero F, Zimmermann F (2016) Camera trapping for wildlife research. Pelagic Publishing, Exeter

Roy DP, Ju J, Mbow C et al (2010) Accessing free Landsat data via the Internet: Africa's challenge. Remote Sens Lett 1(2):111–117

Samways MJ (2015) Future-proofing insect diversity. Curr Opin Insect Sci 12:71–78
Samways MJ, Simaika JP (2016) Manual of freshwater assessment for South Africa: dragonfly biotic index. South African National Biodiversity Institute (SANBI), Pretoria
Samways MJ, Sharratt NJ, Simaika JP (2011) Effect of alien riparian vegetation and its removal on a highly endemic river macroinvertebrate community. Biol Invasions 13(6):1305–1324
Schaafsma M, Morse-Jones S, Posen P et al (2014) The importance of local forest benefits: economic valuation of non-timber forest products in the Eastern Arc Mountains in Tanzania. Global Environ Chang 24:295–305
Schowalter TD, Noriega JA, Tscharntke T (2018) Insect effects on ecosystem services. Basic Appl Ecol 26:1–7
Secades C, O'Connor B, Brown C, Walpole M (2014) Earth observation for biodiversity monitoring: a review of current approaches and future opportunities for tracking progress towards the Aichi biodiversity targets. CBD technical series no. 72. Secretariat of the Convention on Biological Diversity, Montreal
Secretariat of the Convention on Biological Diversity (2014) Global biodiversity outlook 4. Montréal
Segan DB, Bottrill MC, Baxter PWJ, Possingham HP (2011) Using conservation evidence to guide management. Conserv Biol 25(1):200–202
Seto KC, Güneralp B, Hutyra LR (2012) Global forecasts of urban expansion to 2030 and direct impacts on biodiversity and carbon pools. Proc Natl Acad Sci 109(40):16083–16088
Simaika JP, Samways MJ (2008) An easy-to-use index of ecological integrity for prioritizing freshwater sites and for assessing habitat quality. Biodivers Conserv 18(5):1171–1185
Simaika JP, Samways MJ (2011) Comparative assessment of indices of freshwater habitat conditions using different invertebrate taxon sets. Ecol Indic 11(2):370–378
Sitati NW, Ipara H (2012) Indigenous ecological knowledge of a human-elephant interaction in Transmara District, Kenya: implications for research and management. Adv Anthropol 2(3):107–111
Sitati NW, Walpole MJ (2006) Assessing farm-based measures for mitigating human-elephant conflict in Transmara District, Kenya. Oryx 40(3):279–286
Sitati NW, Walpole MJ, Leader-Williams N (2005) Factors affecting susceptibility of farms to crop raiding by African elephants: using a predictive model to mitigate conflict. J Appl Ecol 42(6):1175–1182
Spiegel SJ (2017) EIAs, power and political ecology: situating resource struggles and the technopolitics of small-scale mining. Geoforum 87:95–107
Stephenson PJ (2017) Small mammal monitoring: why we know so little about the Afrotheria. Afrotherian Conserv 13:34–42
Stephenson (2018) A global effort to improve species monitoring for conservation. Oryx 52(3):412–413
Stephenson PJ (2019) Integrating remote sensing into wildlife monitoring for conservation. Environ Conserv 46(3):181–183
Stephenson PJ, Malima C, Tchamba M et al (2007) Human-elephant conflict: WWF case studies from Cameroon and Tanzania. In: Walpole M, Linkie M (eds) Mitigating human-elephant conflict: case studies from Africa and Asia. Fauna & Flora International, Cambridge, pp 63–72
Stephenson PJ, Burgess ND, Jungmann L et al (2015a) Overcoming the challenges to conservation monitoring: integrating data from in situ reporting and global data sets to measure impact and performance. Biodiversity 16(2–3):68–85
Stephenson PJ, O'Connor S, Reidhead W, Loh J (2015b) Using biodiversity indicators for conservation. Oryx 49(3):396
Stephenson PJ, Soarimalala V, Goodman SM (2016a) *Limnogale mergulus*. The IUCN Red List of threatened species 2016: e.T11979A97189690. Downloaded on 01 May 2018
Stephenson PJ, Soarimalala V, Goodman SM (2016b) Microgale jenkinsae. The IUCN Red List of threatened species 2016: e.T62015A97192062. Downloaded on 01 May 2018

Stephenson PJ, Bowles-Newark N, Regan E et al (2017a) Unblocking the flow of biodiversity data for decision-making in Africa. Biol Conserv 213:335–340

Stephenson PJ, Brooks TM, Butchart SHM et al (2017b) Priorities for big biodiversity data. Front Ecol Environ 15(3):124–125

Stephenson PJ, Monadjem A, Decher J (2018) Uplisting a threatened small mammal: the Nimba otter-shrew of West Africa. Oryx 52(4):609

Stephenson PJ, Stengel C (2020) An inventory of biodiversity data sources for conservation monitoring. PLoS One 15(12):e0242923

Stropp J, Ladle RJ, Malhadu ACM et al (2016) Mapping ignorance: 300 years of collecting flowering plants in Africa. Glob Ecol Biogeogr 25(9):1085–1096

Stuart-Hill G, Diggle R, Munali B et al (2005) The event book system: a community-based natural resource monitoring system from Namibia. Biodivers Conserv 14(11):2611–2631

Swanson A, Kosmala M, Lintott C et al (2015) Snapshot Serengeti: high-frequency annotated camera trap images of 40 mammalian species in an African savannah. Scientific Data 2(1):150026

Tallis H, Kareiva P, Marvier M, Chang A (2008) An ecosystem services framework to support both practical conservation and economic development. PNAS 105(28):9457–9464

Tarboton W, Tarboton M (2015) Guide to the dragonflies & damselflies of South Africa. Random House, Struik

Thapa I, Butchart SHM, Gurung H et al (2016) Using information on ecosystem services in Nepal to inform biodiversity conservation and local to national decision-making. Oryx 50(1):147–155

Troudet J, Grandcolas P, Blin A et al (2017) Taxonomic bias in biodiversity data and societal preferences. Sci Rep 7(1):9132

Underwood FM, Burn RW, Milliken T (2013) Dissecting the illegal ivory trade: an analysis of ivory seizures data. PLoS One 8(10):e76539

Valentini A, Taberlet P, Miaud C et al (2016) Next-generation monitoring of aquatic biodiversity using environmental DNA metabarcoding. Mol Ecol 25(4):929–942

Van Swaay C, Regan E, Ling M et al (2015) Guidelines for standardised global butterfly monitoring. GEO BON technical series 1. Group on Earth Observations Biodiversity Observation Network, Leipzig

Velasco D, García-Llorente M, Alonso B et al (2015) Biodiversity conservation research challenges in the 21st century: a review of publishing trends in 2000 and 2011. Environ Sci Pol 54:90–96

Vimal R (2017) Monitoring for conservation in African tropical national parks: an agenda towards policy-relevant science. Biol Conserv 214:127–135

Waeber PO, Wilmé L, Mercier JR et al (2016) How effective have thirty years of internationally-driven conservation and development efforts been in Madagascar? PLoS One 11(8):e0161115

Wakoli EN, Sitati N (2012) Analysis of temporal and spatial patterns of elephant attacks on humans and elephant mortality in Transmara District, Kenya. Greener J Environ Manage Pub Safety 1(1):27–37

Walpole M, Sitati N, Stewart-Cox B et al (2006) Mitigating human-elephant conflict in Africa: a lesson-learning and network development meeting. Pachyderm 41:95–99

Webb TJ, Mindel BL (2015) Global patterns of extinction risk in marine and non-marine systems. Curr Biol 25(4):506–511

Weru S (2016) Wildlife protection and trafficking assessment in Kenya: drivers and trends of transnational wildlife crime in Kenya and its role as a transit point for trafficked species in East Africa. TRAFFIC, Cambridge

Wetzel FT, Saarenmaa H, Regan E et al (2015) The roles and contributions of biodiversity observation networks (BONs) in better tracking progress to 2020 biodiversity targets: a European case study. Biodiversity 16(2-3):137–149

Wich SA, Koh LP (2018) Conservation drones: mapping and monitoring biodiversity. Oxford University Press, Oxford

Wilhelm-Rechmann A, Cowling RM (2011) Framing biodiversity conservation for decision makers: insights from four South African municipalities. Conserv Lett 4(1):73–80

Wongbusarakum S, Heenan A (2019) Integrated Monitoring with SocMon/SEM-Pasifika: Principles and Process: Global Coral Reef Monitoring Network (GCRMN) Socio-economic Monitoring for Coastal Management (SocMon) Methodological Updates. GCRMN SocMon, NOAA Coral Reef Conservation Program, Silver Springs

Wotton SR, Eaton MA, Sheehan D et al (2020) Developing biodiversity indicators for African birds. Oryx 54(1):62–73

WWF (ed) (2007) Management effectiveness tracking tool: reporting progress at protected area sites, 2nd edn. WWF International, Gland

WWF (2014) Living planet report 2014. WWF International, Gland

Yevide AS, Wu B, Khan AS et al (2016) Bibliometric analysis of ecosystem monitoring-related research in Africa: implications for ecological stewardship and scientific collaboration. Int J Sust Dev World 23(5):412–422

Young JC, Waylen KA, Sarkki S et al (2014) Improving the science-policy dialogue to meet the challenge of biodiversity conservation: having conversations rather than talking at one-another. Biodivers Conserv 23(2):387–404

Zarfl C, Lumsdon AE, Berlekamp J et al (2015) A global boom in hydropower dam construction. Aquat Sci 77(1):161–170

Chapter 12
The Marine Conservation Landscape in Europe: Knowledge Support to Policy Implementation and Conservation Action

Rita Araújo, Ana Cristina Cardoso, Victoria Tornero, and Nikos Zampoukas

Contents

R. Araújo (✉) · A. C. Cardoso · V. Tornero
European Commission, Joint Research Centre (JRC), Ispra, Italy
e-mail: rita.araujo@ec.europa.eu; ana-cristina.cardoso@ec.europa.eu; victoria.tornero@ec.europa.eu

N. Zampoukas
European Commission, DG-RTD, Brussels, Belgium
e-mail: nikos.zampoukas@ec.europa.eu

C. C. Ferreira, C. F. C. Klütsch (eds.), *Closing the Knowledge-Implementation Gap in Conservation Science*, Wildlife Research Monographs 4,
https://doi.org/10.1007/978-3-030-81085-6_12

Acronyms

AFS Convention	International Convention on the Control of Harmful Anti-fouling Systems on Ships
AMAP	Arctic Monitoring and Assessment Programme
ATEMP	AMAP's Trends and Effects Monitoring Programme
Baltic Sea Pharma	Platform to reduce pharmaceuticals in the Baltic environment
Black Sea Commission	Convention for the Protection of the Black Sea against pollution
BSAP	Black Sea Strategic Action Plan
CG PHARMA	HELCOM Correspondence Group on Pharmaceuticals
CLRTAP	Geneva Convention on long-range transboundary air pollution
COR GEST	Correspondence Group on GES and Targets in the Mediterranean
COR MON	Correspondence Group on Monitoring in the Mediterranean
DAIMON	Decision Aid for Marine Munitions
EcAp	Ecosystem approach
ECHA	European Chemicals Agency
ECOSTAT	WFD Working Group Ecological Status
EEA	European Environment Agency
EFSA	European Food Safety Agency
Eionet	European Environmental Information and Observation Network
EMBLAS-Plus	Improving Environmental Monitoring in the Black Sea—Special Measures
EMODnet	European Marine Observation and Data Network
EMSA	European Maritime Safety Agency
EN-HZ	HELCOM expert network on hazardous substances
ESAS	Advisory Group on the Environmental Safety Aspects of Shipping in the Black Sea
EUSBSR	EU Strategy for the Baltic Sea Region
EWG OWR	HELCOM Expert Working Group on Oiled Wildlife Response
GES	Good Environmental Status

GESAMP	Joint Group of Experts on the Scientific Aspects of Marine Environmental Protection
EHS	GESAMP Working Group on environmental hazards of harmful substances carried by ships
WG 42	GESAMP Working Group on impacts of wastes and other matter in the marine environment from mining operations
HELCOM	Convention on the Protection of the Marine Environment in the Baltic Sea Area
ICES	International Council for the Exploration of the Sea
ICG 4PE	OSPAR RSC's subsidiary Intersessional Correspondence Group Delivering the Fourth Periodic Evaluation
ICG CTZ	OSPAR RSC's subsidiary Intersessional Correspondence Group Close to Zero
ICG EAC	OSPAR RSC's subsidiary Intersessional Correspondence Group Environmental Assessment Criteria
ICG MOD	OSPAR RSC's subsidiary Intersessional Correspondence Group MODelling
IMAP	Integrated Monitoring and Assessment Programme for the Mediterranean
IMO	International Maritime Organization
INPUT	OSPAR's working group on Inputs to the Marine Environment
JPI Oceans	Joint Programming Initiative Healthy and Productive Seas and Oceans
LBS	Advisory Group on Control of Pollution from Land-Based Sources in the Black Sea
London Convention	Convention on the Prevention of Marine Pollution by Dumping of Wastes and Other Matter
MAP	Mediterranean Action Plan
MARPOL	International Convention for the prevention of Pollution from Ships
MCWG	ICES's Marine Chemistry Working Group
MED POL	Mediterranean Pollution Assessment and Control Programme
MEPC	IMO's Marine Environment Protection Committee

MIME	OSPAR's working group on Monitoring and on Trends and Effects of Substances in the Marine Environment
MORS EG	HELCOM expert group on monitoring of radioactive substances
MSFD	Marine Strategy Framework Directive
NORMAN	Network of reference laboratories, research centres and related organisations for monitoring of emerging environmental substances
OIC	OSPAR's Offshore Industry Committee
OSPAR	Convention for the Protection of the Marine Environment in the North-East Atlantic
HASEC	OSPAR's Hazardous Substances and Eutrophication Committee
PA Hazards	Policy Area Hazards
PMA	Advisory Group on the Pollution Monitoring and Assessment in the Black Sea
PPR	IMO's Sub-Committee on Pollution Prevention and Response
PRESSURE	HELCOM working group on reduction of pressures from the Baltic Sea catchment area
REACH	Registration, Evaluation, Authorisation, and Restriction of Chemical substances
REMPEC	Regional Marine Pollution Emergency Response Centre for the Mediterranean Sea
RESPONSE	HELCOM working group Response
RSC	OSPAR's Radioactive Substances Committee
SUBMERGED	HELCOM Expert Group on Environmental Risks of Hazardous Submerged Objects
Barcelona Convention	Convention for the Protection of Marine Environment and the Coastal Region of the Mediterranean
WFD	Water Framework Directive
WG Chemicals	Working Group Chemicals
WG MARITIME	HELCOM working group Maritime
WG STATE and CONSERVATION	HELCOM working group on state of the environment and nature conservation
WGBEC	ICES's Working Group on Biological Effects of Contaminants

WGMS	ICES's Working Group on Marine Sediments in Relation to Pollution
WISE Marine	Marine Information System for Europe

12.1 Biodiversity in Europe: Relevance, Instruments for Governance and Knowledge Base

Biodiversity loss impacts ecosystem functions and services at different levels and has implications for human life in terms of food provision, regulation of nature service uses, social and economic interactions and recreation (TEEB 2010). Conservation efforts in Europe, similarly to other regions around the globe, have been focused on setting the instruments and priorities for:

1. The management of habitat degradation and species protection.
2. Sustainable exploitation of natural resources.
3. Control of alien species introduction and pollution impacts.
4. Monitoring and mitigation of climate change impacts.

Europe holds a high diversity of wild animals and plants, some of them endemic. These species are protected under several policy instruments, including the Habitats Directive (92/43/ECC). The latter established the EU Natura 2000 network that represents the largest coordinated network of protected sites in the world (Maes et al. 2012), covering over 18% of land and 6% of the EU countries' surface area, respectively.

The most important environmental framework at the European level is the EU Biodiversity Strategy that aims at halting or significantly reducing biodiversity loss and degradation of ecosystem services in the EU by 2020, also contributing to diminishing global biodiversity loss. This Strategy is mainly based on two legal pillars: the Habitats Directive and the Birds Directive (79/409/EEC)—collectively called the Nature Directives—and establishes six main targets, each supported by a set of actions, to achieve important European conservation objectives by 2020:

1. Fully implement the Birds and Habitats Directives.
2. Maintain and restore ecosystems and their services.
3. Increase the contribution of agriculture and forestry to maintain and enhance biodiversity.
4. Ensure the sustainable use of fisheries resources.
5. Control invasive alien species (IAS).
6. Contribute to avert global biodiversity loss.

However, the mid-term review of the EU 2020 Biodiversity Strategy (EU COM (2015) p. 478) showed that, in spite of noticeable progress in biodiversity conservation at the EU level, biodiversity loss is continuing (mostly caused by habitat degradation), highlighting the need for additional and substantial measures to revert this trend. This applies also to marine species and ecosystems that continue declining across Europe's regional seas. Many of these measures need to be based on cross-sectorial, articulated efforts from concerned stakeholders and based on the best available scientific evidence, fitted to the implementation needs. The EU Action Plan for nature, people, and the economy that followed this mid-term review mandated an improvement of the implementation of the Nature Directives to boost their contribution towards reaching the EU's biodiversity targets for 2020. The Action Plan focuses on four priority areas and comprises 15 actions to be carried out by 2020.

Marine resources are considered an important source of livelihood and economic income, providing different ecosystem services (e.g., bioremediation, food, and recreation) and contributing significantly to the global primary production (Charrier et al. 2017). The European seas cover 5.7 million km^2 and include several regional seas (Fig. 12.1) with 82% of the EU Member states having a coastline and with the maritime area under EU jurisdiction being larger than the total area of the EU. Almost half of the population in Europe is concentrated in the maritime areas which account for almost half of its Gross Domestic Product (GDP). The European Marine Strategy Framework Directive (EMSFD, 2008/56/EC, https://www.eea.europa.eu/data-and-maps/data/europe-seas#tab-gis-data) is the overarching framework for community action in the field of marine environmental policy in Europe.

The governance and management framework for European marine waters has a central role in achieving a good conservation status and reversing the decreasing trend in marine biodiversity. Beyond the MSFD, marine governance in Europe involves different legal instruments, organizations and strategies. Some are exemplified below.

Marine Protected Areas (MPAs) Protected areas are valuable tools to reduce the pressures on biodiversity by enabling long-term protection and recovery of target organisms, habitats, and ecosystem services. The expansion of the marine protected area network has been partly driven by the need to meet European and international marine conservation targets and sustainable socioeconomic growth based on marine activities. Subsequently, the number and total area of MPAs has increased rapidly in recent years with most MPAs concentrated in intertidal and coastal waters. In 2012, 5.9% of the EU waters (within 200 nautical miles) were covered by MPAs, corresponding to 7725 sites and a total area of 338,623 km^2 (EEA report 2015). For example, in the Mediterranean, MPAs almost doubled both in number and in area between 2008 and 2012 (Gabrié et al. 2012) and altogether Portugal, Spain, and France have 134 MPAs covering 227.2 km^2 (Batista and Cabral 2016). However, in Europe the coverage of MPAs still needs to be extended and the existing management plans for these areas improved (Batista and Cabral 2016).

Regional Sea Conventions Regional Sea Conventions (RSC) have an important role in European regional seas governance, representing coordinated regional cooperation structures aimed at protecting the coastal and marine environment. There are four European Regional Sea Conventions as follows:

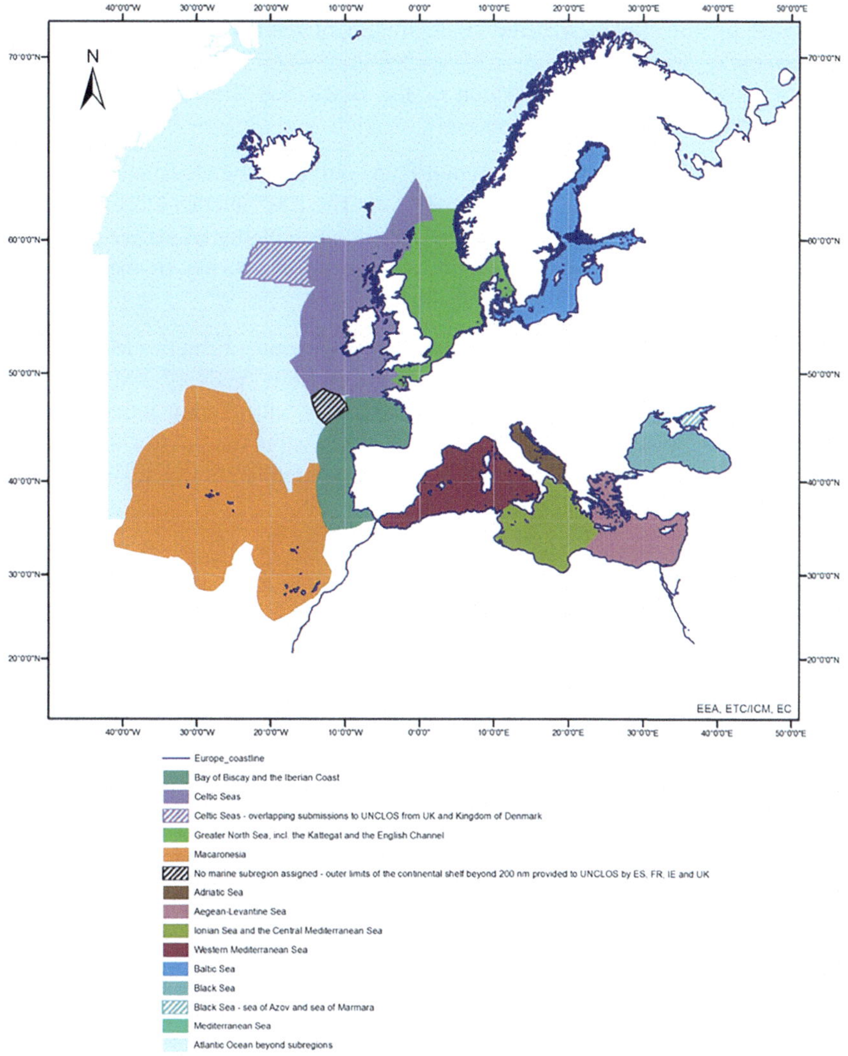

Fig. 12.1 Marine regions and sub-regions considered under the European Marine Strategy Framework Directive (the overarching framework for community action in the field of marine environmental policy in Europe) (2008/56/EC, https://www.eea.europa.eu/data-and-maps/data/europe-seas#tab-gis-data)

- The Convention for the Protection of the Marine Environment in the North-East Atlantic (OSPAR: https://www.ospar.org/convention).
- The Convention on the Protection of the Marine Environment in the Baltic Sea (HELCOM: http://www.helcom.fi/about-us/convention).

- The Convention for the Protection of the Marine Environment and the Coastal Region of the Mediterranean (Barcelona Convention: http://web.unep.org/unepmap/).
- The Convention for the Protection of the Black Sea (Bucharest Convention: http://www.blacksea-commission.org/).

The RSC periodically agree on Action Plans outlining the challenges, activities and goals to be developed at the regional level. The RSC interact closely with the European Commission developing the tools for implementation of marine policies at the regional level and providing support to contracting parties in meeting their national obligations towards the marine environment.

Regional Fisheries Management Organizations Regional Fisheries Management Organizations (RFMOs) are international organizations advising and managing fisheries in a specific area and formed by countries with fishing interests in the respective area. Some of them only deal with tunas while others deal with all fish stocks in the area. Some RFMOs only focus on international waters. Examples of RFMOs with activities in European seas are:

- *The International Commission for the Conservation of Atlantic Tunas* (ICCAT: https://www.iccat.int/en/): ICCAT compiles fisheries statistics for tuna and tuna-like species in the Atlantic ocean and its adjacent seas.
- *The North-East Atlantic Fisheries Commission* (NEAFC: https://www.neafc.org/about): NEAFC manages the fisheries resources in the North-East Atlantic with Denmark (Faroe Islands and Greenland), EU, Iceland, Norway, and the Russian Federation as contracting parties.
- *North Atlantic Salmon Conservation Organization* (NASCO: http://www.nasco.int/about.html): NASCO takes action in the conservation and management of the Atlantic salmon fisheries.
- *General Fisheries Commission for the Mediterranean* (GFCM: http://www.fao.org/gfcm/background/about/en/): GFCM works on the conservation of living marine resources as well as the sustainable development of aquaculture in the Mediterranean and in the Black Sea.

12.1.1 European Policy Initiatives and Legal Framework Related to the Conservation of the Marine Environment

There are several policy initiatives related to the EU sectorial priorities that have implications for the conservation of the marine environment. The economic sector connected to marine bio-based products and biotechnology is receiving increasing attention and support. Several initiatives (Bioeconomy Strategy, Blue Growth Strategy, and Circular Economy Action Plan) aim to design the frameworks for the sustainable development of these activities because of their potential to impact the

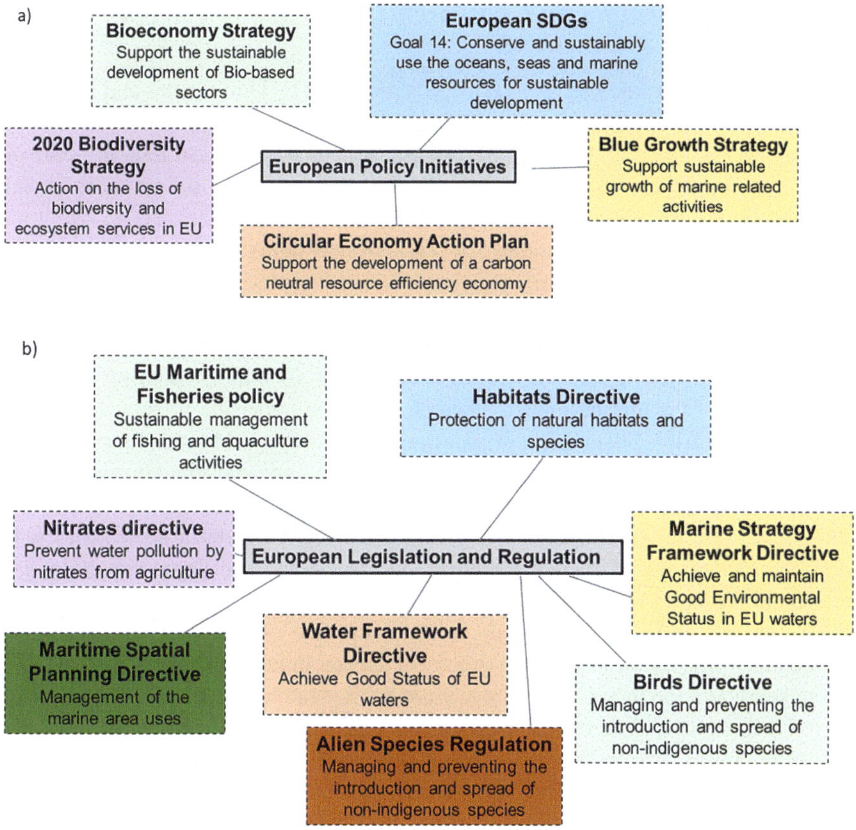

Fig. 12.2 European policy initiatives (**a**) and EU pieces of legislation (**b**) relevant for the governance of the marine environment and their relations to marine conservation in Europe

conservation of marine ecosystems (Fig. 12.2a). These initiatives include also the protection measures to be developed under the Biodiversity Strategy. Additionally, the EU has committed to incorporate the United Nations adopted 17 Sustainable Development Goals (SDGs). These include Goal 14 (Life Below Water) on the conservation and sustainable use of the oceans, seas, and marine resources for sustainable development, addressing marine pollution and the sustainable management of fisheries and aquaculture.

Several EU pieces of legislation also address marine conservation-related topics like the protection of species and habitats (e.g., Habitats Directive, Birds Directive, Regulation (EU) 1143/2014 on invasive alien species), the maintenance of the environmental quality (e.g., Water Framework Directive (2000/60/EC), Marine Strategy Framework Directive, Nitrates Directive (91/676/EEC)), the sustainable management of marine-related economic activities (e.g., EU Common Fisheries Policy (Regulation (EU) No 1380/2013), and the Maritime Spatial Planning Directive (2014/89/EU)) (Fig. 12.2b).

12.1.2 The Scientific and Research Landscape in Europe

Most of the scientific knowledge in Europe is produced by Universities and Research Institutes dedicated both to fundamental and to applied research (European Commission 2017; Powell and Dusdal 2017). Several of these institutions are part of thematic networks and hubs that translate research into innovation, promote global data sharing or are organized in research infrastructure consortia.

The EU framework programmes for research and innovation (https://www.eda.europa.eu/procurement-biz/information/codeda-regulationaba/eu-framework-programme-for-research-and-innovation) span over a 7-year timeframe and fund projects in EU and other countries. The current framework programme, Horizon H2020, started in 2014 and has a budget of around 70 billion euros for different actions. Most of the marine-related research is funded under the societal Challenge 2 on Food Security, Sustainable Agriculture and Forestry, Marine, Maritime and Inland Water Research and the Bioeconomy. Under societal challenges, the EU funds multi-partner collaborative projects bringing together at least three independent entities and the participation of several dozens of partners in these projects is common. Demonstrated impact, together with excellence and implementation, are the criteria based on which proposals are selected.

EU funding still constitutes only a small fraction of the total investment in research and innovation in Europe. In a 2013 speech to key research policy-makers and stakeholders the then EU Commissioner for Research, Innovation and Science, Máire Geoghegan-Quinn, mentioned that national research programmes account for 88% of the public research investments in Europe (Geoghegan-Quinn 2013). Consequently, the alignment of national research programmes across the EU is crucial. The Joint Programming Initiative Healthy and Productive Seas and Oceans (JPI Oceans) (http://www.jpi-oceans.eu/) tries to do just that. JPI Oceans is an intergovernmental platform, open to all EU Member States and Associated Countries with the participation of international partners on actions of mutual interest. JPI Oceans promotes the development of joint research programs based on participating countries' contributions with the aim of fostering cooperative initiatives and optimizing resources. The main challenges addressed by this initiative are related to the marine environmental status and spatial planning, mitigation of climate change and sustainability of anthropogenic activities and maritime economy.

The available knowledge to support policy implementation in marine conservation has increased through the mandatory periodic reporting by Member States on the status of different components of the marine environment under the related EU legislation, the coordination of initiatives at national and European level and the associated research. However, obtaining reliable and comprehensive data on the different areas of the marine environment is still a challenge. Pan-European organizations such as the European Marine Board (EMB: http://www.marineboard.eu/) function as advisory bodies and help bridge the gap between science and policy in marine research and technology.

12.2 The Knowledge-Implementation Pipeline in Europe

Providing conservation policy and management with the best available knowledge is fundamental for the successful, meaningful, and impactful implementation of evidence-based measures at the legislative, political, and practical levels. Mechanisms for an effective communication and knowledge transfer from evidence producers to stakeholders implementing conservation policy and actions have been the focus of intense debate over the last decade (Hulme 2014; Cvitanovic et al. 2016). It has been argued that a proportion of the knowledge produced has poor value or significance to directly support decision-making and therefore it is being seldom used by practitioners (Anderson 2014; Bertuol-Garcia et al. 2018). Even when directly relevant, the ways of conveying that knowledge are frequently not effective (Bainbridge 2014).

The reasons pointed out for this arthritic knowledge transfer in conservation science can be summarized in different categories of arguments. The most important barriers identified in the literature are: (1) the *accessibility to scientific information* by stakeholders (Pullin et al. 2004; Dicks et al. 2014; Anderson 2014; Bainbridge 2014); (2) the *insufficient motivation and awareness of researchers to conduct applied research* tailored by implementation priorities (Cook et al. 2013; Balme et al. 2014; Hulme 2014); (3) the *lack of alignment between the temporal and spatial coverage of research studies and action needs* (Knight et al. 2008; Cook et al. 2013) and (4) the *lack of willingness of stakeholders to integrate the available evidence into their decision-making frameworks* (Ntshotsho et al. 2015).

In the marine context in Europe, an exception to this general pattern is the stock assessment and related maximum sustainable yield estimation performed by organizations such as the International Council for the Exploration of the Sea and the General Fisheries Commission for the Mediterranean and delivered to the European Commission or to RFMOs. This suggests that when there is a clear legislative requirement, scientific advice can more easily find its way into policy implementation. However, even in the case of fisheries assessment and management, the suggested advice is not always taken up by policy-makers and there is often a profound inertia in preventing the integration of new knowledge, adjusted models, and innovative approaches (Stephenson et al. 2017).

Studies evaluating the effectiveness of the knowledge-implementation flow in conservation are relatively rare in the peer-reviewed literature. However, a number of successful and unsuccessful cases of knowledge uptake leading to action are available (Table 12.1) reflecting negative and positive interactions between knowledge producers and end users. Success cases refer to a positive impact of managers' involvement in the articulation between research and implementation. Another example is the positive impact in public dissemination and communication between researchers and stakeholders when conservation policies are based on the best available knowledge. Negative interactions refer mainly to the failure in bridging research results and implementation needs and the inability of practitioners to integrate research results into practice.

Table 12.1 Non-exhaustive list of examples of published scientific studies evaluating the effectiveness of knowledge transfer between producers and users in conservation science in Europe

Scope and reference	Goals	Main findings
Protected areas in Sweden, Sweden (*Angelstam et al. 2011*)	Evaluation of the policy implementation process between 1991–2010 and assessment of ecological knowledge use for conservation planning	Positive transfer and application of available knowledge in conservation
Marine protected areas in Scotland, UK (*Bainbridge 2014*)	Integration of scientific knowledge into policy making by providing user-friendly formats to present available data	Positive translation of available knowledge into conservation
Conservation science of *Triturus cristatus* in England, UK (*Griffiths 2004*)	Comparison of the number of papers published in conservation science and the number of mitigation projects conducted in response to developmental threats	Negative perception by practitioners of the relevance of research outcomes and by researchers of the use of produced knowledge in management decisions
Habitat conservation policy as part of the Habitats Directive in Europe, European Union (*Jeanmougin et al. 2017*)	Evaluation of how the conservation policy contributes to conservation: is it science-based, operational, and legitimate?	Negative evaluation of the application of available knowledge on the implementation of the Habitats Directive
BONUS programme: policy driven joint Baltic Sea research programme (2007-2020), EU Baltic Sea States (*Snoeijs-Leijonmalm et al. 2017*)	Assessment of the potential of the BONUS projects to address the challenges faced by the Baltic ecosystems and bibliometric analysis of BONUS funded papers	Positive effect on reducing fragmentation in research funding and policy and contribution of BONUS products and participants to public policies and relevance
Upupa epops (hoopoe) demographic recovery in the Swiss Alpes, Switzerland (*Arlettaz et al. 2010*)	Recovery response of *U. epops* populations after application of tailored conservation actions based on evidence-based conservation guidelines	Positive outcome of the practical involvement of researchers in the implementation of conservation measures by stakeholders
UK Research Council's Rural Economy and Land Use (RELU) Programme, UK (*Phillipson et al. 2012*)	Evaluation of stakeholders' engagement level and impact in 38 of RELU's research projects	Although only happening in part of the cases, positive impact of stakeholders' involvement at the knowledge production stage (i.e., input to the research projects)
Management plans from major conservation organizations in the UK, UK (*Pullin et al. 2004*)	Examination of the process of Nature Reserve Management Plans formulated by several conservation organizations	Majority of management plan stakeholders do not systematically consider available scientific evidence and do not disseminate the outcomes of their actions
Effectiveness of research-based knowledge for the implementation of genetic conservation programs in EU-28 (Pérez-Espona et al. 2017)	Compilation of the peer-reviewed publications addressing conservation genetics and analysis by target species	Research outputs relevant for conservation and management of European species are abundant but fail to focus on species of conservation interest

The studies listed were related to the group of barriers identified in the main text. Note: Green cells highlight positive impacts

The following sections illustrate examples of supporting mechanisms, platforms, networking activities and initiatives to promote the production and uptake of knowledge on marine conservation at the European level. All these initiatives have a focus on the EU-28 level although many are also covering other countries. This exercise does not intend to exhaustively list all of the ongoing EU initiatives in marine conservation nor does it necessarily reflect the landscape at the country level.

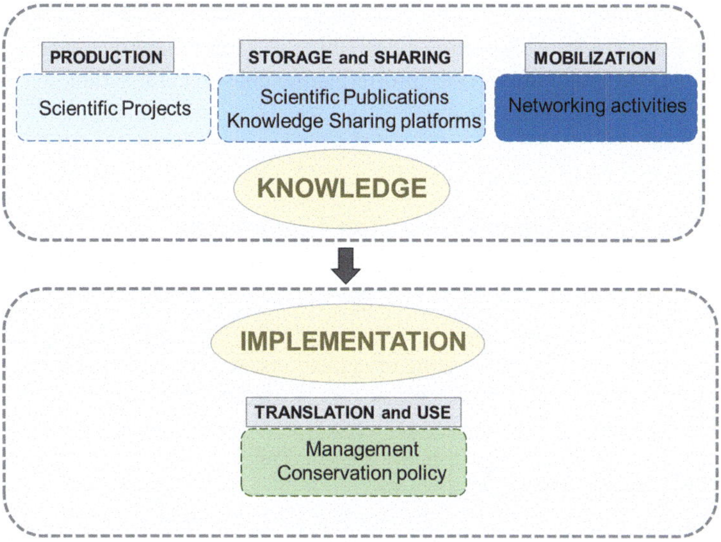

Fig. 12.3 Framework used in this chapter to assess the knowledge production-implementation flow in marine conservation in Europe

Instead, it aims to provide a snapshot of how the knowledge-implementation gap in Europe is being bridged by exemplifying ongoing relevant activities in Europe, discussing the potential constraints to implementation in relation to the general topics identified in the literature and identifying best practices.

12.3 The Knowledge Production-Implementation Flow in Marine Conservation in Europe

This section will characterize the knowledge production-implementation flow in marine conservation in Europe and its main gaps and strengths. Two main components are considered (Fig. 12.3):

1. *Generation and dissemination of knowledge by the scientific community*: This component includes (a) the production of scientific knowledge, as measured by the number of scientific projects supported by EU research funding in marine conservation, (b) the knowledge storage and sharing as illustrated by the scientific production (peer-reviewed publications) and sharing platforms (raw data, expertise or relevant information for marine conservation), and (c) the knowledge mobilization related to initiatives promoting networking activities in the field of marine conservation.

2. *Use of available evidence by decision- and policy-makers*: This component will be illustrated by four case studies describing different initiatives where knowledge was used and combined to inform marine conservation action.

12.3.1 The Knowledge Production Component

Scientific projects that generate knowledge encompass a range of products from simple graduate thesis to multimillion Euro projects implemented by multinational consortia with dozens of collaborating organizations. As an example, the EU-funded AtlantOS project on ocean observations brings together 67 partners from 13 EU Countries and six non-EU countries and territories to develop a sustainable, efficient, and fit-for-purpose Integrated Atlantic Ocean Observing System (https://cordis.europa.eu/project/rcn/193188/factsheet/en).

Tracking all marine-related projects in Europe is challenging, particularly considering different national languages. Therefore, this section focuses on research supported by EU funding streams which are more easily accessible. Additionally, EU research policy (supported by its own funding instruments) is considered to shape the entire European research landscape, in particular with instruments such as the ERA-NETSs (http://ec.europa.eu/research/participants/docs/h2020-funding-guide/cross-cutting-issues/era-net_en.htm), which incentivize participating countries to align funding of their national research programmes with projects of common European interest. The current marine ERA-NET is the BlueBio project (https://bluebioeconomy.eu/) that brings together 16 European countries and includes several calls and other activities covering all aspects of the blue bioeconomy.

Among the EU funding streams, the last two research and innovation framework programmes, the 7th Framework Programme (FP7) and Horizon 2020 (H2020), have provided by far the main support for research at the European level. However, some information on other EU funding streams that have a research component are included, such as the LIFE programme (https://ec.europa.eu/easme/en/life) and the grants provided by the European Maritime and Fisheries Fund (EMFF, https://ec.europa.eu/fisheries/cfp/emff_en).

FP7, the European Union's Research and Innovation funding programme for 2007–2013, had a budget of 50 million Euros and funded 25,778 research projects. A search done using only three keywords (marine, ocean, sea) in the Community Research and Development Information Service (CORDIS, https://cordis.europa.eu), the European Commission's primary source of results from the projects funded by the EU's framework programmes,, retrieved a total of 471 projects with a total budget of around 960 million Euros. This represents roughly 1.8% of the total number of EU-funded projects and 1.9% of the EU budget.

Refining this list of 471 projects by adding any one of the keywords in Table 12.2 resulted in a total of 402 projects specifically related to marine conservation with a total budget of around 875 million Euros. This roughly represents 85.3% of all

Table 12.2 Keywords used to screen marine conservation FP7 and H2020 projects in CORDIS

Acidification	Cephalopod	Demersal	Eutrophication	Molluscs	Plastics	Seabird
Algae	Climate	Diatom	Fish	Oceanographic	Pollution	Seagrass
Bacteria	Conservation	Dolphin	Habitat	Pelagic	Remediation	Seal
Benthic	Copepod	Ecology	Invasive	Physiology	Reproduction	Shellfish
Biodiversity	Crustacean	Ecosystem	Litter	Phytoplankton	Seabed	Whale
						Zooplankton

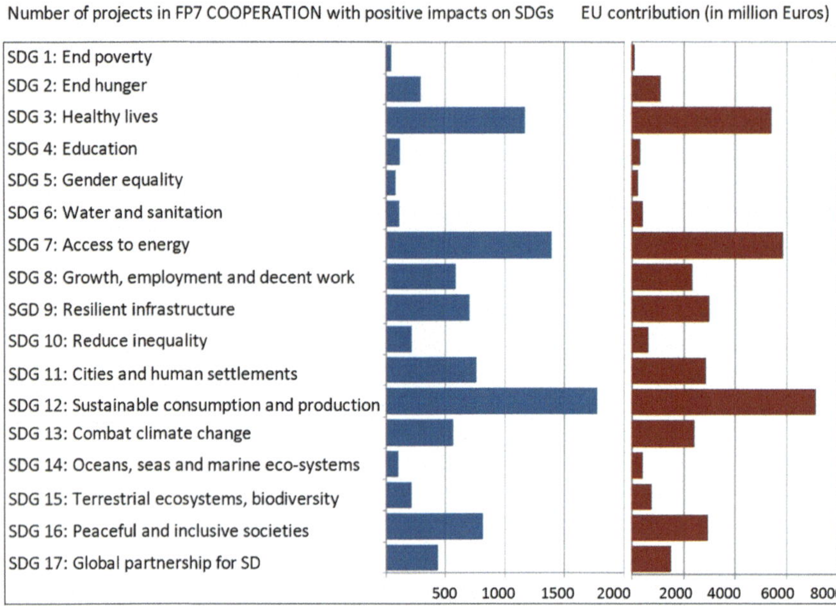

Fig. 12.4 Number of projects in FP7-Cooperation with expected positive impacts in the different SDGs and respective monetary EU contribution (Fresco et al. 2015)

marine projects and 91.1% of the marine-related funding but 1.6% of all FP7 projects and 1.7% of the total FP7 funding.

In accordance with the Fig. 12.4, the evaluation of FP7 once the programme ended (Fresco et al. 2015) revealed that FP7 cooperation projects related to SDG14 (the most relevant for marine conservation) are rather few in number and in funding compared to other SDGs, e.g., related to energy, health, and terrestrial ecosystems (Fig. 12.4).

FP7's successor for 2014–2020, H2020, has a budget of around 70 million Euros. A similar search in CORDIS revealed that, in the five first years of H2020, 463 marine projects were funded totaling around 900 million Euros. Of these, 244 projects are related to marine conservation and their total budget is approx. 500 million Euros, i.e., roughly 52.7% and 55.6% of all marine EU-funded projects. This could suggest a decrease in funding of marine conservation projects but because H2020 is not completed yet this figure should be taken with caution.

However, specifically with respect to blue sky research funded under the Excellence pillar by the European Research Council (ERC) and according to the interim evaluation of H2020 (European Commission 2017), of the 19 key hot research fronts in which ERC grantees are working, two are explicitly related to marine conservation (microplastic pollution in the marine environment and carbon cycle of inland waters and the ocean) and another two are implicitly related to marine conservation (biodiversity loss and its impact on ecosystem functions and ecosystem services and

HOT RESEARCH FRONTS

1. Outbreak, prevention and control of microbial contamination of fresh produce
2. Mechanism of plant innate immunity
3. Microplastic pollution in the marine environment
4. Biodiversity loss and its impact on ecosystem functions and ecosystem services
5. Global warming hiatus
6. Carbon cycle of inland waters and the ocean
7. Clinical trials of direct-acting antivirals (DAAs) for hepatitis C infections
8. Immune checkpoint inhibitors anti-PD-1 antibodies in melanoma immunotherapy
9. The molecular mechanism for origin, development and differentiation of macrophage
10. Differentiation, function, and metabolism of T cells
11. Phosphors for white LEDs
12. Sodium-ion batteries
13. Galactic center gamma-ray excess
14. Property and application of monolayer/few-layer black phosphorus
15. Observations of the cosmic microwave background (CMB) by Planck
16. Baryon acoustic oscillation (BAO) related research based on sky survey missions like SDSS
17. The internet of things, cloud manufacturing and related information technology services
18. Research on measurement-device-independent quantum key distribution
19. DEA (Data Envelopment Analysis) based assessment of environmental and energy efficiency

EMERGING RESEARCH FRONTS

1. Effects of systemic insecticides (neonicotinoids and fipronil) on non-target organisms and environment
2. Elemental composition of the North Atlantic Ocean and Southern Ocean
3. Principles of chromatin looping and evolution of chromosomal domain architecture
4. Research fronts on perovskite
5. Experimental realisation of fractional Chern insulators
6. Studies of Comet 67P/Churyumov-Gerasimenko by Rosetta

Source: ERCEA, Thematic assessment of the European Research Council (see Annex 2), 2017

Fig. 12.5 Key hot and emerging research fronts in which ERC grantees are working under H2020 (European Commission 2017)

global warming hiatus) (Fig. 12.5). Additionally, of the six emerging research fronts in which ERC grantees are working, one is explicitly marine related (Fig. 12.5).

The LIFE programme is the EU's funding instrument for the environment and climate action. The current funding period 2014–2020 has a budget of 3.4 billion Euros. This is not a research programme per se but it includes science-related activities that have a clear relation to nature conservation and environmental protection. A search in the LIFE database (http://ec.europa.eu/environment/life/project/Projects/index.cfm) for the years 2014–2018 showed that of a total of 755 LIFE projects, 73 (9.7%) were related to marine conservation.

The European Maritime and Fisheries Fund (EMFF) is the fund for the EU's maritime and fisheries policies for 2014–2020. A part of it is directly managed by the European Commission and funds studies that may include science-related activities. A search in the dedicated fund webpage (https://ec.europa.eu/easme/en/emff-projects) showed that until 2018 the programme funded 78 projects with 24 of them (30.8%) directly related to marine conservation while the rest are more related to other blue economy activities (e.g., maritime surveillance).

The overall analysis of these results indicates that marine-related topics contributed to less than 10% of all the funding programs considered, with the exception of the EMFF which is, of course, a full marine programme. However, marine science is particularly prevalent in EU-funded blue sky research. Within the marine research area, marine conservation contributed to more than 50% of the funded projects except for the EMFF that is more targeted to maritime economy.

12.3.2 Knowledge Storage and Sharing: Scientific Publications

For this section, only studies published in peer-reviewed scientific journals and indexed in publication databases were considered. The scope of the analysis included research conducted from 2010 to mid-2018 and publication details were gathered by searching the ISI Web of Knowledge, using the terms "marine *and* conservation", "marine *and* ecology", "marine *and* management" and "marine *and* policy".

Although research published as grey literature might contain relevant information, these documents were not included in the analysis due to low accessibility and consistency which impaired a comprehensive sampling of these works. Nevertheless, we consider that the analysis of the peer-reviewed literature provides a comprehensive picture of the scientific production in the field of marine conservation in Europe.

We grouped the results by geographical location of the study, as ascertained by the authors' affiliation (inside/outside Europe; country in Europe), publication year (2000–2018), research areas (following the categories established in ISI web of knowledge), and publication quartile. The majority of marine research comes from outside Europe, with European-based studies representing 41% of the studies retained by the search (Fig. 12.6a). In Europe, these publications were scattered among 39 countries. Most of the countries (62%) published less than 500 papers while only 5 countries produced more than 1000 publications between 2010 and 2018 (Fig. 12.6b).

The number of publications in the field of marine conservation in Europe gradually increased from 2010 to 2015 and has, since then, stabilized at around 1500 publications per year (Fig. 12.7a). The number of publications has decreased slightly in 2018 but that is likely a reflection of only half of the year being considered in the search. These publications are distributed among several different research subdisciplines but 10 of these research subdisciplines contribute to 69% of the total number of published papers. The top three subdisciplines are marine freshwater biology, ecology, and environmental sciences, each one corresponding to more than 10% of the papers (Fig. 12.7b). The distribution per quartile of the publications (ranking of the publications based on their factor in a given topic category) was analyzed considering the journals with more than 80 papers published between 2010 and 2018, corresponding to 91% of the total number of publications. The number of journals considered was 39 out of the total number of 197. These publications were distributed among quartiles Q1, Q2, and Q3 of the distribution of the scientific journal rankings (SJRs) with 86% of the publications falling under the first quartile (Fig. 12.7c).

The scientific productivity measured by the number of peer-reviewed publications in the field of marine conservation in Europe is slightly under half of the publications in this field worldwide. Although the impact of these publications might extend way beyond the local scope, these results show to some extent the potential of

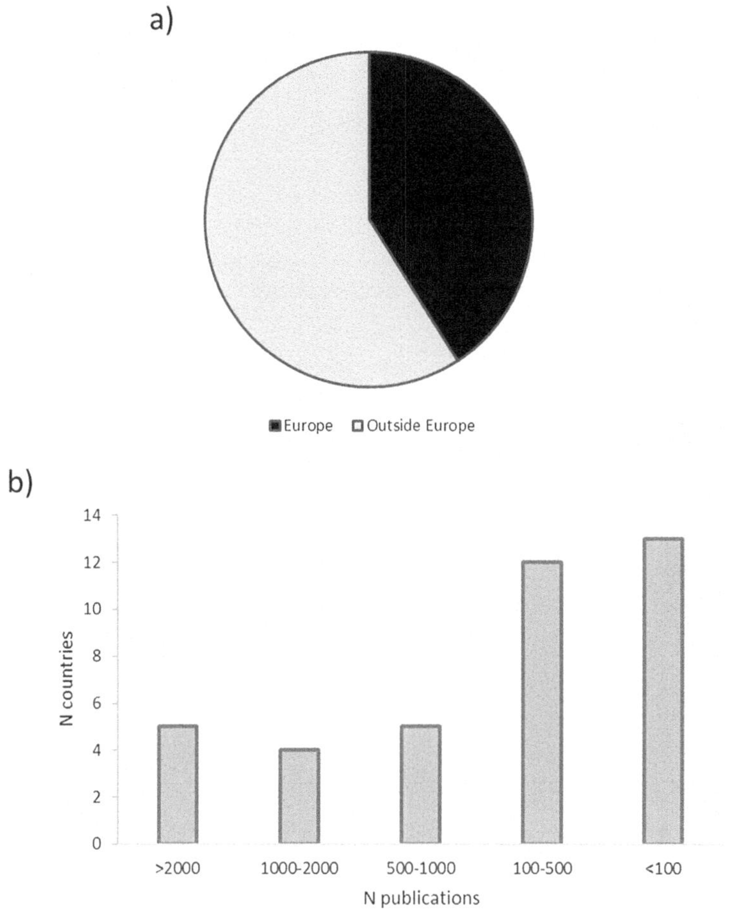

Fig. 12.6 (**a**) Proportion of publications per region (Europe/Outside Europe) as per the search criteria used; (**b**) Distribution of the number of publications by a number of European countries publishing in marine conservation

the knowledge provided by the European research community to support implementation in marine conservation. Some issues highlighted as barriers in the previous section of this chapter, such as the mismatch between implementation needs and scientific research topics or the contribution of fundamental research to the total scientific production, are only evaluated in general terms in this exercise. However, more technological areas, for example engineering or science technology, represented the minority of the publications. Thus, it is expected that a considerable part of the peer-reviewed published literature will potentially contribute to inform knowledge-based management plans or policy initiatives. It is clear that the impact factor of the published research is a relevant issue as demonstrated by the almost complete dominance of scientific journals from quartile 1 in the publication list. This

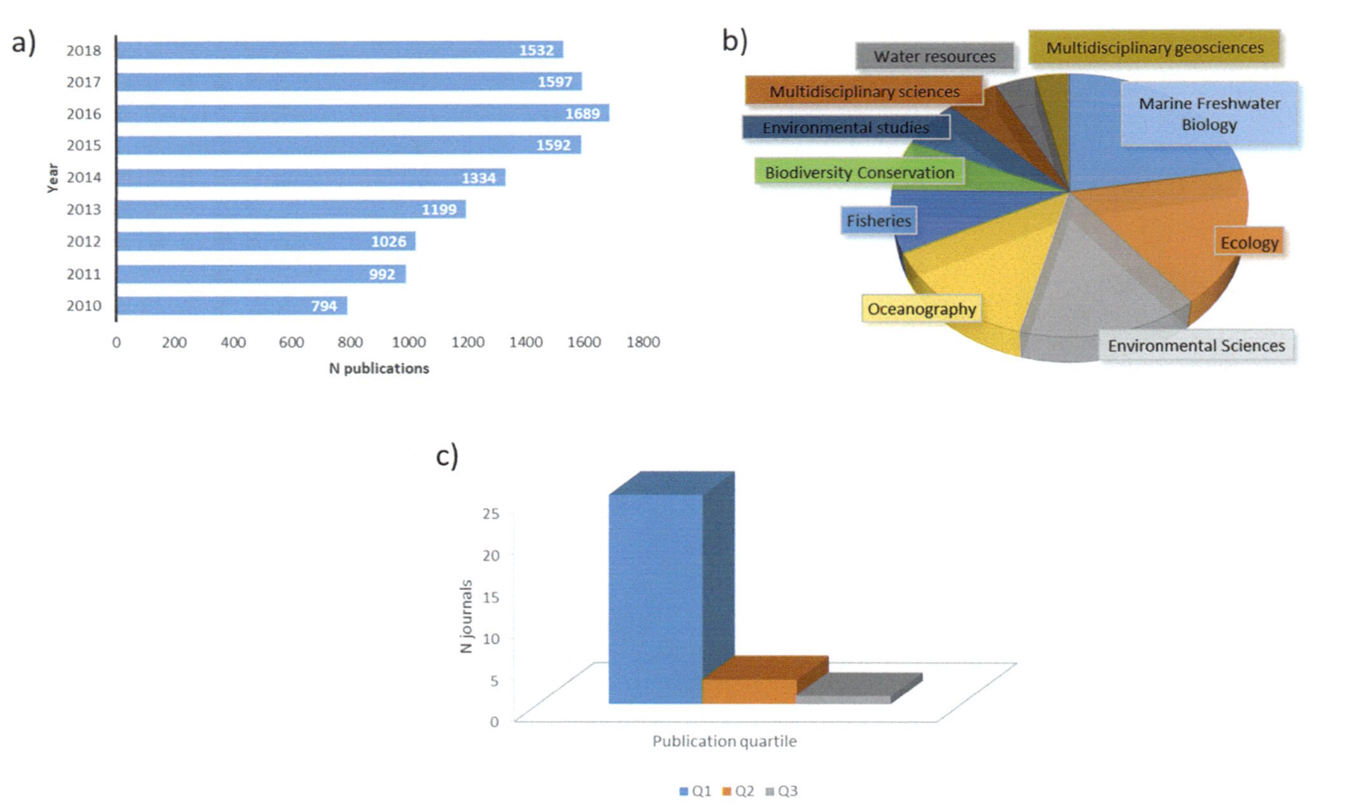

Fig. 12.7 (**a**) Temporal trend (early 2010 to mid-2018) in the number of publications related to marine conservation in Europe; (**b**) Distribution of number of publications by each of the 10 research subdisciplines (as defined in ISI web of knowledge) with highest number of publications (2010–2018); (**c**) Quartile distribution of the journals with more than 80 publications related to marine conservation (2010–2018)

trend is understandably related to the pressure researchers face to publish in high impact scientific journals. The impact factor is the most commonly used metric of research quality for career development and job performance. Frequently, applied research conducted at a local scale (potentially the most useful for conservation action) is not as suitable for publication in high impact journals as conceptual, wide scoped, or fundamental research. This was one of the main constraints identified previously in this chapter to an effective knowledge transfer and seems to be also an issue in marine conservation science in Europe. However, the high number of publications in quartile 1 also reflects the potential impact and broad interest in this topic. Additionally, the temporal trend in the yearly number of publications showing a steady increase from 2010 to 2016 reflects the development of this research field and was very timely to feed the many initiatives evolving at the European level related to the marine environment such as the Blue Growth or the Biodiversity Strategies, and the Marine Strategy Framework Directive.

12.3.3 Knowledge Storage and Sharing: Platforms and Initiatives

An important bridge to narrow the knowledge-implementation gap consists in ensuring that published data, expertise, or infrastructures are organized to provide centralized access to information and knowledge. Here, examples of web platforms or consortia sharing marine data and knowledge at the European level were mapped and listed. For each platform, the type of data, objective, source and collection process are described, when available (Table 12.3).

There are a variety of platforms at the European level aiming to cover, in a coordinated and centralized way, different areas relevant to marine conservation. These platforms include raw data (environmental and biological), information on infrastructure and research topics networks or target specific groups of species (e.g., alien species, Table 12.3). Although some of these data sharing platforms encompass terrestrial, freshwater and marine records, and their geographical scope is not only European, there are already several examples of platforms exclusively dedicated to curate marine data. This coverage of areas and topics at the European level represents an important effort to optimize resources and it greatly increases the quality and impact of knowledge transfer when informing policy sectors or facilitating access to harmonized data by users. This facilitation will certainly positively impact the effectiveness and adequacy of political initiatives. These platforms also streamline the work of researchers using cross-sectorial data with wide geographical coverage. One of the barriers previously identified was the difficulty of practitioners to access relevant published information. These platforms are an opportunity to facilitate access to information in a coordinated way. It was not assessed in this work if the thematic and infrastructure networks have the necessary support to be maintained in the long-term or if there is a geographical balance in accessibility or equal

Table 12.3 Examples of knowledge sharing platforms working with European information also related to marine conservation

Data sharing platform	Description
EMODnet (European Marine Observation and Data Network: www.emodnet.eu)	*EMODnet* is a network of organizations supported by the EU's integrated maritime policy with the objective of providing easily accessible, reliable, and accurate information on different topics related to the marine environment that is publicly available to marine data users. The website provides information on a variety of topics including bathymetry, geology, seabed habitats, chemistry, biology, physics, and human activities
EASIN (European Alien Species Information Network: https://easin.jrc.ec.europa.eu/)	*EASIN* is a scientific network developed by the European Commission's Joint Research Centre with the objective of providing support to European Alien Species policies. EASIN constitutes the central platform of the official information system foreseen under Article 25 of the EU Regulation 1143/2014 on Invasive Alien Species supporting its implementation. EASIN data also serve the broader scientific community. By collecting available information from projects and publications at the European and global level this platform provides easy access to data on Alien Species, including marine ones, reported in Europe
EurOcean (European Centre for information on Marine Science and Technology: http://www.eurocean.org/)	*EurOcean* is a non-governmental organization aiming to promote initiatives supporting the Blue Growth and the implementation of the European Maritime Policy. It facilitates the knowledge transference in the field of marine sciences and technologies among different stakeholders. EurOcean databases cover three main domains: • Marine research infrastructures: Marine research infrastructures database includes information on infrastructures and equipment used to collect ocean data. • Marine knowledge management: the Marine knowledge gate is an online repository cataloguing marine research projects. • Science and technology communications: with activities promoting ocean literacy and awareness.
GBIF (Global Biodiversity Information Facility: https://www.gbif.org/)	*GBIF* is an intergovernmental collaboration between countries and international organizations joining efforts to advance free and open access to biodiversity data. It includes 40,415 datasets from 1231 publishing institutions. Twenty European countries participate in this

(continued)

Table 12.3 (continued)

Data sharing platform	Description
	initiative accounting for 43% of the 1,011,821,225 records of the database. Some of these are marine records.
OBIS (The Ocean Biogeographic Information System: http://iobis.org/)	*OBIS* is a free and open-access data and information portal on marine life with 45 million observations of nearly 1,200,000 species. The information provided is based on the collaboration with the scientific community and refers to the biodiversity and biogeography of marine species including also environmental parameters (physical and chemical).
AQUANIS (The information system on aquatic non-indigenous and cryptogenic species: http://www.corpi.ku.lt/databases/index.php/aquanis/)	*AQUANIS* is an information system on aquatic non-indigenous and cryptogenic species present in marine, brackish, and coastal freshwater in Europe and neighbouring regions. It includes information on the biology of the species, introduction pathways, potential impacts and geographical information on species distribution.
WORMS (The World Register of Marine Species: http://www.marinespecies.org/about.php)	*WORMS* provides a comprehensive list of names of marine species to facilitate the consistency of taxonomic designations. The content of the lists is controlled by taxonomic and thematic experts that combine information from other marine species lists and provide additional details on bibliographic references and biogeographic data.
WRIMS (The World Register of Introduced Marine Species: http://www.marinespecies.org/introduced/)	*WRIMS* reports which species included in WORMS have been introduced by human activities to geographic areas outside of their native distributional range.
ICES (The International Council for the Exploration of the Sea; https://www.ices.dk/Pages/default.aspx)	*ICES* is an intergovernmental marine science organization gathering expertise from a very extended network of scientists from different marine fields and covering a high extension of maritime areas. It provides evidence on the state and sustainable use of the seas and oceans. ICES coordinates the work of 150 expert groups covering fundamental and applied science on different marine trophic levels. ICES also manages datasets and provides advice to policy bodies on marine science related topics.
MARS network (The European Network of Marine Stations) https://www.marinestations.org/	*MARS network* is a foundation connecting European marine research institutes and marine stations in a forum to discuss topics related to fundamental research and inform policies related to marine conservation.

(continued)

Table 12.3 (continued)

Data sharing platform	Description
EMBRC-ERIC (The European Marine Biological Research Centre) http://www.embrc.eu/	*EMBRC-ERIC* is a European research infrastructure consortium in the field of fundamental and applied marine biology and ecology. The consortium develops initiatives to facilitate the mobility and access to research facilities and services in different areas of expertise as well as coordination of educational programmes.
EuroGOOS (The European Global Observing System) http://eurogoos.eu/	*EuroGOOS* is an international association of national governmental agencies, private companies, and research organisations, including members from 18 European countries, aiming at providing coordinated operational oceanographic services. EuroGOOS provides data and collaborates with other pan-European data portals.
WISE Marine (The Marine Information System for Europe) https://water.europa.eu/	*WISE Marine* is a portal and infrastructure for sharing European-level information on marine topics in support of ocean governance and ecosystem-based management. This initiative is a partnership between the European Commission and the European Environmental Agency and provides information on EU water policies, access to datasets and maps, and ongoing water related research activities.

opportunities for all the European countries. These are aspects that will influence the impact of these initiatives at the European level. Another important topic, when considering data sharing platforms, is to understand if, regardless of the apparently already reasonable coverage of topics, the comprehensiveness, coverage and quality of data related to each topic is high. Of paramount importance to the quality and effectiveness of these platforms is their ability to transfer data from knowledge production to knowledge visualization and sharing.

To assess the main constraints to the transfer and integration of available and published data to data sharing platforms, a short questionnaire was sent to the contact points of 16 platforms from which seven replied. The questionnaire comprised four questions:

1. Which are the main data sources used by your platform to collect information?
2. How easy is it to thoroughly map and access the information needed to include in your platform?
3. Is the information found adequate to meet your needs regarding content and format?

4. Which are the main constraints encountered when accessing the information you need?

The outcomes of this survey highlight that the main knowledge sources used by data sharing platforms are scientific publications, monitoring datasets, and project databases (Fig. 12.8a). Project reports were the least used source of information (probably because they are more difficult to access; Fig. 12.8a). Most of the respondents identified difficulties in mapping the relevant information needed although the access to this information was only moderately difficult (Fig. 12.8b). The content of the information sources was considered appropriate to the needs of the platform, but the format was often inadequate (Fig. 12.8c). The main constraints identified to access the available information were the fragmentation of information and confidentiality issues (Fig. 12.8d). The difficulty to find the information needed was not identified by any of the respondents as a constraint (Fig. 12.8d), suggesting that finding information is easy but accessing it often becomes a limitation.

The results of the questionnaire show that scientific knowledge that is potentially useful for marine conservation action in Europe is moderately difficult to access by data gathering platforms. Publications on applied studies, with potential relevance for management and conservation policy, are often contained in non-peer review literature (e.g., reports and other grey literature). The documents most frequently reporting results from local studies are often difficult to track and to extract relevant information since they are commonly written in national languages. Additionally, data (e.g., environmental) collected by individual institutions might be difficult to gather in a coordinated, harmonized, and comparable way because of geographically different monitoring schemes. Some regions have much more developed monitoring schemes, that include different environmental parameters and variables as well as temporal timeframes assessed. Platforms providing data have to overcome these challenges by developing strategies and tools to harmonize the information displayed, maximizing the knowledge support to research, management, and policy initiatives. One of the main issues raised are the different formats used for knowledge storage which leads to difficulties in extracting information. This might contribute to the fact that project reports are the least used source of information by data platforms. Fragmentation of the information and confidentiality issues were also identified as constraints to access the information. This fragmentation probably is due to the different developmental states of science across European countries. To conclude, the information needed is easy to find but difficult to map and extract.

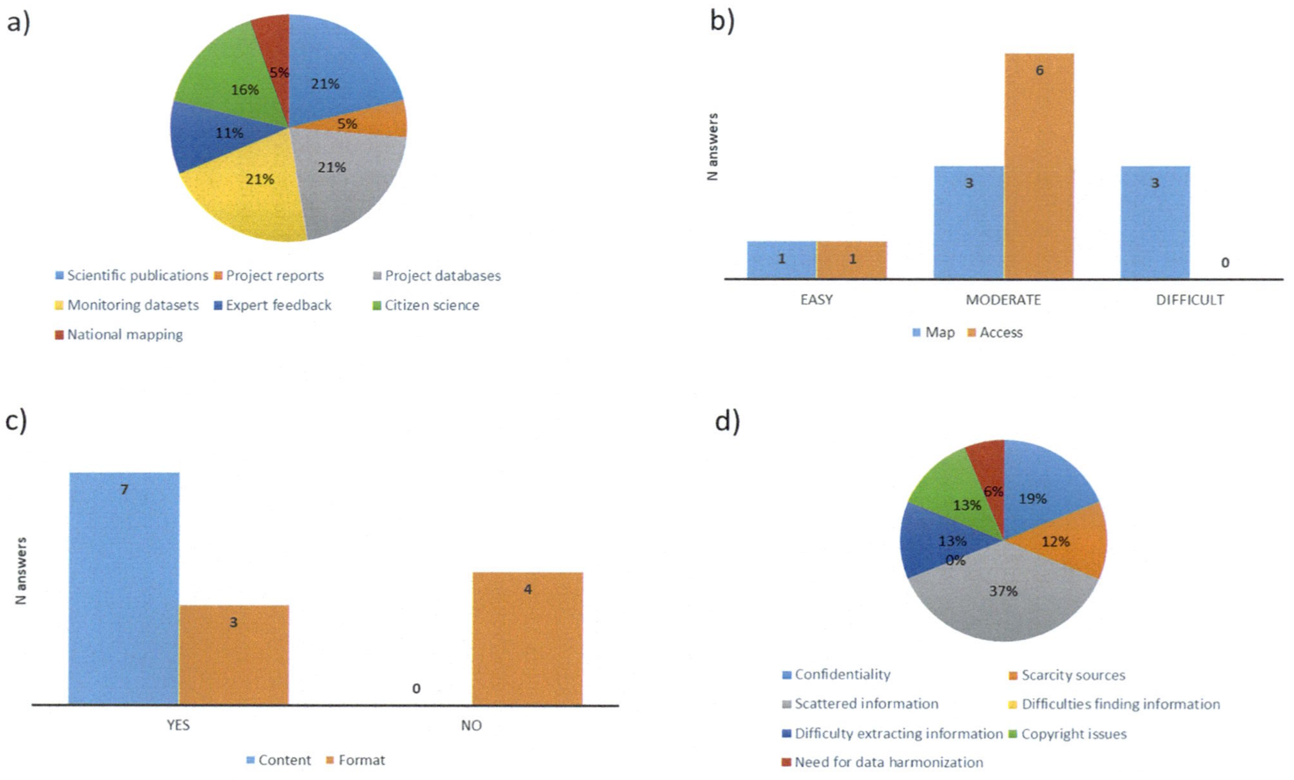

Fig. 12.8 Replies from contact points of data sharing platforms to the survey. (**a**) Data sources used; (**b**) Facility to map and access information; (**c**) Adequacy of content and format; (**d**) Constraints

12.3.4 Knowledge Mobilization

There are several EU-level initiatives promoting networking and facilitating knowledge exchange and transfer as well as capacity building, supporting either exclusively marine activities or scientific activities in which marine topics are also included. Below, we detail some of these initiatives.

The *Projects for Policy (P4P) initiative* (https://ec.europa.eu/info/research-and-innovation/strategy/support-policy-making/scientific-support-eu-policies/p4p_en) aims to use research and innovation project results to shape policy-making. The European Commission identifies policy areas which deserve particular attention, analyses the related knowledge which comes from research and innovation programmes and delivers recommendations in the form of P4P reports to reach out to partners and stakeholders and contribute to a highly impactful policy making process. One of the first P4P reports is on blue economy (https://publications.europa.eu/en/publication-detail/-/publication/ada65c0f-aef9-11e7-837e-01aa75ed71a1/language-en/format-PDF/source-69927165).

COST-European Cooperation in Science and Technology The *COST Association* (http://www.cost.eu/) supports collaborative transnational networking activities, through the financial support to *COST Actions*, covering all scientific and technological domains across Europe. The objective of these COST Actions is to promote impactful scientific developments contributing to Europe's development in Research and Innovation. This is achieved by building researchers' capacity, promoting networking and opportunities for knowledge exchange, and increasing the knowledge transfer between stakeholders from different sectors. COST Actions cover a range of networking activities such as workshops, conferences, training schools, short-term scientific missions, and dissemination activities and include also the participation of non-EU28 countries.

As an attempt to illustrate the contribution of these Actions to marine conservation knowledge mobilization in Europe, the EurOcean database (http://www.eurocean.org/) was used to analyze the share of COST Actions dedicated to marine conservation between 2010 and 2018. The EurOcean database categorizes the projects by disciplines, activities, and themes. Table 12.4 lists the project tags and keywords selected for the search.

Table 12.4 Project tags and keywords used to screen marine conservation projects at the European level in the EurOcean database

Keywords		Project tags	
Conservation	Ecology	Seaweed and other sea-based food harvesting	
Physiology	Remediation	Environmental impact assessment	
Global warming	Ecosystem	Survey and monitoring (not research related)	
Climate change	Benthic	Marine technology and responsible research and innovation	
Fisheries	Phytoplankton		
Pollution	Eutrophication	Marine and coastal tourism	Recreational activities
Biodiversity	Ecosystem services	Protection of habitats	Marine research
Biological invasions	Habitat	Biotechnology	Marine pollution
Ocean acidification	Algae	Climate change	Carbon capture and storage
Bacteria	Seabirds	Marine litter	Marine aquatic products
Diatoms	Mammals	Fisheries	
Zooplankton	Contaminants	Aquaculture	
Litter	Plastics	Chemical oceanography	
Demography	Phenotypic traits	Biological oceanography	
Reproduction	Pelagic	Physical oceanography	
Fauna	Trophic webs	Marine geology	
Deep sea		Environment	

Marine conservation represents a small part (less than 5%) of the COST Actions supported from 2010 to 2018 (Fig. 12.9a). Taking into consideration that a high variety of topics is covered by COST initiatives it is already significant that this number of Actions in marine conservation is supported. Given that each Action has a duration of 4 years there were likely several of these initiatives in marine conservation running in parallel between 2010 and 2018. These Actions involved researchers from more than 30 countries with most of the participating countries being involved in more than 15 COST Actions during this period (Fig. 12.9b).

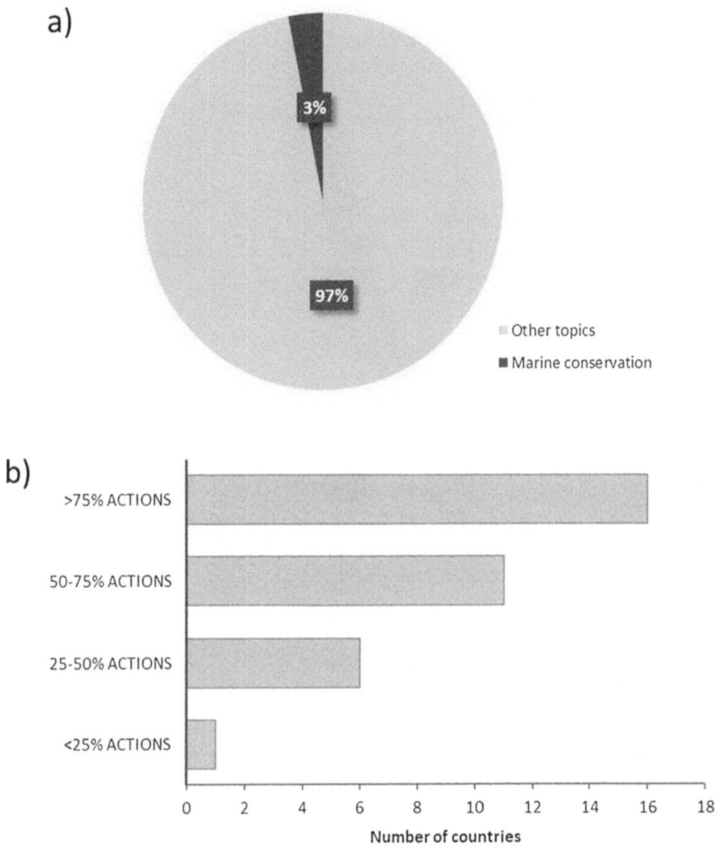

Fig. 12.9 (**a**) Proportion of COST Actions supported from 2010 to 2018 in the field of marine conservation; (**b**) Number of countries participating in the COST Actions funding Marine Conservation from 2010 to 2018

This kind of initiatives has an important potential impact in facilitating the access to research facilities and technologies, support capacity building especially for less intensive research countries and foster the participation of these countries in research consortia.

An example of an ongoing COST Action in the topic of marine conservation is described in Box 12.1.

Box 12.1 Description of the MarCons COST Action (www.marcons-cost.eu)

MarCons

Advancing marine conservation in the European and contiguous seas

This COST Action aims to bridge the gap between conservation science and policy makers by promoting regional coordination and transboundary conservation, proposing specific conservation actions, considering climate change and biological invasions and providing guidance to a more effective spatial management of marine areas

- **WG1: FROM DESCRIPTION TO PREDICTION OF CUMULATIVE HUMAN IMPACTS ON MARINE ECOSYSTEMS TO INFORM CONSERVATION PLANNING:** advance the understanding of cumulative impacts of human stressors (e.g. related to climate change, on marine biodiversity and ecosystems) considering their spatial and temporal heterogeneity and their impact on marine species, communities and ecosystems.
- **WG2: PRESENT CHALLENGES AND LIMITATIONS IN MARINE CONSERVATION:** to address relevant gaps in marine conservation and to identify the most critical actions to significantly improve the protection of the marine biodiversity of the region.
- **WG3: MARINE CONSERVATION AND BIOLOGICAL INVASIONS:** to offer methodological approaches and tools for designing networks of MPAs that are more resilient to biological invasions and to propose management actions to mitigate the impacts of invasive species on biodiversity and ecosystem services.
- **WG4: INTEGRATED CONSERVATION PLANNING ACROSS TERRESTRIAL, FRESHWATER AND MARINE ECOSYSTEMS:** to synthesize existing knowledge on integrated conservation planning across terrestrial, freshwater and marine ecosystems aiming at biodiversity conservation and the sustainability of ecosystem services, and to develop methodological tools for improving this integration within the framework of ecosystem- based management.
- **WG5: REGIONAL COORDINATION AND TRANSBOUNDARY CONSERVATION:** To explore the conditions for the development of successful transboundary conservation in a region of multiple states.
- **WG6: GOVERNANCE OF MARINE PROTECTED AREAS:** to evaluate how different geopolitically and biogeographically representative examples of MPAs are governed across the European and contiguous seas and analyze governance approaches in relation to effectiveness in achieving conservation objectives.

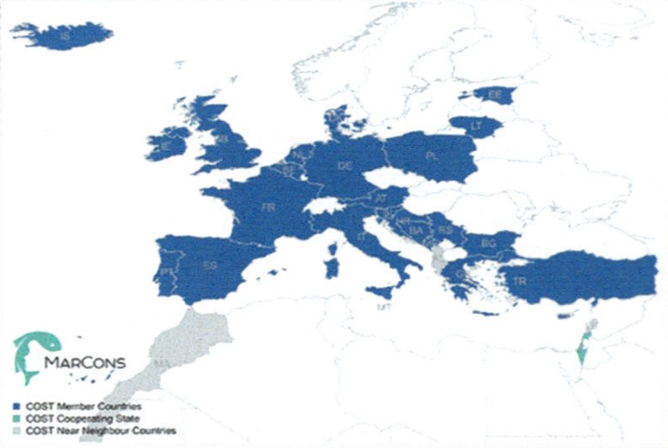

MarCons will contribute towards the targets of a series of European policies and international conventions by providing methods and tools for their implementation, and proposing ways to address some of their requirements and obligations: Framework Directive for Maritime Spatial Planning, Habitats and Birds Directives, the Marine Strategy Framework Directive, the Regulation on the prevention and management of the introduction and spread of invasive alien species, the Water Framework Directive, the Biodiversity Strategy, the Convention on Biological Diversity, the Barcelona Convention, the OSPAR Convention, the Bern Convention, the Helsinki Convention and The Convention on the Protection of the Black Sea Against Pollution.

Towards improving the knowledge-implementation flow, MarCons has established links and investigated possibilities for joint meetings with existing networks of stakeholders, such as MedPAN, the network of Mediterranean MPAs, and ICES working groups. Policy briefs will be translated to local languages and will be disseminated in native languages in an accessible form for managers, policy makers and the general public. A final conference will be organized to disseminate all the outputs of this Action to stakeholders and the wider public. MarCons participants have already started to apply the developed frameworks and tools in their countries.

Box 12.1 exemplifies how these initiatives work in practice promoting transboundary synergies between a high number of European countries and stakeholders to coordinate efforts in marine conservation. This coordination is particularly important for the marine environment where national borders do not apply and thus actions taken in national waters will potentially impact the neighbouring countries. Additionally, these initiatives promote the gathering of the critical mass and expertise needed to produce and provide advice in a coordinated and harmonized way for the wide European level.

EUROMARINE-European Marine Research Network *EuroMarine* (www.euromarinenetwork.eu) was launched in 2014 and is based on the connection between former European Networks of Excellence on marine sciences: the European Network of excellence for Ocean ecosystems analysis (EUR-OCEANS), the Marine Genomics Europe, and the Marine Biodiversity and Ecosystem Functioning EU Network of excellence (MarBEF). The objective of this consortium is to promote marine research activities providing expertise, facilitating knowledge exchange, and increasing capacity building in the marine research-related topics.

This consortium annually supports networking and capacity building activities (workshops, sponsorship of congress attendance etc.) through internal calls for proposals aiming to advance knowledge on emerging important scientific topics in marine sciences.

An example of a recent Euromarine initiative is the Marine Forest for Stakeholders (Marforstake: https://www.euromarinenetwork.eu/activities/marine-forests-stakeholders) workshop that gathered stakeholders connected to marine conservation to discuss the research needs in relation to marine conservation in Europe (Fig. 12.10). From this workshop a collaborative paper addressing the topic at the European level was prepared that is currently under development.

This kind of program exemplifies the advantages of developing initiatives at the European level fostering the participation of a representative group of experts and shortening potential financial constraints for participation. It gives the opportunity to build concerted opinions and approaches with wide European significance reducing fragmentation in political advice.

12.3.5 Knowledge Translation and Use

In this section, we describe four case studies illustrating different processes of gathering and transferring knowledge at the EU level. This knowledge was used to inform specific EU policy initiatives or priorities related to different aspects of the marine environment conservation.

Case study 1 provides a detailed and extended description of marine contaminants, a topic of concern at the EU level. It illustrates the role of the different initiatives, governance structures and regulatory and management approaches that can aid in the implementation of the European Marine Strategy Framework

Fig. 12.10 Description of the Marforstake workshop (2017) funded in the framework of EuroMarine calls

Directive. This case study intends to show the complexity of the network of actors and interactions that influence, provide knowledge and interact to support, in a coordinated way (ideally), well-developed knowledge-based policies in Europe.

The other case studies refer to examples of specific knowledge production initiatives developed to meet direct policy needs. *Case study 2* refers to a scenario where advice from the scientific community is requested to answer a particular question directly driven by the policy sector. *Case study 3* describes the development of a database to support specific needs related to the implementation of an EU regulatory framework. Finally, *case study 4* describes the process of collection of very specific information to support EU level initiatives on an emerging marine-related sector. In this case, the knowledge available was not of sufficient quality to support informed actions on the topic.

12.3.5.1 Case Study 1: Integrating Knowledge from Scientific Community, Regional/European Bodies, Stakeholders and International Organizations for Policy Implementation Support: The Marine Strategy Framework Directive (MSFD) Descriptors on Chemical Contaminants

Background Different anthropogenic pressures, including chemical pollution, can affect the health of our seas and oceans. Attaining and bringing together the necessary knowledge to identify the most relevant contaminants, their potential environmental impacts, and best approaches for their monitoring and assessment, pose a great challenge for managers and researchers all around the world. Here, we provide an overview of major players (across Europe and beyond), whose efforts for capturing, understanding, and improving knowledge on chemical pollution-related matters might be transferred into policy support.

Policy context The MSFD, aims at achieving Good Environmental Status (GES) in all EU marine waters by 2020. The MSFD includes a set of 11 qualitative descriptors that describe and help understand what GES means in practice.

Two of these descriptors deal specifically with chemical contaminants: D8 aims at concentrations of contaminants not giving rise to pollution effects and D9 refers to contaminant levels in edible tissues of fish and other seafood compared to human health threshold values (MSFD Commission Decision on criteria and methodological standards on GES, 2017/848/EU).

Knowledge Sources European legislation, specialized agencies, committees, and working groups from regional frameworks, international conventions and other international agreements as well as dedicated research programmes can act as source of knowledge and information of relevance for MSFD implementation regarding chemical contaminants.

EU Legislation

Water Framework Directive The MSFD has a very close link with the Water Framework Directive (WFD). Under the Common Implementation Strategy (CIS) of the WFD, the work supporting the prioritization of substances and the technical aspects of the chemical status assessment takes place within the Working Group Chemicals. This group collaborates closely with the WFD Working Group Ecological Status (ECOSTAT) to link the chemical and ecological status of surface waters.

Food legislation Regarding MSFD D9, member states shall consider the contaminants and maximum levels in food established in the Commission Regulation (EC) No 1881/2006. In this context, the European Food Safety Agency (EFSA) has a crucial role by providing scientific advice and risk assessments on a wide range of chemicals. This work is carried out by the EFSA's Panel on Contaminants in the Food Chain.

Registration, Evaluation, Authorisation, and Restriction of Chemical substances (REACH) Regulation (EC No 1907/2006) This regulation aims at improving the protection of human health and the environment through the better and earlier identification of the intrinsic properties of chemical substances. The European Chemicals Agency (ECHA) helps companies to comply with REACH, advances the safe use of chemicals, provides information on chemicals, and addresses chemicals of concern.

Regional Sea Conventions

The MSFD also includes provisions for cooperation at (sub)regional level on issues like the identification of additional relevant contaminants and threshold value establishment. To this end, EU Member States can benefit from the established regional cooperation structures (the RSC), which aim to protect the marine environment and bring together Member States and neighbouring countries in the shared marine basins.

OSPAR The work related to hazardous substances is implemented through the OSPAR's Hazardous Substances and Eutrophication Committee (HASEC), along with its subsidiary working groups on Monitoring and on Trends and Effects of Substances in the Marine Environment (MIME) and Inputs to the Marine Environment (INPUT).

The OSPAR's Offshore Industry Committee (OIC) collects and assesses data on the use and discharge of offshore chemicals, accidental spills, and emissions to air.

The OSPAR's Radioactive Substances Committee (RSC) carries out periodic evaluations to reduce discharges of radioactive substances to the North-East Atlantic. There are four RSC's subsidiaries Intersessional Correspondence Groups: Delivering the Fourth Periodic Evaluation (ICG 4PE), Close to Zero (ICG CTZ), Environmental Assessment Criteria (ICG EAC), and MODelling of additional concentrations of NORM in seawater from discharges of produced water from the offshore oil and gas sector (ICG MOD).

HELCOM The Working group on reduction of pressures from the Baltic Sea catchment area (PRESSURE) provides the technical basis to the work on inputs of hazardous substances from both diffuse and point sources on land. This includes the HELCOM Correspondence Group on Pharmaceuticals (CG PHARMA), which provides scientific background for the management of pharmaceuticals and their impacts in the environment.

The working group on the state of the environment and nature conservation (WG STATE and CONSERVATION) covers monitoring and assessment functions as well as issues related to nature conservation and biodiversity protection.

The work related to hazardous substances is supported by the HELCOM expert network on hazardous substances (EN-HZ), which serves as a discussion platform that provides expert advice to HELCOM working groups. Moreover, the expert group on monitoring of radioactive substances in the Baltic Sea (HELCOM MORS EG) focuses on the monitoring and assessment of radioactive substances.

Regarding pollution events, the Maritime Working Group (WG MARITIME) works to prevent any deliberate operational discharges as well as accidental pollution from ships. The Response Working Group (RESPONSE) works to ensure swift and right joint response to maritime pollution incidents. Within RESPONSE, the Expert Working Group on Oiled Wildlife Response (EWG OWR) acts as a forum for the exchange of information on progress and best practices in oiled wildlife response, while the Expert Group on Environmental Risks of Hazardous Submerged Objects (SUBMERGED) compiles and assesses information about hazardous objects, including chemical munitions dumped in the Baltic Sea.

The Barcelona Convention The Mediterranean Action Plan (MAP) is the institutional framework for cooperation in addressing common challenges of marine environmental degradation in the Mediterranean. The main MAP components in relation to chemical pollution are the Mediterranean Pollution Assessment and Control Programme (MED POL), which aims at preventing and eliminating land-based pollution and the Regional Marine Pollution Emergency Response Centre for the Mediterranean Sea (REMPEC), which targets prevention and reduction of pollution from ships.

The Integrated Monitoring and Assessment Programme (IMAP) is based on the ecosystem approach (EcAp), which is the guiding principle to MAP Work Programme. The EcAp process builds on the expert level discussions of the EcAp Coordination Group, including the Correspondence Groups on GES and Targets (COR GEST) and the Correspondence Group on Monitoring (COR MON) on Pollution.

The Black Sea Commission The Advisory Groups to the Black Sea Commission are its main source of expertise, information and support to implementation of the Black Sea Strategic Action Plan (BSAP). The Advisory Group on the Pollution Monitoring and Assessment (PMA) establishes a regionally coordinated network of National Status and Trends monitoring programmes; the Advisory Group on Control of Pollution from Land-Based Sources (LBS) provides technical support for the assessment and control of discharges of pollution from land-based sources; and the Advisory Group on the Environmental Safety Aspects of Shipping (ESAS) coordinates the regional approach to emergency response.

In addition, the Working Group on the WFD assists the Black Sea Commission in promoting the principles of the WFD. The European Union and the United Nations Development Programme (EU-UNDP) EMBLAS-Plus (Improving Environmental Monitoring in the Black Sea—Special Measures) aims to help improve protection of the Black Sea environment.

Multilateral Environmental Agreements

The Stockholm Convention (http://www.pops.int/): To lead to gradual decrease of the presence of persistent organic pollutants (POP) in the environment.

The Rotterdam Convention (http://www.pic.int/): To promote shared responsibility and cooperative efforts among Parties in the international trade of certain hazardous chemicals and contribute to the environmentally sound use of those hazardous chemicals.

The Basel Convention (http://www.basel.int/): To protect human health and the environment against the adverse effects of hazardous wastes.

The Minamata Convention on mercury (http://www.mercuryconvention.org/): To protect human health and the environment from the adverse effects of mercury.

The Geneva Convention on long-range transboundary air pollution (CLRTAP; http://www.unece.org/env/lrtap/welcome.html.html): To reduce air pollution.

The International Convention for the prevention of Pollution from Ships (MARPOL; http://www.imo.org/en/About/Conventions/ListOfConventions/Pages/International-Convention-for-the-Prevention-of-Pollution-from-Ships-(MARPOL).aspx): To prevent pollution by oil from ships.

Convention on the Prevention of Marine Pollution by Dumping of Wastes and Other Matter (London Convention and Protocol; http://www.imo.org/en/OurWork/Environment/LCLP/Pages/default.aspx): To control all sources of marine pollution and prevent pollution of the sea through regulation of dumping into the sea of waste materials.

International Convention on the Control of Harmful Anti-fouling Systems on Ships (AFS Convention; http://www.imo.org/en/About/Conventions/ListOfConventions/Pages/International-Convention-on-the-Control-of-Harmful-Anti-fouling-Systems-on-Ships-(AFS).aspx): To prohibit the use of harmful organotin compounds in anti-fouling paints used on ships and prevent the potential future use of other harmful substances in anti-fouling systems.

The Bonn Agreement (https://www.bonnagreement.org/): To combat pollution in the North Sea Area from maritime disasters and chronic pollution from ships and offshore installations.

The Lisbon Agreement (https://www.wipo.int/treaties/en/registration/lisbon/): To protect the North-East Atlantic against pollution.

Other Relevant European and International Bodies

The European Environment Agency (EEA; https://www.eea.europa.eu/) provides independent information on the environment for policy makers as well as the general public.

The European Maritime Safety Agency (EMSA; http://www.emsa.europa.eu/) aims at reducing the risk of maritime accidents and marine pollution from ships.

The International Maritime Organization (IMO; http://www.imo.org) has responsibility for the safety and security of shipping and the prevention of marine and atmospheric pollution by ships. The IMO's Marine Environment Protection Committee (MEPC), initially focused on the prevention of marine pollution by oil, resulting in the adoption of MARPOL. MEPC is aided by a number of IMO's Sub-Committees, such as the Sub-Committee on Pollution Prevention and Response (PPR).

IMO relies on the work by affiliated bodies and programmes like the Joint Group of Experts on the Scientific Aspects of Marine Environmental Protection (GESAMP). At present, IMO is the lead agency for four active Working Groups in GESAMP, including:

- WG 1 (EHS Working Group) to examine data for evaluating the environmental hazards of harmful substances carried by ships.
- WG 42 to provide independent advice on impacts of wastes and other matter in the marine environment from mining operations.

The Arctic Monitoring and Assessment Programme (AMAP; https://www.amap.no/) focuses on the monitoring and assessment of the status of the Arctic region with respect to pollution and climate change issues. Within AMAP, the Trends and Effects Monitoring Programme (ATEMP) is a harmonized programme for monitoring the trends and effects of contaminants.

The International Council for the Exploration of the Sea (ICES) (Table 12.3) coordinates the work of many expert groups, including:

- Marine Chemistry Working Group (MCWG), which focuses on the status and fate of pollutants in marine ecosystems.
- Working Group on Biological Effects of Contaminants (WGBEC), which examines the biological effects of contaminants in the marine environment and helps identify research and monitoring needs.
- Working Group on Marine Sediments in Relation to Pollution (WGMS), which conducts work on sediment-related science and advice.

The Policy Area Hazards within the EU Strategy for the Baltic Sea Region (EUSBSR) (PA Hazards) is a platform of cooperation between policy and science to reduce the use and emissions of hazardous substances to the Baltic Sea. Within PA Hazards, there are several relevant ongoing flagship projects, e.g.,:

- Baltic Sea Pharma Platform to reduce pharmaceuticals in the Baltic environment.
- DAIMON to evaluate the impacts of dumped ammunition.
- CHANGE to reduce the supply of toxic compounds from antifouling paints in leisure boats.

Research

A challenge in the implementation of the MSFD is to achieve the necessary knowledge upon which integrated management can build the tools for assessing progress towards GES. Many research initiatives can contribute to the development of knowledge and improvement of understanding of the elements that define the status of the marine environment, including those developed under. Horizon Europe, JPI Oceans, LIFE programme, and the specific DG ENV/MSFD proposals to support MSFD implementation.

Platforms for Data Collection and Sharing Information

WISE Marine (Table 12.3) shows the information and knowledge gathered or derived through the MSFD.

The European Environmental Information and Observation Network (Eionet; https://www.eionet.europa.eu/): Partnership network of the EEA to gather data on several topics related to the environment.

EMODnet (Table 12.3): The portal EMODnet Chemistry provides access to marine chemistry data sets and data products related to eutrophication and contaminants.

Network of reference laboratories, research centres and related organisations for monitoring of emerging environmental substances (NORMAN; https://www.norman-network.net/): Network to enhance the collection and exchange of data on emerging environmental substances.

Knowledge Implementation and Dissemination This "Marine Contaminants Landscape" (Fig. 12.11) intends to help EU national authorities, researchers, and stakeholders understand ongoing processes and interactions relevant for the assessment and monitoring of chemical contaminants in European marine waters. More collaborative efforts between the different stakeholders and expert groups are necessary for effective transfer of the knowledge into policy support and identification of opportunities to multiply and synergize efforts. With this purpose, the JRC of the European Commission established the MSFD Expert Network on Contaminants, a network of experts to exchange information and support EU Member States in MSFD implementation, while also providing interactions with the Regional Sea Conventions and other relevant platforms or frameworks.

12.3.5.2 Case Study 2: Scientific Knowledge to Inform Policy Priorities: *Food from the Oceans Initiative*

Background The "Food from the Oceans" report was published in 2017 by SAPEA (Horizon 2020—funded Science Advice for Policy by European Academies) as an evidence review report on the potential of the oceans to supply the expected global increase in food demand. The specific question to be answered was *"How can more food and biomass be obtained from the oceans in a way that does not deprive future generations of their benefits?"*

The work developed to answer this question involved working groups including experts from a range of specialization fields. The report produced (High level group

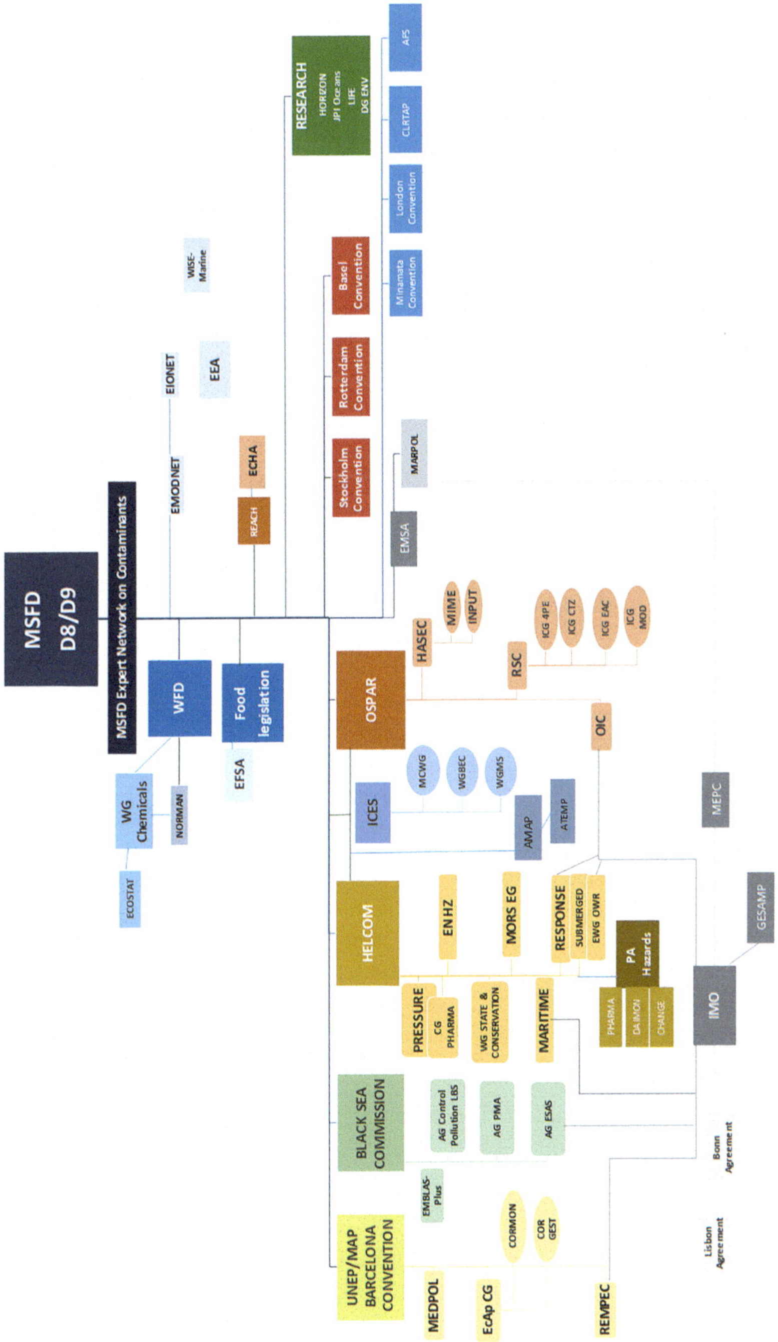

Fig. 12.11 Marine contaminants landscape. Note: List of acronyms is provided at the beginning of the chapter

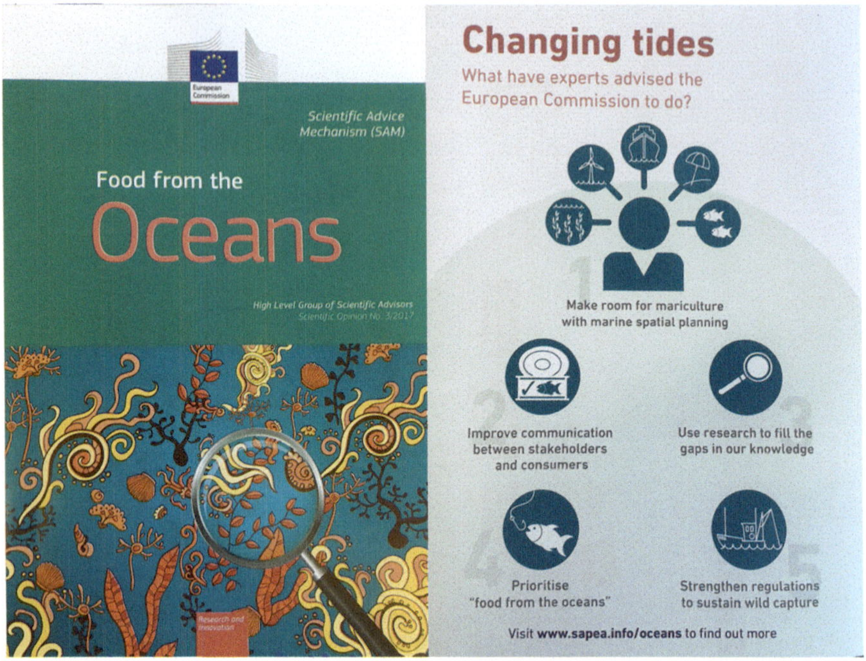

Fig. 12.12 Evidence review report and a part of the dissemination brochure of the Food from the Oceans scientific opinion

of scientific advisers 2017) establishes the state of the art on the topic of sustainable food extraction from the ocean and sets a group of recommendations on how to guarantee the sustainable exploitation of marine resources for food.

Policy Context Requested by Commissioner Karmenu Vella (Environment, Maritime Affairs and Fisheries), this report will inform the preparation of the future European Maritime and Fisheries Fund. Among the recommendations of the report is the expansion of aquaculture production, with potential implications for the Maritime Spatial Planning Directive and the Common Fisheries Policy. Finally, this information is relevant for the Blue Growth and Bioeconomy strategies and the Circular Economy Action Plan.

Knowledge Sources Consulted The scientific opinion from the Food from the Oceans Initiative is based on publicly available scientific evidence and literature (including grey literature), workshops, and consultation with the scientific community.

Knowledge Implementation and Dissemination The scientific opinion was published as an evidence review report and was showcased in different initiatives at the European Commission, European Parliament, and stakeholders and expert thematic meetings (Fig. 12.12).

12.3.5.3 Case Study 3: International Network on Alien Species for Research and Support to Policy: EASIN

Background EASIN was officially started by the JRC in 2012 with the objective of providing a single point of access to scientific information and georeferenced distribution data on alien species occurring in Europe, for the effective support of policies and scientific research on biological invasions in Europe (Katsanevakis et al. 2012, 2015). This network, interconnecting existing databases at national, European and global level, currently integrates information on approximately 14,000 alien species (of which approximately 1400 are marine), which can be searched and mapped online. The network indexes all information needed to:

(1) Efficiently link to existing online databases and retrieve spatial information for alien species distributions in Europe; (2) Access more detailed information in other sources, such as research articles, factsheets, and webpages; (3) Analyze spatial and temporal trends and patterns of biological invasions. Among the information compiled are species taxonomy, synonyms, common names, biology, year of first introduction, pathways, impacts, and occurrence records.

Policy Context In general, EASIN work technically and scientifically supports the European Commission and EU Member States on biodiversity-related policies (including the relevant assessment needs of the MSFD). Specifically, EASIN constitutes the central platform of the official information system foreseen under Article 25 of the EU Regulation 1143/2014 on Invasive Alien Species (IAS) supporting its implementation.

Knowledge Sources Consulted EASIN NOTSYS, scientific peer-reviewed publications, grey literature (e.g., reports from Member States), species occurrence and distribution databases (e.g., GBIF; Global Invasive Species information network, http://www.gisinetwork.org/; Ellenic Network on Aquatic Invasive Species, https://elnais.hcmr.gr/; The Mediterranean Science Commission; http://www.ciesm.org/; Marine Mediterranean Invasive Alien Species; http://www.mamias.org/) and citizen science records.

Knowledge Implementation and Dissemination The knowledge generated by EASIN activities is disseminated and implemented by using different channels:

Direct support to policy implementation, for example:

- Through written advice entailing official collaboration with the MS (Tsiamis et al. 2017, 2019a, b), provide baseline information on the EU geographical distribution of the IAS of Union concern, which among other things, can provide useful information to MS obligations under the EU Regulation on IAS, and factual basis for the review of the application of the IAS Regulation. Tsiamis et al. (2019b) provide refined national baseline inventories of alien species in the context of the MSFD.
- Through providing access to specific datasets: Country-level EASIN data packages for IAS of union concern, obeying to the requirements of EU Regulation

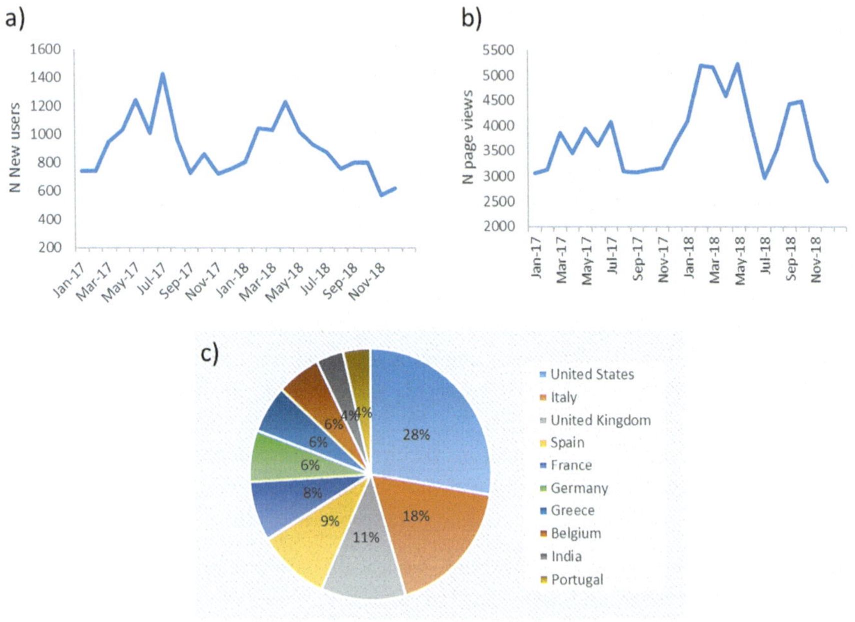

Fig. 12.13 Statistics of the EASIN platform in the period between January 2017 and December 2018 for the number of (**a**) new users, (**b**) number of page views and (**c**) country distribution of users

2017/1454 and Directive 2007/2/EC (INSPIRE), in support to MS reporting under Article 24 of the IAS Regulation (https://easin.jrc.ec.europa.eu/easin/Services/Reporting).

– Through a notification system (NOTSYS): A dedicated tool facilitating a timely comprehensive notification of detection of IAS of Union concern, and related eradication measures, as well as allowing an effective communication between the EC and MS.

Public use of the EASIN platform The EASIN web platform provides web services, search and mapping tools through which EASIN data and information can be accessed. The site shows a monthly number of new users ranging from approximately 600 to 1400 (Fig. 12.13a) and a number of page views ranging from approximately 2700 to 5100 (Fig. 12.13b). Considering the total number of users, Europe is the main user but individually there is a big share of USA and India occupying the first and third positions (Fig. 12.13c).

Scientific Publications EASIN data and information proved to be suitable for scientific research. The JRC EASIN team alone is author or co-author of 19 papers in peer-reviewed journals since 2012 analyzing mainly spatial and temporal trends, and patterns of biological invasions in Europe based on EASIN data. Special focus was placed in the analysis of marine biological invasions in terms of numbers of species introduced, their distribution, pathways and gateways of introduction and origins (Katsanevakis et al. 2013a, b, c; Nunes et al. 2014; Tsiamis et al. 2018) patterns and impacts (Katsanevakis et al. 2014a, b), due to the specific requirements on alien species in the MSFD. The knowledge generated is the basis to policy support contributions in the context of the MSFD.

12.3.5.4 Case Study 4: Knowledge Collection from an Emergent Industry Sector to Support Policy Development: The Algae Industry Dataset in EMODnet Human Activities

Background The algae industry dataset was collected in the framework of the JRC's Biomass assessment study (https://ec.europa.eu/knowledge4policy/projects-activities/jrc-biomass-assessment-study_en) to complement the available data on algae production in Europe. These data were very fragmented and of insufficient quality to produce the needed robust and overarching analysis of the status and potential of the European algae sector.

Policy Context The algae biomass production is an important resource for the European Strategies on Bioeconomy and Blue Growth. These initiatives aim to boost the development of the bio-based sectors while assuring the sustainability of the natural resources and exploitation methods. The development of this sector has also several implications and potential impacts for:

- EU regulations related to the quality of the marine environment: Habitats directive, MSFD, WFD.
- EU regulations related with the introduction and control of alien species into marine territory: Alien species regulation.
- EU regulations establishing the rules for the coordinated and sustainable management of the uses in the marine environment: Maritime spatial planning directive, common fisheries policy, Environmental impact assessment directive (2011/92/EU).

The knowledge gathered in the algae industry directory can support or provide insights to this framework of EU regulations.

Knowledge Sources Consulted The algae producing companies in Europe were mapped, both considering seaweeds and microalgae, and information on the location of the production facilities and production method collected. The underlying information consulted was the database shared by the European Algae Biomass Association (EABA) and information collected from individual stakeholders (researchers, managers, and industry) from different countries.

Knowledge Implementation and Dissemination A total of 200 companies were mapped, the information needed and available retrieved from the companies webpages and the companies contacted to confirm the collected information.

Fig. 12.14 Description of the process of data collection, display and example of products for the EMODnet algae production portal

The database was launched in November 2018 and is available on the EMODnet portal (http://www.emodnet-humanactivities.eu/view-data.php) for consultation and download. The information from the database was already used to perform data analysis included in presentations and expert group discussions (e.g., workshop on algae production in Europe) and publications (e.g., Blue Economy Report 2018–2019) (Fig. 12.14). This information will be regularly updated and new relevant categories will be added.

However, as already referred in the previous sections of this chapter, during the work performed in this case study several obstacles to gather data with the needed coverage and reliability were found. Examples are the limitations related to confidentiality issues, difficulties from data providers to understand the impact of good quality local data on the wide EU level analysis, difficulties from data providers to picture how the collected data would be reflected in impactful initiatives to the companies daily life, lack of time, and no perception of reward.

12.4 Conclusions

The marine conservation landscape in Europe harnesses on several instruments to promote knowledge production and exchange, data sharing, networking activities and scientific advice in support of the European level initiatives and legal frameworks related to this topic. Marine conservation is in the spotlight due to the documented global decrease in biodiversity which includes biodiversity decreases in the marine environment. Further, the extent of marine areas in Europe and the socioeconomic and ecological importance of marine resources warrant EU-level coordinated conservation implementation. The exploitation of new economic and biotechnological possibilities based on the marine environment must be in line with the maintenance of the good conservation status of marine ecosystems. This principle is the drive to several initiatives promoting the sustainable development of marine related activities and conservation measures targeting marine communities and the ecosystem services they provide. The information presented and discussed in this chapter shows that Europe has a good scientific production in the marine conservation field that, although not always regionally balanced, has the potential to adequately inform policy and management frameworks. It was not explicitly evaluated if the application of research outcomes in marine conservation is well developed in Europe and covers the topics and geographical extension requested by implementation measures. The survey conducted on knowledge sharing platforms provides some hints on this topic reporting on the non-adequacy of the data format and on scattered and difficult to map information for some cases. However, most of the data included in these databases are extracted from scientific publications and projects datasets which shows how the knowledge produced is being collected and incorporated in tools to facilitate implementation, coordination and harmonization at the European level. Additionally, the initiatives described in the case studies section provide practical examples of how scientific knowledge can be transferred directly to scientific advice or support initiatives and provide information concerning the entire European area, not only targeting the policy sector but also to be used by researchers and managers. The network of interactions between so many platforms, initiatives, and consortia can be complex to manage; implying a huge coordination effort to efficiently integrate the available but fragmented information from different geographical regions. This is due to the fact that national conservation priorities differ across European countries which consequently lead to regional differences in the development status of particular conservation areas. The initiatives supporting coordination efforts in a given sub-area of marine conservation can help to avoid duplication of initiatives, improve harmonization, facilitate synergies, and promote capacity building across European regions. This coordination will obviously positively impact the quality of data available and the knowledge advice on specific relevant topics for implementation.

References

Anderson P (2014) Bridging the gap between applied ecological science and practical implementation in peatland restoration. J Appl Ecol 51(5):1148–1152

Angelstam P, Andersson K, Axelsson R, Elbakidze M, Jonsson BG, Roberge JM (2011) Protecting forest areas for biodiversity in Sweden 1991-2010: the policy implementation process and outcomes on the ground. Silva Fennica 45(5):1111–1133

Arlettaz R, Schaub M, Fournier J, Reichlin TS, Sierro A, Watson JEM, Braunisch V (2010) From publications to public actions: when conservation biologists bridge the gap between research and implementation. Bioscience 60(10):835–842

Bainbridge I (2014) How can ecologists make conservation policy more evidence based? Ideas and examples from a devolved perspective. J Appl Ecol 51(5):1153–1158

Balme GA, Lindsey PA, Swanepoel LH, Hunter LTB (2014) Failure of research to address the rangewide conservation needs of large carnivores: leopards in South Africa as a case study. Conserv Lett 7(1):3–11

Batista MI, Cabral HN (2016) An overview of marine protected areas in SW Europe: factors contributing to their management effectiveness. Ocean Coastal Manage 132:15–23

Bertuol-Garcia D, Morsello C, El-Hani CN, Pardini R (2018) A conceptual framework for understanding the perspectives on the causes of the science-practice gap in ecology and conservation. Biol Rev 93(2):1032–1055

Charrier B, Abreu MH, Araujo R, Bruhn A, Coates JC, De Clerck O, Katsaros C, Robaina RR, Wichard T (2017) Furthering knowledge of seaweed growth and development to facilitate sustainable aquaculture. New Phytol 216(4):967–975

COM (2015) 478. Report from the Commission to the European Parliaments and the Council. The mid-term review of the EU Biodiversity Strategy to 2020

Commission Decision (EU) 2017/848 of 17 May 2017 laying down criteria and methodological standards on good environmental status of marine waters and specifications and standardised methods for monitoring and assessment, and repealing Decision 2010/477/EU

Commission Regulation (EC) No 1881/2006 of 19 December 2006 setting maximum levels for certain contaminants in foodstuffs

Cook CN, Mascia MB, Schwartz MW, Possingham HP, Fuller RA (2013) Achieving conservation science that bridges the knowledge-action boundary. Conserv Biol 27(4):669–678

Cvitanovic C, McDonald J, Hobday AJ (2016) From science to action: principles for undertaking environmental research that enables knowledge exchange and evidence-based decision-making. J Environ Manag 183(3):864–874

Dicks LV, Walsh JC, Sutherland WJ (2014) Organising evidence for environmental management decisions: a '4S' hierarchy. Trends Ecol Evol 29(11):607–613

Directive 2000/60/EC of the European parliament and of the council of 23 October 2000 establishing a framework for Community action in the field of water policy

Directive 2008/56/EC of the European parliament and of the council of 17 June 2008 establishing a framework for community action in the field of marine environmental policy (Marine Strategy Framework Directive)

Directive 2013/39/EU of the European parliament and of the council of 12 August 2013 amending Directives 2000/60/EC and 2008/105/EC as regards priority substances in the field of water policy

EEA report 3/2015 (2015) Marine protected areas in Europe's seas: an overview and perspectives for the future. European Environmental Agency. ISSN 1977-8449. 37 pp

European Commission (2017) Commission staff working document, Interim Evaluation of Horizon 2020

Fresco LO, Martinuzzi A, Anvret M (2015) Commitment and coherence. Ex-post-evaluation of the 7th EU framework programme (2007–2013). ISBN: 978-92-79-69546-9

Gabrié C, Lagabrielle E, Bissery C, Crochelet E, Meola B, Webster C, Claudet J, Chassanite A, Marinesque S, Robert P, Goutx MCQ (2012) The status of marine protected areas in the Mediterranean Sea. RAC/SPA. In MedPAN Collection 256 pp

Geoghegan-Quinn M (2013) Speech: Joint Programming Conference European Commission Speech 13/170

Griffiths RA (2004) Mismatches between conservation science and practice. Trends Ecol Evol 19(11):564-565

High level group of scientific advisors (2017) Food from the oceans: how can more food and biomass be obtained from the oceans in a way that does not deprive future generations of their benefits? Publications Office of the European Union, Luxembourg. ISBN: 978-92-79-67730-4

Hulme PE (2014) Bridging the knowing-doing gap: know-who, know-what, know-why, know-how and know-when. J Appl Ecol 51(5):1131–1136

Jeanmougin M, Dehais C, Meinard Y (2017) Mismatch between habitat science and habitat directive: lessons from the French (counter) example. Conserv Lett 10(5):634–644

Katsanevakis S, Bogucarskis K, Gatto F, Vandekerkhove J, Deriu I, Cardoso AC (2012) Building the European Alien Species Information Network (EASIN): a novel approach for the exploration of distributed alien species data. BioInvasions Rec 1(4):235–245

Katsanevakis S, Gatto F, Zenetos A, Cardoso AC (2013a) How many marine aliens in Europe? Manage Biol Invasions 4(1):37–42

Katsanevakis S, Zenetos A, Belchior C, Cardoso AC (2013b) Invading European Seas: assessing pathways of introduction of marine aliens. Ocean Coast Manag 76:64–74

Katsanevakis S, Zenetos A, Poursanidis D, Nunes AL, Deriu I, Bogucarskis K, Cardoso AC (2013c) ELNAIS meets EASIN: distribution of marine alien species in Greece using EASIN mapping services and ELNAIS spatial data. Mediterr Mar Sci 14(1):95–98

Katsanevakis S, Coll M, Piroddi C, Steenbeek J, Ben Rais Lasram F, Zenetos A, Cardoso AC (2014a) Invading the Mediterranean Sea: biodiversity patterns shaped by human activities. Front Mar Sci 1:32

Katsanevakis S, Wallentinus I, Zenetos A, Leppäkoski E, Çinar ME, Oztürk B, Grabowski M, Golani D, Cardoso AC (2014b) Impacts of marine invasive alien species on ecosystem services and biodiversity: a pan-European review. Aquat Invasions 9(4):391–423

Katsanevakis S, Deriu I, D'Amico F, Nunes AL, Sanchez SP, Crocetta F, Arianoutsou M, Bazos I, Christopoulou A, Curto G, Delipetrou P, Kokkoris Y, Panov VE, Rabitsch W, Roques A, Scalera R, Shirley SM, Tricarico E, Vannini A, Zenetos A, Zervou S, Zikos A, Cardoso AC (2015) European Alien Species Information Network (EASIN): supporting European policies and scientific research. Manage Biol Invasions 6(2):147–157

Knight AT, Cowling RM, Rouget M, Balmford A, Lombard AT, Campbell BM (2008) Knowing but not doing: selecting priority conservation areas and the research-implementation gap. Conserv Biol 22(3):610–617

Maes J, Paracchini ML, Zulian G, Dunbar MB, Alkemade R (2012) Synergies and trade-offs between ecosystem service supply, biodiversity and habitat conservation status in Europe. Biol Conserv 155:1–12

Ntshotsho P, Prozesky HE, Esler KJ, Reyers B (2015) What drives the use of scientific evidence in decision making? The case of the South African working for water program. Biol Conserv 184: 136–144

Nunes AL, Katsanevakis S, Zenetos A, Cardoso AC (2014) Gateways to alien invasions in the European Seas. Aquat Invasions 9(2):133–144

Pérez-Espona S, ConGRESS Consortium (2017) Conservation genetics in the European Union-biases, gaps and future directions. Biol Conserv 209:130–136

Phillipson J, Lowe P, Proctor A, Ruto E (2012) Stakeholder engagement and knowledge exchange in environmental research. J Environ Manag 95(1):56–65

Powell JJW, Dusdal J (2017) Science production in Germany, France, Belgium and Luxembourg: comparing the contributions of research universities and institutes to science, technology, engineering, mathematics and health. Minerva 55(4):413–434

Pullin AS, Knight TM, Stone DA, Charman K (2004) Do conservation managers use scientific evidence to support their decision-making? Biol Conserv 119:245–252

Regulation (EC) No 1907/2006 of the European parliament and of the council of 18 December 2006 concerning the Registration, Evaluation, Authorisation and Restriction of Chemicals (REACH), establishing a European Chemicals Agency, amending Directive 1999/45/EC and repealing Council Regulation (EEC) No 793/93 and Commission Regulation (EC) No 1488/94 as well as Council Directive 76/769/EEC and Commission Directives 91/155/EEC, 93/67/EEC, 93/105/EC and 2000/21/EC

Snoeijs-Leijonmalm P, Barnard S, Elliot M, Andrusaitis A, Kononen K, Sirola M (2017) Towards better integration of environmental science in society: lessons from BONUS, the joint Baltic Sea environmental research and development programme. Environ Sci Policy 78:193–209

Stephenson RL, Benson AJ, Brooks K, Charles A, Degnbol P, Dichmont CM, Kraan M, Pascoe S, Paul SD, Rindorf A, Wiber M (2017) Practical steps toward integrating economic, social and institutional elements in fisheries policy and management. ICES J Mar Sci 74(7):1981–1989

TEEB (2010) The economics of ecosystems and biodiversity: ecological and economic foundation. Earthscan, Cambridge

Tsiamis K, Gervasini E, Deriu I, D'amico F, Nunes A, Addamo A, Cardoso AC (2017) Baseline distribution of invasive alien species of Union concern. Ispra (Italy): Publications Office of the European Union; EUR 28596 EN

Tsiamis K, Zenetos A, Deriu I, Gervasini E, Cardoso AC (2018) The native distribution range of the European marine non-indigenous species. Aquat Invasions 13(2):187–198

Tsiamis K, Gervasini E, Deriu I, D'amico F, Katsanevakis S, Cardoso AC (2019a) Baseline distribution of species listed in the 1st update of Invasive Alien Species of Union concern. Ispra (Italy): Publications Office of the European Union; EUR 29675 EN

Tsiamis K, Palialexis A, Stefanova K, Gladan ŽN, Skejić S, Despalatović M, Cvitković I, Dragičević B, Dulčić J, Vidjak O, Bojanić N, Žuljević A, Aplikioti M, Argyrou M, Josephides M, Michailidis N, Jakobsen HH, Staehr PA, Ojaveer H, Lehtiniemi M, Massé C, Zenetos A, Castriota L, LiviS, Mazziotti C, Schembri PJ, Evans J, Bartolo AG, Kabuta SH, Smolders S, Knegtering E, Gittenberger A, Gruszka P, Kraśniewski W, Bartilotti C, Tuaty-Guerra M, Canning-Clode J, Costa AC, Parente MI, Botelho AZ, Micael J, Miodonski JV, Carreira GP, Lopes V, Chainho P, Barberá C, Naddafi R, Florin A-B, Barry P, Stebbing PD, Cardoso AC (2019b) Non-indigenous species refined national baseline inventories: a synthesis in the context of the European Union's Marine Strategy Framework Directive. Marine Poll Bull 145:429–435

Chapter 13
Translating Research into Wildlife Conservation Actions Through Multi-sector Collaboration in Tropical Asia

Finbarr G. Horgan and Enoka P. Kudavidanage

Contents

F. G. Horgan (✉)
EcoLaVerna Integral Restoration Ecology, Co. Cork, Ireland

Universidad Católica del Maule, Facultad de Ciencias Agrarias y Forestales, Escuela de Agronomía, Curicó, Chile

Environment and Sustainable Resource Management, University College Dublin, Dublin, Ireland
e-mail: f.horgan@ecolaverna.org

E. P. Kudavidanage
EcoLaVerna Integral Restoration Ecology, Co. Cork, Ireland

Tropical Ecosystem Research Network, Piliyandala, Sri Lanka

Department of Natural Resources, Faculty of Applied Sciences, Sabaragamuwa University, Belihuloya, Sri Lanka

C. C. Ferreira, C. F. C. Klütsch (eds.), *Closing the Knowledge-Implementation Gap in Conservation Science*, Wildlife Research Monographs 4,
https://doi.org/10.1007/978-3-030-81085-6_13

13.1 Introduction: Asia in Context

Over the last 60 years, much of Asia has transformed from fragile economies largely based on postcolonial plantation systems to rapidly developing industry-based economies (Hayami 2001; Imai et al. 2018). However, the plantation approach to land management has continued in many regions, particularly where governments support the conversion of high-diversity tropical forests to agricultural or timber-producing monocultures either to incentivize transmigration and development or as concessions to industry (Law et al. 2015; MONRE – Laos 2016; Dhiaulhaq et al. 2018; Varkkey et al. 2018). Sixty years ago, tropical Asia was predominantly covered by native forests; however, since that time, most of these forests have disappeared (Fig. 13.1) and destruction of the remaining forest fragments continues to be a serious problem (Wilcove et al. 2013; Imai et al. 2018; Varkkey et al. 2018). Several nations of tropical Asia are among the most populous on the planet or have among the highest population densities of continental and large island/archipelago nations (UN DESA 2018). Despite continuing population growth, fertility rates have generally declined in recent decades (UN DESA 2018); however, industrial and agricultural productivity have continued to increase dramatically. For example, agricultural productivity (in terms of hectares of agricultural production) in South East Asia increased by between 15 and 19% each decade from 1960 to 2010 (Fig. 13.2). Industrial growth in India and China has further fueled the intensification of agriculture by supplying cheaper inputs and demanding a greater supply of raw materials (Bosworth and Collins 2008; Horgan 2017). These trends have resulted in what Wilcove et al. (2013) have termed Navjot's nightmare (after Prof. Navjot Sodhi), where deforestation for timber production, combined with agricultural expansion, have caused a biodiversity disaster throughout much of tropical Asia (Sodhi et al. 2004; Sodhi and Brook 2008).

As Asian economies grow, government and industry revenues are re-invested into infrastructural and industrial developments that further impact natural resources. The depletion of natural resources and, in particular deforestation, can set in motion a cascade of events that becomes difficult to contain, including the invasion of forests by hunters and foragers seeking lucrative ingredients for traditional medicines or collecting plants and wildlife destined for distant cities (Gray et al. 2018). Imbalances in the natural regulation of ecosystem processes caused by changing landscapes, excessive hunting, and declining numbers of apex predators create "silent forests" (i.e., forests largely devoid of wildlife). Furthermore, a lack of available resources, including forage, water, or prey for wildlife in depleted natural areas can increase the frequency of human–wildlife conflicts (Ickes 2001; Wallach et al. 2015; Luskin et al. 2017; EAZA 2018; Thinley et al. 2018). The management of wildlife

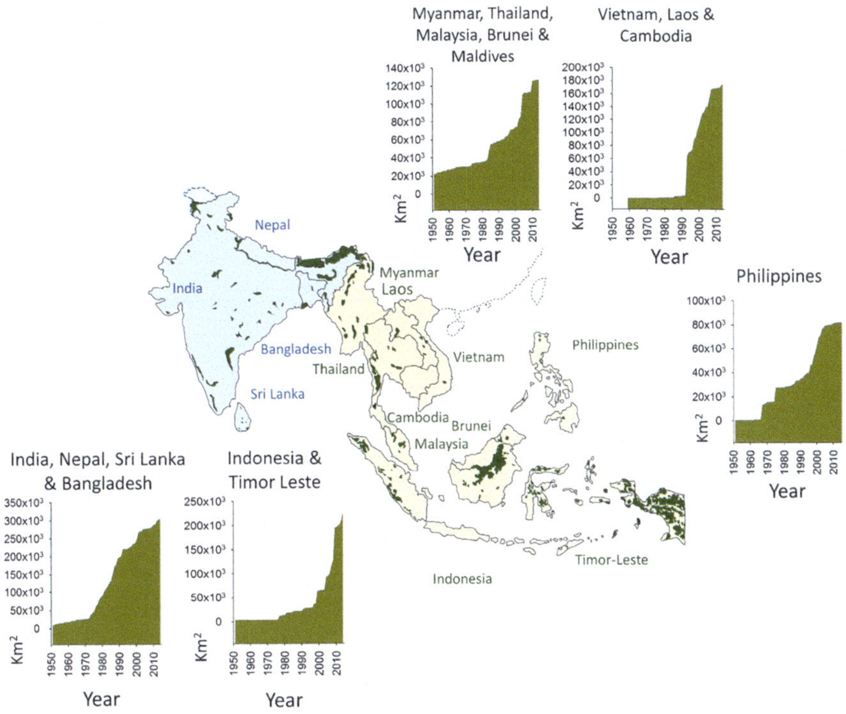

Fig. 13.1 Map of tropical South Asia (blue) and South East Asia (yellow) indicating areas of remaining intact forest cover (dark green). This does not include other forest types (e.g., secondary growth, logged forests, or plantations). Inset: Graphs indicate the accumulation of protected areas (in km^2) since 1950 in different regions within tropical Asia. The figure and graphs were drawn using open-access data available from IUCN (2018)

and compensation for damage to homes and crops are key issues for Natural Resource and Wildlife Departments throughout the region (Pechacek et al. 2013; Gogoi 2018). Inadequate compensation for damage can further drive resentment against wildlife and against conservation practitioners (Pechacek et al. 2013; Suba et al. 2017). Disentangling the economic, ecological, and social factors that determine the success or failure of conservation actions is therefore understandably complex (Fig. 13.3) and requires increasingly transdisciplinary approaches to identify socially acceptable and culturally appropriate environmental solutions with lasting impacts.

Asian governments do not dispute that natural resources require adequate protection and management and that agriculture and industry must grow in an economically and environmentally sustainable fashion. Several governments have made headway by creating conservation policies, designating protected areas (Fig. 13.1), or ratifying international conventions such as the Convention on Biological Diversity (CBD), the Collaborative Programme on Reducing Emissions from Deforestation and Forest Degradation in Developing Countries (REDD+), or the Convention on International Trade in Endangered Species of Wild Fauna and Flora (CITES)

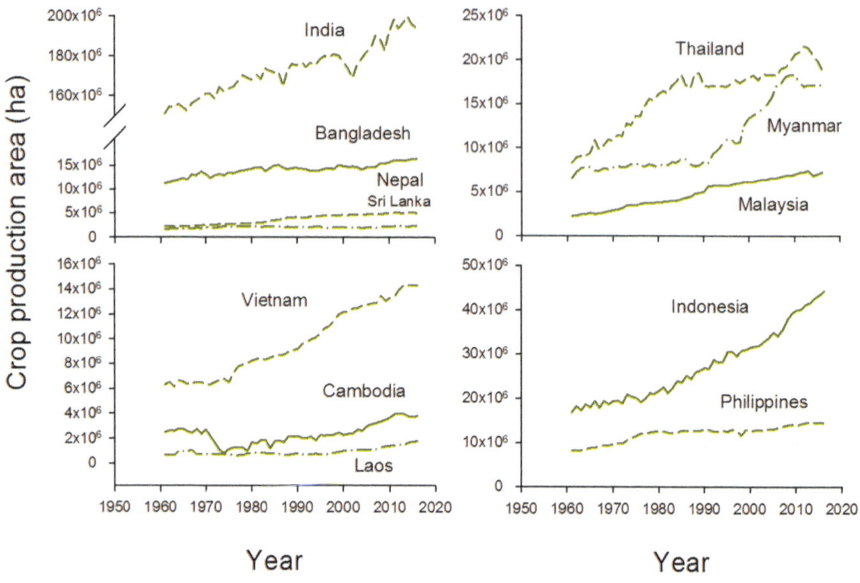

Fig. 13.2 Agricultural production in Asia (area in hectares harvested for all crops) since 1961 (graphs were drawn using open-access data from FAO (2018)—note that values include double and triple cropping in some regions)

(CBD 2018; CITES 2018; IUCN 2018; UN-REDD 2018). Furthermore, several Asian governments and Asian-based institutions participate in frameworks, platforms, and initiatives such as the Intergovernmental Science-Policy Platform on Biodiversity and Ecosystem Services (IPBES) or the International Partnership for the Satoyama Initiative (IPSI) that aim to promote biodiversity conservation (IPBES 2018; IPSI 2018). However, the governments of tropical Asian countries greatly prioritize economic growth resulting in often contradictory policies and legislation. For example, mandates to develop the science and infrastructure necessary to increase agricultural productivity to meet Asia's projected food security needs are often at loggerheads with conservation policies aimed at reducing deforestation and protecting vulnerable habitats (e.g., Government of Nepal 2014; Horgan et al. 2018; Horgan and Kudavidanage 2020). How such contradictions are resolved will depend on investments into reasoned and informed dialogue; a dialogue that can only be achieved if adequate attention, investment, and support are directed toward knowledge capture and the application of that knowledge to policy and practice for both development and conservation.

In this chapter, we examine mechanisms by which science has best supported policy and conservation actions in tropical Asia. We assess the development and status of conservation research in Asia and examine the capacity of tropical Asian countries to gain reliable scientific knowledge and to put that knowledge into conservation practice. We also assess the proposed future directions for research and capacity building in biodiversity and resource management as expressed in the

Fig. 13.3 Tropical Asia is a region of rich ethnic, cultural, religious, sociopolitical, and natural diversity; making conservation science an urgent necessity as well as a logistic challenge. In Sri Lanka, traditional villages that included "wewa" (lake), "dageba" (pagoda in the temple), "gama" (village), and "kumbura" (paddy field) promoted sustainability through respect for religion and nature (photo by F. Horgan)

National Biodiversity Strategies and Action Plans (NBSAPs) prepared by Asian governments. Based on an analysis of case studies, we assess the roles of different stakeholders during research and implementation and highlight the need for coordinated multidisciplinary and multi-sector collaboration in conservation projects. Finally, we make recommendations toward building a research capacity that is crafted for the particulars of tropical Asia by recognizing Asia's diversity of economies, political systems, languages, cultures and religions, and taking into account some key attributes that are frequently linked to project success. Our analysis is limited to tropical Asia and focuses mainly on terrestrial ecosystems.

13.2 Research–Action Partnerships and the Need for Multi-sector Collaboration

Conservation science must lead to concrete conservation actions. Actions can result from the accumulation of knowledge until a threshold where informed interventions become possible [value-of-information for environmental decision-making (e.g., Moore and Runge 2012)], or from careful monitoring that guides projects through adaptive management (Bixler et al. 2015; Gosselin et al. 2018). However, given the immense pressures on Asia's natural resources, conservation actions have often been urgent procedural responses (protocols) to current predicaments without the necessary research or monitoring that allows reasonable predictions of outcomes or impacts (Spangenberg et al. 2015). For example, in many cases the designation of conservation areas in Asia has been made without knowledge of the home ranges or habitat requirements of iconic fauna (Fernando et al. 2012; Jathanna et al. 2015; Sarker et al. 2016), or without consulting local communities—often eventually leading to conflicts, or at least distrust, between local communities and conservation programs (Jadhav and Barua 2012; Perez 2018).

It is increasingly apparent that the successful management of habitats, species, and resources requires collaborative research and implementation teams that include academics, government and nongovernment organizations (NGOs), local communities and their social, political and religious leaders, and—occasionally—experts in social and economic development related to health, tourism, or industry (Novellino and Dressler 2010; Jadhav and Barua 2012; Bixler et al. 2015; Rozylowicz et al. 2017; Dressel et al. 2018; Gosselin et al. 2018). Whereas such collaborators will normally have shared conservation goals, they are each subject to proximate incentives and restrictions that can sometimes reduce the efficiency of workflows. For example, academics, applied conservation scientists, and other conservation practitioners seek different types of knowledge because they are evaluated according to different metrics. Such differences in research goals can represent potential barriers to effective collaboration. Some typical advantages and constraints associated with different types of collaborators are outlined in Fig. 13.4 and explained below.

13.2.1 Asia's Universities

There are over 3300 universities located throughout tropical Asia (UniRank 2018). Most universities follow Western models for administration, research, and teaching developed in Europe and North America (although several institutes grounded in Asian philosophies occur throughout the region and especially in India). Under the Western models, academics are evaluated according to their research output as determined by the number of peer-reviewed scientific articles published in high-impact journals and the number of times their articles are cited by academic peers. The use of such metrics strongly influences the direction of research and the value

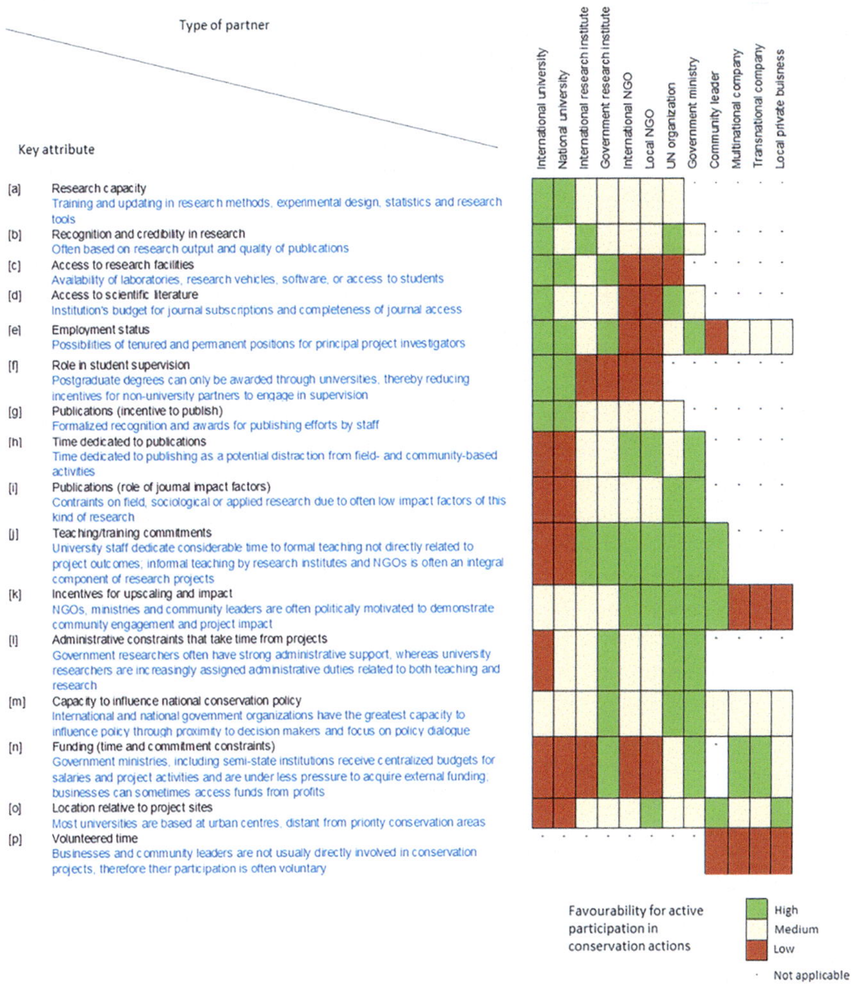

Fig. 13.4 An assessment of the key attributes of 12 different types of partners involved in conservation projects. Partners are categorized based on typical advantages and constraints (favorability) associated with each attribute for the successful completion of conservation projects that integrate scientific knowledge and concrete actions. Explanations for categorizing attributes as high, medium, or low favorability are indicated in blue font; examples are based on Asian contexts drawn from the authors' experiences of working in multidisciplinary teams. The figure indicates that successful conservation initiatives will achieve the greatest benefits from collaborations between a range of international and regional partners across a range of sectors. For example, whereas academic and research institutions can support initiatives through access to research facilities and postgraduate students, they are limited by distances and knowledge of field sites as well as administrative constraints on field work and community access. Such constraints can be overcome through successful partnerships with national ministries, NGOs, or local communities. Further explanations are presented in the main text

given by institutes to their academics—often with little credit for active engagement in concrete conservation actions (Salazar-Vallejo and Carrera-Parra 1998; Ohmer and Bishop 2011). Furthermore, higher citation impacts can be gained from opinion pieces, literature reviews, meta-analyses, and desktop studies (Statzner and Resh 2010; Fox et al. 2016), which reduces the relative gains from more applied, field-based research and from the type of regionally focused studies that are often required to support local conservation actions (Ríos-Saldaña et al. 2018). Shifting donor priorities and increasingly restrictive funding in Asia, as in other regions (Bakker et al. 2010; Verissimo et al. 2014) will also make it difficult for academics to make long-term commitments to conservation projects or to see projects through to final conservation actions. One approach to redress the balance is for academic institutes to formally recognize project contributions and impacts (Meyer et al. 2010). Many Asian universities promote "social projection" among staff and students, particularly students preparing final year theses. Social projection is also increasingly promoted as a modular component to standard undergraduate courses, for example, by introducing mini-projects where students gain hands-on experience in conservation issues. Such an emphasis encourages academics and students to contribute toward meaningful actions. Academics can gain further institutional merits based on public recognition of their roles in conservation actions, on media attention to projects, or where they receive prestigious awards. However, such accolades may be heavily influenced by the nature of the action or target species (i.e., charismatic flagship fauna), the composition of the project team (e.g., charismatic research professor), proximity to urban centers and media outlets, or through wise use of communication opportunities.

13.2.2 Government Services

Conservation practitioners include a diversity of civil servants, including government scientists, forest and fisheries technicians, policy makers, or park administrators, wardens, and ground-level staff. In Asia, these are typically part of a hierarchical structure that ultimately reports back to central governments. In some cases, civil servants are directly appointed by governments, changing under each new administration, and are subject to censorship and pressures to support government agendas (Moon and Hwang 2013). For Asian scientists not based at universities, scientific publications may enhance credibility (Loh et al. 2018); however, the efforts of publishing articles in high-impact, international journals are seldom adequately awarded, particularly in countries that use non-Latin scripts or where English is not an official language (Mukherjee 2018). Nevertheless, where government scientists can freely engage in research, they may have greater potential than universities to convert their findings into concrete conservation actions. This is because they devote little time to non-research activities (such as formal teaching) and are often supported by government funding, without pressures to seek funding from external donors. Perhaps most importantly, the salaries and positions of government researchers are guaranteed (Moon and Hwang 2013), thus facilitating long-term involvement in specific conservation projects. However, even among government researchers, progress can be limited because of preferences among postings and duties. For example, we noted that whereas priority conservation areas are mainly located in east, south, and north-central Sri Lanka, Wildlife Department staff predominantly select postings to western areas where they have better opportunities for promotions and access for their families to educational, health, and other facilities (Kudavidanage and Horgan, unpublished data).

13.2.3 International Research Organizations

Part of the mandate of international research and development organizations such as the Centre for International Forestry Research (CIFOR; based in Indonesia), the International Crops Research Institute for the Semi-Arid Tropics (ICRISAT; based in India), or the ASEAN Centre for Biodiversity (ACB; based in the Philippines) is to convert science to policy and actions for development. However, irregular funding and frequent changes in leadership and in research priorities can hamper progress. Furthermore, because research leaders are employed on short-term contracts (usually 3–5 years) and core funds are not generally allocated to project activities, individual researchers usually make only short-term commitments to specific projects (Leeuwis et al. 2018). On the other hand, some institutes, such as those that make up the Consultative Group on International Agricultural Research (CGIAR) mainly focus on commodity-based research and are increasingly encouraged to create links with the private sector (Manicad 1999), diverting resources away from environmental conservation toward product-based research and sometimes reinforcing exploitative national agricultural and forestry policies.

13.2.4 Nongovernment Organizations

Although sometimes involved in research and data collection, nongovernment organizations (NGOs) play a largely distinct role in conservation compared to academics and other researchers. NGOs are generally tied to specific animal and plant species, regions, communities, or issues (such as animal trafficking, poaching, or deforestation) and generally depend on external project funding from donors (although some may include paying services, or they may represent the corporate responsibility arm of some private company: Harada and Wiyono 2014; KimDung et al. 2016). Many NGOs do not publish results in scientific journals (they tend to rely on social media outlets), and they are evaluated almost entirely by donors, although they must also maintain a good rapport with their target communities (Ahsan et al. 2009; Novellino and Dressler 2010; Harada and Wiyono 2014). Nevertheless, NGOs play a vital role as knowledge "gatekeepers" that bring

Fig. 13.5 Researchers from a multi-sector team that includes international and national universities and a local NGO conduct surveys with villagers in Sri Lanka. Tropical Asia includes the most populous nations on the planet. Understanding the knowledge, attitudes, and practices of rural communities is essential to guarantee successful conservation actions (photo by E. Kudavidange)

science, innovation, and information to communities, interpreting the information and helping translate it into concrete actions (Shukla and Gardner 2006; Fig. 13.5). NGOs are also important in lobbying governments to create, adapt, or change policy. International NGOs, in particular, have played a tremendous role in bringing public attention to conservation issues in Asia by focusing on flagship species such as orangutans (*Pongo* spp.), rhinoceros (Rhinocerotidae), tigers (*Panthera tigris*), leopards (*Panthera pardus*), and elephants (*Elephas maximus*) or to issues such as animal trafficking and poaching. Furthermore, NGOs increasingly play a role as facilitators of research by supporting university researchers and students in the field. However, because of their diversity of functions and internal pressures to ensure that they appear credible, accurate assessments of NGO impacts can sometimes be difficult (Wahlén 2014).

13.2.5 Private Sector

The private sector is increasingly apparent in conservation projects in Asia. This is partly because funders—particularly government institutions and ministries—increasingly demand that researchers seek private sector participation (Manicad 1999; Bakker et al. 2010). Furthermore, many companies realize the importance of green labelling to improve their market shares. However, it can be difficult to successfully integrate industry and conservation science because the objectives of the two are seldom of mutual immediate advantage. Recognition of responsible actions is often a sufficient incentive for private sector participation in conservation projects (Svensson et al. 2008; Thompson 2018). However, because private sector partners are generally both collaborators and funders, they can unduly influence the direction of projects, particularly where they dictate intellectual property rights on data and publishing, or demand entitlements from donations and investments (e.g., demanding priority accommodation in the national parks that they support or access to parks during closed seasons: Kudavidanage, personal observation). During open funding calls, participation in conservation projects may be less attractive for Asian industry partners compared to participation in, for example, medical and technology research. Asian conservation biologists must therefore become increasingly innovative if they are to extract direct economic returns on research investments to incentivize industry involvement (e.g., Robinson 2012). Nevertheless, incorporating collaborators from the private sector can be useful to bring about lasting change where the private partner contributes successfully to conservation actions by marketing traditional food products, operating environmentally responsible hotels and tourist facilities, or otherwise supporting conservation actions through funding or in-kind support. For example, the herbs-based medical industry Sidomuncul is reported to contribute between US$10–15 millions annually to biodiversity-related issues including research, training, and sustainable land management in Indonesia (BAPPENAS 2016).

13.2.6 Building Multi-sector Teams that Complement Capacities

During collaboration, partners from different sectors can help each other to overcome individual obstacles and thereby increase the transdisciplinary capacity of each institute as well as ensuring a multidisciplinary approach to the final project. Whereas the greatest form of collaboration will involve an equitable distribution of research funds as well as shared, visible acknowledgment on scientific publications and through popular media outlets (TV, radio, social networks), other in-kind incentives can also improve collaboration. For example, international universities might offer collaborators access to research facilities or services, including access to libraries and training opportunities. Government scientists with access to project funds might also gain better terms in the supervision of graduate students that avoids simply tagging principal academic supervisors to research publications. Funders might encourage partnerships between academics and NGOs, rather than simply focusing on building partnerships with the private sector. Academic partners can facilitate and build capacity that expands key attributes of NGOs, strengthening attributes [a-d] and [f-g] in Fig. 13.4, whereas NGOs can facilitate engagement by academics with rural communities and remote conservation projects, strengthening attributes [k-o] in Fig. 13.4. Such steps would help partners engaged in implementation to gain increased credibility and allow universities to develop metrics that assess the participation and impact of their academics in concrete actions.

In the following sections, we use published articles to gain insights into the participation by different sectors in knowledge production related to conservation science in tropical Asia. We then examine a series of Asian case studies to determine the mechanisms by which multi-sector teams have successfully translated this knowledge into conservation actions.

13.3 Knowledge Production in Asia

13.3.1 Sources of Conservation Knowledge in Tropical Asia

To assess the role of Asian scientists in generating conservation knowledge, and to assess the nature and extent of collaboration during recent conservation projects in the region, we examined articles published from January 2000 to June 2018 in ten conservation journals. The journals were *Conservation Biology*, *Biological Conservation*, *Animal Conservation*, *Environmental Conservation*, *Biodiversity and Conservation*, *Conservation Genetics*, *Oryx*, *Conservation Letters*, *Tropical Conservation Science*, and *Insect Conservation and Diversity*. During this time, these journals published a total of 21,197 papers. We used the ISI Web of Science to perform a geographical breakdown (by continent and country) of author affiliations for these articles using combinations of country names in the "address" prompt as search criteria. Over 12% of the papers ($n = 2544$) included institutions from Asia (Fig. 13.6a). Although this number is equivalent to the number from Oceania and greater than that from Central and South America or from Africa, in terms of per capita output, Asian institutions were underrepresented (i.e., 0.67 papers per million population for Asia compared to 80.09 for Oceania, 4.24 for Central and South America, and 1.61 for Africa). Furthermore, among the Asian institutions represented in these papers, 46% were from India and China, with few publications focused on conservation topics from the remaining countries (Fig. 13.6b), or few with contributors from institutions in any of these remaining countries (Fig. 13.6c). We did not assess the quality of the papers, or their potential to contribute to concrete actions, but assumed that representation by authors in papers will be correlated with the research capacities of the different institutions and their respective nations. Although there are several local journals produced in the region, most are not included in international or interlibrary databases. Furthermore, Asian authors contribute large numbers of articles (over 75%) to a growing number of predatory journals (Xia et al. 2015); however, many of these fail to meet international standards for experimental design, analysis, and reporting (Beall 2015). Assuming that peer-reviewed publications are an indicator of higher quality research outputs, our rapid assessment suggests that institutions in many tropical Asian countries are likely to be knowledge deficient during the design and implementation of conservation strategies. Indeed, this has been acknowledged in many of the National Biodiversity Strategies and Action Plans (NBSAPs) prepared by Asian governments (see below).

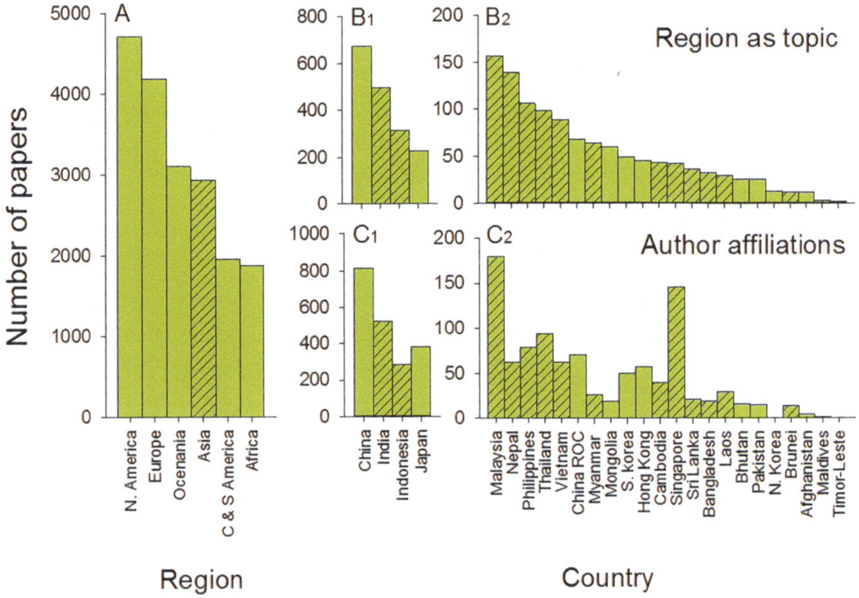

Fig. 13.6 The number of papers published in ten conservation journals (see main text) between January 2000 and June 2018 according to (**a**) geographical breakdown (by continent) of author affiliations, ($B_{1,2}$) Asian countries targeted by the published research, i.e., the geographical focus of the research, or ($C_{1,2}$) with an Asian affiliation (by country). Shaded bars in (**b**) and (**c**) are tropical Asian countries. Note differences in scales between B_1 and B_2, and C_1 and C_2

To investigate aspects of collaboration and local capacity, we recorded which institutions participated in each research paper. We then checked the authors, where possible, to assess whether these were ex-patriots or nationals based on published information, mainly through institution webpages, and we categorized the institutes into one of five categories. These were: (a) national educational institutes (national universities and colleges) and international educational institutes based in Asia (e.g., Monash and Nottingham Universities in Malaysia, or the French Institute of Pondicherry in India); (b) National government research institutes and semi-state research institutes (e.g., Indian Institute of Science, national parks, zoos, and botanic gardens) including government institutions such as wildlife and natural resource ministries; (c) International NGOs (e.g., the World Wildlife Fund [WWF], Fauna and Flora International [FFI], or Wildlife Conservation Society [WCS]), and international projects (Turtle Survival Alliance, Leuser Conservation Forum, Roundtable on Sustainable Palm Oil [RSPO]); (d) local NGOs; and (e) private companies.

A total of 1443 institutions participated in the studies (from 1361 papers). Over 63% of these were educational institutes or government and semi-state research institutes. However, across countries, the prominence in research by different types of institutions varied. In Bangladesh and Sri Lanka, university scientists played a predominant role in conservation projects (authoring 61% and 57% of papers,

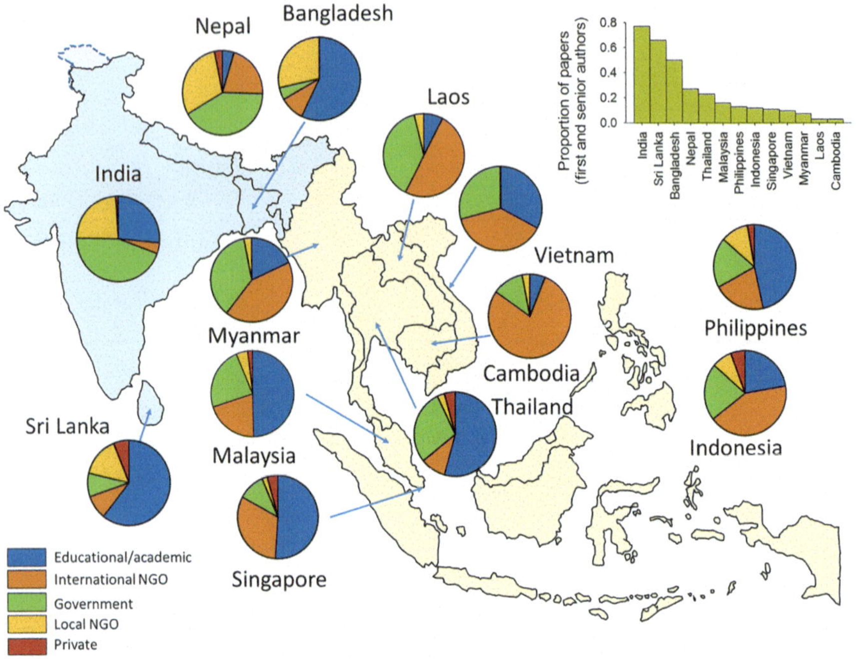

Fig. 13.7 The proportional contributions by researchers from different types of institutions in Asia to conservation papers published in ten international journals between 2000 and 2018 (see also Fig. 13.4 $C_{1,2}$). The inset graph indicates the proportion of conservation papers with affiliates in each country that had national (not ex-patriot) first or senior authors

respectively) with government institutions playing a large role in India (45%) and Nepal (41%). Local NGOs contributed between 15% (Sri Lanka) and 30% (Nepal) of papers from South Asia. Furthermore, authors from Bangladeshi, Indian, and Sri Lankan institutions frequently served as first or senior authors on corresponding conservation papers (Fig. 13.7 [inset]). International NGOs played a relatively minor role in conservation publications from Bangladesh, India, and Sri Lanka (<10%), but were more prominent in research from Nepal (21%). Overall, these trends suggest that the countries in South Asia have developed a strong local capacity to participate in national conservation research (albeit with few papers from Bangladesh, Nepal, and Sri Lanka).

The relative institutional participation in conservation research among South East Asian countries varied considerably. Universities accounted for most of the affiliated research institutions in Thailand (54%), Singapore (51%), Malaysia (50%), and the Philippines (47%). In the remaining countries, international NGOs and research organizations played a prominent role in published conservation research, particularly in Cambodia where nearly 80% of the organizations named on papers were international conservation NGOs. Local authors from South East Asian institutions rarely served as first or senior authors, although ex-patriot authors based at high-ranking universities such as the National University of Singapore, at international universities and international research centers or NGOs (e.g., CIFOR, WorldFish, WWF, or WCS) often generated considerable output (these are not included in Fig. 13.7 [inset]). These trends indicate an important role for international organizations in generating knowledge and developing the research capacity of emerging economies such as Cambodia, Laos, and Myanmar.

In general, there has been little participation by the private sector in recent conservation publications—although wildlife authors often gain patronage from private sector partners (e.g., Dilmah Conservation 2018). Private sector partners that have participated in research toward publications included consultancy companies engaged in environmental research (particularly in Indonesia), tourism operators or tourist resorts, geographic information and imaging companies, or biotech companies that provide technical expertise in tracking illegal wildlife and plant products (see below). Private sector involvement has also included industry bodies such as banks, oil prospectors, and bird and reptile hatcheries. Among the government partners named on publications from tropical Asia, a large number also have industry interests as their principal mandate—e.g., forestry, fisheries, and agricultural research and development institutes and ministries. No authors from the private sector were listed on studies from Cambodia, Laos, Myanmar, or Vietnam.

We examined the relative participation by institutions from different Asian countries in projects either geographically focused only on national conservation issues or with a broader international focus. To assess relative geographical focus among national researchers we divided the number of conservation papers related to each country (Fig. 13.7b) by the number of published papers with at least one national institute from that country involved in the research (Fig. 13.7c). This simple metric, as plotted in Fig. 13.8, separates countries with nationally based scientists participating in international conservation projects (<1 on y-axis) from those countries with national scientists engaged only in research on local, national conservation issues ($=1$ on y-axis), or countries with a significant amount of conservation research conducted in-country by nonnational institutions (>1 on y-axis). The analysis indicated Brunei, Indonesia, Malaysia, and Singapore as countries with scientists participating in international conservation projects. This was often due to prominent internationally recognized institutes such as CIFOR in Indonesia, a number of Malaysian universities, and the National University of Singapore (Fig. 13.8). However, most countries from the region were largely focused only on national issues, and many were the focus of international researchers without any local, national collaborators involved in research outputs (i.e., Philippines, Bangladesh, Maldives, Myanmar, Nepal, Timor-Leste, and Vietnam: Fig. 13.8). In some cases, these papers included analyses of remote sensing images that are available internationally or large reviews focused on international conservation issues. When the relative contributions of national institutions to regional conservation publications are plotted against national Gross Domestic Product (GDP), countries with a nominal GDP below US $5000 were seen to rely on international scientists to generate much of their conservation knowledge—often without any participation or apparent co-authorship from researchers at local institutions (≥ 1; Fig. 13.8). This simple

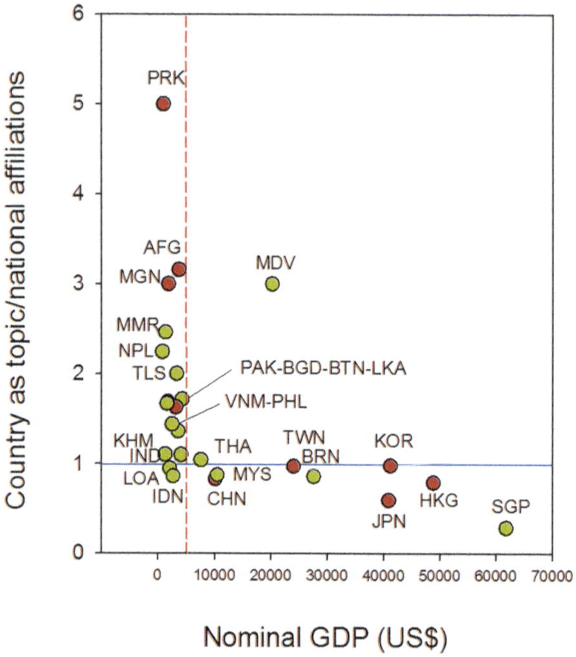

Fig. 13.8 Bi-plot indicating the involvement of researchers from Asian institutions in national or international research (y-axis) against national nominal GDP from each country (x-axis). Countries indicated at $y = 1$ (blue line) have national representation on all papers investigating corresponding national conservation issues. Countries below the blue line (<1) have national representation in international conservation studies. Countries above the blue line (> 1) are the focus of international research often without national participants. The dashed red line indicates a nominal GDP of US $5000. Green points are tropical South and South East Asian countries. Data are from papers published in ten conservation journals between 2010 and 2018 (see main text for details). Countries are identified using ISO alpha-3 country codes (for codes see URL: unstats.un.org/unsd/tradekb/knowledgebase/country-code)

analysis indicates that low revenues in the developing countries of Asia continue to represent a major barrier to knowledge production and research capacity for conservation in the region.

13.3.2 *Future Directions for Research and Funding*

Our analysis suggests that greater investment is needed to develop conservation research capacity in most Asian countries—or, where current research investments are relatively high, that recipients are adequately evaluated for their contributions to publishing or implementation (i.e., that some national funds are converted to accessible knowledge). However, while economies are still limited, and GDPs are relatively low, many governments will simply not have sufficient revenue to appreciably enhance conservation science. This has been acknowledged in the NBSAPs of Asia's tropical countries, with each country identifying clear research needs, including needs for capacity building and for research funding. Among the areas frequently mentioned as priority areas for research in these documents are database management, molecular biology, and the application of molecular tools (see Chap. 3; Klütsch and Laikre 2021), and expertise in modelling and geographic information systems. However, with little funding for research, governments, and institutes may need to carefully reconsider what types of research and data will have the greatest and most immediate impact on conservation actions. With high costs to some tools and methods (e.g., costs of chemical reagents or specialized laboratories) the cost–benefit ratios of conservation-oriented research should be considered in a realistic and regionally appropriate manner. For example, among the NBSAPs there is little mention of research in the social sciences or humanities, including human behavior, or in communications and marketing strategies. Such research will be essential to develop sound conservation strategies with lasting impact (see next section). Other novel strategies that are lacking from the documents include the development of citizen science approaches to data collection (see Chap. 4; Phillips et al. 2021). On mentioning the valuation of biodiversity, there is also a lack of emphasis on life cycle assessments (LCA) for agricultural and other products. Although the development of national capacities in the science and tools that precisely monitor environmental changes is desirable, experimental field ecology using participatory methods and focused on concrete solutions to biodiversity loss merits greater emphasis in the NBSAPs.

Adequate funding will be essential to support the research required to improve biodiversity management and to implement conservation strategies in tropical Asia. Several countries have presented budget commitments to support their NBSAPs. In some cases, such as Brunei (BRIC—Brunei 2016) and Singapore (NPB Singapore 2015), these are detailed accounts of estimates related to specific actions. In many of the remaining documents, discussion of the financial aspects of knowledge production includes any research that is linked to resource management, including research and development budgets for agriculture, forestry, and fisheries (Government of Nepal 2014; DENR – Philippines 2015; BAPPENAS 2016; MOEF 2016). However, without clear goals to develop more sustainable extraction industries, these budgets and related policies only permit a "business-as-usual" approach to development with little pressure or incentive to change the direction of research. Furthermore, apart from building capacity through training courses, curricula development, and university degrees, financial requirements should recognize the need for stable employment of scientists and practitioners to maintain young professionals working nationally.

As indicated from the breakdown of research institutions on published papers from Asia, support from international NGOs and development organizations will probably continue to play an important role in generating knowledge in the region, as has been the case with Cambodia, Laos, and Myanmar. International funding organizations should be encouraged to further contribute to conservation research and actions in low GDP countries, particularly where there is a strong requirement for capacity building; but they must also monitor impacts and base support on merit rather than national prominence (e.g., avoiding repeated funding of unproductive but prominent organizations instead of smaller active ones). The presence of international institutes such as CIFOR, ICRISAT, WorldFish, and the World Agroforestry Institute (based in the Philippines) and international NGOs, such as WWF, FFI, and WCS, tends to improve national scientific output. Although many of the principal authors on papers from these institutions are non-nationals, several papers include national authors working at the institutions or collaborating with them. Perhaps more importantly, international institutions can act as capacity builders by offering bursaries or research internships for national scientists and students to conduct conservation research in their native regions. International collaboration will therefore remain an important component of conservation research in much of tropical Asia and should be actively encouraged during project development.

13.4 Knowledge Mobilization and Conservation Actions: Asian Case Studies

Conservation of wildlife and natural areas occurs across three main categories of land or species use. These are:

1. *Protected lands and species*, without possibilities for direct resource use. In many cases protected areas are designated according to the presence of iconic or flagship species.
2. *Natural areas*, where limited resource use is permitted, and some level of anthropogenic disturbance is tolerated. These include wild areas with developed ecotourism industries and extractivist reserves with traditional or indigenous farming, fishing, or forestry operations.
3. *Production areas*, ranging from mixed farm/forest landscapes to intensive monocultures that are inhabited by animals and plants of conservation interest.

In this section, we present case studies from Asia related to resource management under each of these categories and examine how knowledge is successfully transferred through the production-to-action pipeline. We also examine the roles played by different sectors in influencing the final outcomes of conservation interventions.

13.4.1 Category 1: Protected Lands and Species

13.4.1.1 Establishing Protected Areas

As indicated in Fig. 13.1, the total number and size of protected areas have dramatically increased in Asia since the 1992 Earth Summit. Although these trends are impressive, realities on the ground often reveal only modest impacts. Many of Asia's protected areas are simply tracts of the original vegetation, garden parks, or marine habitats that have been maintained due to difficult topographies or limited access (Dewi et al. 2013), and often do not represent optimal designs for species conservation (Fernando et al. 2012; Jathanna et al. 2015; Sarker et al. 2016). Many are effectively unprotected (Masum et al. 2016). For example, the Philippines has already protected 11% of its territory with the intention to reach the 17% recommended by Aichi target 11 before 2020. However, Mallari et al. (2016) note that 64% of key biodiversity areas in the Philippines remain outside the system of national parks and reserves, and several designated areas do not match the requirements for key conservation species. Frequent disparities between priority areas that conservation biologists highlight for protection, and the extent or nature of areas that are officially included under national park systems, indicate apparent systemic inefficiencies in the knowledge-to-action pipeline related to the designation of parks. Furthermore, failures to adequately manage parks and protect their resources often reflect a lack of attention to the concerns and aspirations of local communities.

Researchers have played an important role in supporting governments to define boundaries for protected areas and to identify landscapes and habitats with optimal conditions for conserving the greatest number of animal and plant species (Hossain et al. 2014, Sarker et al. 2016) or with significant geodiversity or geoheritage value (e.g., Nam Ou Valley, Laos: Kiernan 2013). However, even where significant detailed knowledge has been gathered by researchers and clear action plans defined, this is not always translated into optimal designs for establishing protected areas. For example, Thi et al. (2017) report how high-quality scientific solutions were largely overlooked in favor of weaker research during the establishment of Ngoc Son Ngo Luong Nature Reserve in Vietnam because of better integration between decision-makers and the researchers involved with the latter research. This indicates a growing requirement for researchers to actively lobby for their proposed actions, which demands established, often long-term relations between the researchers and policy makers. Based on their observations, Thi et al. (2017) highlight the need to not only improve such professional integration between researchers and policy makers, but to also develop local conservation capacity (e.g., regional conservation officers and government employees) if science is to better influence decisions.

In many cases, a low success in natural area protection has been due to the poor integration of local communities during planning and actions (Gunatilleke et al. 1993; Zingerli 2005; Hossain et al. 2018; Perez 2018). Protected areas are expected to improve landscapes and lifestyles for local communities that were previously affected by unregulated resource use—often by individuals and businesses not from the local region. People living adjacent to areas zoned for protection often predict advantages from conservation areas (Gier et al. 2017), particularly where they are guaranteed some regulated access to the areas and their resources, or where they become stakeholders in ecotourism or park operations (Clements et al. 2014). As such, the designation of protected areas can represent a win-win for conservation biologists and for local communities that consolidates cooperation and interest in the success of the project. However, as one of the most densely populated parts of the world, the designation of protected areas in Asia, rather than representing an opportunity for local economies, frequently leads to consternation from local

Fig. 13.9 Asian elephants (*Elephas maximus*) grazing on early succession plants near a roadside in Sri Lanka. A preference for plants from disturbed habitats brings elephants and humans into close proximity elevating human–elephant conflicts (photo by E. Kudavidanage)

communities, particularly where key stakeholders have been overlooked or actively excluded from planning consultations, or where local communities suffer potential damages from wildlife near the parks (Fig. 13.9). Such top-down approaches to establishing protected areas are increasingly recognized as ineffective due to resentment from local communities. Indigenous communities dwelling inside protected areas are often most affected, particularly where restrictions on use of the parks' natural resources are in conflict with indigenous peoples' rights to self-determination. For example, Perez (2018) noted that the Indigenous Peoples' Rights Act and the National Integrated Protected Areas System of the Philippines are often incompatible and require further multi-sector consultation to redraft existing laws and regulations and to improve relations between natural resource managers and indigenous communities. In several other cases, the establishment of protected areas has caused resentment by failing to recognize the importance of forest resources in the livelihoods of local people (Gunatilleke et al. 1993; Zingerli 2005; Bennett and Dearden 2014). Such exclusion of local communities, and limitations on their extractivist activities in natural areas, can reduce livelihood portfolios and intensify commodity production outside the protected areas (Dewi et al. 2013; Milne 2015; Dressler et al. 2016). This in turn will reduce the habitat value for wild species of production areas outside the parks.

Finally, an adequate understanding of governance structures will help avoid discontent and resentment of conservation actions among local communities during the designation and management of natural areas. A number of studies have indicated that large conservation projects, particularly those with a high level of financial support, often serve to solidify existing power structures to the detriment of poor, rural communities (Zingerli 2005; Singh 2009; Milne 2015; Pasgaard 2015; Chervier et al. 2016; Gillespie 2018). Milne (2015) suggests that government agencies in Cambodia have used conservation strategies and financing to perpetuate lucrative extraction industries. Similar issues have been associated with reforestation projects, including large-scale projects such as the 5 million hectare Reforestation Project in Vietnam (McElwee 2009; Le et al. 2014). On the other hand, governments with limited access to funding are increasingly moving toward collaborative governance with local communities and stakeholders in protecting natural areas (Mallari et al. 2016; de Koning et al. 2017). Solving such systemic problems is generally outside the realm of possibilities for most researchers or conservation practitioners, and because governance systems are dynamic and political structures are sometimes unpredictable, it can be difficult to incorporate key decision makers or strategic community leaders into realistic medium to long-term action plans. These difficulties can only be overcome through long-term attention to capacity building and the establishment of new cultural norms at relevant institutions (Rao et al. 2014). Successful translation of knowledge to conservation actions, therefore, requires not only a common language for communication between researchers, policy makers, and other stakeholders, but also a greater understanding of the political and cultural systems that determine the reception of, and response to management plans.

13.4.1.2 Engaging Local Communities in the Protection of Endangered Species

The success of parks in maintaining populations of iconic species in Asia has been largely determined by investments in enforcement; however, the importance of cooperation with local communities at each stage of the knowledge-to-action pipeline is increasingly recognized. For example, Nepal's success in reducing Indian rhinoceros (*Rhinoceros unicornis*) poaching, although largely due to effective policing, has benefitted from the active support of local communities who receive shares in park revenues (Martin et al. 2013). Effective communication with thousands of local people living near national parks of the benefits they receive from these parks has consolidated their support and facilitated their contribution to park vigilance and rhino protection (Martin et al. 2013). In contrast to the success of Nepalese rhino conservation, researchers have documented continuing high levels of tiger poaching in Bangladesh (Aziz et al. 2017; Hossain et al. 2018). For example, protection of tigers by park rangers in the Sundarbans Reserve Forest has been largely ineffective. Local communities receive little benefit from the reserve and government support for tourism and infrastructure development near the reserve has been minimal.

Consequently, the local communities play little role in vigilance or park protection (Aziz et al. 2017; Hossain et al. 2018). Whereas the creation of protected areas has been an important step toward tiger conservation, communication of the need for conservation (particularly considering that tigers are considered dangerous to humans) has been ineffective and plans failed to recognize the need for local community support or for the necessary incentives to gain that support. In a similar case, the protected status of the banteng (*Bos javanicus birmanicus*) and designated parks for its conservation have failed to curtail poaching in Vietnam largely due to poor governance (Pedrono et al. 2009). Pedrono et al. (2009) have described the areas for protection of Vietnamese banteng as "paper parks" with ineffective patrolling by poorly paid park staff and little consequence for displaying poached trophies. In effect, the monetary gains for lawbreakers outweigh any compensation for community participation in conservation actions. In contrast, in the Philippines, populations of a similar native cattle species, the tamaraw (*Bubalus mindorensis*), are recovering following education campaigns with local communities and community participation during conservation actions. This includes monitoring of tamaraw populations by locals during community events, which foments community appreciation of conservation goals, as well as providing data for wildlife researchers (Ishihara et al. 2015). These examples indicate that the successful conservation of protected species in Asia often depends, not only on incorporating high-quality research into action plans, but also on the effective inclusion of local communities as stakeholders in actions either through active participation in conservation efforts or through recognized economic incentives derived from the conservation actions.

The management of hilsa shad (*Tenualosa ilisha*) is a good example of an effective knowledge mobilization into action that successfully integrated researcher and local stakeholder participation during planning and adequately incorporated community concerns into action plans, thereby achieving effective community participation. Hilsa is a fish species of tremendous economic and cultural significance to Bangladesh (Sahoo et al. 2018) (Fig. 13.10). Declines in hilsa populations were noted by fishermen and government scientists in the 1990s. Scientists identified key spawning habitats for the fish through dialog with local fishermen (this was later verified and georeferenced in a habitat database: Hossain et al. 2014). The Bangladeshi government has used this information to establish strict hilsa sanctuaries; however, they also recognized the need to compensate local communities for the ban on fishing in the sanctuaries by providing rice to affected households (Bladon et al. 2018). Field monitoring indicates that government actions have led to an increase in hilsa populations. This has been acknowledged by local fishermen who attribute higher catches and larger fish to establishment of the sanctuaries (Islam et al. 2016; Porras et al. 2017; Bladon et al. 2018). Although there have been biases in the distribution of compensation and the greatest hardships have been suffered by the poorest households (Porras et al. 2017; Bladon et al. 2018), nevertheless, compensation—a form of payment for ecosystem services—has helped fishermen recognize that their stakeholder positions were duly considered by the government during conservation planning and actions.

Fig. 13.10 A hilsa fisherman sells his day's catch at Kushtia, Bangladesh. Cooperation between the Government of Bangladesh and hilsa fishermen has played an important role in recovering populations of this economically and culturally significant fish (photo by Shahnewaj Uddin Priom)

The role of compensation to fishermen, in this case, demonstrates the understanding by researchers and planners (informed through consultation with local communities) that short-term necessities often outweigh the long-term benefits of conservation actions in driving local stakeholder participation. Such attention to payments for ecosystem services has proved successful in community-based conservation actions across Asia (Clements et al. 2013; Claassen et al. 2017); however, payments are difficult to distribute equitably (Clements et al. 2013) and it is largely unknown how stakeholder attitudes might change when subsidies are withdrawn, are reduced, or become irregular.

13.4.2 Category 2: Natural Areas with Limited Resource Use

13.4.2.1 Engaging Ecotourism Stakeholders in Conservation Actions

One of the key objectives of ecotourism is to encourage local stakeholder interest in protecting natural areas and their associated fauna and flora. National parks often draw tourists to otherwise remote areas where there are few opportunities for income generation. Local communities generally regard tourism positively, indeed prior to tourism development, communities can be overly optimistic about the potential social and economic impacts of tourism (Gier et al. 2017). High expectations can make it difficult to create a balance between the benefits of tourism for local

Fig. 13.11 Elephants (*Elephas maximus*) at Yala National Park. Yala is one of the most visited wildlife parks in Sri Lanka. Animals are followed by a heavy parade of safari jeeps with jeep drivers using mobile phones to communicate to each other the locations of elephants. While tourists enjoy these encounters, heavy traffic disturbs animals, sometimes evoking aggression and cutting off herds from their water and food sources (photo by E. Kudavidange)

communities and for wildlife. For example, studies indicate that wildlife tourism and visitors to protected areas can affect the resource-use and behavior of wild animals, with unknown long-term impacts (Jathanna et al. 2015; Alwis et al. 2016; Araujo et al. 2017). Scientists have been instrumental in monitoring these effects and in recommending or improving management plans, including determining effective pricing guidelines to best protect natural resources while enhancing the tourists' wildlife experience (Rathnayake 2016a, b). In the case of national parks, such recommendations can be quickly implemented through government institutions, with corresponding adjustments to visitor quotas. However, this can also generate resistance from local businesses that benefit from high visitor numbers. Therefore, adequate provisioning for projected losses of revenue to local businesses must also be incorporated into management plans. Without adequate provisioning for stakeholders, government directives aimed at reducing the negative impacts of ecotourism development on wildlife can be counter-productive (Newsome 2013; Buultjens et al. 2016). Such considerations may explain why some national parks continue to allow high tourist numbers and congestion of park roadways, despite clear recognition of the negative impacts on wildlife and depreciation of the tourist experience (Newsome 2013; Kudavidanage, unpublished data; Fig. 13.11).

Stakeholders such as hoteliers and safari operators are playing an increasing role in managing reserves in Asia (e.g., marine reserves in South East Asia: Svensson et al. 2010), as well as funding, organizing, or assisting in research for conservation (e.g., Romero-Brito et al. 2016). In some cases, high levels of shareholder participation result from anticipated or realized benefits related to increases in the tourists' satisfaction with their wildlife experience. For example, scuba divers have a lower impact on Philippine reefs where dive operators comply with '*Green Fins*'—an environmentally responsible diving program that monitors adherence by operators to a code of conduct and promotes participating members (Roche et al. 2016). Although participation in such conservation actions brings direct economic benefits for private sector stakeholders, in many cases (perhaps the most successful ones) participation has also resulted from conscientious decisions by business owners to support local communities and a sense of responsibility for the environment (Svensson et al. 2008; Romero-Brito et al. 2016; Thompson 2018). The conscientious investment of personnel, facilities, time, and money by the tourism sector represents a successful and sustainable mechanism of paying for ecosystem services. However, such payments may be more suited to relatively small-scale local businesses than to large corporations. Thompson (2018) indicates that large multinational and transnational corporations that might be expected to make greater contributions to conservation projects have in general contributed little. Large corporations are likely to be more reticent to participate in large-scale, complicated schemes that are difficult to understand (e.g., including difficult concepts such as "ecosystem services") or that require long-term financial commitments (Thompson 2018). Successful knowledge-to-action in the case of ecotourism, therefore, requires a good understanding of the incentives and limitations underlying participation by stakeholders in conservation actions.

13.4.3 Category 3: Production Areas and Captive Breeding

13.4.3.1 Obstacles to Knowledge Transfer in the Conservation of Traditional High-Diversity Agricultural Landscapes

Rice agroecosystems constitute the largest land-use system in tropical Asia (Horgan 2017). Intensification of agricultural production, particularly the production of rice, is proposed as a mechanism to spare wild areas while at the same time achieving Asia's formidable food security targets (i.e., land sparing approach; Cassman 1999; Neumann et al. 2010; Seck et al. 2012; Tran et al. 2018). However, this recommendation has overlooked the importance of production landscapes as habitat that supports a high diversity of animal and plant species. For example, traditional, low-intensity rice production systems are associated with high biodiversity (Catling 1993; Bambaradeniya et al. 2004; Settele et al. 2018) as well as preserving aspects of local cultural heritage (e.g., terraced rice, deep-water rice, maavee rice: Castonguay et al. 2016; Horgan et al. 2018). Such rice landscapes can act as buffers for natural areas including nature reserves; indeed, traditional systems such as terraced rice are part of a broader landscape system that includes woodlots on higher slopes (land

Fig. 13.12 Traditional rice production landscapes such as this one, at Bangaan, Ifugao, in the Philippines, maintain a high diversity of flora and fauna by avoiding pesticides and promoting forest conservation at higher elevations (photo by F. Horgan)

sharing; Castonguay et al. 2016; Horgan et al. 2018; Settele et al. 2018: Fig. 13.12). Although rice terraces have benefited from intense anthropological, sociological, and ecological study (Castonguay et al. 2016; Dominik et al. 2017), other traditional agricultural systems, such as maavee rice, have received relatively little research attention (Horgan et al. 2018).

Records of maavee rice paddies in Sri Lanka date back to the 1930s, but paddy areas have decreased dramatically due to land drainage, flood protection, and the cultivation of modern rice varieties. Information on the nature of peat soils and the damaging consequences of drainage for agricultural use (which increases soil acidity) has been available for decades; however, decision makers under pressure to modernize rice production apparently overlooked this information. Consequently, flood control has reduced rice yields and caused widespread losses to soil fertility as well as depleting available habitats for water birds and other wetland species in areas that were formerly devoted to maavee production (Horgan et al. 2018). This cascade of events has been repeated in several other lowland peat areas throughout Asia. Evidence suggests that the destruction of such natural wetlands is the result of key stakeholders repeatedly opting to modernize farming methods driven by government policies around food security and agricultural intensification (Horgan et al. 2018; Horgan and Kudavidanage 2020). Such incongruity between available knowledge that promotes conservation on the one hand, and development policies promoting landscape changes on the other, is a major hindrance to achieving sustainable productivity and conservation goals. Conservation of the remaining maavee paddies in Sri Lanka depends on securing their economic sustainability (Horgan et al. 2018) and will likely benefit from better links to private sector stakeholders interested in marketing traditional and low input rice varieties, or from developing community capacity to better process and market the unique maavee product (traditional rice grain of high nutrient value). Sri Lanka's NBSAP does mention the importance of traditional maavee rice systems and the need for their conservation (MMDE 2016), but the government has not yet invested in the research required to develop environmentally and economically sustainable conservation strategies for this traditional rice system. Currently, Sri Lanka's Rice Research and Development Institute (RRDI) remains focused on increasing national rice yields through intensification methods with most of the work on maavee and other traditional rice systems conducted by smaller horticulture research stations with limited funds (communications with Sri Lankan rice researchers).

13.4.3.2 The Role of Knowledge Transfer in the Success of Ecological Engineering

Over the last decade, research into the potential for ecological engineering to reduce pesticide use in intensified rice production systems has increased dramatically. Much of this research has been conducted in China, Vietnam, Thailand, and the Philippines (Westphal et al. 2015; Gurr et al. 2016; Horgan et al. 2017). By 2014, almost 3500 hectares of rice paddies in Tien Giang Province, Vietnam, had established flower strips as an ecological engineering approach to conserving the natural enemies of rice pests and ecological engineering was included as a strategy for sustainable rice production in regional agricultural policy (van Chien, personal communication). In China, several rice villages have similarly established flower or vegetable strips to conserve functional biodiversity (Gurr et al. 2016). However, in Thailand and the Philippines, ecological engineering has not moved beyond agricultural research stations. These contrasting outcomes offer an interesting possibility to examine how links between researchers and implementers affected the relative successes of the knowledge-to-action pipelines in the different countries.

In the case of Vietnam, implementation of ecological engineering in rice began largely before demonstration trials had been completed or assessed. Much of the success of the project was therefore due to effectively communicating only ideas and concepts to decision-makers at the same time that research was underway (Westphal et al. 2015). In effect, the project's success was largely due to several years of prior experience by key researchers in communicating and implementing pesticide reduction strategies in Vietnam (Huan et al. 2005). Strong ties with local Vietnamese and Chinese researchers, together with a solid understanding of agricultural policy in these countries, facilitated the mainstreaming of ecological engineering among regional decision makers. Furthermore, it is noteworthy that the centralized political systems of Vietnam and China contributed to the rapid success of implementation in these countries, with key government individuals identified to include ecological engineering in the relevant policy. In contrast, in the Philippines, research took precedence over implementation, with detailed monitoring of field experiments conducted across a number of research centers (Horgan et al. 2016, 2017, 2019; Vu et al. 2018). Positive preliminary results, including unanticipated benefits for wildlife (birds and pollinators; Horgan et al. 2016, 2017), as well as a close relationship with the government institutes that funded and participated in the project, led to favorable reviews from the government. However, eventual structural and budget changes at the main research institutes meant that the project did not advance beyond the research experience. These examples indicate that concurrent research and implementation through participatory methods are likely to have the greatest impact on conservation actions. However, the approach includes risks because positive results from interventions cannot be guaranteed. In contrast, the need for detailed research before implementation can run the risk of waning interest and funding over time or changes in research teams; thus, limiting the progression from research to concrete actions.

13.4.3.3 Knowledge Transfer and Captive Breeding

The captive breeding of animals to tentatively reduce effects on wild populations is somewhat analogous to land sparing. While captive breeding for marketing is popular in Asia, captive breeding with concrete plans for reintroduction (ex situ conservation) is relatively rare. There is much debate in the scientific literature about the potential impact of captive breeding and regulated trade in animal products on wild populations (Thorbjarnarson et al. 2000; Drury 2009; Lyons and Natusch 2011; Collins et al. 2016; Tensen 2016). The main argument among proponents is that a regulated industry will make wild harvesting economically unviable, whereas opponents argue that the availability of animal products only encourages opportunistic wild harvesting because wild caught and farmed products can be merged at market.

In Asia, a large number of farms breed wild animals for food, pet, and other industries. These include facilities that rear songbirds (Fig. 13.13), turtles and snakes, butterflies and other exotic insects, porcupines, and bears (Jepson and Ladle 2005; Drury 2009; Brooks et al. 2010; Van der Heyden 2011; Livingstone and Shepherd 2016; Aust et al. 2017).

Evidence for the impact of wildlife farming on wild populations in Asia has mainly come from stakeholder interviews and reveals a range of possible impacts. For example, despite government regulations aimed at curbing wild harvests and a threefold increase in the numbers of bears (*Ursus thibetanus*) farmed in Laos, current prices for bear bile are likely to further incentivize poaching (Livingstone and Shepherd 2016). Similarly, porcupine (*Hystrix brachyuran*) farms have failed to reduce demands for wild caught meat and farm managers continue to purchase wild animals from poachers (Brooks et al. 2010). Based on interviews with consumers and nonconsumers of wild meat in Vietnam, Drury (2009) suggests that the

Fig. 13.13 Java has a strong tradition of keeping songbirds in ornate cages thus fueling Indonesia's cage bird industry and depleting bird populations in rural landscapes (photo by F. Horgan)

availability of meat from farmed wildlife will increase consumer demand, and that this demand, combined with a belief that farmed meat is inferior, could increase pressure on wild animal populations.

In contrast to the above examples, Aust et al. (2017) suggest that snake (*Naja* spp., *Ptyas mucosus*, *Ophiophagus hannah*) farming in Vietnam has reduced pressures on wild populations because snake farmers select species that are fast growing, early maturing, and have a high reproductive capacity. Similar positive effects on wild populations have been attributed to butterfly farms (Van der Heyden 2011). However, the scale of research required to examine real impacts has been difficult to achieve and indirect effects on wild populations are not generally considered. For example, Thorburn (2014) reports that to meet demands for swiftlet (*Aerodramus* spp.) nests, one of the most expensive animal products in Asia, investors in Vietnam, Thailand, Malaysia, and Indonesia erect special buildings that mimic limestone caves. Even though this decreases pressures on natural caves and wild birds, evidence suggests that swiftlet eggs are now so widely traded that the species' distribution range has shifted. Caution should therefore be exercised when interpreting the results of research into the effects of captive breeding. With such a polemic topic, it is perhaps inevitable that the knowledge used to guide breeding as a conservation strategy will be carefully selected by stakeholders to accommodate their own political or economic incentives.

Considering the apparent extent of captive breeding in Asia, the topic has received surprisingly little research attention. Much of the current research is directed toward policing the trade and assisting wildlife officers and law enforcement in crime investigations. Researchers have helped develop new tools with potential to ensure that products from wild stock are not laundered through legal farms and markets. For example, molecular analyses (DNA barcoding and diagnostic kits) can be used to ensure that meats and other products from wildlife farms are derived only from legally traded species (Asis et al. 2016). Furthermore, analysis of isotopes or the elemental composition of proteins could potentially differentiate between wild and farmed animals (Van Schingen et al. 2016), reducing any potential gains from including illegally sourced products into legal supply chains. Despite the promises, these technologies are limited without adequate attention to the social and political context of the wildlife trade, and without sufficient collaboration with or incentives for relevant communities. Furthermore, such technologies are often expensive, particularly for developing Asian countries, and often require access to well-equipped laboratories, making them impractical in many cases. Although researchers, particularly social scientists and economists, have been instrumental in monitoring the impact of some captive breeding operations, there is still much work to be done. To date, NGOs, including animal welfare groups, have perhaps had the greatest impact on the industry by applying national and international pressure to improve regulations and enforcement, and by focusing on issues of animal welfare rather than perceived conservation benefits (Nurse 2016).

13.4.4 Some Common Features of Successful Conservation Projects

The above case studies indicate that researchers and conservation practitioners can have the greatest impact on conservation policies and actions where they acknowledge power and governance structures, community aspirations, and the traditional, cultural, and economic values attributed to wildlife and natural areas (Baynes et al. 2015). Among the cases with the greatest successes, governments, NGOs, and researchers have benefitted from close cooperation with local/target communities in data collection, problem solving, and implementation. In each of the Asian case studies that we examined, sufficient knowledge has been available to guide conservation actions. This knowledge, produced by scientists from universities, government institutes, and/or local NGOs, variously offered potential solutions to key conservation problems as responses to environmental, social, or economic concerns (Fig. 13.14, [b-d]). However, the availability of information and action plans that addressed each of these three concerns appears not to have determined the final success of many of the projects (although action plans that incorporated all three factors tended to be more successful: Fig. 13.14 [cases 1, 3, 7, and 11]). On the other hand, key project attributes that appeared to govern the success of most of the

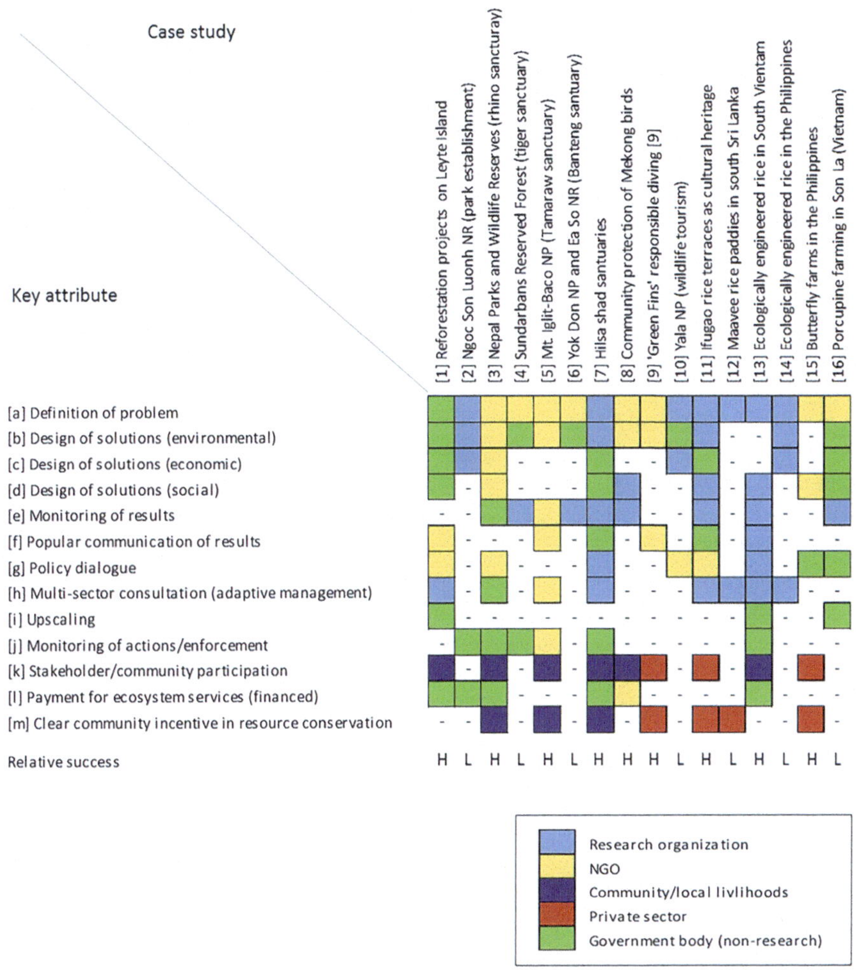

Fig. 13.14 Key project attributes [a-m] associated with 16 case studies of resource management from tropical Asia. Details of each study are presented in the text. The relative success of each study is indicated as high (H) or low (L) at the base of the figure

projects included multi-sector consultations, shareholder and/or community participation, and clear community incentives (Fig. 13.14 [h, k, m]). Although not a feature of all successful projects, those projects that included payments for ecosystem services were also predominantly successful (Fig. 13.14 [l]). Projects with apparently low success, often lacked these same key project attributes, particularly those related to stakeholder and community participation or payments for ecosystem service (Fig. 13.14 [cases 2, 4, 6, 10, 12, 14, 16]). Successes attributed to payments for services and community incentives indicate one of the painful realities of conservation knowledge-to-action in developing countries. That is that knowledge, no matter how detailed or significant, can become irrelevant in the face of individual and community hardships. This point is perhaps most apparent in the recognition by the Bangladeshi Government that fishermen who had most to gain from the establishment of hilsa sanctuaries, nevertheless required food support during fishing bans to facilitate the project. Furthermore, we noted that where conservation actions are likely to affect stakeholder businesses, a lack of clear messages from researchers or policy writers hinders the translation of knowledge into actions. Two current issues, each with vastly opposing schools of thought, are the land sharing-land sparing and the captive breeding debates. Although each may be intellectually stimulating and the focus of an ever-increasing number of academic papers, a lack of clarity around the issues may represent a real obstacle to effective conservation actions, particularly given the range of possible social, economic, and environmental landscapes to which these debates are relevant across Asia.

Conflicting national policies represent a further and considerable obstacle to progress in translating knowledge to conservation actions. Taking agriculture and particularly rice production as an example, Asian nations have clear policies to increase productivity to meet food security targets with national rice research and development authorities firmly focused on conventional, high input production methods (Cassman 1999; Horgan and Crisol 2013; Spangenberg et al. 2015; Tran et al. 2018). As clearly outlined in the Nepalese NBSAP, such policies are in direct conflict with biodiversity action plans, thus indicating a need for greater coordination on legislation and actions (Government of Nepal 2014). Myanmar and the Philippines have targeted an increase of 10% in agricultural areas dedicated to all types of biodiversity-friendly agriculture before 2020 and 2028, respectively (DENR – Philippines 2015; Forest Department – Myanmar 2015). However, in the case of agricultural production in Asia, much of the direction of research is influenced by international, commodity-based research centers focused on the intensification of agriculture (Renkow and Byerlee 2010; Herdt 2012). Such approaches have been justified by regarding intensification as a means of reducing land clearing and limiting the expansion of agricultural frontiers (land sparing: Phalan et al. 2014). However, evidence for such an effect, particularly in Asia is lacking, indeed much of the evidence is contrary to the land-sparing model (Phalan et al. 2014; Law et al. 2015; Dressler et al. 2016; Johansson et al. 2016; Vongvisouk et al. 2016). Such key issues will need to be clarified to better direct integrated agriculture and conservation strategies, particularly where productive land sharing strategies already exist (Horgan 2017). While such key issues remain unresolved, even coordinated policies can be confounded and the sharing of budget resources between conservation and development potentially counter-productive.

13.5 Conclusions

Researchers have played a major role in defining Asia's environmental problems and in outlining priorities for conservation. In many cases, global, regional, and national trends in deforestation, biodiversity loss, wildlife trading, and other environmental problems have been used to influence the environmental policies of tropical Asian nations. This is most notably expressed in Asia's NBSAPs. However, researchers are often limited because they prioritize communication with peers through academic reports and publishing. Furthermore, conservation science in Asia has lagged behind other world regions and many nations continue to rely heavily on research support from international nongovernment scientific, development, and advocacy organizations. The low GDPs of many of Asia's nations suggest that such international support will continue to play a prominent role in defining the direction of conservation research in the region. However, Asian nations must balance between the tremendous benefits from international collaboration and maintaining research sovereignty by defining directions for research that are appropriate to their specific cultures, economies, and needs. The emergence of new academic journals (e.g., *Journal of Asia-Pacific Biodiversity*, *Asian Journal of Conservation Biology*, *Asian Journal of Biodiversity*) focused specifically on the Asian region may help to promote regionally centered research; however, further efforts should be made to redress the current lack of representation by Asian scientists on the editorial boards of international journals (Campos-Arceiz et al. 2018). Furthermore, efforts should be taken to avoid any erosion of research quality resulting from the proliferation of predatory journals in the region.

To better translate research findings into concrete actions, Asian researchers must ensure that informed roadmaps to implementation are prepared prior to project initiation, and that key institutions are identified and included as part of transdisciplinary and multidisciplinary project teams. Our analysis of the current roles and pressures on professionals in different sectors demonstrates how multi-sector partnerships are necessary to incorporate the maximum range of key attributes required for effective knowledge-to-action pipelines. Our examination of case studies across a gradient of resource use categories suggests that successful conservation programs have progressed beyond problem definition and toward concrete actions by engaging larger interdisciplinary teams that include government scientists and policy makers. Successful projects also worked closely with target communities (including local communities and the private sector) and often provided tangible incentives as payments for ecosystem services. Our interpretation of case studies from Asia suggests that the completeness of knowledge (i.e., including problem definition, and environmental, social, or economic solutions) is often less important than the participation of key stakeholders in driving project success. We suggest that effective conservation actions are also more likely to emerge where knowledge and legislation from researchers and policy makers are clear and without contradictions, this will require greater communication between stakeholders, and, in the case of policy makers, improved intergovernmental communication.

Despite awareness of the need for multidisciplinary approaches to conservation knowledge production and implementation, regional policies continue to focus on the types of science that achieve high citation impacts. Through a review of the region's most recent NBSAPs, we noted a clear emphasis on tool-driven research including molecular, biotechnological, and remote sensing technologies as well as big data approaches to elucidating the drivers of environmental deterioration (i.e., all focused on describing problems). Whereas such tools are useful, there has been a general lack of attention to the field-based experimental and participatory studies required to design, test, and implement solutions. Future research teams will require anthropologists, political scientists, social scientists, environmental economists, as well as communications and marketing experts to take projects beyond problem definition by working with regional governments and local communities to bring about change. Large interdisciplinary teams with shared goals but often different incentives, although complex and sometimes unwieldy, will be necessary to work through the often conflicting goals for development and resource management as apparent in national policies. Although there is a need to train transdisciplinary researchers and professionals for conservation programs, it will also be important to balance between transdisciplinary and classical disciplines to ensure that key skills are not lost or, alternatively, that generalist researchers are not disadvantaged because they remain outside traditional roles. Finally, to ensure commitments to the long-term and large-scale projects essential for nature conservation, it will be necessary for governments to provide attractive employment opportunities, security of tenure, and security of funding for young professionals throughout tropical Asia.

References

Ahsan DA, DelValls TA, Blasco J (2009) The relationship of national and international environmental NGOs in Bangladesh and their role in wetland conservation. Int J Environ Res 3(1):23–34

Alwis NS, Perera P, Dayawansa NP (2016) Response of tropical avifauna to visitor recreational disturbances: a case study from the Sinharaja World Heritage Forest, Sri Lanka. Avian Res 7(1):15

Araujo G, Vivier F, Labaja JJ, Hartley D, Ponzo A (2017) Assessing the impacts of tourism on the world's largest fish *Rhincodon typus* at Panaon Island, Southern Leyte, Philippines. Aquat Conserv 27(5):986–994

Asis AMJM, Lacsamana JKM, Santos MD (2016) Illegal trade of regulated and protected aquatic species in the Philippines detected by DNA barcoding. Mitochondrial DNA A 27(1):659–666

Aust PW, Van Tri N, Natusch DJD, Alexander GJ (2017) Asian snake farms: conservation curse or sustainable enterprise? Oryx 51(3):498–505

Aziz MA, Tollington S, Barlow A, Goodrich J, Shamsuddoha M, Islam MA, Groombridge JJ (2017) Investigating patterns of tiger and prey poaching in the Bangladesh Sundarbans: implications for improved management. Glob Ecol Conserv 9:70–81

Bakker VJ, Baum JK, Brodie JF, Salomon AK, Dickson BG, Gibbs HK, Jensen OP, McIntyre PB (2010) The changing landscape of conservation science funding in the United States. Conserv Lett 3(6):435–444

Bambaradeniya CNB, Edirisinghe JP, De Silva DN, Gunatilleke CVS, Ranawana KB, Wijekoon S (2004) Biodiversity associated with an irrigated rice agro-ecosystem in Sri Lanka. Biodivers Conserv 13:1715–1753

BAPPENAS (2016) Indonesian biodiversity strategy and action plan 2015–2020. Ministry of National Development Planning. Government of Indonesia, Jakarta

Baynes J, Herbohn J, Smith C, Fisher R, Bray D (2015) Key factors which influence the success of community forestry in developing countries. Global Environ Chang 35:226–238

Beall J (2015) Predatory journals and the breakdown of research cultures. Inf Dev 31(5):473-476

Bennett NJ, Dearden P (2014) Why local people do not support conservation: community perceptions of marine protected area livelihood impacts, governance and management in Thailand. Mar Policy 44:107–116

Bixler RP, Dell'Angelo J, Mfune O, Roba H (2015) The political ecology of participatory conservation: institutions and discourse. J Polit Ecol 22(1):164–182

Bladon AJ, Mohammed EY, Hossain B, Kibria L, Ali L, Milner-Gulland EJ (2018) Evaluating the ecological and social targeting of a compensation scheme in Bangladesh. PLoS One 13(6): e0197809

Bosworth B, Collins SM (2008) Accounting for growth: comparing China and India. J Econ Perspect 22(1):45–66

BRIC Brunei (2016) National biological resources (biodiversity) policy and strategic plan of action. Biodiversity Research and Innovation Centre. Ministry of Industry and Primary Resources, Brunei

Brooks EGE, Roberton SI, Bell DJ (2010) The conservation impact of commercial wildlife farming of porcupines in Vietnam. Biol Conserv 143:2808–2814

Buultjens J, Ratnayke I, Gnanapala A (2016) Whale watching in Sri Lanka: perceptions of sustainability. Tour Manag Perspect 18:125–133

Campos-Arceiz A, Primack ARB, Miller-Rushing AJ, Maron M (2018) Striking underrepresentation of biodiversity-rich regions among editors of conservation journals. Biol Conserv 220:330–333

Cassman KG (1999) Ecological intensification of cereal production systems: yield potential, soil quality, and precision agriculture. PNAS 96(11):5952–5959

Castonguay AC, Burkhard B, Müller F, Horgan FG, Settele J (2016) Resilience and adaptability of rice terrace social-ecological systems: a case study of a local community's perception in Banaue, Philippines. Ecol Soc 21(2):15

Catling D (1993) Rice in deep water. Macmillan, London

CBD (2018) National biodiversity strategies and action plans (NBSAPs). Convention on Biological Diversity. https://www.cbd.int/nbsap/. Accessed 15 June 2018

Chervier C, Peresse A, Millet-Amrani S, Meral P (2016) Changement institutionnel et paiements pour services environnementaux au Cambodge: l'intérêt de l'approche Commonsienne. Développement Durable et Territoires 7:1

CITES (2018) List of parties to the convention. Convention on International Trade in Endangered Species of Wild Fauna and Flora, https://www.cites.org/eng/disc/parties/index.php. Accessed 15 June 2018

Claassen AH, Sok K, Arnold TW, Cuthbert FJ (2017) Effectiveness of direct payments to increase reproductive success of sandbar-nesting river birds in Cambodia. Bird Conserv Int 27(4):495–511

Clements T, Rainey H, An D, Rours V, Tan S, Thong S, Sutherland WJ, Milner-Gulland EJ (2013) An evaluation of the effectiveness of a direct payment for biodiversity conservation: the bird nest protection program in the northern plains of Cambodia. Biol Conserv 157:50–59

Clements T, Suon S, Wilkie DS, Milner-Gulland EJ (2014) Impacts of protected areas on local livelihoods in Cambodia. World Dev 64(Suppl 1):S125–S134

Collins A, Fraser G, Snowball J (2016) Issues and concerns in developing regulated markets for endangered species products: the case of rhinoceros horns. Camb J Econ 40(6):1669–1686

de Koning M, Nguyen T, Lockwood M, Sengchanthavong S, Phommasane S (2017) Collaborative governance of protected areas: success factors and prospects for Hin Nam No National Protected Area, Central Laos. Conserv Soc 15(1):87–99

DENR – Philippines (2015) Philippine biodiversity strategy and action plan 2015–2028. Biodiversity Management Bureau, Department of Environment and Natural Resources, Quezon City

Dewi S, van Noordwijk M, Ekadinata A, Pfund JL (2013) Protected areas within multifunctional landscapes: squeezing out intermediate land use intensities in the tropics? Land Use Policy 30(1):38–56

Dhiaulhaq A, McCarthy JF, Yasmi Y (2018) Resolving industrial plantation conflicts in Indonesia: can mediation deliver? Forest Policy Econ 91:64–72
Dilmah Conservation (2018). https://www.dilmahconservation.org. Accessed 12 Sept 2018
Dominik C, Seppelt R, Horgan FG, Marquez L, Settele J, Václavík T (2017) Regional-scale effects override the influence of fine-scale landscape heterogeneity on rice arthropod communities. Agric Ecosyst Environ 246:269–278
Dressel S, Ericsson G, Sandström C (2018) Mapping social-ecological systems to understand the challenges underlying wildlife management. Environ Sci Pol 84:105–112
Dressler W, de Koning J, Montefrio M, Firn J (2016) Land sharing not sparing in the "green economy": the role of livelihood bricolage in conservation and development in the Philippines. Geoforum 76:75–89
Drury R (2009) Reducing urban demand for wild animals in Vietnam: examining the potential of wildlife farming as a conservation tool. Conserv Lett 2(6):263–270
EAZA (2018) Silent forest Asian bird crisis. European Association of Zoos and Aquaria. https:// wwwsilentforesteu. Accessed 15 June 2018
FAO (2018) FAOSTAT. Food and Agriculture Organization of the United Nations. http://www.fao. org/faostat/en/?#data. Accessed 15 June 2018
Fernando P, Leimgruber P, Prasad T, Pastorini J (2012) Problem-elephant translocation: translocating the problem and the elephant? PLoS One 7(12):e50917
Forest Department – Myanmar (2015) National biodiversity strategy and action plan 2015–2020. Myanmar. Forest Department
Fox CW, Paine CET, Sauterey B (2016) Citations increase with manuscript length, author number, and references cited in ecology journals. Ecol Evol 6(21):7717–7726
Gier L, Christie P, Amolo R (2017) Community perceptions of scuba dive tourism development in Bien Unido, Bohol Island, Philippines. J Coast Conserv 21(1):153–166
Gillespie J (2018) Wetland conservation and legal layering: managing Cambodia's great lake. Geogr J 184(1):31–40
Gogoi M (2018) Emotional coping among communities affected by wildlife-caused damage in north-east India: opportunities for building tolerance and improving conservation outcomes. Oryx 52(2):214–219
Gosselin F, Cordonnier T, Bilger I, Jappiot M, Chauvin C, Gosselin M (2018) Ecological research and environmental management: we need different interfaces based on different knowledge types. J Environ Manag 218:388–401
Government of Nepal (2014) Nepal – National biodiversity strategy and action plan 2014–2020. Ministry of Forests and Soil Conservation, Kathmandu
Gray TNE, Hughes AC, Laurance WF, Long B, Lynam AJ, O'Kelly H, Ripple WJ, Seng T, Scotson L, Wilkinson NM (2018) The wildlife snaring crisis: an insidious and pervasive threat to biodiversity in Southeast Asia. Biodivers Conserv 27:1031–1037
Gunatilleke IAUN, Gunatilleke CVS, Abeygunawardena P (1993) Interdisciplinary research towards management of non-timber forest resources in lowland rain-forests of Sri Lanka. Econ Bot 47:282–290
Gurr GM, Lu ZX, Zheng XS, Xu HX, Zhu PY, Chen GH, Yao XM, Cheng J, Zhu ZR, Catindig JL, Villareal S, Chien HV, Cuong LQ, Channoo C, Chengwattana N, Lan LP, Hai LH, Chaiwong J, Nicol HI, Perovic DJ, Wratten SD, Heong KL (2016) Multi-country evidence that crop diversification promotes ecological intensification of agriculture. Nat Plants 2:16014
Harada K, Wiyono (2014) Certification of a community-based forest enterprise for improving institutional management and household income: a case from Southeast Sulawesi, Indonesia. Small-Scale For 13:47–64
Hayami Y (2001) Ecology, history, and development: a perspective from rural Southeast Asia. The World Bank Res Obser 16(2):169–198
Herdt RW (2012) People, institutions, and technology: a personal view of the role of foundations in international agricultural research and development 1960-2010. Food Policy 37(2):179–190

Horgan FG (2017) Integrated pest management for sustainable rice cultivation: a holistic approach. In: Sasaki T (ed) Achieving sustainable cultivation of rice—cultivation, pest and disease management. Burleigh Dodds Science Publishing, Cambridge, pp 309–342

Horgan FG, Crisol E (2013) Hybrid rice and insect herbivores in Asia. Entomol Exp Appl 148(1):1–19

Horgan FG, Kudavidanage EP (2020) Farming on the edge: farmer training to mitigate human-wildlife conflict at an agricultural frontier in south Sri Lanka. Crop Prot 127:104981

Horgan FG, Ramal AF, Bernal CC, Villegas JM, Stuart AM, Almazan MLP (2016) Applying ecological engineering for sustainable and resilient rice production systems. Procedia Food Sci 6:7–15

Horgan FG, Ramal AF, Villegas JM, Almazan MLP, Bernal CC, Jamoralin A, Pasang JM, Orboc G, Agreda V, Arroyo C (2017) Ecological engineering with high diversity vegetation patches enhances bird activity and ecosystem services in Philippine rice fields. Reg Environ Chang 17: 1355–1367

Horgan FG, Kudavidanage EP, Weragodaarachchi A, Ramp D (2018) Traditional 'maavee' rice production in Sri Lanka: environmental, economic and social pressures revealed through stakeholder interviews. Paddy Water Environ 16(2):225–241

Horgan FG, Crisol-Martínez E, Stuart AM, Bernal CC, de Cima Martín E, Almazan MLP, Ramal AF (2019) Effects of vegetation strips, fertilizer levels and varietal resistance on the integrated management of arthropod biodiversity in a tropical rice ecosystem. Insects 10(10):328

Hossain MS, Sarker S, Chowdhury SR, Sharifuzzaman SM (2014) Discovering spawning ground of Hilsa shad (*Tenualosa ilisha*) in the coastal waters of Bangladesh. Ecol Model 282:59–68

Hossain ANM, Lynam AJ, Ngoprasert D, Barlow A, Greenwood Barlow C, Savini T (2018) Identifying landscape factors affecting tiger decline in the Bangladesh Sundarbans. Global Ecol Conserv 13:e00382

Huan NH, Thiet LV, Chien HV, Heong KL (2005) Farmers' participatory evaluation of reducing pesticides, fertilizers and seed rates in rice farming in the Mekong Delta, Vietnam. Crop Prot 24(5):457–464

Ickes K (2001) Hyper-abundance of native wild pigs (*Sus scrofa*) in a lowland dipterocarp rain forest of peninsular Malaysia. Biotropica 33(4):682–690

Imai N, Furukawa T, Tsujino R, Kitamura S, Yumoto T (2018) Factors affecting forest area change in Southeast Asia during 1980-2010. PLoS One 13(5):e0197391

IPBES (2018) Members. Intergovernmental Science-Policy Platform on Biodiversity and Ecosystem Services. https://www.ipbes.net/members. Accessed 15 June 2018

IPSI (2018) Partnership. International Partnership for the Satoyama Initiative. http://satoyama-initiative.org/partnership/ipsi_members. Accessed 15 June 2018

Ishihara S, Boyles RM, Matsubayashi H, del Barrio AN, Cebrian MR, Ishida A, Lapitan RM, Atabay EP, Cruz LC, Kanai Y (2015) Long-term community-based monitoring of tamaraw *Bubalus mindorensis* on Mindoro Island, Philippines. Oryx 49(2):352–359

Islam MM, Islam N, Sunny AR, Jentoft S, Ullah MH, Sharifuzzaman SM (2016) Fishers' perceptions of the performance of Hilsa shad (*Tenualosa ilisha*) sanctuaries in Bangladesh. Ocean Coast Manage 130:309–316

IUCN (2018) World database on protected areas. International Union for Conservation of Nature. https://www.iucn.org/theme/protected-areas/our-work/world-database-protected-areas. Accessed 15 June 2018

Jadhav S, Barua M (2012) The elephant vanishes: impact of human-elephant conflict on people's wellbeing. Health Place 18(6):1356–1365

Jathanna D, Karanth KU, Kumar NS, Karanth KK, Goswami VR (2015) Patterns and determinants of habitat occupancy by the Asian elephant in the Western Ghats of Karnataka, India. PLoS One 10(7):e0133233

Jepson P, Ladle RJ (2005) Bird-keeping in Indonesia: conservation impacts and the potential for substitution-based conservation responses. Oryx 39(4):442–448

Johansson O, Rauset GR, Samelius G, McCarthy T, Andren H, Tumursukh L, Mishra C (2016) Land sharing is essential for snow leopard conservation. Biol Conserv 203:1–7

Kiernan K (2013) The nature conservation, geotourism and poverty reduction nexus in developing countries: a case study from the Lao PDR. Geoheritage 5:207–225

KimDung N, Bush SR, Mol APJ (2016) NGOs as bridging organizations in managing nature protection in Vietnam. J Environ Dev 25(2):191–218

Klütsch CFC, Laikre L (2021) Closing the conservation genetics gap: integrating genetic knowledge in conservation management to ensure evolutionary potential. In: Ferreira CC, Klütsch CFC (eds) Closing the knowledge-implementation gap in conservation science – evidence transfer across spatiotemporal scales and different stakeholders. Springer, New York, pp 51–82

Law EA, Meijaard E, Bryan BA, Mallawaarachchi T, Koh LP, Wilson KA (2015) Better land-use allocation outperforms land sparing and land sharing approaches to conservation in Central Kalimantan, Indonesia. Biol Conserv 186:276–286

Le HD, Smith C, Herbohn J (2014) What drives the success of reforestation projects in tropical developing countries? The case of the Philippines. Global Environ Chang 24:334–348

Leeuwis C, Klerkx L, Schut M (2018) Reforming the research policy and impact culture in the CGIAR: integrating science and systemic capacity development. Glob Food Secur 16:17–21

Livingstone E, Shepherd CR (2016) Bear farms in Lao PDR expand illegally and fail to conserve wild bears. Oryx 50(1):176–184

Loh TL, Larson ER, David SR, de Souza LS, Gericke R, Gryzbek M, Kough AS, Willink PW, Knapp CR (2018) Quantifying the contribution of zoos and aquariums to peer-reviewed scientific research. Facets 3(1):287–299

Luskin MS, Brashares JS, Ickes K, Sun IF, Fletcher C, Wright SJ, Potts MD (2017) Cross-boundary subsidy cascades from oil palm degrade distant tropical forests. Nat Commun 8(1):2231

Lyons JA, Natusch DJD (2011) Wildlife laundering through breeding farms: illegal harvest, population declines and a means of regulating the trade of green pythons (*Morelia viridis*) from Indonesia. Biol Conserv 144:3073–3081

Mallari NAD, Collar NJ, McGowan PJK, Marsden SJ (2016) Philippine protected areas are not meeting the biodiversity coverage and management effectiveness requirements of Aichi target 11. Ambio 45:313–322

Manicad G (1999) CGIAR and the private sector: public good versus proprietary technology in agricultural research. Biotechnol Dev Monit 37:8–13

Martin E, Martin C, Vigne L (2013) Successful reduction in rhino poaching in Nepal. Pachyderm 54:66–74

Masum KM, Islam MN, Saha N, Hasan MZ, Mansor A (2016) Assessment of land grabbing from protected forest areas of Bhawal National Park in Bangladesh. Landsc Res 41(3):330–343

McElwee P (2009) Reforesting "bare hills" in Vietnam: social and environmental consequences of the 5 million hectare reforestation program. Ambio 38:325–333

Meyer JL, Frumhoff PC, Hamburg SP, de la Rosa C (2010) Above the din but in the fray: environmental scientists as effective advocates. Front Ecol Environ 8(6):299–305

Milne S (2015) Cambodia's unofficial regime of extraction: illicit logging in the shadow of transnational governance and investment. Crit Asian Stud 47(2):200–228

MMDE – Sri Lanka (2016) National biodiversity strategic action plan 2016–2022. Ministry of Mahaweli Development and Environment, Colombo

MOEF (2016) National biodiversity strategy and action plan of Bangladesh 2016–2021. Ministry of Environment and Forests, Dhaka

MONRE - Laos (2016) National biodiversity strategy and action plan 2016–2025. Ministry of Natural Resources and Environment. Vientiane, Lao PDR

Moon MJ, Hwang C (2013) The state of civil service systems in the Asia-Pacific region: a comparative perspective. Rev Public Pers Adm 33:121–139

Moore JL, Runge MC (2012) Combining structured decision making and value-of-information analyses to identify robust management strategies. Conserv Biol 26(5):810–820

Mukherjee B (2018) Measuring scientific value of Indian journals beyond impact factor: a case with physics-chemistry-biology journals. Curr Sci 114(8):1613–1623
Neumann K, Verburg PH, Stehfest E, Müller C (2010) The yield gap of global grain production: a spatial analysis. Agric Syst 103(5):316–326
Newsome D (2013) An 'ecotourist's' recent experience in Sri Lanka. J Ecotour 12(3):210–220
Novellino D, Dressler WH (2010) The role of "hybrid" NGOs in the conservation and development of Palawan Island, the Philippines. Soc Nat Resour 23:165–180
NPB Singapore (2015) Singapore fifth national report to the convention on biological diversity (2010–2014). National Parks Board Singapore, Singapore
Nurse A (2016) A global movement: NGOs and the policing of international wildlife trafficking. JTOCS 2:50–61
Ohmer ME, Bishop PJ (2011) Citation rate and perceived subject bias in the amphibian-decline literature. Conserv Biol 25:195–199
Pasgaard M (2015) Lost in translation? How project actors shape REDD plus policy and outcomes in Cambodia. Asia Pac Viewp 56:111–127
Pechacek P, Li G, Li JS, Wang W, Wu XP, Xu J (2013) Compensation payments for downsides generated by protected areas. Ambio 42:90–99
Pedrono M, Ha MT, Chouteau P, Vallejo F (2009) Status and distribution of the endangered banteng *Bos javanicus birmanicus* in Vietnam: a conservation tragedy. Oryx 43:618–625
Perez PL (2018) Living with the problem of national parks: indigenous critique of Philippine environmental policy. Thesis Eleven 145:58–76
Phalan B, Green R, Balmford A (2014) Closing yield gaps: perils and possibilities for biodiversity conservation. Philos T R Soc B 369:20120285
Phillips TB, Parker A, Bowser A, Hakley M (2021) Publicly generated data: the role of Citizen-Science for knowledge production, action, and public engagement. In: Ferreira CC, Klütsch CFC (eds) Closing the knowledge-implementation gap in conservation science – evidence transfer across spatiotemporal scales and different stakeholders. Springer, New York, pp 83–107
Porras I, Mohammed EY, Ali L, Ali MS, Hossain MB (2017) Power, profits and payments for ecosystem services in Hilsa fisheries in Bangladesh: a value chain analysis. Mar Policy 84:60–68
Rao M, Johnson A, Spence K, Sypasong A, Bynum N, Sterling E, Phimminith T, Praxaysombath B (2014) Building capacity for protected area management in Lao PDR. Environ Manag 53:715–727
Rathnayake RMW (2016a) Pricing the enjoyment of 'elephant watching' at the Minneriya National Park in Sri Lanka: an analysis using CVM. Tour Manag Perspect 18:26–33
Rathnayake RMW (2016b) Willingness to pay for a novel visitor experience: ecotourism planning at Kawdulla National Park in Sri Lanka. Tour Plan Dev 13(1):37–51
Renkow M, Byerlee D (2010) The impacts of CGIAR research: a review of recent evidence. Food Policy 35(5):391–402
Ríos-Saldaña CA, Delibes-Mateos M, Ferreira CC (2018) Are fieldwork studies being relegated to second place in conservation science? Glob Ecol Conserv 14:e00389
Robinson JG (2012) Common and conflicting interests in the engagements between conservation organizations and corporations. Conserv Biol 26:967–977
Roche RC, Harvey CV, Harvey JJ, Kavanagh AP, McDonald M, Stein-Rostaing VR, Turner JR (2016) Recreational diving impacts on coral reefs and the adoption of environmentally responsible practices within the SCUBA diving industry. Environ Manag 58:107–116
Romero-Brito TP, Buckley RC, Byrne J (2016) NGO partnerships in using ecotourism for conservation: systematic review and meta-analysis. PLoS One 11(11):e0166919
Rozylowicz L, Nita A, Manolache S, Ciocanea CM, Popescu VD (2017) Recipe for success: a network perspective of partnership in nature conservation. J Nat Conserv 38:21–29
Sahoo AK, Wahab MA, Phillips M, Rahman A, Padiyar A, Puvanendran V, Bangera R, Belton B, De DK, Meena DK, Behera BK, Sharma AP, Bhaumik U, Mohanty BP, Choudhury SR, Mohan

CV (2018) Breeding and culture status of Hilsa (*Tenualosa ilisha*, Ham. 1822) in South Asia: a review. Rev Aquacult 10(1):96–110

Salazar-Vallejo SI, Carrera-Parra LF (1998) Biological taxonomy, impact factor and curricular evaluation for the 21st century. Interciencia 23(5):293–305

Sarker SK, Reeve R, Thompson J, Paul NK, Matthiopoulos J (2016) Are we failing to protect threatened mangroves in the Sundarbans world heritage ecosystem? Sci Rep 6:21234

Seck PA, Diagne A, Mohanty S, Wopereis MCS (2012) Crops that feed the world 7: Rice. Food Secur 4:7–24

Settele J, Heong KL, Kühn I, Klotz S, Spangenberg JH, Arida G, Beaurepaire A, Beck S, Bergmeier E, Burkhard B, Brandl R, Bustamante JV, Butler A, Cabbigat J, Le XC, Catindig JLA, Ho VC, Le QC, Dang KB, Escalada M, Dominik C, Franzen M, Fried O, Görg C, Grescho V, Grossmann S, Gurr GM, Hadi BAR, Le HH, Harpke A, Hass AL, Hirneisen N, Horgan FG, Hotes S, Isoda Y, Jahn R, Kettle H, Klotzbücher A, Klotzbücher T, Langerwisch F, Loke WH, Lin YP, Lu ZX, Lum KY, Magcale-Macandog DB, Marion G, Marquez L, Muller F, Nguyen M, Nguyen QA, Nguyen VS, Ott J, Penev L, Pham HT, Radermacher N, Rodriguez-Labajos B, Sann C, Sattler C, Schadler M, Scheu S, Schmidt A, Schrader J, Schweiger O, Seppelt R, Soitong K, Stoev P, Stoll-Kleemann S, Tekken V, Thonicke K, Tilliger B, Tobias K, Trisyono YA, Dao TT, Tscharntke T, Le QT, Turke M, Václavík T, Vetterlein D, Villareal S, Vu KC, Vu Q, Weisser WW, Westphal C, Zhu ZR, Wiemers M (2018) Rice ecosystem services in South-east Asia. Paddy Water Environ 16:211–224

Shukla S, Gardner J (2006) Local knowledge in community-based approaches to medicinal plant conservation: lessons from India. J Ethnobiol Ethnomed 2:20

Singh S (2009) Governing anti-conservation sentiments: forest politics in Laos. Hum Ecol 37:749–760

Sodhi NS, Brook BW (2008) Fragile southeast Asian biotas. Biol Conserv 141:883–884

Sodhi NS, Koh LP, Brook BW, Ng PKL (2004) Southeast Asian biodiversity: an impending disaster. Trends Ecol Evol 19(12):654–660

Spangenberg JH, Douguet JM, Settele J, Heong KL (2015) Escaping the lock-in of continuous insecticide spraying in rice: developing an integrated ecological and socio-political DPSIR analysis. Ecol Model 295:188–195

Statzner B, Resh VH (2010) Negative changes in the scientific publication process in ecology: potential causes and consequences. Freshw Biol 55(12):2639–2653

Suba RB, van der Ploeg J, van't Zelfde M, Lau YW, Wissingh TF, Kustiawan W, de Snoo GR, de Iongh HH (2017) Rapid expansion of oil palm is leading to human-elephant conflicts in North Kalimantan Province of Indonesia. Trop Conserv Sci 10:1–10

Svensson P, Rodwell LD, Attrill MJ (2008) Hotel managed marine reserves: a willingness to pay survey. Ocean Coast Manage 51(2):854–861

Svensson P, Rodwell LD, Attrill MJ (2010) The perceptions of local fishermen towards a hotel managed marine reserve in Vietnam. Ocean Coast Manage 53(3):114–122

Tensen L (2016) Under what circumstances can wildlife farming benefit species conservation? Global Ecol Conserv 6:286–298

Thi HD, Krott M, Böcher M (2017) The success of scientific support for biodiversity conservation policy: the case of Ngoc Son Ngo Luong nature reserve in Vietnam. J Nat Conserv 38:3–10

Thinley P, Rajaratnam R, Lassoie JP, Morreale SJ, Curtis PD, Vernes K, Leki L, Phuntsho S, Dorji T, Dorji P (2018) The ecological benefit of tigers (*Panthera tigris*) to farmers in reducing crop and livestock losses in the eastern Himalayas: implications for conservation of large apex predators. Biol Conserv 219:119–125

Thompson BS (2018) Institutional challenges for corporate participation in payments for ecosystem services (PES): insights from Southeast Asia. Sustain Sci 13:919–935

Thorbjarnarson J, Platt SG, Khaing UST (2000) A population survey of the estuarine crocodile in the Ayeyarwady Delta, Myanmar. Oryx 34(4):317–324

Thorburn C (2014) The edible birds' nest boom in Indonesia and South-east Asia: a nested political ecology. Food Cult Soc 17(4):535–553

Tran DD, van Halsema G, Hellegers PJGJ, Ludwig F, Wyatt A (2018) Questioning triple rice intensification on the Vietnamese mekong delta floodplains: an environmental and economic analysis of current land-use trends and alternatives. J Environ Manag 217:429–441

UN DESA (2018) World population prospects 2017. United Nations Department of Economic and Social Affairs. https://esa.un.org/unpd/wpp. Accessed 15 June 2018

UniRank (2018) Universities in Asia – Asian Universities by Country. https://www.4icu.org/Asia. Accessed 15 June 2018

UN-REDD (2018) UN-REDD Programme. United Nations Collaborative Programme on Reducing Emissions from Deforestation and Forest Degradation in Developing Countries. http://www.un-redd.org. Accessed 15 June 2018

Van der Heyden T (2011) Local and effective: two projects of butterfly farming in Cambodia and Tanzania (Insecta: Lepidoptera). SHILAP Revta Lepid 39:267–270

Van Schingen M, Ziegler T, Boner M, Streit B, Nguyen TQ, Crook V, Ziegler S (2016) Can isotope markers differentiate between wild and captive reptile populations? A case study based on crocodile lizards (*Shinisaurus crocodilurus*) from Vietnam. Global Ecol Conserv 6:232–241

Varkkey H, Tyson A, Choiruzzad SAB (2018) Palm oil intensification and expansion in Indonesia and Malaysia: environmental and socio-political factors influencing policy. Forest Policy Econ 92:148–159

Veríssimo D, MacMillan DC, Smith RJ, Crees J, Davies ZG (2014) Has climate change taken prominence over biodiversity conservation? Bioscience 64(7):625–629

Vongvisouk T, Broegaard RB, Mertz O, Thongmanivong S (2016) Rush for cash crops and forest protection: neither land sparing nor land sharing. Land Use Policy 55:182–192

Vu Q, Ramal AF, Villegas JM, Jamoralin A, Bernal CC, Pasang JM, Almazan MLP, Ramp D, Settele J, Horgan FG (2018) Enhancing the parasitism of insect herbivores through diversification of habitat in Philippine rice fields. Paddy Water Environ 16:379–390

Wahlén CB (2014) Constructing conservation impact: understanding monitoring and evaluation in conservation NGOs. Conserv Soc 12(1):77–88

Wallach AD, Ripple WJ, Carroll SP (2015) Novel trophic cascades: apex predators enable coexistence. Trends Ecol Evol 30(3):146–153

Westphal C, Vidal S, Horgan FG, Gurr GM, Escalada M, Van Chien H, Tscharntke T, Heong KL, Settele J (2015) Promoting multiple ecosystem services with flower strips and participatory approaches in rice production landscapes. Basic Appl Ecol 16(8):681–689

Wilcove DS, Giam X, Edwards DP, Fisher B, Koh LP (2013) Navjot's nightmare revisited: logging, agriculture, and biodiversity in Southeast Asia. Trends Ecol Evol 28(9):531–540

Xia J, Harmon JL, Connolly KG, Anderson MR, Howard HA (2015) Who publishes in 'predatory' journals? J Assoc Inf Sci Tech 66(7):1406–1417

Zingerli C (2005) Colliding understandings of biodiversity conservation in Vietnam: global claims, national interests, and local struggles. Soc Nat Resour 18(8):733–747

Chapter 14
Environmental Knowledge Exchange in Australia and Oceania: How Researchers and Practitioners Are Working Together to Bring Change

Alexandra Ruth Knight

Contents

A. R. Knight (✉)
School of Environmental Science and Institute for Land, Water and Society, Charles Sturt University, Port Macquarie, NSW, Australia
e-mail: aknight@csu.edu.au

C. C. Ferreira, C. F. C. Klütsch (eds.), *Closing the Knowledge-Implementation Gap in Conservation Science*, Wildlife Research Monographs 4,
https://doi.org/10.1007/978-3-030-81085-6_14

14.1 Introduction

14.1.1 Oceania and Its Diversity

Oceania is that vast region encompassing Australia, New Zealand, Papua New Guinea, the islands of Melanesia, Polynesia and Micronesia and over 62 million km^2 of the Pacific Oceanscape (IUCN 2017a) that surrounds them. The region incorporates not only a great range of terrestrial and freshwater diversity, but also significant marine ecosystems, including the Great Barrier Reef and part of the Coral Triangle. A culturally rich and diverse region, human occupation is estimated to range from 65,000 years in Australia to only 800 years in Aotearoa/New Zealand (Braje et al. 2017; Douglas and Ballard 2008). Two distinct regions occur, with Australia and New Zealand considered as developed economies, and the Pacific island countries and territories (PICT) faced with the significant challenges associated with severe poverty and structural impediments to sustainable development (United Nations 2018). Of the world's 36 biodiversity hotspots (Conservation International Foundation 2018), eight occur in Oceania, and both Australia and Papua New Guinea are considered megadiverse (Mittermeier 1997) (Fig. 14.1).

Within this context of rich biodiversity, conservation challenges abound. Oceania has a high record of extinctions; for instance, of Australia's 316 unique land mammal species, 28 have become extinct in the last 200 years (Woinarski et al. 2015). Recent investigations suggest that Australia and Papua New Guinea are two

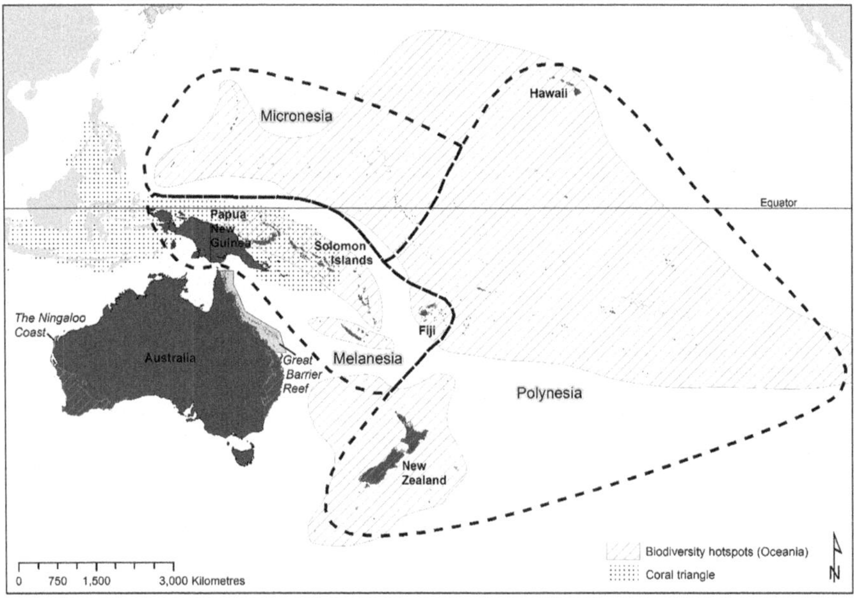

Fig. 14.1 Location of Oceania showing biodiversity hotspots (Source: CSU SPAN)

of only seven countries contributing 60% of the total global biodiversity decline (Waldron et al. 2017). Ongoing threats to biodiversity include habitat clearing, invasive species, lack of appropriate biosecurity measures, the spread of disease, over-exploitation, climate change and associated sea-level rise. The recognition of rich diversity, together with the immediacy of threats, has brought together environmental scientists, practitioners and local people in a number of relationships and partnerships, which have built increased knowledge of how to implement change.

In this chapter, I investigate examples of the relationship between scientific evidence and conservation action from Oceania and contrast findings across scales and approaches. Inevitably and appropriately, the role of indigenous knowledges and environmental management are also discussed, as more and more the importance of different types of knowledge and suitable local responses to environmental challenges are acknowledged. I explore the successes and failures faced by scientists, practitioners and policy makers in the region, as documented in the literature from the region, and recommend future approaches.

14.1.2 The Impetus for Evidence-Informed Decision-Making in Oceania

The call for evidence-informed decision-making and better relationships between environmental researchers and implementers has a long history in Oceania. There is a widespread desire that knowledge generated from research should be incorporated into practice and used in management situations. Discussion on the best way to undertake this is ongoing (van Kerkhoff and Pilbeam 2017), however, there is a well-established view that better collaboration between scientists and decision-makers will lead to better implementation of new knowledge (Chapman et al. 2017).

In Australia, especially within the field of freshwater ecology, there is a strong legacy of building a relationship between scientists and environmental managers and policy makers. This has been championed by one of our prominent water scientists, Peter Cullen, who saw that actively engaging science in policy development was essential for success (Cullen 2006) and also recognized that scientists themselves have strong values (Cullen 2006). Impetus from scientists including Peter Cullen resulted in the foundation of the Wentworth Group of Concerned Scientists in 2002 (Cullen 2004). This group of leading environmental scientists initially advocated for fundamental environmental reform to improve environmental water, undertake landscape repair, value ecosystem services, and eliminate hidden environmental subsidies (Cullen et al. 2002). They have provided ongoing submissions to the Australian government on environmental matters and contributed to public debate, particularly on issues surrounding water allocation (Wentworth Group of Concerned Scientists 2017). The development of this highly visible group of scientists, openly influencing environmental policy, has not only had impact on particular issues, but also provided a licence for and encouraged open input of environmental scientists into management and policy. In 2015, another group of Australian scientists, the Australian Coral Reef Society, undertook concerted action on the 'coal *versus* coral' war resulting in the reversal of a decision to dump dredged sediment into the Great Barrier Reef (Hamylton 2018).

In Melanesia, Micronesia and Polynesia, the impetus for stronger relationships between scientists and practitioners has grown with the increased awareness of threats to biodiversity and sustainability resulting from climate change (Kingsford and Watson 2011; Kumar and Tehrany 2017) and increasing human populations (Butler et al. 2014). Loss of terrestrial and freshwater biodiversity on island ecosystems has been earmarked as an increasing problem (Jupiter et al. 2014b), with a number of collaborative programs instigated in 2008 to begin to address these issues (Critical Ecosystem Partnership Fund 2013). As well as having a strong basis in science, these programs rely on collaborative approaches using traditional local knowledge to instigate conservation actions. The Secretariat of the Pacific Regional Environment Program, a 25-nation consortium established in 1993, has played a key role in developing and implementing conservation-based programs throughout the region (SPREP 2017). Both ensuring implementation of scientific findings and incorporating local knowledge and management have been accelerated with the intervention of international non-government organizations, including Conservation International (CI) and the International Union for the Conservation of Nature (IUCN), and through the significant international funding opportunities provided through organizations such as the Global Environment Facility.

Global interest has also stimulated collaborative work between governments, culturally diverse peoples and scientists in the marine and coastal environments of Oceania. World Heritage listing (Fig. 14.2) has ensured that concern regarding the degradation of the Great Barrier Reef is in the spotlight, and embedding science into management has been one response of the Australian government (Commonwealth of Australia 2015). The Coral Triangle has been rightly a focus for marine science, traditional knowledge and practice to the interface, with multiple international organizations, who espouse the importance of science in management, involved.

14.1.3 Current Approaches to Bridging the Gap Between Scientists and Practitioners

Improved communication between practitioners and scientists has been seen as a prerequisite for effective conservation, and there has been a call for all parties to take measures to do so, whether they be ecologists and landholders discussing frogs (Carr and Hazell 2006); wetland ecologists and managers (Ryder et al. 2010); landscape-scale research teams (Chapman et al. 2017); invasive species managers and researchers (Masters et al. 2018); or, river basin managers exploring the barriers to implementation and relationships between scientists and policy makers (Webb et al. 2010; Weber et al. 2011). New approaches to science communication have been

Fig. 14.2 (**a**) The Wet Tropics of Queensland World Heritage area stretches along the northeast coast of Australia, protecting important coastal and terrestrial ecosystems adjacent to the Great Barrier Reef. (**b**) *Papurana daemeli*, Australian wood frog, is found through northern Australia, Papua New Guinea and nearby islands. Photos: Geoff Heard

developed, and a number of organizations offer specialized services in communication (e.g., Australian science media center https://www.smc.org.au/ and Econnect Communication—"Bringing Science to Life" www.econnect.com.au) (also see O'Connell and McKinnon 2021).

Beyond the call for better communication between researchers and/or practitioners, sophisticated knowledge exchange models and practices have developed that focus on the flow and integration of different types of knowledge, including:

- collaboration,
- co-production,
- embedding of scientists in practice-based organizations,
- employment of specialist knowledge brokers,
- practitioners with strong science training "practitioner–researchers", and
- the development of boundary-spanning organizations.

Collaborative work approaches allow knowledge to be built and shared between experts and practitioners (Thomas 2003; Roux et al. 2006; Gawne et al. 2010; Raymond et al. 2010). Recently, Australian wetland ecologists have entered the natural resource policy discussion advancing the case for collaboration between and across disciplines to produce shared knowledge that meets the needs of users (Ryder et al. 2010, p. 826). They support Roux et al. (2006), who suggest that scientists need to enter collaborative learning partnerships with managers rather than continuing in roles of dispassionate experts (Roux et al. 2006). In this context, the need for ecological research to move away from objective positions, where the researcher is detached from the research outcomes, towards an approach within which the researcher is responding to and involving her/himself with immediate problems has been argued and sometimes accepted (e.g., Ryder et al. 2010, cf. Fazey et al. 2018).

Co-production involves managers and researchers actively sharing knowledge to form new and integrated understandings and management approaches (van Kerkhoff and Pilbeam 2017). Embedding refers to having a scientist working within a practice-based organization, or a practitioner working within a research-focussed organization (Cook et al. 2013). Knowledge brokers are specialized employees, usually within scientific institutions, who focus on facilitating sharing knowledge and building organizational capacity in knowledge exchange. Boundary organizations are separate entities to research teams or implementation organizations which facilitate knowledge sharing amongst multiple stakeholders. Each of these models of knowledge exchange offers different levels of participation between researchers and managers.

Despite the willingness for evidence-informed action and the emergence of a stronger framework for knowledge exchange, major breakdowns in the link between science and action resulting in irreplaceable loss of biodiversity still occur. The extinction of the endemic Australian vertebrate, the Christmas Island Pipistrelle (*Pipistrellus murrayi*), in 2009, is one of the most recent examples of a complete failure from government to act, despite adequate scientific knowledge to intervene being available (Martin et al. 2012; Ng et al. 2014; Woinarski et al. 2017). This extinction was quickly followed by the confirmation of the extinction of the Christmas Island forest skink (*Emoia nativitatis*) and the Bramble Cay melomys (*Melomys rubicola*), in 2014. For all island species, it has been asserted that the extinctions were able to be anticipated and prevented (Woinarski et al. 2017).

14.1.4 Concepts and Definitions

Language and concepts of "knowledge exchange," a field which focuses on overcoming the constraints between knowledge and action, have not yet been broadly adopted in the literature of Oceania's conservation biologists and environmental practitioners. However, these concepts have been used by those working with Indigenous peoples across the region (e.g., Rose 2008; Howitt et al. 2013; Dobbs et al. 2016; Austin et al. 2017). The adoption of a common language has the potential to enable a clearer and more relevant discussion of the uses and benefits of different sorts of knowledge and the application of and barriers to effective evidence-informed science. This chapter relies on language and concepts of knowledge exchange applied to environmental management by Fazey et al. (2013), Phipps and Shapson (2009), and Raymond et al. (2010) and acknowledges the importance of experiential, evidence-based, expert, local, tacit, Indigenous and traditional knowledge in building conservation successes.

14.2 Knowledge Exchange that Benefits Terrestrial and Freshwater Ecosystems in Melanesia, Micronesia, and Polynesia

Pacific Island countries and territories provide a unique context for conservation efforts and collaboration between scientists and communities due to the complexities of conservation challenges where international interests influence local outcomes (Fitzherbert et al. 2008; Cramb and Curry 2012); the impacts of climate change are severe (Kingsford and Watson 2011); and, diverse cultures with differing world views and attitudes towards nature exist (Jupiter 2017). The region is highly vulnerable to climate change and so is not only reliant on global aid to address environmental challenges, but also dependent on a global response to climate change for survival. The solidarity between Pacific Island Countries supports a multi-country approach to conservation with many regional initiatives underway (see Jupiter et al. 2014b for a detailed list of initiatives and partnerships). Conservation challenges in the region have been enumerated and discussed (Kingsford et al. 2009; Kingsford and Watson 2011; Meyer 2014; Kumar and Tehrany 2017; Taylor 2017), but evaluating the effectiveness of knowledge exchange activities between scientists and practitioners is largely unexplored in the peer-reviewed literature.

With limited peer-reviewed investigations available, information about the input of scientists into the management of the terrestrial and freshwater ecosystems of the Oceania island nations exists largely in the project and annual reports prepared by environment agencies and international aid organizations. These reports provide important information about existing and past programs. While evaluation of the effectiveness of knowledge exchange activities and programs is rare, reports often contain sections on lessons learned. These reports can be difficult to find and also often contain limited information, and have limited life spans. Changing internet platforms and arrangements of different organizations also contribute to the difficulty of finding and using this type of knowledge (e.g., Duffy 2011, p. 194).

Important sources of lessons learned include the rather hard to find "Biodiversity Conservation Lessons Learned" Technical Series published by CI developed in 2013 at the cessation of the Critical Ecosystem Partnership Fund (CEPF) funding in Polynesia-Micronesia (currently available at https://www.thelittledesigncompany.co.nz/bio-cons-lessons-learned-tech-series/intro.htm) and the Food and Agriculture Organization of the United Nations' report evaluating nature conservation success in the South Pacific Islands (Scherl and Hahn 2018).

These reports include detailed research results, as well as practical 'lessons learned' to aid future projects. One recommendation is that 'recovery groups' are a good model for increasing capacity at conservation work. Recovery groups are an example of a collaborative approach to knowledge exchange, where researchers, managers, practitioners, and Indigenous people come together to share knowledge, plan, implement and reflect on good practices for improving the conditions for threatened species or ecosystems.

Suggested improvements include the needs for:

- early involvement of government agencies in research projects,
- utilizing local experience and expertise to develop partnerships,
- conservation NGOs and institutions to communicate their results to the scientific community and relevant stakeholders and local residents at research sites,
- continuity in the local person working on the project,
- inclusion of more local participants, and
- spacing out workshops rather than overloading the local community at one time (Patrick and Edwards 2011; Watling 2011; Whistler 2011)

The IUCN plays an important role in Oceania, particularly with the PICTs. The IUCN encourages the provision and use of the latest science to its stakeholders and members and recognizes and proclaims as an enabling principle that the nexus between indigenous local knowledge (ILK) and modern conservation science is vital (IUCN 2017a). Annual reports of IUCN Oceania reveal a variety of activities that support this approach, including gathering basic data describing species and their distributions, learning exchange visits, natural resource management symposiums, provision of open resources, and capacity building activities (IUCN 2017b).

The Secretariat for the Pacific Regional Environmental Program (SPREP) is a 25-nation consortium with an important role in developing and undertaking conservation programs in Oceania. It focuses on four core priority areas: climate change resilience, island and ocean ecosystems, environmental governance and waste management and pollution control (www.sprep.org 2018). SPREP provides a knowledge hub, ensuring that technical and scientific information, as well as traditional knowledge, is available to members. SPREP meetings provide an opportunity for direct exchange information about important scientific findings, for instance, at the twenty-eighth meeting held in 2017, SPREP exchanged information with all member nations about the impact of plastic bags on marine life, including the micro-plastics which bio-accumulate, based on recently published research findings (SPREP 2017, p.26). SPREP is also a facilitator of major research collaborations. Recently, SPREP and the University of Newcastle, Australia have formalized an agreement to develop a research node that provides for five Ph.D. research scholarships for Pacific island nationals focussed on waste, pollution, and ecosystem management (University of Newcastle and SPREP 2018). With only one university, the University of the South Pacific, regionally based, international agreements for research activity are important. The University of the South Pacific is supported by 12 Pacific Island Countries and undertakes research in environment, sustainable development and climate change.

Conservation successes, such as community-managed protected areas and invasive species eradication (Keppel et al. 2014), are documented in few peer-reviewed articles. Kingsford et al. (2009) point out that conservation research in Oceania mostly occurs in developed countries, with the vast majority of publications focussed on Australia, New Zealand, and the Hawaiian Islands. In addition, research is under resourced and not well coordinated. Recent literature which makes some commentary, either explicitly or as secondary to the main research material, regarding knowledge exchange in terrestrial and freshwater biodiversity of Polynesia, Melanesia and Micronesia is detailed in the table below (Table 14.1).

These publications offer big-picture solutions to evidence-based implementation of conservation measures, insight into particular case studies, and some novel and challenging recommendations for new approaches. Common themes emerge throughout these publications, many of which resonate with the lessons learned in the grey literature. Of particular importance are:

- the lack of data and knowledge developed from western scientific methods,
- the importance of early engagement with local people and organizations,
- the need for longer timeframes to develop relationships between researchers and implementers,
- the need for continuity of Indigenous staff, researchers, and implementers working on a particular project or program,
- the use of local expert knowledge,
- the importance for place-based implementation and local decisions,

Table 14.1 Selected key terrestrial and freshwater biodiversity studies that provide commentary on knowledge exchange and effectiveness in Polynesia, Melanesia, and Micronesia

	Location	Issue/case study	Knowledge exchange issues	Knowledge exchange recommendations	References
1	Solomon Islands Australia Fiji	Marine protected areas, fisheries and nutrition in Oceania. "Care for Country" initiative. Managing small-island river basins for aquatic biodiversity, water-borne disease prevention and psychosocial wellbeing	Need to break down boundaries of "silos" and challenge the status quo	Consider the relationships between biodiversity conservation and human health and wellbeing. Place-based decisions negotiated locally. Transparency of information upon which decisions are made. Draw upon different types of knowledges.	Jenkins et al. (2018)
2	Melanesia	Case studies of establishing protected areas, conservation agreements, eco-tourism initiatives, and research-action arenas	Failure to understand differences between western and indigenous world views and issues around land and marine tenure	Clarify expectations early during project planning. Identify local champions. Benefits should be achievable and clearly articulated. Interventions need to be culturally sensitive and build on customary knowledge.	Jupiter (2017)
3	Pacific Island countries and territories	The major threats to biodiversity are reviewed, including habitat loss, overexploitation, invasive species, pollution, disease, human-forced climate change. Bright spots of implementation are identified	Patchy implementation of biodiversity initiatives due to lack of resources, limited data and political will	Provides detailed recommendations including: Knowledge-sharing networks, Community-based management, Increased engagement with local communities to promote wise stewardship and local environmental monitoring, and dissemination of best practice guidelines for management	Jupiter et al. (2014b)

(continued)

Table 14.1 (continued)

	Location	Issue/case study	Knowledge exchange issues	Knowledge exchange recommendations	References
				through learning networks. Implementation of integrated island management including synergistic benefits of ecosystem management (e.g., climate adaptation, disaster risk reduction, improved health)	
4	Pacific Island countries and territories	Integrated island management (IIM) proposed as a new approach to planning and implementation of ecosystem management. Evaluates 36 case studies	Lack of coordinated and linked communities	Building links across groups—communities, business, industry and government to develop common goals for conserving biodiversity, maintaining ecosystem services and securing human health and well-being. Ten guiding principles are developed (Table 14.2, p. 196). Adaptive management is required.	Jupiter et al. (2014a)
5	Solomon Islands	Explanation and detailed case study of developing a cultural landscape approach to conservation which incorporates the indigenous perceptions of landscape in cases where The Nature Conservancy has been working on biodiversity projects	Problem of overcoming long-term community buy-in to biodiversity. Disjunct between the values of NGOs and of the indigenous communities with whom they work. Need for conservation practitioners working in indigenous settings to recognize and respect the landscape values of the	Using a cultural landscape approach can generate support for environmental objectives of NGOs. Gaining strong community support for biodiversity requires a cultural heritage component.	Walter and Hamilton (2014)

(continued)

Table 14.1 (continued)

	Location	Issue/case study	Knowledge exchange issues	Knowledge exchange recommendations	References
			custodians of those places		
6	South Pacific independent island nations	Compares approaches of different types of conservation agencies and their resourcing	Need for better capacity building and information exchange Better collaboration, with coordinated and integrated approaches Need for understanding of cultural and social arrangements	Long-term funding for projects is required. Need for greater involvement of Indigenous communities	Keppel et al. (2012b)
7	Oceania Pacific islands	Effects of unprecedented anthropogenic climate change on biodiversity of Pacific island states, especially atolls	Need for useable information and technical advice. Lack of data. Knowledge generated does not reach decision-makers in local communities	Involving local people will be more effective in conserving biodiversity.	Duffy (2011)
8	Kahua, Solomon Islands	Improving community livelihood and well-being without negative social and environmental consequences	Integration of different knowledges is challenging	Research embedded in the application by engaging local communities in the research process and working with grass roots organizations is required. External knowledge of western researchers provides a different perspective on the problem. This kind of research can have an ongoing legacy—communities undertake implementation when their confidence in their own knowledge and capacities is built during the research process.	Raymond et al. (2010)

- the need to acknowledge, respect and include traditional and local customs, knowledge and values, and
- working and developing linkages across interdisciplinary boundaries.

Beyond these constraints and recommendations, Kingsley et al. (2015), Jenkins et al. (2018), and Jupiter et al. (2014a) challenge both environmental scientists and practitioners when they ask that we move beyond disciplinary boundaries, break down silos and consider the relationship between human health and well-being and the environment, both in research design and implementation. Walter and Hamilton (2014) also provide some provoking insight for biodiversity research and implementation when they recommend that the starting point for sustainable, long-term conservation intervention is through mapping and modeling cultural landscapes. Novel insights into knowledge exchange and successful implementation case studies challenge us to work harder at overcoming the knowledge-implementation gap. Educated consideration of and reflection on context, culture and social-ecological systems provide the opportunity to improve approaches.

14.3 Knowledge Exchange in Oceania's Marine and Coastal Ecosystems

Australia's Great Barrier Reef is the world's largest coral reef system, listed as World Heritage, and contributes around $6 billion to the Australian economy annually (Hoegh-Guldberg nd), yet like most coral reefs around the world (Guest et al. 2018), is in serious decline. Despite management being underpinned by a massive and long-standing scientific program, including the highly successful Cooperative Research Centre (Reef CRC), which undertook an integrated program of applied research and development, education training and extension (Woodley et al. 2006; Cvitanovic et al. 2014a; Australian Government Department of the Environment 2015), political and economic decisions are still made that are in direct contradiction to sound management practice.

Sarah Hamylton, an established reef researcher focused on mapping, monitoring and modeling the effects of climate change on reefs (Hamylton 2013, 2014, 2017b) recently spoke out about how scientists can create change (Hamylton 2018), stating that it is a responsibility of scientists to take a leading role in creating change. Her focus is on working with social scientists to explore interdisciplinary approaches and developing powerful communication and creative ways to emphasize the value of landscapes (Hamylton 2017a, 2018), in particular, art-science collaborations including drawings and story-telling. Hamylton has also commented on the influence of the Homeward Bound leadership course for science, technology, engineering, mathematics, and medicine (STEMM) women, which has as a goal developing collaborations for impactful outcomes (Hamylton and Balez 2018). In addition, she advocates for the role of professional societies in implementing evidence-based science,

Fig. 14.3 Waterfall Bay, a popular tourist destination in Tasmania, occurs within the Waterfall-Fortescue Marine Conservation Area. The reserve contains spectacular sea cliffs, as well as rocky reef habitats, giant string kelp forests, and sea caves containing complex invertebrate assemblages

discussing the role of the Australian Coral Reef Society (2018) in influencing the management of the Great Barrier Reef (Hamylton 2018).

Coral reefs are not the only high priority for marine conservation in Oceania: the Pacific Ocean is home to globally important tuna fisheries largely occurring in waters under the jurisdiction of PICTs (Cordonnery 2005). Overexploitation is a significant threat to fish stocks (Adams et al. 2016) which provide an important source of economic development. Quentin Hanich, also based in Australia, is another scientist working to ensure evidence-based decision making is undertaken in the area of marine conservation and fisheries management (Hanich et al. 2010, 2015, 2018; Hanich 2012; Hanich and Ota 2013; Charlton et al. 2016; Gourlie et al. 2018; Wabnitz et al. 2018). His approach is to directly advise international organizations and governments and work through his membership of the Pacific Sector Working group, the IUCN Fisheries Experts Group and the IUCN World Commission on Protected Areas (Hanich 2018).

Marine conservation requires complex governance arrangements between countries to develop integrated ocean management (Vince et al. 2017). Marine protected areas (Fig. 14.3) and linking networks are considered an important approach to marine conservation. Oceans and seas surrounding the coastline of Australia contain the world heritage marine areas of the Great Barrier Reef and Ningaloo. Oceania also contains a large proportion of the Coral Triangle, a highly important center of marine diversity and a global priority for conservation (Green et al. 2011). Developing

Table 14.2 Selected key marine and coastal research that provides commentary on knowledge exchange and effectiveness in Oceania

	Location	Issue/case study	Knowledge exchange issues	Knowledge exchange recommendations	References
1	Pacific locally managed marine areas (LMMA)	Diffusion of knowledge about LMMAs—more established in Fiji than Solomon Islands	Limited capacity building and networking due to distances and restricted resources.	Interventions should be simple, readily observable and consistent with social beliefs and values and context. Theories of innovation diffusion provide insight into the success of measures.	Mascia and Mills (2018)
2	Four cases in Micronesia	Conservation practitioners are using new knowledge in the form of monitoring data to advance marine conservation in an adaptive management framework	Adaptive management relies on knowledge exchange, yet that is lacking. Formal and informal communication pathways are lacking.	Monitoring should be conducted in a manner relevant to the social and ecological systems and integrated into the decision-making process. Conservation practitioners and scientists, in these cases, integrated culturally appropriate stakeholder engagement throughout all phases of the adaptive management cycle. Providing more details on how monitoring and management activities are linked at similar spatial scales and across similar time frames can enhance the application of knowledge. The strength of communicating messages in a	Montambault et al. (2015)

(continued)

Table 14.2 (continued)

	Location	Issue/case study	Knowledge exchange issues	Knowledge exchange recommendations	References
				locally appropriate way.	
3	The coral Triangle (PNG)	Effective management of Marine Protected Areas (MPA)	Need for access to better data. Poor information and technical expertise. Presence of scientifically rigorous plan does not mean MPA is successfully implemented.	Adaptive planning needed. Social and learning networks outside of scientifically designed networks are effective in sharing knowledge. Need for simplicity. Local advocates and champions.	Green et al. (2011)
4	Pacific region Oceans	Challenges in implementing the Pacific Islands Regional Ocean Policy	Purely scientific approaches to resource management have limited success. The role of customary resources owners and traditional knowledge needs to be considered. Need for improved access to scientific information regarding coastal processes and ecosystems to support decision-making. Capacity building is a major issue. Disjunct between aid donors interest and power and that of PICTs.	Partnerships are required between international and regional organizations. Capacity building in enhancing knowledge management needs addressing. Procedures for information and data collection, maintenance and sharing need to be established. Networks need a systematic approach. All stakeholders, including local communities need to be involved in decision-making. Need for establishment and protection of traditional knowledge and intellectual property rights.	Cordonnery (2005)

(continued)

Table 14.2 (continued)

	Location	Issue/case study	Knowledge exchange issues	Knowledge exchange recommendations	References
5	Australia's Ningaloo marine park and coastal region	Evaluating Ningaloo research program	Very little evaluation of knowledge exchange practices has occurred. There is a need to investigate the long-term benefits of knowledge exchange.	Evaluating knowledge exchange practices is important. In this case: Science has improved the trust of community members in marine managers and improved social and environmental outcomes, Need for targeted engagement programs to improve trust and enhance participatory marine governance, and Need for citizen science programs.	Cvitanovic et al. (2018)
6	Australia's Ningaloo Marine Park and coastal region	Knowledge brokering to improve management of Ningaloo Marine Park	Need to consider the role of relational and reflective knowledge as well as explicit knowledge. More evaluation of the effectiveness of knowledge brokering is needed.	Prolonged stakeholder engagement is required to improve stakeholder knowledge.	Chapman et al. (2017)
7	Australian marine protected area networks	Interviews with MPA practitioners regarding quantitative condition assessment and evaluation of management effectiveness	Limitations on quantitative assessments include: Lack of agency capacity (time and money), Knowledge gaps, and Need for improved cooperation between different levels of government	Using new decision support tools. Promoting more management relevant science.	Addison et al. (2017)

(continued)

Table 14.2 (continued)

	Location	Issue/case study	Knowledge exchange issues	Knowledge exchange recommendations	References
8	Coral dominated marine protected areas in Australia	Lack of primary scientific literature used in marine management planning	Peer-reviewed literature is not the predominant source of information in management plans. Access, timing, and clarity of recommendations influence the lack of use.	Need for knowledge brokers, layperson summaries, and other novel approaches.	Cvitanovic et al. (2014a)
9	Australia Marine Protected Areas	Climate change impacts on biodiversity, species extinctions and ecosystem function in marine parks	Ineffective communication between scientists and decision-makers limits the integration of science into management.	Need for new measures of success for researchers that include uptake of knowledge. Scientists need to provide systematic reviews. Science should provide pragmatic recommendations. Bias must be disclosed. Evidence-based rather than theoretical and conceptual research is more trusted by managers.	Cvitanovic et al. (2014b)

participatory management programs incorporating the best available science has been a focus for marine conservation efforts. Specific research into the knowledge exchange mechanisms underpinning these efforts has been the focus of a number of recent investigations (Table 14.2).

Discussion of effective knowledge exchange and implementation of evidence-informed conservation in the Pacific is focussed on the importance of the social-ecological context of both research and implementation. Acknowledging the importance of traditional knowledge and the intellectual property associated with it (Table 14.2, #1–4) is as important as recognizing filling knowledge gaps associated with western scientific processes. As with terrestrial and freshwater conservation interventions, the need to acknowledge, respect and include traditional and local customs, knowledge and values is fundamental to success. The use of adaptive management processes is considered important in improving evidence-based decision-making (Table 14.1, #3). Knowledge gaps and access to already developed knowledge are still key problems (Table 14.1, #4). Knowledge hubs, meetings, workshops and networking events can play a key role in overcoming these difficulties if these mechanisms are well understood and managed, however, there is little evaluation of the effectiveness of these measures. Importantly, time for collaborative reflection on what works is recognized as useful time.

In Australia, the focus is on investigating successful models and particular approaches to sharing knowledge. Cvitanovic et al. (2015) provide a wide-ranging review of knowledge exchange mechanisms between scientists and practitioners. While knowledge gaps are still acknowledged as a problem for evidence-informed management, the literature supports the need for layperson summaries and reviews of scientific information, the development of adaptive management programs and the need for pragmatic and specific solutions to management problems. The use of knowledge brokers to facilitate engagement and relationship building has been trialed in coastal management in Australia, with mixed success (Table 14.2 #5). The necessity of longer-term programs and time-relevant science is fundamental to success. Acknowledgment, respect and incorporation of the different types of knowledge (particularly tacit knowledge) that lead to effective conservation management is crucial. Importantly, the literature also focuses on the more human characteristics of implementing evidence-informed management, in particular the development of support and trust between managers, scientists, local and indigenous peoples, and other stakeholders.

14.4 Knowledge Exchange Practices Supporting Biodiversity Conservation in Australia and New Zealand

14.4.1 Landcare, Natural Resource Management and the Development of Capacity Building Activities in Regional Australia

Natural resource management has been regionalized in Australia with 56 regional agencies and over 5400 Landcare and Coastcare groups undertaking conservation activities (Commonwealth of Australia 2017). Australian Landcare, founded in 1986, is a nationally accepted and supported community-based and largely volunteer movement that works to build capacity and protect and care for the natural environment. Investment in Landcare has increased local community participation in environmental activities, education and training and the building of social capital, with landholders being more environmentally conscious, an outcome which has not been quantitatively evaluated (Hamparsum et al. 2016) (Fig. 14.4). Recent reviews have focussed on the engagement of the community in Landcare and the opportunities it provides for learning (Curtis et al. 2014; Hamparsum et al. 2016; Commonwealth of Australia 2017).

Fig. 14.4 Bethanga Landcare Group is one of many groups in South-eastern Australia which is focussing on revegetating over-cleared Grassy Box Woodland communities. Volunteers gather on National Tree Day each year to replant cleared areas. (Photo: Alexandra Knight)

Given its prevalence and widespread recognition, Landcare provides an important opportunity for evidence-informed conservation action, including biodiversity management on private land. Community capacity building through knowledge exchange (Fig. 14.5) has been a strong and accepted component of Landcare and regional natural resource management activities. Environmental managers use the scientific links and knowledge they build to develop community events focusing on both raising awareness of biodiversity conservation and improving abilities to benefit biodiversity in the broader community, for instance, developing Brolga (*Grus rubicunda)* habitat management guidelines and having wetland planting days at Brolga sites (Herring 2018). University academics, ecological consultants, and other researchers are frequent guest speakers at Landcare events. Field-based knowledge-sharing events are considered particularly important in developing understanding and ensuring that knowledge is utilized (Knight 2015).

One of the seven key attributes for natural resource management agencies in Australia is to build sustainable action based on synthesized knowledge gathered from multiple sources, including western scientific and Indigenous knowledge (Vella et al. 2017, p. 9). Despite the energy spent by both ecologists and practitioners in developing local scale conservation projects and the knowledge that natural resource management groups are catalysts for change at the local scale, there has been a little measurement of the effectiveness of knowledge exchange or evidence-informed programs (Vella et al. 2017).

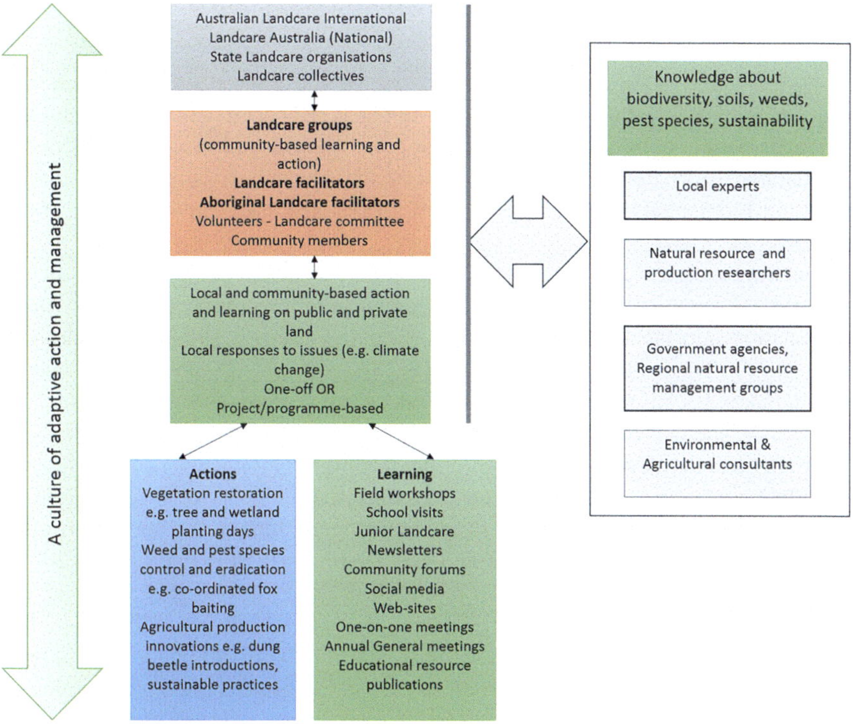

Fig. 14.5 Knowledge exchanges between Landcare and research organizations and other knowledge holders occur at international, national, state, regional, and local levels within a culture of adaptive action and management. Community-based, local Landcare groups are central to success

In addition to undertaking on-the-ground conservation management activities, many Landcare practitioners and environmental managers act as knowledge brokers for the broader community (Pettit et al. 2011). Despite very limited access to peer-reviewed journal articles (Pullin et al. 2004; Pullin and Knight 2005; Cook et al. 2010; Cvitanovic et al. 2014a; Gossa et al. 2014), committed and passionate environmental managers develop links with the scientific community, subscribe to Twitter feeds (Knight 2015), review conference proceedings and pass on the knowledge they have gleaned in informal conversations and formal knowledge-sharing events initiated and run by them. In coordinated efforts, organizations such as Natural Resource Management Research and Innovation network (NRM RAIN) specifically aim "to improve communication and knowledge sharing, encourage collaboration and to seek new opportunities for research and science partnerships" (NRM RAIN 2015).

14.4.2 Threatened Species and Threat Abatement

In Australia and New Zealand, a large part of the conservation research and on-ground effort is aimed at recovering threatened species and abating threats, with a particular emphasis on invasive species control and eradication. Despite considerable effort, both Australia and New Zealand have been largely unsuccessful in halting biodiversity decline (Craig et al. 2000; Seabrook-Davison et al. 2009; Woinarski et al. 2015).

The improvement in the conservation status of Gould's Petrel (*Pterodroma leucoptera leucoptera*) is one of a few documented recovery programs where interrogation into the key elements for success has been undertaken (Priddel and Carlile 2009). The context for action was a key to success, with actions being able to be undertaken because the threats were localized, there were no nested threats, and the recovery site was an island (Priddel and Carlile 2009). Context was not the only important factor, with the authors concluding that the strong linkage between science and implementation was a key factor including the elements of a foundation of robust ecological research, adaptive management, good monitoring and reporting, a multidisciplinary approach, and not being risk averse (Priddel and Carlile 2009). Priddel and Carlile (2009) drill down into each of these factors and emphasize the need for field ecologists to be employed by conservation agencies. While there has been a decline of research-based conservation staff in conservation agencies in both Australia and New Zealand, this is an approach that has been adopted by a number of non-government organizations undertaking important conservation work in Australia, as discussed below. Priddel and Carlile (2009) also comment on the importance of continuity of expertise, a key factor that has received little attention in the peer-reviewed literature.

Garnett et al. (2018) provide a volume that gives insight into 24 cases of "successful" threatened species recovery in Australia and provides insight into careful management. The case studies provided respond to the editors' questions about the characteristics of the species that help recovery efforts, the contribution of the community to recovery, individuals' commitment and leadership, policy and governance, funding and costs, and the ongoing effort needed. Many of the case studies provide great insight into the evidence-based collaboration and knowledge-sharing efforts that lead to success. No explicit questions were asked regarding the use of research or best available science or how that knowledge was produced, translated or exchanged, yet two of the editors' seven conclusions regarding successful recovery relate strongly to environmental knowledge exchange theory. These are that: evidence about species biology and threats is the basis for the recovery effort, which is also responsive to change, and that narratives about the species and its recovery are told well. These recommendations suggest some key questions for effective knowledge exchange and evidence-informed action:

- Can recovery actions be undertaken based on generalized scientific knowledge (for instance, control of invasive species and habitat rehabilitation), or is specific research investigation required?

- What scale (particularly temporal) of monitoring is required that will trigger appropriate changes in management actions? (e.g., Cook et al. 2016).
- What mechanisms are most effective in sharing knowledge well that lead to changes in management?
- What level of community involvement is required for success, and what level of resourcing is required to achieve that level?
- Is the application of western scientific and Indigenous cultural knowledge streams required to achieve success?

Constraints on effectively implementing evidence-informed conservation include human foibles. In a detailed investigation into the long-standing recovery programs of the Bridled nailtail wallaby *(Onychogalea fraemata)* and the Eastern bristlebird *(Dasyornis brachypterus)*, Guerrero et al. (2017) provide a revealing qualitative analysis of the implementation processes of recovery programs. They identified two common factors underlying success across the two programs: the commitment of people involved and the availability of supporting research (p. 22). The difficulties of the collaborative process included:

- Inflexible influential individuals hampering collaborative processes
- Attachment of individuals to particular sites
- Personal agendas which cause conflict and focus resources that do not reflect the needs of the species
- Personal biases hampering adaptive management and willingness to try new things, and
- Attitudinal differences between those in the field and those working in remote offices

(Guerrero et al. 2017, pp. 36–37).

Complex collaborations can fail when expectations of success do not occur, as is often the case in threatened species programs. Forthright discussion of the difficulties of collaborative action is uncommon in the literature regarding the successful implementation of biodiversity conservation, with a focus on the more positive aspects of collaboration highlighted. Yet these considerations are real, and they hamper effective knowledge exchange as well as on-ground success. Guidance in and application of effective participative processes and collaboration is required, both for researchers and practitioners (Allen et al. 2014).

Another often overlooked key consideration in effecting successful evidence-informed management is the importance of constructing knowledge and mutual understanding "in-the-field." For instance, in a qualitative study undertaken in south-eastern Australia, Knight (2015) gained insight from environmental practitioners in how new scientific findings could be mobilized to conserve a rare and little-known amphibian, Sloane's Froglet, *Crinia sloanei*. Group discussions held at management sites built up sound corporate knowledge of the site. Decisions made "in-the-field" also led directly to the use of new scientific knowledge. The process of coming together in field situations was considered powerful in ensuring appropriate implementation practices, and particularly when collaborating with Indigenous people and landholders, absolutely essential. "In-the-field" discussions were not only useful for knowledge sharing and influencing knowledge utilization, they also led to the recognition of the possibility that current practice was ill-informed and was powerful in leading to changes in project or policy direction.

In contrast to these mostly positive examples are those mentioned in the introduction to this chapter, the anticipated extinction of the island species Christmas Island Pipistrelle, the Christmas Island forest skink, and the Bramble Cay melomys (Woinarski et al. 2017), all within the last 10 years. Webb et al. (2016) provide an example of another island-dwelling species, the King Island Scrubtit (*Acanthornis magnus greenianus*) that is in imminent danger of extinction. In this case, a past lack of scientific information (Webb et al. 2016) is combined with a lack of management action. Based on field research undertaken in 2012 and 2015, it is argued that now there is enough information to act in an informed manner, but there needs to be better information sharing and collaboration. Webb et al. (2016) recommend that the development of a national action group or recovery team is critical to overcoming the paralysis and inaction. As recently as May 2018, the plight of the Scrubtit has been highlighted in the press, with inaction still dominating any recovery effort (Cox 2018).

New Zealand also has stories of success and failure. There is a strong history of implementing conservation science with successful programs in controlling invasive species (Allen et al. 2014) and threatened species recovery, especially in island situations (Taylor et al. 2005; Jones and Merton 2012; Russell et al. 2015; Longson et al. 2017). As in Australia, much of this work has been led by recovery groups or threatened species teams which bring together scientists and practitioners. Ewen et al. (2013) provide insight into the recovery team approach using a case study of the threatened Hihi (*Notiomystis cincta*). They confirm that where threatened species challenges are difficult and case-specific (p. 283), recovery teams provide suitable management approaches and a very important channel for evidence-informed action (p. 284).

As many conservation threats are common across species and ecosystems, methods of abating them may be applied across the broader landscape. Internet knowledge hubs and informal communication media enable this practice. Following New Zealand's success in invasive species control and island fauna recovery, a new multistrand program to eradicate introduced predators (Department of Conservation Te Papa Atawhai 2017), which clearly relies on evidence-based implementation and technological innovation, is being undertaken. In order to undertake this extraordinary vision, ecologists are encouraged to move beyond usual boundaries and collaborate with social scientists, economists and policymakers (Russell et al. 2015, p. 523).

In a New Zealand-based case study of transdisciplinary pest management, Allen et al. (2014) address the methods of building integrated regional and landscape programs from site-specific approaches. They provide detailed advice on how to build effective participatory programs through adaptive management and action research. Important aspects underlying effective evidence-based conservation are discussed, including:

- facilitating processes where disciplines and stakeholders can engage respectfully on equal terms,
- encouraging reflection on approaches that foster collaborative production of knowledge,
- problem-solving and dispute resolution,
- using multiple approaches to engage the range of decision-makers at various levels,
- building a culture of trust and respect, and
- the need to integrate solutions with cultural aspirations of Indigenous peoples.

14.4.3 Regional, Landscape and Protected Area Approaches to Evidence-Informed Management in Australia

Many environmental managers implementing biodiversity conservation measures in Australia and New Zealand work as national park rangers, managers of privately managed conservation reserves, and project officers in Landcare and regional natural resource management agencies. Day-to-day management activities, for instance, running rabbit and fox control programs or negotiating revegetation agreements with private landholders, are often undertaken without direct input from university academics or other research scientists, yet provide a major contribution to protecting and enhancing biodiversity. Often practitioners have a university education in environmental science, many with Honours degrees, and so have some experience of environmental research. Other implementers are also well-seasoned researchers (e.g., Murphy and Shea 2015; Croft et al. 2016) undertaking research or monitoring and evaluation programs in their own organizations. Tacit, experiential, Indigenous and observational knowledge together with monitoring and evaluation and research input form the foundation for program implementation. These activities are undertaken without the benefit of large, multi-partner collaborative research implementation measures. Yet, protected area management efforts play a key role in conservation success in Australia (Taylor et al. 2011). It seems unlikely that management actions are not evidence-informed, so how are these practitioners accessing relevant information?

Conservation practitioners, while acknowledging the importance of electronic mechanisms for knowledge exchange, use their own networks and the conversations that take place within those networks as their most important source of new scientific knowledge (Knight 2015). Practitioners have great confidence that these networks provide them with a sound and comprehensive knowledge source. In this context, the ability and confidence of managers to approach researchers is vital to knowledge exchange. Practitioners often focus on established networks and on contacting people that they already know, even using networks established many years before at university. Networks can be quite small. Forums and other formal knowledge exchange networks are also important ways of facilitating knowledge sharing and utilization. In particular, place-based forums, which provide the opportunity for researchers and practitioners to directly swap narratives are vital for mobilizing scientific information.

The role of the formal knowledge broker (or science broker) placed within research organizations (Cvitanovic et al. 2017; Reed et al. 2018) has become more important in biodiversity conservation in Australia. This responds to the growing understanding that enabling uptake of knowledge requires an organizational approach and one that is embedded throughout the research process (Andrews 2012). Australia's lack of success in halting biodiversity decline has provided the impetus for the development of the National Environmental Science's Program (NESP) Threatened Species Recovery Hub. The Hub is convened by ten research organizations, supported by a broader collaboration with more than 20 research organizations and implementers, and espouses an overt goal of delivering evidence for biodiversity conservation. Multiple science communication methods are undertaken by the Hub, which employs science communicators and knowledge brokers. Knowledge brokers perform a key role in Australia's NESP, which aims to produce accessible results that inform decisions (Reef and Rainforest Research Centre 2015), facilitating the flow of knowledge sharing; increasing the involvement of stakeholders in research and practice activities; increasing the influence of science on policy; and, building institutional capacity in sharing knowledge.

Along with knowledge brokers, boundary-spanning activities and organizations such as the Great Southern Science Council (GSSC 2018) are playing an increasing role in evidence-informed management. Limited information about the effectiveness of knowledge brokers and boundary-spanning organizations in building evidence-based implementation in Australia is currently available (Pettit et al. 2011; Cvitanovic et al. 2017).

Non-government organizations managing privately owned lands for nature conservation now hold a key and vital role in developing and applying conservation solutions in Australia. Organizations such as Bush Heritage Australia, Arid Recovery, Australian Wildlife Conservancy (Freudenberger 2016), and The Nature Conservancy (The Nature Conservancy Australia 2018) purchase and manage land for conservation. Management with a strong evidence base is at the core of these organizations. Arid Recovery, an applied restoration ecology initiative in inland South Australia with successes in the reintroduction of threatened mammals, has a sound history of applied and published scientific inquiry, directly employs ecologists and has a distinguished scientific advisory panel (Arid Recovery 2016). In addition, it seeks and supports research collaborations with a number of Universities and provides opportunities for post-graduate research projects. Bush Heritage Australia also has a practice of employing ecologists who work directly alongside field staff to implement biodiversity conservation on the 6 million hectares they manage for conservation across Australia. In addition, their science program articulates a vision and current actions which underpin evidence-based conservation, including:

- post-graduate research scholarships,
- internships,

- science reports synthesizing current conservation thinking,
- scientist in residence programs,
- transformational change forums,
- science fellowships, and
- field stations that are centers of learning (Bush Heritage Australia 2015).

14.4.4 Water Reform and Freshwater Ecological Systems

Conflict over water resources and associated biodiversity losses are a major focus for biodiversity conservation in Australia and New Zealand. Mounting evidence of the worldwide failure of conventional command and control approaches to ensure the environmentally sound and equitable use of water has prompted interest in a new water management paradigm. This way of thinking of water emphasizes sustainability, water security and adaptive capacity (Allan et al. 2013) and privileges learning, governance, dialogic problem solving, and adaptive management (Medema et al. 2008).

Adaptive management processes used in water management have the potential to enhance knowledge exchange and co-learning, yet the outcomes in practice are varied. When discussing environmental flows, Webb et al. (2018) noted three groups of stakeholders are essential for adaptive management of environmental flows (Fig. 14.6) but acknowledge that not all are represented in many programs.

Allan and Watts (2018) provide an example of multiple forms of information exchange occurring around environmental flows in the Edward-Wakool area of NSW. They conclude by encouraging policymakers, managers, and community members involved in adaptive management to share their procedural knowledge—how to do adaptive management—as well as biophysical information. This means supporting, through funding, people to actively seek and share information and document the progress of the shared learning that is occurring. Schoeman et al. (2014) note that adaptive management and Ecosystem-Based Approaches, such as those underpinning the Ramsar Convention and the Millennium Ecosystem Assessment, see management as part of complex ecological systems that require and promote knowledge exchange among multiple stakeholders. In New Zealand, the focus is on participatory practices and the barriers to collaborative processes (e.g., Cradock-Henry et al. 2017). In an example of an environmental stalemate, Weber et al. (2011) offer an enlightening discussion of knowledge exchange, management impasses, and scientist and community values in their case study of water resource management in the Selwyn watershed, Canterbury in New Zealand. They argue for the adoption of the collaborative approach of civic science as described by Lee (1993). However, this kind of approach can only be undertaken where trust and willingness to cooperate are central.

Enquiries into effective water management in Australia and New Zealand highlight the highly complex nature of participatory management processes and the difficulties in implementing evidence-informed management (e.g., Frame 2018).

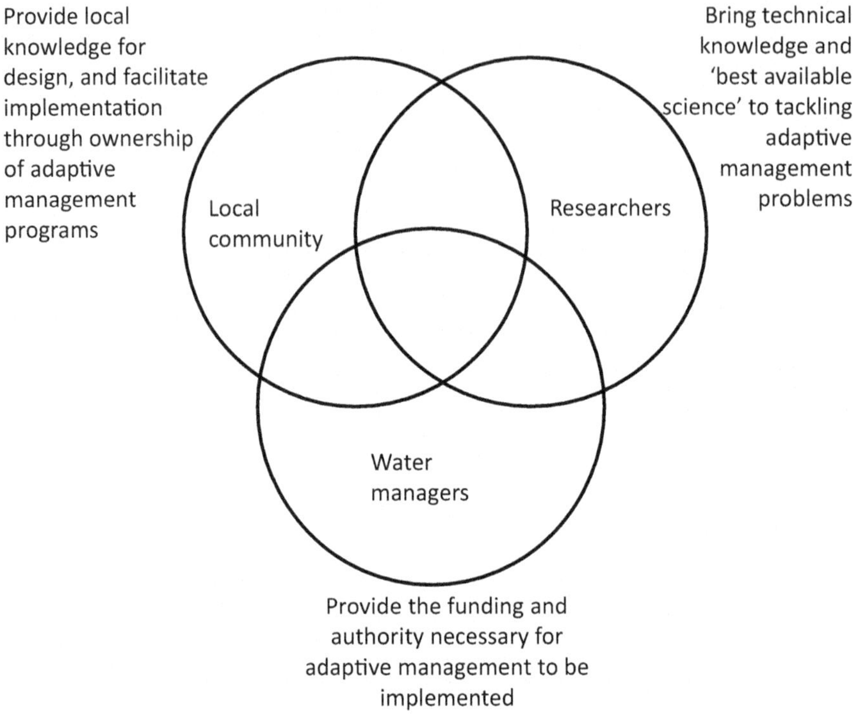

Fig. 14.6 Three different groups of stakeholders are essential for successful adaptive management. The figure depicts different regions of interaction among the local community, researchers, and water managers. Also included is a brief note on what each stakeholder group brings to adaptive management (after Webb et al. 2018)

Endeavors to use scientific knowledge alongside other local and experiential knowledge in participative processes have not always been inclusive or had effective outcomes (Weber et al. 2011; Duncan 2013). Adaptive management approaches that incorporate tracking and evaluating social learning as well as responding to biophysical monitoring and evaluation are rare. However, adaptive management does appear to be increasing in application and effectiveness, providing opportunities to evaluate how learning and knowledge-sharing approaches can work (Allan and Watts 2018; Hart and Butcher 2018; Webb et al. 2018). The analysis of water management initiatives with their focus on processes and social-ecological complexity provides excellent learning opportunities for other biodiversity conservation interventions.

14.5 What Works?: Effective Processes for Knowledge Exchange in Oceania

Undertaking evidence-informed conservation involves, at the very least, two-way processes of knowledge exchange in which researchers and practitioners build an understanding of what can and should work. More likely, is that effective and successful evidence-informed conservation requires multiple knowledge building and sharing processes that are inclusive of traditional, Indigenous and local peoples' knowledge, understandings, customs and perspectives, as well as those of researchers and practitioners. Uni-directional knowledge transfer where limited advice on the application of research findings is provided solely at the conclusion of a peer-reviewed publication is unlikely to be effective. Nor is information dissemination via brochures and social media likely to be enough to improve knowledge uptake. Working together is key.

The frequent call by western scientists for "evidence-based" interventions and initiatives has been superseded by an "evidence-informed" approach, which takes other sources of knowledge as well as practical and external constraints into consideration. Many researchers and other owners and providers of knowledge are working together with implementers in Oceania to achieve conservation successes.

Combinations of approaches with varying levels of collaboration and complexity lead to effective and successful implementation. Key common factors that emerge are listed below.

1. Acknowledgement, respect and application of the different types of knowledge that contribute to successful conservation.
2. An ability to understand, reflect on and communicate the social-ecological context of the conservation research and initiative.
3. Local and place-based collaborative research and interventions as well as knowledge-sharing events.
4. Longer term timeframes with continuity of personnel. Perseverance and persistence.
5. Well-managed collaborative and participative processes including early identification (Colvin et al. 2016) and involvement of stakeholders.
6. Commitment and willingness to advocate.

While the use of citizen science has not been appraised in this chapter (but see Phillips et al. 2021), it is highly likely that the many initiatives now underway can provide better ways for researchers to understand the social-ecological context of their research and learn to involve stakeholders better.

The processes used to underpin evidence-informed management currently being undertaken by non-government organizations such as Bush Heritage Australia and Arid Recovery on their own conservation lands show considerable success and are worth emulating. A key component of success is embedding scientists with management (Hulme 2014; Freudenberger 2016). Embedding scientists in resource management agencies is a vital way of ensuring that science will inform management (Cook et al. 2013; Addison et al. 2017). The loss of embedded scientists in both Australian conservation agencies and the lead government conservation agency in New Zealand is a major constraint on their ability to undertake informed action, reducing both their scientific and technical capabilities (Ewen et al. 2013).

Scientists embedded in management organizations are able to develop strong relationships with practitioners enabling both the co-production and sharing of knowledge. While not all researchers working outside of academia are field-based, many non-government organizations employ ecologists situated in regional or field positions. Place-based positions allow ecologists to work directly alongside field staff and build mutual understanding. This practice of embedding scientists in place-based conservation initiatives alongside the local community and field staff is particularly effective in ensuring not only evidence-informed implementation is undertaken but also that it is accepted by local communities. This approach resonates strongly with the suggestion made by practitioners that discussions of practice undertaken in the field build knowledge and promote changes in decisions (Knight 2015). It also resonates with the barriers to effective action and differences in priorities highlighted by Guerrero et al. (2017) between practitioners based in the field and those remote from daily activities. The recommendation for combined field-based research and decision-making is important throughout Oceania, with Pacific Island learnings emphasizing the importance of researchers communicating their findings with local people at the research sites and undertaking place-based decision-making and initiatives. While many ecological practitioners and researchers remain situated in major cities, place-based initiatives still provide examples of effective ways to work, for instance, taking extra time in the field to not only gather data but also have conversations with managers, researchers, and the community. Another option is to organize and participate in place-based forums such as those initiated by regional natural resource management agencies (e.g., GB CMA 2015).

14.6 Where Is Improvement Required?

14.6.1 Advocacy

Traditionally, many scientists have not been involved in decision-making for policy or management practices, as advocates for particular management actions or about particular conservation issues or even in explaining the results of their research to the public (Lach et al. 2003; Garrard et al. 2016). Some scientists fear that their credibility suffers when they participate in advocacy roles (Lach et al. 2003). Yet researchers are passionate and committed people, particularly in their area of expertise. There is not necessarily a contradiction between being an objective scientist and an advocate (Noss 2007). Advocacy can and does make a difference to the use of scientific evidence in conservation. While it is acknowledged there could be institutional restrictions on advocacy, scientists need to speak up.

Opportunities to change decisions that ensure evidence-informed conservation is implemented occur at many levels. Despite best practice place-based efforts, Landcare interventions, multi-stranded operations of global conservation organizations, the downward trajectory of nature and biodiversity continues. While a multitude of fundamentally important actions are undertaken at local levels, in Oceania, there continues to be a lack of connection between environmental science and government policy. A recent example of this disconnect is the feral horse conflict in eastern Australia. Ecologists have clearly determined that introduced feral horses are severely damaging important protected environments, yet policy decisions fly in the face of reason, further diminishing nature, diminishing the input of science and depleting researchers' will to continue contributing to policy (Hannam 2018; Driscoll et al. 2019; Knight 2019).

Researchers and practitioners need to make collective efforts to ensure policy uptake of conservation sciences. They need to build broader-scale pictures and narratives from the localized activities that are occurring and provide this information to decision-makers in convincing ways. Uncertainty about the nature and state of knowledge and the apparent disagreement of ecologists about key concepts has been seen as a major concern for the utilization of knowledge in environmental policy (Knight 2015).

Professional societies can play a significant role in advocacy. The example given in the introduction of the Australian Coral Reef Society's input changing decisions on reef management is a case in point (Hamylton 2018). Scott et al. (2008) provide detailed advice on how professional societies can play a key role and conclude that professional societies have an important role in bringing policy-relevant science to lobbyists and decision-makers. Professional societies in Oceania need to improve in this area. In addition, particularly in Australia, institutional and systematic arrangements that enable good scientists to provide sound policy advice to government need to be developed and respected.

14.6.2 Education

Guidance in and application of effective participative processes and collaboration is required, both for researchers and implementers. A great deal of information and expertise is already available in this discipline (e.g., the International Association for Public Participation iap2.org.au, and books such as Williamson and DeSouza 2007). For researchers, this guidance needs to start at the undergraduate level so that along with ecological theory, they receive training in Indigenous knowledge systems, environmental ethics, engagement processes and the complexities of social-ecological contexts.

14.6.3 Publications

Researchers often provide brief information on the relevance and management implications of new findings in the discussion section of their articles. On the occasions where these articles are available to practitioners, feedback indicates that the application section is not detailed enough (Ewen et al. 2013). Editors of applied conservation journals along with article reviewers, have a responsibility to ensure that the "application" section of articles is articulated clearly and with detail.

Several journals, e.g., *Pacific Conservation Biology,* provide a section of "field notes." Journals can provide an equivalent section to 'field notes' that provides insight into knowledge exchange case studies to share better information about processes that work. For instance, the *Journal of Applied Ecology* provides practitioners' perspectives (Hulme 2014) and *Ecological Management and Restoration* (Wiley Online Library 2017) provides features stories about projects and actions which are often co-authored by practitioners.

A great deal of excellent "lessons learned" exists in reports. Even more exists in Landcare magazines and natural resource management newsletters. Knowledge hubs can play a key role in making this information available. For instance, in Australia *Knowledge for Purpose: managing research for uptake—a guide to a knowledge and adoption program* (Andrews 2012) provides excellent guidelines for researchers and implementers and is currently found here https://www.environment.gov.au/system/files/resources/7fee4ddb-e1df-4d13-85f0-e0091a95d80f/files/knowledge-purpose.pdf .

14.6.4 Learning with and from Practitioners and Implementers

Reviewing the peer-reviewed evidence regarding how knowledge exchange and evidence-informed management can work and be improved is less than half the story. A very few of the publications cited in this chapter provide implementers'/practitioners' viewpoints (cf. Goggin et al. 2015). Listening to practitioners' stories and attending non-academic forums for information swapping allows us to hear how practitioners are implementing evidence-informed management, the processes they use, constraints they face and how other sorts of knowledge are incorporated and respected in decision-making.

Hulme (2014) points out that many practitioners apply different types of knowledge, which may be intuitive and difficult to define, and that scientists need to consider practitioners' tacit knowledge (p. 1192). Scientists need to take into account that many practitioners undertake management interventions based on their own research, and considerable field naturalist experience (e.g., Parker et al. 2010), as well as the considered results of in-house monitoring and evaluation programs (e.g., OEH 2011).

14.6.5 Evaluation

There is little reported evaluation of the effectiveness of environmental knowledge exchange practices in Oceania. What evaluation has been undertaken is difficult to locate. The question remains as to how much of the continued decline of biodiversity in Oceania is an indication that knowledge exchange is not working or an indication that other factors are responsible. For instance, cultural, economic, land ownership and social situations all are involved in the success and failure of conservation initiatives (Keppel et al. 2012a, b). The continued decline suggests that knowledge sharing with the broader community has failed to impact society's values and priorities and associated government policy and funding decisions.

The concern regarding the lack of evaluation of knowledge exchange processes has been well-articulated in the New Zealand context of pest management. Allen et al. (2014) consider that collaborative and social components are often not explicit in research proposals and published results. To address this, the scientific community needs to use robust and relevant practices to make sure that these areas are not just adequately peer-reviewed but also skilfully implemented (p. 430).

In the Australian water context, Allan and Watts (2018) provide guidance in evaluation approaches when they reveal the iterative nature of learnings occurring throughout all stages of adaptive management processes. To improve practice and policy, there is an urgent need to evaluate the effectiveness of knowledge exchange processes as well as the outcomes of the knowledge exchange and the success or failure of the conservation initiative.

14.7 Conclusions

There is an expansive and complex landscape of researchers and other generators, owners, and providers of knowledge working together with implementers in Oceania to achieve conservation successes. In Australia and New Zealand, enough information is available to undertake effective environmental management and address many key ecological issues (Morton et al. 2009). Elsewhere in Oceania, particularly in the marine and terrestrial ecosystems of the Pacific Islands and oceans, knowledge gaps require urgent filling, yet action does and should take place.

Successful conservation initiatives are informed by western scientific knowledge, local, Indigenous and traditional knowledge and tacit and experiential knowledge of researchers, practitioners, and the community. Conflicts and tensions occur between these different types of knowledge and between the different values and visions of those researching and implementing conservation and land management practices. Integrating other sorts of knowledge with scientific knowledge represents a deep change in our understanding and approach to knowledge governance as other types of knowledge become accepted as authoritative and reliable (Duncan 2013, p. 206).

Research processes and implementation and integration practices can be improved (Bammer 2013). Cross-sectoral approaches warrant further investigation, for instance, approaches that consider health and the environment. They have the potential to draw upon different types of knowledge in place-based responses, which carefully and hopefully engage people in learning and action (Jenkins et al. 2018, p. 131).

Working together requires effort and can be difficult. Working together needs time, resourcing, commitment and the willingness to understand complex social-ecological and cultural contexts. Using established pathways such as conservation action planning, recovery teams, adaptive management, and knowledge brokerage can be very effective. Accommodating differing approaches, such as the cultural landscape approach trialed by Walter and Hamilton (2014) in the Solomon Islands, requires willingness as well as conceptual flexibility and agility. Further investigation which reveals the nitty-gritty of successful approaches to evidence-informed conservation in Oceania is warranted and will enlighten the processes of transformation that scientists, practitioners and the community seek.

References

Adams VM, Spindler RE, Kingsford RT (2016) Thinking globally, acting locally - conservation lessons from Oceania Pacific. Conserv Biol 22(2):85–89

Addison PFE, Flander LB, Cook CN (2017) Towards quantitative condition assessment of biodiversity outcomes: insights from Australian marine protected areas. J Environ Manage 198(Part 1):183–191

Allan C, Watts RJ (2018) Revealing adaptive management of environmental flows. Environ Manage 61:520–533

Allan C, Xia J, Pahl-Wostl C (2013) Climate change and water security: challenges for adaptive water management. Curr Opin Environ Sustain 5(6):625–632

Allen W et al (2014) Bridging disciplines, knowledge systems and cultures in pest management. Environ Manage 53:429–440

Andrews K (2012) Knowledge for purpose: managing research for uptake – a guide to a knowledge and adoption program. Department of Sustainability, Environment, Water, Population and Communities, Canberra

Arid Recovery (2016) Arid Recovery Annual Report 2015–2016

Austin B, Vigilante T, Cowell S, Dutton I, Djanghara D, Mangolomara S et al (2017) The Uunguu monitoring and evaluation committee: intercultural governance of a land and sea management Programme in the Kimberley, Australia. Ecol Manage Restor 18(2):124–133

Australian Coral Reef Society (2018) About the Society. Australian Coral Reef Society. https://australiancoralreefsociety.org/about/. Accessed 21 June 2018

Australian Government Department of the Environment (2015) Science and the Great Barrier Reef

Bammer G (2013) Disciplining interdisciplinarity: integration and implementation sciences for researching complex real-world problems. ANU Press, Canberra

Braje TJ, Leppard TP, Fitzpatrick SM, Erlandson JM (2017) Archaeology, historical ecology and anthropogenic island ecosystems. Environ Conserv 44(3):286–297

Bush Heritage Australia (2015) Science for healthy country. Sci Res Plan 2015–2025

Butler JRA, Skewes T, Mitchell D, Pontio M, Hills T (2014) Stakeholder perceptions of ecosystem service declines in Milne Bay, Papua New Guinea: is human population a more critical driver than climate change? Mar Policy 46:1–13
Carr A, Hazell D (2006) Talking frogs: the role of communication in ecological research on private land. Biodivers Conserv 15:3177–3191
Chapman K, Boschetti F, Fulton E, Horwitz P, Jones T, Scherrer P, Syme G (2017) Knowledge that acts: evaluating the outcomes of a knowledge brokering intervention in Western Australia's Ningaloo region environmental management 60:896–907
Charlton KE, Russell J, Gorman E, Hanich Q, Delisle A, Campbell B, Bell J (2016) Fish, food security and health in Pacific Island countries and territories: a systematic literature review. BMC Public Health 16:285
Colvin RM, Witt GB, Lacey J (2016) Approaches to identifying stakeholders in environmental management: Insights from practitioners to go beyond the 'usual suspects'. Land Use Policy 52: 266–276
Commonwealth of Australia (2015) Reef 2050 Long-Term Sustainability Plan
Commonwealth of Australia (2017) The report on the review of the National Landcare Program
Conservation International Foundation (2018) Critical ecosystem partnership fund. Explore the biodiversity hotspots. https://wwwcepfnet/our-work/biodiversity-hotspots. Accessed 6 June 2018
Cook CN, Hockings M, Carter RW (2010) Conservation in the dark? The information used to support management decisions. Front Ecol Environ 8(4):181–186
Cook CN, Mascia MB, Schwartz MW, Possingham HP, Fuller RA (2013) Achieving conservation science that bridges the knowledge–action boundary. Conserv Biol 27:669–678
Cook CN, de Bie K, Keith DA, Addison PFE (2016) Decision triggers are a critical part of evidence-based conservation. Biol Conserv 195:46–51
Cordonnery L (2005) Implementing the Pacific Islands Regional Ocean Policy. Vict Univ Wellington Law Rev 36:723–731
Cox L (2018) King Island birds face extinction without urgent federal help, residents say. Guardian News and Media Limited. https://www.theguardian.com/australia-news/2018/may/24/king-island-birds-face-extinction-without-urgent-federal-help-residents-say
Cradock-Henry NA, Greenhalgh S, Brown P, Sinner J (2017) Factors influencing successful collaboration for freshwater management in Aotearoa, New Zealand. Ecol Soc 22(2):14
Craig J et al (2000) Conservation issues in New Zealand. Annu Rev Ecol Syst 31:61–78
Cramb R, Curry GN (2012) Oil palm and rural livelihoods in the A sia–Pacific region: an overview. Asia Pac Viewp 53(3):223–239
Critical Ecosystem Partnership Fund (2013) Critical ecosystem partnership fund final evaluation conference for the Polynesia-micronesia hotspot. April 15–18, 2013 Suva-Fiji
Croft P, Hunter JT, Reid N (2016) Forgotten fauna: habitat attributes of long-unburnt open forests and woodlands dictate a rethink of fire management theory and practice. Forest Ecol Manage 366:166–174
Cullen P (2004) Turning the tide: how does science change public policy? Paper presented at the CSIRO land and water seminar world water day address, Adelaide, 22 March 2004
Cullen P (2006) Science and politics – speaking truth to power. Paper presented at the North American Benthological society annual conference, Anchorage, Alaska, June 2006
Cullen P, Flannery T, Harding R, Morton SR, Possingham HP, Saunders DA, Thom B, Williams J, Young MD, Cosier P, Boully L (2002) Blueprint for a living continent: a way forward from the Wentworth group of concerned scientists. WWF Australia, Sydney
Curtis A et al (2014) The great experiment with devolved NRM governance: lessons from community engagement in Australia and New Zealand since the 1980s. Austral J Environ Manage 21(2):175–199
Cvitanovic C, Fulton CJ, Wilson SK, van Kerkhoff L, Cripps IL, Muthiga N (2014a) Utility of primary scientific literature to environmental managers: an international case study on coral-dominated marine protected areas. Ocean Coast Manage 102(Part A):72–78

Cvitanovic C, Marshall NA, Wilson SK, Dobbs K, Hobday AJ (2014b) Perceptions of Australian marine protected area managers regarding the role, importance, and achievability of adaptation for managing the risks of climate change. Ecol Soc:19(4):33

Cvitanovic C, Hobday AJ, van Kerkhoff L, Wilson SK, Dobbs K, Marshall NA (2015) Improving knowledge exchange among scientists and decision-makers to facilitate the adaptive governance of marine resources: a review of knowledge and research needs. Ocean Coastal Manage 112:25–35

Cvitanovic C, Cunningham R, Dowd AM, Howden SM, van Putten EI (2017) Using social network analysis to monitor and assess the effectiveness of knowledge brokers at connecting scientists and decision-makers: an Australian case study. Environ Policy Governance 27(3):256–269

Cvitanovic C et al (2018) Building trust among marine protected area managers and community members through scientific research: insights from the Ningaloo Marine Park, Australia. Mar Policy 93:195–206

Department of Conservation Te Papa Atawhai (2017) Predator Free 2050

Dobbs RJ et al (2016) Collaborative research partnerships inform monitoring and management of aquatic ecosystems by indigenous rangers. Rev Fish Biol Fish 26:711–725

Douglas B, Ballard C (2008) Foreign bodies: Oceania and the science of race 1750–1940. ANU Press, Canberra

Driscoll DA, Worboys GL, Allen H, Banks SC, Beeton NJ, Cherubin RC, et al (2019) Impacts of feral horses in the Australian Alps and evidence-based solutions. Ecol Manag Restor 20(1):63–72

Duffy DC (2011) No room in the ark? Climate change and biodiversity in the Pacific Islands of Oceania. Pac Conserv Biol 17(3):192–200

Duncan R (2013) Converting community knowledge into catchment nutrient limits: a constructivist analysis of a New Zealand collaborative approach to water management. Nat Cult 8(2):205–225

Ewen JG, Adams L, Renwick R (2013) New Zealand species recovery groups and their role in evidence-based conservation. J Appl Ecol 50(2):281–285

Fazey I et al (2013) Knowledge exchange: a review and research agenda for environmental management. Environ Conserv 40(1):19–36

Fazey I, Schäpke N, Caniglia G, Patterson J, Hultman J, van Mierlo B, et al (2018) Ten essentials for action-oriented and second order energy transitions, transformations and climate change research. Energy Res Soc Sci 40:54–70

Fitzherbert EB, Struebig MJ, Morel A, Danielsen F, Brühl CA, Donald PF, Phalan B (2008) How will oil palm expansion affect biodiversity? Trends Ecol Evol 23(10):538–545

Frame B (2018) New Zealand: new futures, new thinking? Futures 100:45–55

Freudenberger D (2016) The vital role of environmental NGOs: trusted brokers in complex markets. Australian National University Press, Canberra

Garnett ST, Woinarski J, Lindenmayer D, Latch P (eds) (2018) Recovering Australian threatened species: a book of hope. Recovering Australian threatened species. CSIRO, Clayton

Garrard GE, Fidler F, Wintle BC, Chee YE, Bekessy SA (2016) Beyond advocacy: making space for conservation scientists in public debate. Conserv Lett 9(3):208–212

Gawne B, Crase L, Watson AS (2010) Can a collaborative focus on solutions improve our capacity to achieve sustainable water management? Mar Freshw Res 61(7):814–820

GB CMA (2015) Goulburn Broken Catchment Research Forum

Goggin CL et al (2015) Working with practitioners:how scientists can develop effective relationships that produce an enduring legacy. Paper presented at the 21st International Congress on Modelling and Simulation, Gold Coast, 29/11–4/12 2015

Gossa C, Fisher M, Milner-Gulland EJ (2014) The research–implementation gap: how practitioners and researchers from developing countries perceive the role of peer-reviewed literature in conservation science. Oryx 49(1):80–87

Gourlie D, Davis R, Govan H, Marshman J, Hanich Q (2018) Performing "a new song": suggested considerations for drafting effective coastal fisheries legislation under climate change. Marine Policy 88:342–349

Green SJ et al (2011) Emerging marine protected area networks in the coral triangle: lessons and way forward. Conserv Soc 9(3):173–188
Guerrero AM, McKenna R, Woinarski JCZ, Pannell DJ, Wilson KA, Garnett ST (2017) Threatened species recovery planning in Australia: learning from two case studies. National Environmental Science Programme, Threatened Species Recovery Hub
Guest JR et al (2018) A framework for identifying and characterising coral reef "oases" against a backdrop of degradation. J Appl Ecol 55(6):2865–2875
Hamparsum J, O'Neil C, Walker D (2016) Landcare's role in buliding adaptive capacity and resilience
Hamylton S (2013) Five practical uses of spatial autocorrelation for studies of coral reef ecology. Mar Ecol Progress Ser 478:15–25
Hamylton S (2014) Will Coral Islands maintain their growth over the next century? A deterministic model of sediment availability at Lady Elliot Island, Great Barrier Reef. PLoS One 9(4):e94067
Hamylton S (2017a) Mapping the islands: how art and science can save the great barrier reef
Hamylton S (2017b) Spatial analysis of coastal environments. Cambridge University Press, Cambridge
Hamylton S (2018) Make your science count. Nature 554(7690):137–138
Hamylton S, Balez R (2018) How a trip to Antarctica became a real-life experiment in decision-making. The Conversation Media Group
Hanich Q (2012) Distributing the Bigeye conservation burden in the Western and Central Pacific Fisheries. Mar Policy 36(2):327–332
Hanich Q (2018) Associate professor Quentin Hanich. University of Woollongong. https://ancors.uow.edu.au/staff/UOW076782.html
Hanich Q, Ota Y (2013) Moving beyond rights-based management: a transparent approach to distributing the conservation burden and benefit in tuna fisheries. Int J Mar Coastal Law 28(1):135–170
Hanich Q, Teo F, Tsamenyi M (2010) A collective approach to Pacific islands fisheries management: moving beyond regional agreements. Mar Policy 34(1):85–91
Hanich Q, Campbell B, Bailey M, Molenaar E (2015) Research into fisheries equity and fairness – addressing conservation burden concerns in transboundary fisheries. Mar Policy 51:302–304
Hanich Q, Wabnitz CCC, Ota Y, Amos M, Donato-Hunt C, Hunt A (2018) Small-scale fisheries under climate change in the Pacific Islands region. Marine Policy 88:279–284
Hannam P (2018) 'Wilful disregard': scientist quits NSW panel over wild horses bill. The Sydney Morning Herald. https://www.smh.com.au/environment/conservation/wilful-disregard-scientist-quits-nsw-panel-over-wild-horses-bill-20180607-p4zk19.html. Accessed 07 June 2018
Hart BT, Butcher R (2018) Commonwealth long-term intervention monitoring project: stage 1 mid-term review and evaluation, report by Water Science Pty Ltd and Water's Edge Consulting for Commonwealth Environmental Water Office
Herring M (2018) Brolga breeding habitat. A guide to managing wetlands on your farm, 2nd edn. http://www.murraywildlife.com.au/wp-content/uploads/2018/09/Brolga-Breeding-Habitat-FINAL.pdf
Hoegh-Guldberg O (nd) Science, solutions and the future: The Great Barrier Reef in a time of change
Howitt R, Lunkapis G, Suchet-Pearson S, Miller F (2013) New geographies of coexistence: reconsidering cultural interfaces in resource and environmental governance. Asia Pac Viewp 54(2):123–125
Hulme PE (2014) EDITORIAL: bridging the knowing–doing gap: know-who, know-what, know-why, know-how and know-when. J Appl Ecol 51:1131–1136
IUCN (2017a) IUCN Oceania Regional Plan 2017–2020
IUCN (2017b) Oceania Regional Office annual report 2017
Jenkins A, Horwitz P, Arabena K (2018) My island home: Place-based integration of conservation and public health in Oceania. Environ Conserv 45(2):125–136

Jones CG, Merton DV (2012) A tale of two islands: the rescue and recovery of endemic birds in New Zealand and Mauritius. In: Ewen JG, Armstrong DP, Parker KA, Seddon PJ (eds) Reintroduction biology: integrating science and management, pp 33–72

Jupiter SD (2017) Culture, kastom and conservation in Melanesia: what happens when worldviews collide? Pac Conserv Biol 23(2):139–145

Jupiter SD et al (2014a) Principles for integrated island management in the tropical Pacific. Pac Conserv Biol 20(2):193–205

Jupiter SD, Mangubhai S, Kingsford RT (2014b) Conservation of biodiversity in the Pacific Islands of Oceania: challenges and opportunities. Pac Conserv Biol 20(2):206–220

Keppel G, Morrison C, Hardcastle J, Rounds IA, Wilmott IK, Hurahura F, Shed P (2012a) Conservation in Pacific Island countries: case studies of successful programs. Parks 18:112–124

Keppel G, Morrison C, Watling D, Tuiwawa MV, Rounds IA (2012b) Conservation in tropical Pacific Island countries: why most current approaches are failing. Conserv Lett 5(4):256–265

Keppel G, Morrison C, Meyer J-Y, Boehmer HJ (2014) Isolated and vulnerable: the history and future of Pacific Island terrestrial biodiversity. Pac Conserv Biol 20(2):136–145

Kingsford RT, Watson JEM (2011) Climate change in oceania – a synthesis of biodiversity impacts and adaptations. Pac Conserv Biol 17(3):270–284

Kingsford RT et al (2009) Major conservation policy issues for biodiversity in oceania. Conserv Biol 23(4):834–840

Kingsley J et al (2015) Exploring ecosystems and health by shifting to a regional focus: perspectives from the Oceania EcoHealth chapter. Int J Environ Res Public Health 12(10):12706–12722

Knight AR (2015) The case for Sloane's Froglet: generating ecological knowledge with the intent to benefit biodiversity. Charles Sturt University

Knight AR (2019) How can the social sciences work with ecology in informing feral horse policy and management in South-Eastern Australia? Ecol Manage Restorat 20(1):9–12

Kumar L, Tehrany MS (2017) Climate change impacts on the threatened terrestrial vertebrates of the Pacific Islands. Sci Rep 7(1):5030

Lach D, List P, Steel B, Shindler B (2003) Advocacy and credibility of ecological scientists in resource decisionmaking: a regional study. BioScience 53(2):170–178

Lee KN (1993) Compass and gyroscope integrating science and politics for the environment. Island Press, Washington, DC

Longson CG, Brejaart R, Baber MJ, Babbitt KJ (2017) Rapid recovery of a population of the cryptic and evolutionarily distinct Hochstetter's frog, *Leiopelma hochstetteri*, in a pest-free environment. Ecol Manage Restorat 18(1):26–31

Martin TG et al (2012) Acting fast helps avoid extinction. Conserv Biol 5(4):274–280

Mascia MB, Mills M (2018) When conservation goes viral: the diffusion of innovative biodiversity conservation policies and practices. Conserv Lett 11(3):e12442

Masters P, Markopoulos N, Florance B, Southgate R (2018) The eradication of fallow deer (*Dama dama*) and feral goats (*Capra hircus*) from Kangaroo Island, South Australia. Austral J Environ Manage 25:86–98

Medema W, McIntosh BS, Jeffrey PJ (2008) From premise to practice a critical assessment of integrated water resources management and adaptive management approaches in the water sector. Ecol Soc 13(2):29

Meyer J-Y (2014) Critical issues and new challenges for research and management of invasive plants in the Pacific Islands. Pac Conserv Biol 20(2):146–164

Mittermeier RA (1997) Megadiversity: earth's biologically wealthiest nations (1st English ed.). CEMEX, México, D.F

Montambault JR et al (2015) Use of monitoring data to support conservation management and policy decisions in Micronesia. Conserv Biol 29(5):1279–1289

Morton SR et al (2009) The big ecological questions inhibiting effective environmental management in Australia. Austral Ecol 34(1):1–9

Murphy MJ, Shea M (2015) Survey of the land snail fauna (Gastropoda: Pulmonata) of Mount Kaputar National Park in northern inland New South Wales, Australia, including a description of the listing of Australia's first legally recognised endangered land snail community. Molluscan Res 35(1):51–64

Ng CF, McCarthy MA, Martin TG, Possingham HP (2014) Determining when to change course in management actions. Conserv Biol 28(6):1617–1625

Noss RF (2007) Values are a good thing in conservation biology. Conserv Biol 21(1):18–20

NRM RAIN (2015) NRM RAIN. About the network. http://nrmrain.org.au/about/. Accessed 29 June 2018

O'Connell C, McKinnon M (2021) Communication: the bridge between knowledge and implementation. In: Ferreira CC, Klütsch CFC (eds) Closing the knowledge-implementation gap in conservation science – evidence transfer across spatiotemporal scales and different stakeholders. Springer, New York, pp 171–197

OEH (2011) NSW threat abatement plan for predation by the red fox

Parker DG, Egan D, Ballestrin ML (2010) Recent observations of the thick-billed Grasswren '*Amytornis textilis modestus*' in New South Wales. Australian Field Ornithology 27(4):159–164

Patrick B, Edwards E (2011) Biodiversity conservation lessons learned technical series 3: capacity building to secure endemic Samoan swallowtail butterfly as a model for valuing and conserving butterflies distinctive in the Polynesia–Micronesia Hotspot

Pettit C et al (2011) Exploring the potential of knowledge brokering to enhance natural resource management: findings from the catchment knowledge exchange project in Victoria Australasian. J Environ Manage 18(4):233–247

Phillips TB, Parker A, Bowser A, Hakley M (2021) Publicly generated data: the role of citizen-science for knowledge production, action, and public engagement. In: Ferreira CC, Klütsch CFC (eds) Closing the knowledge-implementation gap in conservation science – evidence transfer across spatiotemporal scales and different stakeholders. Springer, New York, pp 83–107

Phipps DJ, Shapson S (2009) Knowledge mobilisation builds local research collaborations for social innovation. Evid Policy: J Res Debate Pract 5(3):211–227

Priddel D, Carlile N (2009) Key elements in achieving a successful recovery programme: a discussion illustrated by the Gould's Petrel case study. Ecol Manage Rest 10:S97-S102

Pullin AS, Knight TM (2005) Assessing conservation management's evidence base: a survey of management-plan compilers in the United Kingdom and Australia. Conserv Biol 19(6):1989–1996

Pullin AS, Knight TM, Stone DA, Charman K (2004) Do conservation managers use scientific evidence to support their decision-making? Biol Conserv 119(2):245–252

Raymond CM, Fazey I, Reed MS, Stringer LC, Robinson GM, Evely AC (2010) Integrating local and scientific knowledge for environmental management. J Environ Manage 91(8):1766–1777

Reed SE et al (2018) Roles for scientific societies to engage with conservation policy. Conserv Biol 32(3):513–515

Reef and Rainforest Research Centre (2015) National environmental science programme tropical water quality hub knowledge brokering and communication strategy v 2.0

Rose DB (2008) Love in the time of extinctions. Austral J Anthropol 19(1):81–84

Roux DJ, Rogers KH, Biggs HC, Ashton PJ, Sergeant A (2006) Bridging the science management divide: moving from unidirectional knowledge transfer to knowledge interfacing and sharing. Ecol Soc 11(1):4

Russell JC, Innes JG, Brown PH, Byrom AE (2015) Predator-free New Zealand: conservation country. BioScience 65(5):520–525

Ryder DS, Tomlinson M, Gawne B, Likens GE (2010) Defining and using 'best available science': a policy conundrum for the management of aquatic ecosystems. Mar Freshw Res 61(7):821–828

Scherl LM, Hahn R (2018) GEF-PAS forest and protected area Management in Fiji, Samoa, Vanuatu and Niue. Lessons Learned from five years of Nature Conservation and Development in South Pacific Islands

Schoeman J, Allan C, Finlayson CM (2014) A new paradigm for water? A comparative review of integrated, adaptive and ecosystem-based water management in the Anthropocene. Int J Water Resour Dev 30(3):1–14
Scott JM, Rachlow JL, Lackey RT (2008) The science-policy interface: what is an appropriate role for professional societies? Bioscience 58(9):865–869
Seabrook-Davison MNH, Brunton DH, Ji W (2009) Future direction for the conservation of New Zealand's biodiversity. Pac Conserv Biol 15(3):153–157
SPREP (2017) Report of the twenty-eighth SPREP meeting of officials
Taylor S (2017) Impacts of climatic and oceanic processes on the threatened terrestrial vertebrates of the Pacific region. Geo Res J 13:1–8
Taylor SS, Jamieson IG, Armstrong DP (2005) Successful island reintroductions of New Zealand robins and saddlebacks with small numbers of founders. Anim Conserv 8(4):415–420
Taylor MFJ, Sattler PS, Evans M, Fuller RA, Watson JEM, Possingham HP (2011) What works for threatened species recovery? An empirical evaluation for Australia. Biodivers Conserv 20:767–777
The Nature Conservancy Australia (2018) Impact report:2018
Thomas C (2003) Bureaucratic landscapes: interagency cooperation and the preservation of biodiversity. MIT, Boston
Threatened Species Recovery Hub (nd) About. Threatened species recovery hub. http://www.nespthreatenedspecies.edu.au/about
University of Newcastle, SPREP (2018) Research road map. A joint capacity building initiative of The University of Newcastle and The Secretariat of the Pacific Regional Environment Programme
van Kerkhoff L, Pilbeam V (2017) Understanding socio-cultural dimensions of environmental decision-making: a knowledge governance approach. Environ Sci Policy 73:29–37
Vella K, Cole-Hawthorne R, Hardaker M (2017) The value proposition of regional natural resource management in Queensland final report
Vince J, Brierley E, Stevenson S, Dunstan P (2017) Ocean governance in the South Pacific region: progress and plans for action. Marine Policy 79:40–45
Wabnitz CCC, Cisneros-Montemayor AM, Hanich Q, Ota Y (2018) Ecotourism, climate change and reef fish consumption in Palau: benefits, trade-offs and adaptation strategies. Marine Policy 88:323–332
Waldron A et al (2017) Reductions in global biodiversity loss predicted from conservation spending. Nature 551(7680):364-367
Walter RK, Hamilton RJ (2014) A cultural landscape approach to community-based conservation in Solomon Islands. Ecol Soc 19(4):41
Watling D (2011) Biodiversity conservation lessons learned technical series 1: conservation of the endangered Fiji Flying Fox Mirimiri acrodonta on Taveuni Island, Fiji
Webb JA, Stewardson MJ, Chee YE, Schreiber ESG, Sharpe AK, Jensz MC (2010) Negotiating the turbulent boundary: the challenges of bulding a science-management collaboration for landscape-scale monitoring of environmental flows. Marine Freshwater Res 61(7):798–807
Webb MH, Holdsworth M, Stojanovic D, Terauds A, Bell P, Heinsohn R (2016) Immediate action required to prevent another Australian avian extinction: the King Island Scrubtit. Emu - Austral Ornithol 116(3):223–229
Webb JA, Watts RJ, Allan C, Conallin JC (2018) Adaptive management of environmental flows. Environ Manage 61(3):339–346
Weber EP, Memon A, Painter B (2011) Science, society, and water resources in New Zealand: recognising and overcoming a societal impasse. J Environ Policy Plan 13(1):49–69
Wentworth Group of Concerned Scientists (2017) Review of water reform in the Murray-Darling Basin
Whistler A (2011) Biodiversity conservation lessons learned technical series 2. In: The rare plants of Samoa

Wiley Online Library (2017) Ecological management and restoration. Author Guidelines. Wiley. https://onlinelibrary.wiley.com/page/journal/14428903/homepage/forauthors.html#manuscript. Accessed 30 June 2018

Williamson A, DeSouza R (2007) Researching with communities: grounded perspectives on engaging communities in research. Muddycreek Press, Auckland

Woinarski JCZ, Burbidge AA, Harrison PL (2015) Ongoing unraveling of a continental fauna: decline and extinction of Australian mammals since European settlement. PNAS 112(15):4531–4540

Woinarski JCZ, Garnett ST, Legge SM, Lindenmayer DB (2017) The contribution of policy, law, management, research, and advocacy failings to the recent extinctions of three Australian vertebrate species. Conserv Biol 31(1):13–23

Woodley SJ, Wlliams DM, Harvey T, Jones A (2006) World heritage research: making a difference. CRC Reef Research, Education and Capacity Building 1999e2006. CRC Reef Research Centre, Townsville, Australia.

www.sprep.org (2018) Home. SPREP Pacific Environment. www.sprep.org. Accessed 29 Jun 2018

Part V
Synthesis

Chapter 15
Closing the Gap Between Knowledge and Implementation in Conservation Science: Concluding Remarks

Cornelya F. C. Klütsch and Catarina C. Ferreira

Contents

C. F. C. Klütsch (✉)
Division of Environment and Natural Resources, Norwegian Institute of Bioeconomy Research (NIBIO), Svanvik, Norway
e-mail: cornelya.klutsch@nibio.no

C. C. Ferreira
Department of Conservation Biology, Helmholtz-Zentrum für Umweltforschung – UFZ, Leipzig, Germany

C. C. Ferreira, C. F. C. Klütsch (eds.), *Closing the Knowledge-Implementation Gap in Conservation Science*, Wildlife Research Monographs 4,
https://doi.org/10.1007/978-3-030-81085-6_15

15.1 Valuable Lessons Learned

Life on Earth is facing severe challenges. Human action is leading to deterioration of natural resources and ecosystems, with over 35,000 species threatened with extinction worldwide (IUCN 2020) and widespread declines documented systematically among populations of wild species (WWF 2020). This presents a severe threat to humanity by undermining the capacity of biodiversity to support human livelihoods and well-being. There is, counter-intuitively, an ever-growing body of literature showing that knowing more about this crisis does not lead to more action to abate it. In addition, research has shown that ecological knowledge is often misused during the implementation of policy instruments (for example, when deciding where to best place protected areas) in part driven by a lack of understanding of the mechanisms underlying the policy-making process (Chassé et al. 2020). The consequences of not using scientific evidence when making decisions about conservation interventions can be damaging, both in terms of wasting limited human and financial resources and failing to meet biodiversity objectives (Pettorelli et al. 2020). Therefore, effective use of biodiversity knowledge to elicit successful decision-making and implementation of conservation actions needs political, societal, and behavioural incentives that can only be instigated through good knowledge brokerage and science diplomacy, underpinned by strong collaborative and interdisciplinary endeavours.

15.1.1 The Importance of Multi- and Interdisciplinary Knowledge Sources and Co-production

With this book, we set out to explore the depths of the knowledge–implementation gap in conservation science around the world and from various socio-economic, cultural, and political perspectives. To understand the flow of knowledge from the stages of (co)production to implementation, we dedicated the first part of the book to describing the major knowledge sources available to inform environmental decision- and policy-making processes, including some that have been traditionally underutilized and/or overlooked. One of the main take-aways from the book is that there are major challenges in integrating western science with indigenous and local knowledges, particularly in culturally diverse regions, although this is pivotal to advance conservation knowledge implementation. Multi- and interdisciplinary research integration will require changes in governance and funding schemes to incentivize different stakeholders to participate in co-designed research. This section emphasized the importance of considering multidisciplinary sources of knowledge, which are based on various knowledge generation models (e.g., academic, traditional and local knowledge, and citizen science), when engaging and informing stakeholders (including civil society) on conservation rationales and decisions, while highlighting the value, relevance, and cultural significance of certain streams of knowledge generation. This realization opposes the still widespread paradigm that scientists are the sole knowledge producers and providers and that they do it in a linear or top-down fashion to the public and decision-makers (otherwise known as the information-deficit model; Toomey et al. 2017). Rather, it underscores knowledge production as a collaborative, societal, and multidimensional process that connects different knowledge holders to different publics. Environmental challenges as complex as the ones we are dealing with daily cannot be effectively tackled with a unidimensional lens of knowledge. Indeed, a greater recognition of citizen science and traditional ecological/local knowledge as legitimate conservation knowledge sources is a testament to this need and promotes the democratization of science by opening up the knowledge (co)production process to previously largely excluded societal groups (Turini et al. 2018). The participation of these groups has additionally the potential to facilitate the expansion of environmental democracy and decision-making (Kiss 2014) as well as the integration of human behaviours and social dimensions in the conservation process (Bennett et al. 2017).

15.1.2 Improving Science Communication and Spatial–Temporal Connectivity of Various Societal Groups for Evidence-Informed Decision-Making

One critical element of knowledge brokerage in biodiversity conservation is effective science communication and increasing connectivity between different societal groups. Effective science communication can change perceptions and beliefs and ultimately trigger actions by different audiences, hence being a key element in closing the knowledge–implementation gap. In that spirit, the second part of this book was dedicated to addressing the barriers and identifying opportunities for science communication to reach different audiences with empowering, engaging, and effective messaging as well as the role of boundary and bridging organizations and decision support systems in these efforts.

Challenges around effective science communication are mostly twofold: (1) the lack of time, funding, and career incentives for scientists to engage more regularly in these activities, and (2) that scientists are often more concerned with knowledge transfer in traditional formats rather than adjusting it to different audiences, an aspect linked to a lack of formal training in science communication. Strategies for better science communication include adequate reflection on who the audience is and what the communicator is trying to achieve, and a focus on connecting to the audience (e.g., by storytelling and inclusion of personal experiences) rather than delivering pure facts (through traditional lecturing, for example). Modern communication tools like social media platforms are increasingly utilized as a two-way street of science communication, biodiversity conservation research, and public outreach (Sbragaglia et al. 2020; Toivonen et al. 2019), as news media are able to increase discussion on policy topics on social media platforms and webpage forums by ~63% (King et al. 2017). Thus, news outlets and social media platforms offer a communication channel for public discussions of environmental and conservation issues with its full potential probably yet to be realized by a large proportion of scientists and practitioners in conservation science.

Scientists are also insufficiently engaged in science communication because of an ongoing debate about whether scientists can/should be activists, and/or engaged in political actions (Muir 2020). The conception that science should be an apolitical separate endeavour and that scientists are providers of objective and value-free knowledge leads many scientists to avoid commenting on impacts or applicability of their work beyond scientific publications. In addition, systematic muzzling of government scientists in several regions of the world, typically under conservative governments, has prevented scientists from communicating their research results on issues like climate change, deforestation, shark conservation, etc. (Rapp Learn 2017). In Canada, despite an elected liberal government in 2015 that promised to reverse this situation, in 2017, 53% of government scientists still considered that they could not freely speak about their science to the media (Owens 2018). This example suggests that even in comparatively progressive countries, under such a constraining climate, the proportion of scientists engaging in science communication will remain low. There has been a call to strengthen whistleblower protection laws and proactive measurements to encourage a cultural change in government organizations in Canada (Owens 2018). Similar actions may be desirable at a global scale.

A second key element for improved knowledge flow is increasing connectivity among the different parties that make, and are potentially affected by, conservation decisions. Scale and network boundaries are important components to consider in knowledge networks with bridging organizations playing an important role as multipliers of brokerage efforts, amplifying information flow through space and time. They can also help to link global science to local implementation, by facilitating the establishment of science–management partnerships and, consequently, the conversion of scientific research into local action. This is particularly critical in the case of abstract environmental issues, like climate change, where knowledge mobilization needs to operate within matching scales and relevance to local communities. Decision support systems can also aid in the implementation of conservation actions by taking into account uncertainty and complex socio-environmental factors with competing interests among stakeholders. Decision support systems can increase knowledge flow to decision-makers if they are well-designed and maintained and if they are tailored towards the need of decision-makers in conservation science. However, these systems remain underutilized in conservation science and more research and development is necessary to fully integrate them for this purpose.

15.1.3 A Global Review of the Knowledge–Implementation Gap

The third and last section of the book provided state-of-the-art summaries on the status of the knowledge–implementation gap in different parts of the world and on what solutions are being offered as feasible in these different regional settings to close the gap. These contributions were provided, to the extent possible, by native regional authors to ensure that descriptions of regional initiatives were backed up by strong local expertise. This proved to be a challenge for numerous reasons, among which the prevalence of non-native authors dominating the research field is probably the common denominator. Further, language barriers and hard-to-find expert contact details added to the difficulty of finding local and regional experts. More support in the form of funding (e.g., long-term employment opportunities for young professionals in the field), additional training opportunities and probably most importantly, inclusion of (non-English speaking) native scientists and knowledge holders in scientific research outputs (research papers, conference posters, etc.) are needed to achieve better integration of local perspectives and people at the knowledge production stage. This will also have the added advantage of gaining a better insight into local socio-economic factors inhibiting or competing with conservation action (see also Sect. 15.2.3).

Several common barriers to turning ecological knowledge into conservation action are shared across regional contexts. For example, mismatches between the research scope, its temporal and spatial coverage, and its relevance to address conservation issues are common challenges. A general lack of interest from scientists in policy, low political priority of biodiversity conservation issues, and a low accessibility and uptake of scientific evidence by decision- and policy-makers were additionally systematically pointed out as major contributors to the knowledge–implementation gap worldwide. Little is known about the cost and effectiveness of efforts aimed at narrowing the gap in certain regions of the world, such as Asia, Africa, and South America. In these contexts, regional and global initiatives, as well as international agreements, are critical to align national development goals and biodiversity protection priorities and underpin the development of local initiatives (for example, capacity building and development of standards) that support biodiversity mainstreaming efforts.

Other major barriers to the uptake of ecological data in conservation decision-making processes in some parts of the world include a complete absence of knowledge of some taxonomic groups (like insects and plants). Some of these geographic and taxonomic knowledge gaps could be filled with an increased utilization of standardized data collection methods and citizen science projects that would also allow to upscale conservation efforts. In addition, an increasing involvement of the private sector in biodiversity monitoring and the achievement of biodiversity strategic goals are warranted as pointed out in several book chapters and elsewhere (e.g., de Silva et al. 2019; Krause et al. 2021; Smith et al. 2018, 2020). Partnerships between the private sector and other stakeholders have often been problematic, mostly because of conflicting interests related to economic and conservation goals. However, biodiversity loss has been identified as a risk to business in 2005 by the World Economic Forum, and since then moved from 'a potential concern' to 'critical issue' in about 15 years (WEF 2019), suggesting a growing awareness of the interdependencies between business and biodiversity status. Nevertheless, engagement of private companies with strategic biodiversity goals and sustainable development goals varies widely (Smith et al. 2018, 2020), with business representatives struggling to pinpoint how they can include targeted actions into business plans (CBD 2018). This situation feeds further into the conservation knowledge–implementation gap and needs addressing to ensure all parts of society are engaged in solving the biodiversity crisis. The winning formula to get the private sector on board to commit to biodiversity protection will likely be a combination of stronger governmental environmental regulations and ecological knowledge transfer to business by scientists and practitioners, outlining clear links to biodiversity strategies and sustainable development goals as well as the strengthening of Open Access principles that in the case of private companies probably need to be legally enforced. Finally, science-based criteria are needed to measure the private sectors' commitment to halt biodiversity loss and force companies to outline specific, measurable, and time-bound action plans and business practices (de Silva et al. 2019).

While the original goal of this book was to describe both success stories and failures of the knowledge–implementation flow in biodiversity conservation and investigate which strategies are more successful across different geographic, economic, and social settings, it soon became evident that well-documented failures and associated consequences are much harder to find, likely because they are reported less frequently (but see Aganyira et al. 2019; Catalano et al. 2018; Giakoumi et al. 2018; Godet and Devictor 2018). For this reason, the book focused more on the positive messaging of successful attempts to close the knowledge–implementation gap. However, there is much to be learned from analysing and publishing failed conservation (research) initiatives as they could give an insight on the role played by different stakeholders and local communities, as well as improve understanding of interpersonal relationships, in the context of knowledge flow. The one-sided reporting of perceived success stories and reliance on this knowledge can lead to confirmation bias, overconfidence, and false determination of causal relationships, sampling error, and reduced impetus to look for alternative options (Catalano et al. 2018), particularly in understudied contexts. The major social and economic causes for conservation failures identified by Catalano et al. (2019) include communication difficulties between stakeholders and local communities, psychological reactions, and previous (negative) experiences by people involved in a conservation project. Further, economic constraints including lack of funding, mismanagement (e.g., corruption), and donor conflicts as well as missing incentives for conservation actions and lack of alternative income strategies are predominating reasons for conservation failures. Finally, lack of political support and shifting political priorities can lead to offsetting conservation goals. When properly addressed, all these variables have been identified by the contributing authors of this book as potential catalysts for closing the knowledge–implementation gap in conservation science across diverse socio-economic and cultural contexts.

15.2 What's Left to Uncover?

With this book we offer potential cross-cutting broad solutions to close the knowledge–implementation gap in conservation science around the world. While we have covered an important portion of the subject matter, there is a lot still to be explored. The book has focused substantially on how we can further improve the mainstreaming of scientific evidence into the environmental decision- and policy-making arenas, namely through better communication and public engagement. However, a major goal of all of these efforts is ultimately to change human behaviour because only then can we as a society shift to a sustainable bioeconomy (Crowling 2014). Understanding which socio-economic-cultural factors hinder/support the success of conservation actions, and how information flow varies across different social structures, are important aspects that we would like to expand on here a bit more. Moreover, including the social dimension in this conversation is a critical piece to generate additional knowledge about resource use and impacts of conservation actions on different societal groups as well as various human behaviours (de la Torre-Castro et al. 2017; Martin and Hall-Arber 2008).

15.2.1 The Critical Contribution of Social Sciences to Closing the Knowledge–Implementation Gap

Traditionally, the natural sciences have been the main knowledge provider for guiding the implementation of conservation action (Bennett et al. 2017; Moon and Blackman 2014). However, there is an increasing recognition of the importance of social sciences as a complementary field of expertise, alongside other emerging fields like citizen science and traditional ecological and local knowledge, in successful conservation planning and action (Bennett et al. 2017; Moon and Blackman 2014; Moon et al. 2019; Sandbrook et al. 2013). The main barrier for the integration of social sciences into conservation policy and management appears to be a lack of awareness on the side of practitioners and scientists regarding the potential contributions, sub-disciplines, and objectives of the field (Bennett et al. 2017). In a recent review, Bennett et al. (2017) identified 18 sub-disciplines of conservation social sciences that represent social sciences contributing to biodiversity conservation. For instance, conservation or social marketing research investigates the possibility of applying marketing strategies, including concepts and methods, to change the behaviour of target audiences and make it more environmentally friendly (Green et al. 2019; Kidd et al. 2019; Veríssimo 2019; Wright et al. 2015). The recent emergence of social marketing as a research field is the consequence of the realization that awareness-raising initiatives or environmental education programmes alone generally do not result in behavioural changes (Green et al. 2019). Social marketing aims to develop engaging and attractive campaigns that enable behaviour change in society, and a meta-analysis of 84 conservation social marketing campaigns showed that they were successful in changing behaviours by approx. 9% (Green et al. 2019), reinforcing the potential for this subdiscipline to help close the conservation knowledge–implementation gap (David et al. 2019; Tapp and Rundle-Thiele 2016). More broadly, conservation social sciences can investigate and provide local social, economic, cultural, and governance context to increase our understanding of the needs, challenges, successes, and potential impacts of conservation actions (Bennett et al. 2020). They can also help with adaptive co-management practices as well as stakeholder engagement, and hence are instrumental to develop socially equitable and just conservation actions and outcomes (Bennett et al. 2017). Below, we provide a few examples of how social sciences can help to close the knowledge–implementation gap and support long-term societal support of conservation implementation.

15.2.2 Information Flow Analysis for a Better Understanding of Social Structure and Knowledge Transfer in Local Communities

An alternative way to look at information flow is to consider how information travels within a community (i.e., interpersonal communication) and how social structures impact information flow and influence behavioural change (de Lange et al. 2019). Information flow within a society will be affected by the identities, personalities, and relationships of communicators that try to spread a message. At the receiving end, acceptance of the message will depend on the relationship of the two individuals (e.g., an elder communicates to a younger person), the credibility of the source (i.e., trustworthy source of information), power structures, and social norms (de Lange et al. 2019 and references therein). Individual resistance to messaging is connected to several social phenomena, including *uncertainty of outcome*, which is the reluctance to implement change until benefits of the action are clearly visible in society and/or that change is socially acceptable. Hence, understanding these factors better can make messaging more targeted and improve communication efficiency in conservation efforts to solicit behavioural change and adoption of new practices.

Social network analysis (SNA) is an analytical method to assess social structures within communities by considering individual nodes that are connected via links, representing their relationships to other individuals in society, often including a measurement of closeness of relationship (for example, shorter distances for kin relationships than for acquaintances). Improving our understanding of societal information flow may aid in identifying key people with high connectivity to other individuals in the community (i.e., centrality); thereby improving information flow and enhance behavioural change. It may be also important to understand different communication tools as these are context-dependent (e.g., if no access to internet is available, then interpersonal communication might be the sole way for information flow). Identification of individuals that connect two or more societal subgroups may facilitate increased information flow if these individuals are trusted by both groups.

SNA has been used to analyse societal relationships at different levels including organizations from various sectors (i.e., academia, government organizations, NGOs, etc.; Adán et al. 2020; Bixler 2021; Riggs et al. 2020) and more fine-scale analyses investigating and identifying information flows among individual local community members (Arlidge et al. 2020). For example, a network analysis revealed that information flow for sea turtle bycatch information differed significantly from information networks concerning other topics (e.g., fishing gear, fishing location network, etc.), although fishing information-sharing networks were predictive to a certain extent of how information about sea turtle bycatch is shared within the community (Arlidge et al. 2020). This demonstrates that information flow is difficult to predict and that it needs to be determined on a case-by-case basis for best implementation.

15.2.3 Trade-Offs Between Environmental and Socio-Economic Development Goals Impact Conservation Implementation

There is an ongoing debate about the trade-offs between environmental and socio-economic development goals that can determine long-term success or failure of conservation actions and that is not sufficiently considered when planning conservation strategies (Oldekop et al. 2016). Socio-economic dimensions are key drivers of local communities buying into implementation of conservation strategies. This is because acceptance of conservation implementation largely depends on personal considerations, for example, the question of how implementation of conservation measures will impact income and livelihood for different demographics (e.g., gender, rural communities, etc.). Thus, even if knowledge flow among different actors and stakeholders is achieved, competing interests among these stakeholders can hamper conservation actions. Therefore, knowledge about resource users is crucial to increase conservation effectiveness (de la Torre-Castro et al. 2017).

15.2.3.1 Trade-off Between Conservation Impact and Livelihoods of Local Communities

This book largely focussed on the identification of barriers and solutions to closing the knowledge–implementation gap. However, what happens after implementation of conservation actions and are there any negative consequences associated with them? What supports successful long-term implementation and what can we learn from existing initiatives to reduce negative impacts of conservation action? And how do these feed retroactively into and affect the knowledge–

implementation gap? Most of the knowledge on this topic has been generated studying established protected areas (PAs) and their socio-economic impacts and therefore, we will highlight a few examples from those studies below. This knowledge can be used to refine future conservation plans to avoid/mitigate negative consequences and increase acceptance and support of conservation actions by local communities. Ultimately, we can consider the knowledge–implementation gap only truly closed, if the conservation action is supported and respected and if compliance with conservation guidelines is achieved.

The selection of geographic regions for the establishment of PAs may have both positive and negative impacts on local communities (Friedman et al. 2018; Jones et al. 2020; Mizrahi et al. 2018; Ward et al. 2018; Table 15.1). For example, a PA might reduce/prohibit hunting and harvesting thereby potentially lowering sustenance use of the area for the local community. Without mitigation measures like, for example, provision of complementary income and food sources, local communities will be unwilling to follow guidelines and respect the boundaries of the PA, leading to unsuccessful implementation and lowering the efficiency of the PA. Probably as a consequence of that, protected areas are often placed in spaces which are considered economically less valuable; side-stepping the human conflict but also reducing the impact of conservation action, as regions that have a lower human footprint likely have more intact ecosystems (Geldmann et al. 2019; Mizrahi et al. 2018). Qualitative assessments of protected areas and their potential impact on long-term conservation are largely missing at a global scale (Geldmann et al. 2019; see also Chap. 13; Horgan and Kudavidanage 2021 in this book for examples from Asia).

Additional examples are summarized in Table 15.1 to highlight some of the positive and negative consequences of conservation implementations that are

Table 15.1 Some examples of how conservation action can have positively and negatively socio-economic impacts. Importantly, some, if not most, of these can have both positive and negative effects

Positive impacts of conservation actions	Negative impacts of conservation actions	References
Livelihood provision through tourism development	Displacement of local communities	Naidoo et al. (2019)
Participation in governance Empowerment of local natural resource management	Exclusion from decision-making processes and governance	
Protection of historic sites	Access to spiritual sites restricted	
Reduction of poverty	Increase of poverty	Andam et al. (2010), Mammides (2020)
Long-term protection of resources	Access to resources restricted	
Increased income due to, for example, eco-tourism	Unbalanced cost and benefit distribution	Holmes and Cavanagh (2016)
Health benefits like normal child growth		Naidoo et al. (2019)

affecting socio-economic developments of local communities. This list is incomplete and more research on the topic is needed to increase the success of conservation actions.

One possible solution to mitigate the above-mentioned trade-offs (Table 15.1) could be the integration of social impact assessments into adaptive conservation management that evaluate both positive and negative socio-economic, health, cultural, and governance changes to local communities caused by conservation initiatives (Kaplan-Hallam and Bennett 2018). Improving knowledge about potential negative impacts of conservation can help mitigating these negative consequences earlier and in a more targeted way, while positive impacts can be further enhanced to maximize the benefits of conservation. In addition, both positive socio-economic and conservation results are more likely to be achieved when a co-management system with local communities is put in place as this leads to engagement and empowerment of those communities while strict management excluding local communities lead to less efficient conservation and socio-economic developments (Oldekop et al. 2016). Importantly, here again, recognition and acceptance of failures and unexpected outcomes are crucial components for the successful implementation of these adaptive management initiatives. Equally, most of the interpersonal and communication skills mentioned in Sect. 15.1.2 are essential to conduct these types of adaptive management practices.

15.2.3.2 Social Equity

Social equity, a multidimensional concept of ethical issues referring to fairness, is increasingly recognized in the conservation literature and international conservation policy as a pivotal component in closing the knowledge–implementation gap (CBD 2011; Friedman et al. 2018).

Impacts of conservation efforts are unequally distributed in a society and affect local scales more than national or international scales (Adams et al. 2004; Holmes 2007; Oldekop et al. 2016; Ward et al. 2018). Within local societies, the impacts of conservation actions are also felt differently across classes/castes, ethnicities, and gender, with benefits normally accumulating towards affluent members of the society while costs mostly affecting the poorest faction with the least power (Holmes 2007; Dawson et al. 2018). For example, wealthier households can take more risks violating rules (i.e., paying fines) and negotiate permissions for access to PAs, while poorer households cannot, leading to unequal land ownership close to PAs (Dawson et al. 2018). As a consequence, poorer households may resent PAs and associated rules and demand more transparency about land allocation (Dawson et al. 2018). Inclusion of this feedback into adaptive co-management processes would be a step towards a higher level of equity. In turn, this can increase conservation effectiveness.

At larger geographical scales, PAs in Africa and southern Asia led to higher displacement of local communities than in other regions (Oldekop et al. 2016). Similarly, more conflict was associated with PAs in Africa, southern Asia, South-East Asia, and Oceania than other geographical regions (Oldekop et al. 2016).

Recognizing the importance of social equity for successful long-term implementation of conservation actions, the Convention on Biological Diversity states that PAs should, at the minimum, not negatively impact local communities or they should be compensated, but ideally lead to reduction in inequity (CBD 2010). However, a recent analysis suggests that challenges remain in connection to loss of rights over natural resources, inadequate access to mechanisms solving disputes, and lack of transparency of decision-making (Zafra-Calvo et al. 2019). Further, in both community-managed and strict conservation PAs, actions taken to mitigate burdens associated with PAs were considered inadequate to achieve equitably managed PAs (Zafra-Calvo et al. 2019). Finally, geographical differences in transparency and rights existed at the continent level with rights performing best in Europe and worst in Africa and Oceania (Zafra-Calvo et al. 2019). These results suggest that more work is required to fully integrate equity into conservation decision-making processes and implementation. Doing so will further contribute to close the gap as conservation implementation is expected to be more widely accepted if societal groups directly affected by these actions are offered alternative livelihoods and income.

15.2.3.3 Gender

As mentioned above (Sect. 15.2.3.2), different societal groups may be disproportionally affected by conservation action and excluded from decision-making processes; thus, ignoring the potential for these groups to aid in successful implementation of conservation strategies and understanding the way these groups use resources (Cook et al. 2019; de la Torre-Castro et al. 2017). In many rural communities, men are the predominant decision-makers over natural resources (e.g., forests, fisheries, etc.) with women having no or very limited influence (Coleman and Mwangi 2013). Paradoxically, behavioural research into gendered decision-making suggests that women, on average, are seeking more equality and are more willing to share than men (Cook et al. 2019; Eckel and Grossman 1998).

There is a lack of knowledge concerning gender aspects in natural resource management and conservation, but it is increasingly becoming apparent that men and women are holders of different knowledge and expertise (Arora-Jonsson 2014; de la Torre-Castro et al. 2017; Kleiber et al. 2015). For example, in coastal areas of Zanzibar, men know more about fishery-associated ecosystem services than women while women know/value more the oxygen production of coastal forests than men (de la Torre-Castro et al. 2017). This can be at least in part explained by labour divisions between the genders, with men being more involved in fishery activities while women mainly work near the coastline and forest (e.g., seaweed farming and collection of firewood). This introduces also a spatial divide of knowledge that can influence the outcomes of conservation actions if not properly identified and addressed (de la Torre-Castro et al. 2017). Therefore, in order to implement holistic conservation action, it is critical to collect local knowledge of both men and women.

Gender has also emerged as one of the most central stratifying factors influencing vulnerability to climate change and natural resource use (Call and Sellers 2019; Yadav and Lal 2018). Intersectionality of gender and other factors (like poverty, discrimination, geography, race, health status, and education level) need to be considered to gain a better understanding into the impacts of gender issues on the success of conservation actions (Arora-Jonsson 2011; Call and Sellers 2019) and their role in widening the knowledge–implementation gap. This is not to say that men and boys do not face challenges but to emphasize that in order to achieve equity for all members of society it is important to consider specific circumstances of individuals or societal groups like gender.

This non-exhaustive list of examples shows that there is an urgent need to plan conservation and sustainability programmes that consider gender and intersectionality in their designs. Some initiatives to reduce greenhouse emissions have introduced gender quotas with the aim of achieving gender balance to increase equal rights and benefits from decision-making of collective payment for ecosystem services (PES) interventions (Cook et al. 2019). Payment for ecosystem services interventions target collectively owned forest, which is forest owned by communities rather than individual owners. However, collective payments may lead to inequalities in benefit-sharing and more research is needed to understand how this hinders long-term conservation implementation. In a randomized experimental field trial, in which 440 forest users from Peru, Tanzania, and Indonesia participated in a hypothetical scenario of a collective payment that should be distributed among the collective owners, a balanced gender quota of 50% was introduced. This resulted in an increased decision-making towards reduced deforestation by about 50% and increased benefit-sharing in groups that had a gender-balanced group composition in comparison to groups with fewer women.

15.3 Ways Forward

The compilation of perspectives and case studies included in this book has provided an opportunity to assess the different dimensions of the knowledge–implementation gap in conservation science around the globe and identify potential ways to bridge it. Moving forward, it will be important to continue studying the gap in order to understand it increasingly better and to enable more targeted and informed actions to close this gap. This book has offered a baseline and a functional framework for future work on the gap while suggesting a plethora of research directions that could contribute to further close it. Raising awareness of the gap and educating the newer and future generations of conservation scientists in how to address it creatively and across socio-economic and cultural contexts will require substantial investment (financial and human capital) that needs to be prioritized by public ventures and mobilized also by involving the private sector. Ultimately, closing the gap will only be possible if there is enough adequately trained capacity to integrate various co-generated knowledges through a multitude of ways involving different perspectives and package it in a format that is useful and accessible for environmental decision- and policy-making. To build this capacity, cross-sectoral and interdisciplinary role models and mentors, who currently work at various intersections and who can guide the newer generations into effective and honest knowledge brokerage, will be instrumental to motivate more and more individuals to become a part of this movement. This is of particular importance because knowledge generation in conservation science is moving further and further from academia to governmental organizations and NGOs, leading to a diversified landscape that offers new opportunities but also requires new and comprehensive skillsets. In the Anthropocene, the co-creation and brokering of several forms of knowledge and the ability to turn it into narratives with implications for environmental decision-making and policy will have to become the norm if we aspire to overcome the biodiversity crisis.

References

Adams WM, Aveling R, Brockington D, Dickson B, Elliott J, Hutton J, Roe D, Vira B, Wolmer W (2004) Biodiversity conservation and the eradication of poverty. Science 306(5699):1146–1149

Adán A-F, Castillo A, Cárdenas-Tapia M, Moreno-Casasola P, Marín A (2020) Scientific research and information flow in the small-scale fisheries of the central Mexican Pacific coast. Ocean Coast Manag 198:105324

Aganyira K, Kabumbuli R, Muwanika VB, Nampanzira D, Tabuti JRS, Sheil D (2019) Learning from failure: lessons from a forest-based carbon and charcoal project. Int For Rev 21(1):1–10

Andam KS, Ferraro PJ, Sims KRE, Healy A, Holland MB (2010) Protected areas reduced poverty in Costa Rica and Thailand. PNAS 107(22):9996–10001

Arlidge W, Firth J, Alfaro-Shigueto J, Ibanez-Erquiaga B, Mangel J, Squires D, Milner-Gulland EJ (2020) Assessing information-sharing networks within small-scale fisheries and the implications for conservation interventions

Arora-Jonsson S (2011) Virtue and vulnerability: discourses on women, gender and climate change. Glob Environ Chang 21(2):744–751

Arora-Jonsson S (2014) Forty years of gender research and environmental policy: where do we stand? Women's Stud Int Forum 47(Part B):295–308

Bennett NJ, Roth R, Klain SC, Chan K, Christie P, Clark DA, Cullman G, Curran D, Durbin TJ, Epstein G, Greenberg A, Nelseon MP, Sandlos J, Stedman R, Teel TL, Thomas R, Veríssimo D, Wyborn C (2017) Conservation social science: understanding and integrating human dimensions to improve conservation. Biol Conserv 205:93–108

Bennett NJ, Calò A, Di Franco A, Niccolini F, Marzo D, Domina I, Dimitriadis C, Sobrado F, Santoni M-C, Charbonnel E, Trujillo M, Garcia-Charton J, Seddiki L, Cappanera V, Grbin J, Kastelic L, Milazzo M, Guidetti P (2020) Social equity and marine protected areas: perceptions of small-scale fishermen in the Mediterranean Sea. Biol Conserv 244:108531

Bixler RP (2021) The knowledge network: identifying actors and structural dimensions of knowledge transfer. In: Ferreira CC, Klütsch CFC (eds) Closing the knowledge-implementation gap in conservation science - Evidence transfer across spatiotemporal scales and different stakeholders. Springer Publishers, Xx pages

Call M, Sellers S (2019) How does gendered vulnerability shape the adoption and impact of sustainable livelihood interventions in an era of global climate change? Environ Res Lett 14(8):083005

Catalano AS, Redford K, Margoluis R, Knight AT (2018) Black swans, cognition, and the power of learning from failure. Conserv Biol 32(3):584–596

Catalano AS, Lyons-White J, Mills MM, Knight AT (2019) Learning from published project failures in conservation. Biol Conserv 238:108223

Chassé P, Blatrix C, Frascaria-Lacoste N (2020) What is wrong between ecological science and policy? Ecol Lett 23(12):1736–1738

Coleman EA, Mwangi E (2013) Women's participation in forest management: a cross-country analysis. Glob Environ Chang 23(1):193–205

Convention on Biological Diversity (CBD) (2010) Aichi Target 11 – Technical Rationale Extended, COP/10/INF/12/Rev.1. https://www.cbd.int/sp/targets/rationale/target-11/

Convention on Biological Diversity (CBD) (2011) Nagoya protocol on access to genetic resources and the fair and equitable sharing of benefits arising from their utilization to the convention on biological diversity. Secretariat of the Convention on Biological Diversity, Montreal

Convention on Biological Diversity (CBD) (2018) COP/14/INF/31 Engaging business in the development of a post-2020 global biodiversity framework. Prepared for the Conference of the Parties to the Convention on Biological Diversity. Fourteenth meeting. Sharm El-Sheikh, Egypt, 17–29 November 2018. Items 13 and 22 of the provisional agenda. Retrieved from https://www.cbd.int/doc/c/9b08/d19e/1fbeec1724642fe73810e71f/cop-14-inf-31-en.pdf

Cook NJ, Grillos T, Andersson KP (2019) Gender quotas increase the equality and effectiveness of climate policy interventions. Nat Clim Chang 9:330–334

Crowling RM (2014) Let's get serious about human behavior and conservation. Conserv Lett 7(3):147–148

David P, Rundle-Thiele S, Pang B, Knox K, Parkinson J, Hussenoeder F (2019) Engaging the dog owner community in the design of an effective koala aversion program. Soc Mark Q 25(1):55–68

Dawson N, Martin A, Danielsen F (2018) Assessing equity in protected area governance: approaches to promote just and effective conservation. Conserv Lett 11(2):e12388

de la Torre-Castro M, Fröcklin S, Börjesson S, Okupnik J, Jiddawi NS (2017) Gender analysis for better coastal management – increasing our understanding of social-ecological seascapes. Mar Policy 83:62–74

de Lange E, Milner-Gulland EJ, Keane A (2019) Improving environmental interventions by understanding information flows. Trends Ecol Evol 34(11):1034–1047

de Silva GC, Regan EC, Pollard EHB, Addison PFE (2019) The evolution of corporate no net loss and net positive impact biodiversity commitments: understanding appetite and addressing challenges. Bus Strateg Environ 28(7):1481–1495

Eckel CC, Grossman PJ (1998) Are women less selfish than men? Evidence from dictator experiments. Econ J 108(448):726–735

Friedman RS, Law EA, Bennett NJ, Ives CD, Thorn JPR, Wilson KA (2018) How just and just how? A systematic review of social equity in conservation research. Environ Res Lett 13(5):053001

Geldmann J, Manica A, Burgess ND, Coad L, Balmford A (2019) A global-level assessment of the effectiveness of protected areas at resisting anthropogenic pressures. Proc Natl Acad Sci USA 116(46):23209–23215

Giakoumi S, McGowan J, Mills M, Beger M, Bustamante RH, Charles A, Christie P, Fox M, Garcia-Borboroglu P, Gelcich S, Guidetti P, Mackelworth P, Maina JM, McCook L, Micheli F, Morgan LE, Mumby PJ, Reyes LM, White A, Grorud-Colvert K, Possingham HP (2018) Revisiting "success" and "failure" of marine protected areas: a conservation scientist perspective. Front Mar Sci 5:223

Godet L, Devictor V (2018) What conservation does. Trends Ecol Evol 33(10):720–730

Green KM, Crawford BA, Williamson KA, DeWan AA (2019) A meta-analysis of social marketing campaigns to improve global conservation outcomes. Soc Market Q 25(1):69–87

Holmes G (2007) Protection, politics and protest: understanding resistance to conservation. Conserv Soc 5(2):184–201

Holmes G, Cavanagh CJ (2016) A review of the social impacts of neoliberal conservation: formations, inequalities, contestations. Geoforum 75:199–209
Horgan FG, Kudavidanage EP (2021) Translating research into conservation actions through multi-sector collaboration in tropical Asia. In: Ferreira CC, Klütsch CFC (eds) Closing the knowledge-implementation gap in conservation science – evidence transfer across spatiotemporal scales and different stakeholders. Springer, New York, pp 371–411
International Union for the Conservation of Nature (IUCN) (2020). https://www.iucnredlist.org/
Jones N, Graziano M, Dimitrakopoulos PG (2020) Social impacts of European protected areas and policy recommendations. Environ Sci Pol 112:134–140
Kaplan-Hallam M, Bennett NJ (2018) Adaptive social impact management for conservation and environmental management. Conserv Biol 32(2):304–314
Kidd LR, Garrard GE, Bekessy SA, Mills M, Camilleri AR, Fidler F, Fielding KS, Gordon A, Gregg EA, Kusmanoff AM, Louis W, Moon K, Robinson JA, Selinske MJ, Shanahan D, Adams VM (2019) Messaging matters: a systematic review of the conservation messaging literature. Biol Conserv 236:92–99
King G, Schneer B, White A (2017) How the news media activate public expression and influence national agendas. Science 358(6364):776–780
Kiss G (2014) Why should the public participate in environmental decision-making? Periodica Polytech Soc Manage Sci 22(1):13–20
Kleiber D, Harris LM, Vincent ACJ (2015) Gender and small-scale fisheries: a case for counting women and beyond. Fish Fish 16(4):547–562
Krause MS, Droste N, Matzdorf B (2021) What makes businesses commit to nature conservation? Bus Strateg Environ 30(2):741–755
Mammides C (2020) Evidence from eleven countries in four continents suggests that protected areas are not associated with higher poverty rates. Biol Conserv 241:108353
Mizrahi M, Diedrich A, Weeks R, Pressey RL (2018) A systematic review of the socioeconomic factors that influence how marine protected areas impact on ecosystems and livelihoods. Soc Nat Resour 32(1):4–20
Moon K, Blackman D (2014) A guide to understanding social science research for natural scientists. Conserv Biol 28(5):1167–1177
Moon K, Adams VM, Cooke B (2019) Shared personal reflections on the need to broaden the scope of conservation social science. People Nat 1(4):426–434
Muir B (2020) Should scientists be activists? https://www.chemistryworld.com/careers/should-scientists-be-activists/4011293.article
Naidoo R, Gerkey D, Hole D, Pfaff A, Ellis AM, Golden CD, Herrera D, Johnson K, Mulligan M, Ricketts TH, Fisher B (2019) Evaluating the impacts of protected areas on human well-being across the developing world. Sci Adv 5(4):eaav3006.
Oldekop JA, Holmes G, Harris WE, Evans KL (2016) A global assessment of the social and conservation outcomes of protected areas. Conserv Biol 30(1):133–141
Owens B (2018) Half of Canada's government scientists still feel muzzled. https://www.sciencemag.org/news/2018/02/half-canada-s-government-scientists-still-feel-muzzled
Pettorelli N, Brotherton PNM, Davies ZG, Ockendon N, Sutherland WJ, Vickery JA (2020) Successfully translating conservation research into practice and policy: concluding thoughts. In: Sutherland WJ, Brotherton PNM, Davies ZG, Ockenden N, Pettorelli N, Vickery JA (eds) Conservation research, policy and practice (Ecological Reviews, pp 325–328). Cambridge University Press. ISBN: 978-1-108-63821-0
Rapp Learn J (2017) Canadian scientists explain exactly how their government silenced science. https://www.smithsonianmag.com/science-nature/canadian-scientists-open-about-how-their-government-silenced-science-180961942/
Riggs RA, Langston JD, Phann S (2020) Actor network analysis to leverage improvements in conservation and development outcomes in Cambodia. Ecol Soc 25(4):28
Sandbrook C, Adams WM, Büscher B, Vira B (2013) Social research and biodiversity conservation. Conserv Biol 27(6):1487–1490

Sbragaglia V, Correia RA, Coco S, Arlinghaus R (2020) Data mining on YouTube reveals fisher group-specific harvesting patterns and social engagement in recreational anglers and spearfishers. ICES J Mar Sci 77(6):2234–2244

Smith T, Addison P, Smith M, Beagley L (2018) Mainstreaming international biodiversity goals for the private sector: main report & case studies, JNCC report no. 613, JNCC, Peterborough, ISSN 0963–8091. Retrieved from http://jncc.defra.gov.uk/page-7678

Smith T, Beagley L, Bull J, Milner-Gulland EJ, Smith M, Vorhies F, Addison PFE (2020) Biodiversity means business: reframing global biodiversity goals for the private sector. Conserv Lett 13(1):e12690

St. Martin K, Hall-Arber M (2008) The missing layer: geo-technologies, communities, and implications for marine spatial planning. Mar Policy 32(5):779–786

Tapp A, Rundle-Thiele S (2016) Integrating social marketing into a multi-disciplinary environment behaviour change: past present and future. In: Spotswood F (ed) Beyond behaviour change: key issues, interdisciplinary approaches and future directions. University of Bristol; The Policy Press, Bristol, pp 135–156.

Toivonen T, Heikinheimo V, Fink C, Hausmann A, Hiippala T, Järv O, Tenkanen H, Di Minin E (2019) Social media data for conservation science: a methodological overview. Biol Conserv 233:298–315

Toomey AH, Knight AT, Barlow J (2017) Navigating the space between research and implementation in conservation. Conserv Lett 10(5):619–625

Turini T, Dörler D, Richter A, Heigl F, Bonn A (2018) The threefold potential of environmental citizen science – generating knowledge, creating learning opportunities and enabling civic participation. Biol Conserv 225:176–186

Veríssimo D (2019) The past, present, and future of using social marketing to conserve biodiversity. Soc Mark Q 25(1):3–8

Ward C, Stringer LC, Holmes G (2018) Protected area co-management and perceived livelihood impacts. J Environ Manag 228:1–12

World Economic Forum [WEF] (2019) The global risks report 2019, 14th edn. Geneva

Wright AJ, Veríssimo D, Pilfold K, Parsons ECM, Ventre K, Cousins J, Jefferson R, Koldewey H, Llewellyn F, McKinley E (2015) Competitive outreach in the 21st century: why we need conservation marketing. Ocean Coast Manage 115:41–48

WWF (2020) Living planet report 2020 – bending the curve of biodiversity loss. In: Almond REA, Grooten M, Petersen T (eds). WWF, Gland

Yadav SS, Lal R (2018) Vulnerability of women to climate change in arid and semi-arid regions: the case of India and South Asia. J Arid Environ 149:4–17

Zafra-Calvo N, Garmendia E, Pascual U, Palomo I, Gross-Camp N, Brockington D, Cortes-Vazquez J-A, Coolsaet B, Burgess ND (2019) Progress toward equitably managed protected areas in Aichi target 11: a global survey. Bioscience 69(3):191–197

The manufacturer's authorised representative in the EU is Springer Nature Customer Service Centre GmbH, Europaplatz 3, 69115 Heidelberg, Germany. If you have any concerns regarding our products, please contact ProductSafety@springernature.com

Printed and bound by CPI Group (UK) Ltd, Croydon, CR0 4YY
30/09/2026
02225240-0010